BASIC
COAL SCIENCE AND TECHNOLOGY

What Coal Did Today

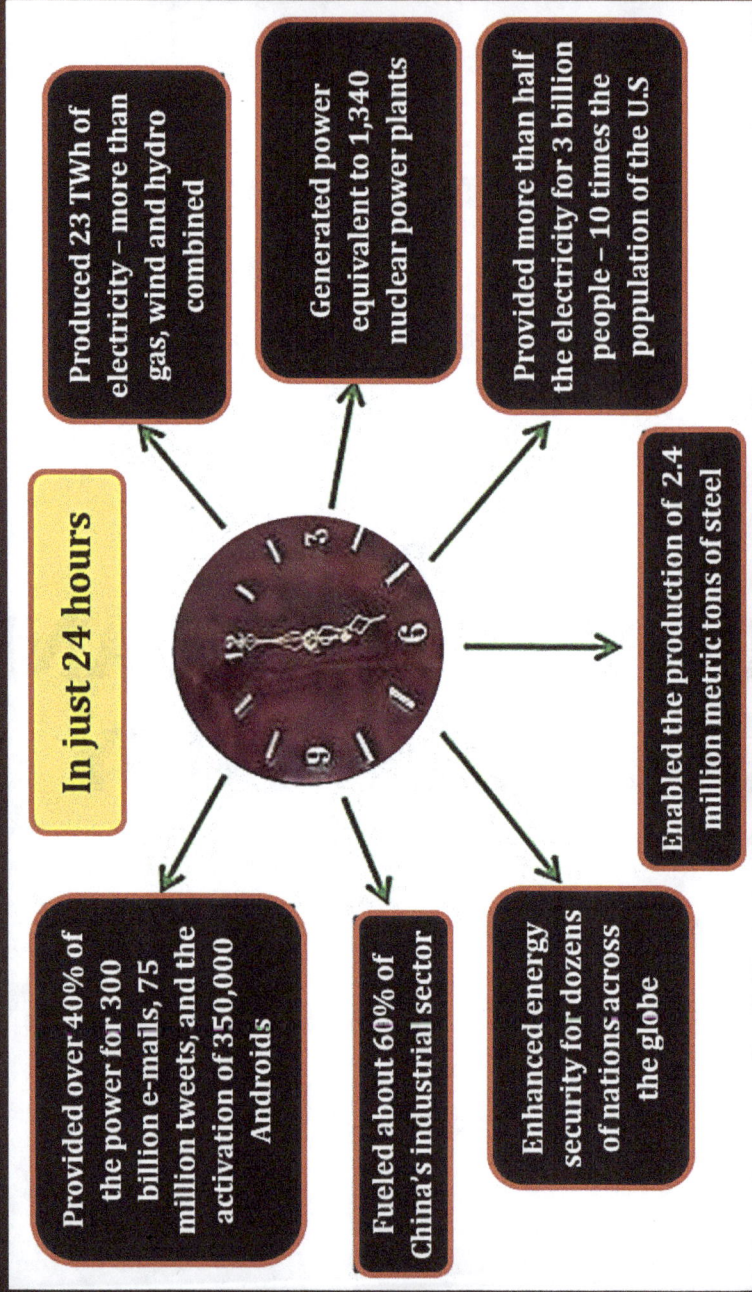

In just 24 hours

Produced 23 TWh of electricity – more than gas, wind and hydro combined

Generated power equivalent to 1,340 nuclear power plants

Provided more than half the electricity for 3 billion people – 10 times the population of the U.S

Enabled the production of 2.4 million metric tons of steel

Provided over 40% of the power for 300 billion e-mails, 75 million tweets, and the activation of 350,000 Androids

Fueled about 60% of China's industrial sector

Enhanced energy security for dozens of nations across the globe

Adapted from iea, 2010 & Science News Today, 2011

BASIC
COAL SCIENCE
AND TECHNOLOGY

Adeniyi A. Afonja

Published by

SineliBooks
2771 East Broad St., Suite 217-150
Mansfield, Texas, U.S.A.

ISBN: 978-0-9985843-0-0

The contents of this work are intended to further general scientific research, understanding, and discussion only and are not intended and should not be relied upon as recommending or promoting a specific method or equipment. The publisher and author make no representations or warranties with respect to the accuracy or completeness of the contents of this work and specifically disclaim all warranties including without limitation any implied warranties of fitness for a particular purpose. In view of ongoing research, development and equipment modifications, changes in governmental regulations, and the constant flow of information relating to the use of equipment and devices, the reader is urged to search for updates on processes and equipment discussed in this work.

Typeset in Calibri/Cambria/11 pt. by McConsult, Mansfield, Texas, U.S.A.
Printed in the United States of America
First Impression 2017
Limited Edition, printed in colour

Contents

Part 1
Coal Science

2   World energy resources. 27

3   Coal in the global energy mix. 47

9   Coal gasification. 227

Acronyms/Abbreviations

ASTM	American Society for Testing and Materials
BAT	Best available technologies
Bcm	Billion cubic metres
BDO	Butanediol
BPT	Best practice technologies
Btu	British thermal unit
BTX	Benzene, toluene, xylene
CBM	Coal bed methane
CCGT	Combined cycle gas turbine
CCS	Carbon capture and storage/sequestration
CFS	Coal-fuel slurry
CIS	Commonwealth of Independent States (Russian Federation)
CMM	Coal mine methane
CO	Carbon monoxide
CO_2	Carbon dioxide
CSG	Coal seam gas
CTL	Coal-to-liquid
CWS	Coal water slurry
DCL	Direct coal liquefaction
DRI	Direct reduction iron
DMC	Dense media cyclone
DME	Dimethyl ether
DMV	Dense media vessel
EAF	Electric arc furnace
EOR	Enhanced oil recovery
ESCII	Energy sector carbon intensity index
ESR	Energy star rating
FBC	Fluidized-bed combustion
F-T	Fischer-Tropsch
GDP	Gross Domestic Product
GHG	Greenhouse gases
Gwh	Gigawatt-hour
GWP	Global warming potentials
HGI	Hardgrove Grindability Index
HVAC	High voltage alternating current
HVDC	High voltage direct current
ICL	Indirect coal liquefaction
ICT	Information and communication technology
IEA	International Energy Agency
IGCC	Integrated gasification combined cycle
IMO	International Maritime Organization
LCA	Life cycle analysis/assessment

MEG	Monoethylene glycol
Mt	Metric tonnes
MTG	Methanol-to-gasoline
NGCC	Natural gas combined cycle
Nox	Nitrogen oxides
PFBC	Pressurized fluidized fuel combustion
PSA	Pressure swing adsorption
PVA	Polyvinyl alcohol
RCF	Radiative climate forcing
SCPF	Supercritical pulverized fuel technologies
SNG	Synthetic natural gas
SOFC	Solid oxide fuel cell
Sox	Sulphur oxides
TCO_2	Tonnes carbon dioxide equivalent
Toe	Tonnes of oil equivalent
TPES	Total Primary Energy Supplies
Twh	Terawatt hour
UGC	Underground Gasification of coal
UK	United Kingdom
UNECE	United Nations Energy Commission for Europe
UNFCCC	United Framework Convention on Climate Change
VAM	Vinyl acetate monomer
VCM	Vinyl chloride monomer
WGS	Water gas shift
ZJ	ZettaJoules (10^{21} Joules)

List of Figures

List of Tables

Preface

Coal is used mainly for iron and steel production and for generating electric power which is the world's fastest-growing form of end-use global energy consumption. Coal currently accounts for around 30% of the global primary energy consumption. Around 80% of the world's iron and steel production depends on coal and the resource is a precursor to a wide range of premium chemicals, gaseous and liquid fuels, and high-technology materials. Research on coal has been intensive for over a hundred years and there are many excellent published books on various aspects of this valuable natural resource. This book differs from previous treatments in several important ways. First, it is a departure from the very specialized approach of most of the existing books. On the contrary, the book attempts to give an overall view of the subject, thereby providing a valuable starting point for a future research career in coal science and technology.

Second, the subjects treated in the book cover a very wide range of topics, from origin of coal through structure-property relationships to coal utilization. Extensive literature on coal research over the last one hundred years have been reviewed in depth and important updates are incorporated in this book. Finally, important topics which are usually given cursory treatment in many excellent books on coal have been treated in considerable depth. These include complex, strong relationships between physico-chemical properties and technological properties of coal, coal weathering and oxidation, cleaner coal technologies, and environmental implications of coal utilization.

Another important aspect of coal treated exhaustively is its potential as a precursor to important high-value engineering materials. These premium materials include industrial graphite, industrial diamond, activated carbon or carbo-molecular sieves, carbon fibre composites, nanocarbons, synthetic natural gas, synthetic gasoline and diesel fuel, and hundreds of valuable industrial chemicals. Coal-derived graphite and tar are now starting materials for the production of a large number of premium materials for the manufacture of aerospace, aircraft, auto, telecommunications and electronic equipment components, industrial equipment parts, sports equipment, polymers, biomedical materials, and a large number of premium chemicals.

Perhaps the most topical issue with coal (and other fossil fuels) today is the extensive atmospheric pollution which emanates from most utilization processes. However, coal will remain a prime primary energy resource for decades to come and, with oil and gas, still account for around 80% of the global primary energy demand in 2040. Development of renewable energy has been rapid in recent years and utilization will grow even faster over the next two or three decades. Also, the world is recovering from the trauma of three nuclear accidents in the past three decades or so and there will we some growth in the deployment of nuclear power plants. However there are formidable problems with widespread utilization of renewable and nuclear energy sources and both will still account for no more than about 20% of the world's total primary energy use in 2040. The reality is that fossil fuels will dominate the primary energy scene in the foreseeable future and the only option is to 'clean up' fossil energy utilization processes. Feasible options for reducing pollution include clean-up of mining, transportation and combustion processes, improved efficiencies of all processes in the upstream, operational and downstream stages in the life cycle chain of fossil fuels, and changes in the fuel mix for power generation, with natural gas utilization rising at the expense of coal. It is estimated that up to 30% reduction in emissions from fossil fuel use could be achieved by 2040.

The topics treated have been selected carefully, based on the author's teaching and research experience in coal science and technology spanning over four decades. Hopefully, the book will serve as a useful starting point for newcomers to coal research as well as a useful update for experienced researchers. There are thirteen chapters in two parts. The first part treats in depth the origin and structure of coal, physical, chemical, thermoplastic and weathering properties, and the role which the resource plays currently in the global primary energy scene, particularly in power generation and iron and steel production. Throughout the chapters in this part, the primary focus is on the complex interrelationship between structure, properties and utilization potentials of coal. Part 2 comprises six chapters on the utilization potentials of coal, from combustion for heat and power production, through carbonization for iron and steelmaking, to conversion to valuable gaseous and liquid fuels, chemicals and premium materials. Chapter 12 deals extensively with the effects of pollution from fossil fuel use on the environment, with emphasis on coal use which accounts for about 41% of the total energy-related anthropogenic emissions. The last chapter discusses in depth the cleaner coal technology options which have become available in the last decade or so and those under development. The potential impact of these new technologies on pollution emissions is also discussed in some depth.

I wish to acknowledge with immense gratitude the invaluable support that I have received over the years in terms of funding, fellowship awards, and access to equipment and facilities. Top on my list is University of Ife, (Now Obafemi Awolowo University), Ile-Ife, Nigeria for giving the grant in 1972 that enabled me to establish the first Coal Research Laboratory in Nigeria. Over the years the laboratory has built up an extensive database on Nigeria's enormous coal reserves as well as many foreign coals. I am also indebted to many institutions and establishments at which I carried out a substantial part of my research, notably, Zaporozhystal Steel Plant, Ukraine (former Soviet Union), Coal Research Establishment, Stoke Orchard, United Kingdom, Northern Carbon Research Laboratories, University of Newcastle Upon Tyne, United Kingdom, National Research Council Coal Research Establishment, Ottawa, Canada, Department of Materials Science and Engineering, Massachusetts Institute of Technology, USA, University of Wisconsin-Whitewater, U.S.A. My visits to these establishments were funded through fellowships, in particular, Commonwealth Research Fellowship (UK), United Nations Development Programme (UNDP) Fellowship, Senior Commonwealth Research Fellowship (UK), Fulbright Fellowship, (USA), Senior Fulbright Fellowship (USA), and National Research Council Fellowship, (Canada).

Finally, I am indebted to Professor Simi Afonja my friend and life partner for nearly six decades not only for her support, keeping the home front stable on my numerous field trips and visits to foreign research establishments, but also for her forbearance and encouragement over the many years I have spent writing this book which is one of three volumes on coal. (The companion volumes are NIGERIAN COALS: TECHNICAL PROPERTIES AND UTILIZATION POTENTIALS, and FOSSIL ENERGY USE AND THE ENVIRONMENT). She also contributed in no small measure to the final manuscripts by reading over many times and making very useful suggestions. I also acknowledge the immense contributions to my research on coal by my numerous former students for over four decades. Many of them are now professors actively involved in coal research.

Adeniyi A. Afonja
Professor Emeritus
July, 2017

Part I

Coal Science

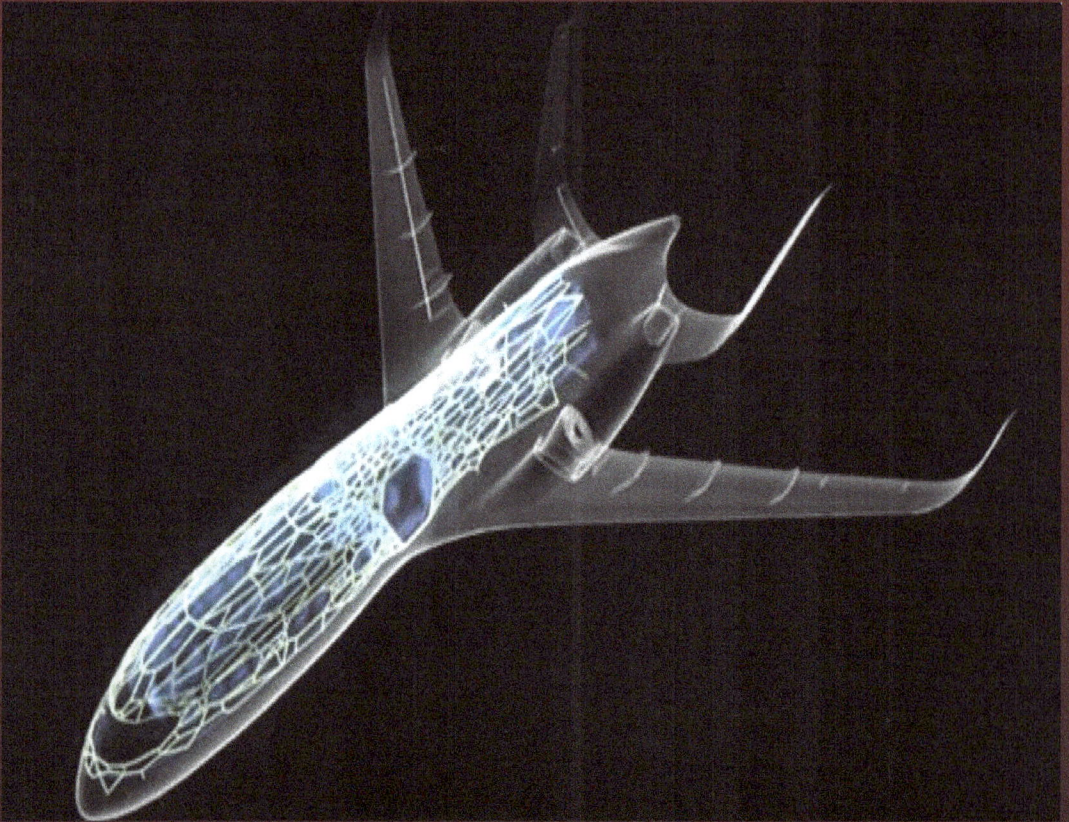

Aircraft of the future

- Fuselage, wings and components manufactured from carbon fibre composites made from coal tar-derived graphite

- Superior aerodynamics

- Up to 20% lighter, translates to around 40 tonnes (40,000 kilograms) for an Airbus 350

- Each kilogram cut in weight saves around $1 million ($40 billion total) over the lifetime of the aircraft

Image from bbc.com/news/business
(28/1/2014)

1   Coal geology, composition and classification

1.1   INTRODUCTION

Coal is a black to brownish black rock-like compound which belongs to the family of mineral, carbonaceous and combustible fuels formed from pre-historic plants and animals buried under the Earth's crust millions of years ago. The other members are oil and gas. All three fuels are believed to have been formed from plant and animal remains in the Carboniferous Period between 350 and 250 million years ago when the Earth's surface was covered with vegetation and swamps. Carbon is the main element in all the three fuels but hydrogen, oxygen, moisture, sulphur and other minerals are also present in varying amounts.

There is ample archaeological and metallurgical evidence that coal obtained from outcrops and shallow depths of the Earth's crust was a significant source of energy for shaping and smelting metals since the times of the Early Man. The Chinese were believed to have been smelting copper using coal sourced from the northeastern part of the country some 3000 years ago, and coal was the primary source of energy that fueled the Industrial Revolution which began around the sixteenth century. Currently the world still depends on coal for nearly a third of the primary energy needs and projections indicate that the fuel will continue to play a prime role in the world's primary energy supply equation in the foreseeable future (WCA, 2004; EIA, 2016).

1.2   COAL ORIGIN AND FORMATION

Fossil fuels were formed from pre-historic plants and animals that lived millions of years ago. When they died, they decomposed and got buried under layers of swamp mud, rock and sand. They decomposed slowly into organic materials and sank to deeper depths, often hundreds of metres and sometimes over a thousand metres deep, subjected to high temperatures and pressures. The masses which initially were made up of carbonaceous material and water lost much of the water content with time and, under conditions of higher temperatures and pressure, became richer in carbon. Changes on the Earth's surface over time due to earthquakes, land slides, migration or drying up of rivers and seas, and ocean incursions caused deposits of soil, sand, clay and other mineral matter to accumulate, burying the carbonaceous materials. Anaerobic conditions (low oxygen) facilitated the breakdown and decomposition of these deposits by micro-organisms.

Whether the buried materials eventually formed coal, oil or gas depended on what combinations of animal and plant debris were present, how long the material was buried, and what conditions of temperature and pressure existed when they were decomposing.

The age in which coal precursors were formed is called the Carboniferous Period on the Geologic time scale, around 350 to 250 million years ago. Some younger coals are believed to have been formed later, probably as late as the Cretaceous Period (65 to 150 million years ago). The chemical composition of these carbon-rich sediments also depended on the conditions in which they were deposited and the types of soil and mineral matter under which they were buried. For example, fuels formed from matter buried under fresh water usually contain less sulphur than those which accumulated under the saline conditions which prevail in seas.

There is considerable scientific evidence in support of the theory that oil and gas also derived from similar processes as coal but the constituents of the original deposits and environmental conditions differed. While coal was formed primarily from plant and vegetable matter, oil and gas were formed from diatomic organisms called *Protoplankton* that lived and died in the water and were buried under ocean or river sediments millions of years ago.

The Protoplankton group of organisms comprises three different functional groups that form a nutrient cycle in the water: *Phytoplankton* are tiny, usually monocellular algae that live near the water surface where there is sufficient light and carbon dioxide to support photosynthesis. *Zooplankton* are small animals, fish, larval stages of larger animals and fish that feed on the Phytoplankton. Like the plants and algae, these species are capable of converting the Sun's energy directly into stored energy. Zooplankton are in turn consumed by small fishes while small fishes themselves are consumed by larger ones.

When the fishes die, bacteria called *Bacterioplankton* break them down and the process cycle continues although probably much more slowly in modern times because of the severe disruption in the ecosystem due to human activities. Thick, viscous oil is formed first but, in deeper, hotter depths, gas also forms. While most coals remain in-situ in the general location of formation, oil and gas tend to migrate and are often found together in geological traps such as porous rock formations called *caprocks* deep beneath the Earth's surface.

The caprocks are dense enough to prevent the oil from seeping to the surface, although some can squeeze through the tiny pores in the rocks under the tremendous pressures that exist in great depths underground, reach the surface and become an energy resource which has been utilized by mankind from the early times.

Oil and gas also migrate sideways and become trapped under dry ground, or the water dries up or migrates leaving dry overburden, which explains why oil and gas are found both under the sea and on dry ground. Modern mining is done by drilling a shaft through many layers of silt and overburden and through the caprocks. The underground pressure is often high enough to force the oil and gas into the shaft but in some situations additional pressure may be required to force the oil upwards.

Coal is a complex black, brownish black, combustible, solid carbonaceous mineral formed from mainly plant debris buried underground in favourable environment millions of years ago when the Earth was covered by steamy, swampy forests. Throughout the life span of these mostly green plants, they captured energy from the Sun, carbon dioxide from the atmosphere and water from the soil to produce through a process called photosynthesis, carbohydrates which are complex compounds that make up plant tissues. The most important element in the plant material is carbon, which gives coal most of its energy.

As plants and trees died, their remains sank to the bottom of the swampy areas, accumulating layer upon layer. As the layers were successively covered, their access to the air became limited and this stopped the full decomposition process, creating a soggy, dense material called *peat.* As seas and swamps receded or dried up, these prehistoric accumulations of plant materials and peat bogs became buried, often to great depths, and went through significant physical and chemical transformations due to the combined effects of high temperatures and pressures (Figure 1.1).

Coal is indeed a generic term for a family of carbonaceous materials in various stages of transformation, from soft *peat* to the most mature and hardest coals. The original peat swamp is transformed through the *coalification* process, and passes through the progressive stages of evolution, from *lignite* through *bituminous* to *anthracite*. Intermediate stages are often classified as *sub-bituminous* and *semi-anthracite*. If the burial conditions are favourable, *graphitization* may also occur and *graphite* is the ultimate product. Graphite, the purest form of coal is also believed to have been formed from buried plant debris but the conditions of burial are different and, whereas coal is largely amorphous, graphite has a partially crystalline structure. Also it has very poor combustion characteristics and is not considered an energy source.

The types and properties of coal formed depend on the nature of the original plant debris, the environmental conditions, and the age which is considered to be of the order of 300 million years or more. It is estimated that about ten metres of prehistoric plant debris was needed to form one metre of mature coal. In effect, a one and a half metre thick coal seam which is usually considered by miners as the minimum that can be mined economically will require the accumulation of about fifteen metres of plant debris, a process which could take thousands of years. Furthermore, not every accumulation will eventually result in the formation of coal. In fact, relatively few get buried under ideal conditions which include a reasonably stable water depth over a very long time. If the water becomes too deep the plants of the swamp will drown and disintegrate. If the water cover is not maintained the plant debris will decay.

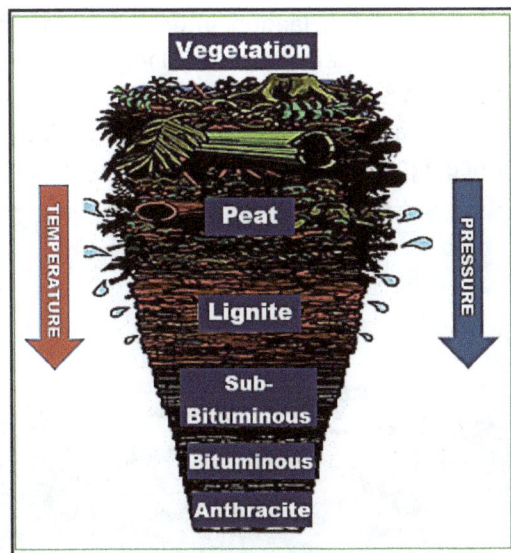

FIGURE 1.1 Stages of coalification. *(Adapted from uky.edu).*

The ideal conditions which promote the biochemical reactions needed to initiate coal formation are highly improbable events, which explains why the conditions for forming coal have occurred only a small number of times in relatively few locations throughout the Earth's history. Carbon is the primary content of coal, ranging between 50% and 96% by weight, the balance being hydrogen, oxygen, moisture and mineral matter in varying proportions depending on the history and geology of the deposit. Coal formation is a continuous process and some of the newest coals are estimated to have been deposited only a million years or so ago. The chemical, biochemical and metamorphic transformations which take place in the conversion of plant debris to coal may be summarized in three stages: *peatification*, *gelification* and *carbonification*.

1.2.1   Peatification

The first stage of coal formation involves the accumulation of suitable plant debris, often with some trapped inorganic mineral matter in conducive aquatic, water-logged environment (Figure 1.2). This process determines to a large extent whether or not the subsequent stages of coalification occur, the rate of carbonification, and the type and quality of coal that is eventually formed. It is estimated that about 3% of the Earth's surface is covered by peat (WEC, 1998) but the percentage would have been much higher in the Carboniferous Period when most of the coals currently being exploited were deposited since there were no human activities that could have destroyed vegetation on a large scale. However, the fact that relatively few peat deposits eventually mature into coal strongly indicates that conducive environmental conditions play a critical role in coal formation.

First, the peat-forming ecosystems, also known as *mire* must be deposited in swampy areas and there must be geological activities such as earth movements and subsidence, volcanic activities, marine transgression and regression which continuously deposit earth and sediments on the mire. Two possible peat forming events have been identified: *terrestialization*, a process by which ponds, lakes, lagoon, inter-distributary bays are replaced by a mire, or *paludification* which is a replacement of dry land by a mire, possibly due to a rising groundwater table or rise in water level as a result of faster seasonal accumulation of water compared with evaporation rate (Thomas, 2013). Peat is relatively impermeable and tends to impede drainage hence deposits may grow to cover wide surface areas.

FIGURE 1.2   A potential site for peat-forming ecosystems.

The peatification process starts with the biochemical degradation of the accumulated plant debris, a process also known as *humification.* Over time, as a result of basin subsidence or deposition of earth and sediments by earth or marine movement over mires, the overburden builds up, temperatures, pressures continue to rise and the peatification process progresses.

Several other factors also play a vital role in peat formation and progressive transformation into high carbon material. Vital variables for peat formation include the type of deposition, the peat-forming plant communities, the nutrient supply, acidity of the environment, bacterial activity, temperature and redox potential (Thomas, 2013).The types of plants that form mires vary widely and greatly influence the ultimate constitution and properties of coal. The nature of the depositional environment, palaeogeography, and the associated hydrogeological conditions also have significant influence on the structure and properties of coal formed in a particular location (Bend, 1992).

Peatification is a biochemical process which begins with the extensive degradation of the plant debris by bacteria, microbial activity, fungi and enzymes. Humic substances form and aromaticity is increased. The process is controlled by various factors including the nature of the plant debris, the hydrological environment, depth, temperature, mobility, salinity and acidity of the surrounding water. The geological history of the peat deposit is also important. Peats that remain in the same location or vicinity of where the forming plants grew tend to mature into low-ash, low sulphur coals, also known as *autochthonous* coals. In contrast, peat that has been transported over considerable distances to new locations as a result of geological, hydro-geological activities or marine invasion which promote active clastic sedimentation often mature into coals which contain considerable amounts of mostly inorganic impurities which may be localized or widespread within seams. High inorganic content, especially pyrites and sulphur tends to degrade the economic value of a coal deposit. Coals formed from displaced peats are known as *allochthonous* coals. Also, repeated transgressive and regressive marine activities at the peat formation stage can cause the infusion of mineral matter, in particular, sulphur, which become part of the aromatic structure of the coal that is ultimately formed, or are finely disseminated in the coal matrix. The tropical climate is believed to favour the formation of laterally extensive, thick, low-sulphur, low-ash peat deposits rather than active clastic sedimentation. Furthermore, erosion and sediment transportation are restricted in the tropical rain forest environment (Thomas, 2002; Gastaldo, 2010). This may explain why many coals found in such locations contain low sulphur.

Penetration of peat into the Earth's crust proceeds gradually. However, the rate of penetration varies from point to point in a particular deposit due to the variable nature and rate of deposit of the overburden, the sporadic nature of peat accumulation and biochemical activity. Hence a series of horizontal seams of coal of different depths are formed. This to a large extent accounts for seam thickness variations and lateral discontinuities that are characteristic of many coal seams. Significant variations of coal properties from point to point may also occur in a seam. The formation of the seams completes the first stage of coal formation and the processes from gelification to carbonification begin, with temperature and pressure being the primary rate controlling factors. The carbonification process may be interrupted from time to time by earth movements, ocean incursions, etc. A typical coal seam is shown in Figure 1.3.

FIGURE 1.3 A Coal deposit with alternating layers of sandstone, siltstone, shale and coal. *(www.elementsunearthed.com).*

1.2.2 Gelification

Extensive chemical, biochemical and physical changes take place in the second stage of coal formation known as the geochemical, or dynamo-chemical or metamorphic stage, promoted by temperature and pressure, beginning with dehydration. The seams become compressed with a reduction in thickness as much as 20 times the original thickness of the peat deposit. The peat formed in the first stage retains much of the structure of the plant debris from which it is formed. The gelification process is a combined chemical and biochemical activity that further degrades the plant structure and homogenizes the peat to the extent that, ultimately the plant cellular structure is no longer recognizable.

1.2.3 Carbonification

Carbonification (or *coalification*) is a geochemical process of progressive transformation of the gelified peat into coal under the influences of temperature, pressure and time. The temperature and pressure build-up is a consequence of progressive deposition of debris on the peat deposit over millions of years. Coalification (*diagenesis, bituminization*) progresses slowly through the stages (ranks) of lignite (brown coal), bituminous coal, and finally anthracite. The carbon content increases progressively from around 50% in the original plant debris to about 95% in anthracite, the H/C ratio and oxygen content decrease while the heating value increases. Some classification systems include two sub-divisions - sub-bituminous and semi-anthracite to describe coals in transition between ranks. Aromatization and cross linkages progress with coalification, hence the higher rank coals have more aromatic carbons than aliphatic carbons. Graphite is sometimes classified as coal because it also originates from organic material and the processes of carbonification are similar. In fact, it can be regarded as the last step in the carbonification process. However, as mentioned earlier, the environmental conditions that facilitate the formation of graphite are different.

Substantial changes in the structure, chemical composition and properties also occur during coalification. For example, carbon content increases progressively towards 100%, volatile matter decreases from around 60% for peat to less than 10% for anthracite, and calorific value more than doubles, increasing from around 15 MJ/kg for peat to 35 MJ/kg or higher for anthracite. Peat has a high moisture content, often over 70% compared with about 1% or less for anthracite.

1.3   COAL RANK AND TYPE

Coal *rank* is the degree or stage of coalification or carbonification, the stage reached in the process of geochemical and metamorphic transformation of carbonaceous material from peat to anthracite and graphite. Although deeper burial and therefore, higher temperatures and pressures imply older age, coals of approximately the same geological age may exhibit a wide range in structure and properties, depending on their geological histories (Schweinfurth, 2009). The often notable variation in properties of coals within the same rank has led to a further classification of coals by *type*, based primarily on the original organic matter from which they derive. The differences between coal rank and type will be discussed in some depth in a later chapter. As mentioned earlier, the main ranks of coal are lignite, bituminous and anthracite but two sub-ranks (sub-bituminous and semi-anthracite) are also commonly used to classify coals which have characteristics that fall between two adjacent ranks.

1.3.1   Peat

Peat is a soft, spongy and often soggy carbonaceous compound not normally considered as coal although it is also used extensively as fuel (Figure 1.4(a). However it is a precursor to coal and its composition and deposition environment determine the type of coal that is eventually formed. Peat is spongy and highly textured, and a good fuel when dry. Peat deposits are widespread and are primary energy sources in many developing countries. Peat is also used extensively in environmental control. It is a highly effective absorbent for fuel and oil spills on land and water and is also used as a conditioner for soil to make it more able to retain and slowly release water.

1.3.2   Lignite

Lignite is the lowest rank of coal, it is soft, spongy, water-saturated and brown to brownish-black in colour (Figure 1.4(b). It constitutes the largest proportion of world reserves of coal and is used extensively as domestic fuel and fuel for electric power generation. It is highly smoky and a rich source of chemicals. The original wood texture is visible in some lignites.

1.3.3   Sub-bituminous coal

The carbonification process proceeds slowly and many coals currently being exploited are in a transition stage between lignite and bituminous coal ranks (Figure 1.4(c). This class of coals also known as sub-bituminous coals has properties such as carbon content and calorific values which are intermediate between lignite and bituminous coals. The coals are brownish to black, harder than lignites but softer than bituminous coals. They are used mainly for power generation and steam raising. They are also a rich source of chemicals because of the high volatile matter content. Lignite and sub-bituminous coal deposits account for the bulk of the world's coal resources and are often classified together as *brown coals*.

1.3.4   Bituminous coal

This class of coals has reached a high degree of coalification. Bituminous coals are hard, black to dark brown, often with well-defined bands of bright and dull material (Figure 1.4(d). This is the most important coal rank, used extensively for steam/power generation and steelmaking. Bituminous coals which have little or no thermo-caking properties are classified as thermal coals and used for steam-raising and power generation. Coals that soften on carbonization and re-solidify on further heating to form strong, porous mass are known as metallurgical caking/coking coals and used primarily for primary metal production. The microstructure and chemical properties of bituminous coals can vary extensively depending on the nature of the precursor plant material and environmental conditions of the peatification process. For example, volatile matter content can vary from around 13% to 35%, hence the sub-classification into high, medium and low volatile bituminous coals.

1.3.5   Anthracite coal

Anthracite is the highest coal rank, hard, dark and glossy (Figure 1.4(e). The carbon content has reached around 95% or higher with very low volatile content and the coal burns almost smokelessly. It is used mainly for producing smokeless fuel for industrial and domestic applications. Deposits are relatively scarce and are often found in great depths or beneath mountains. Anthracite constitutes less than 1% of total world coal utilization and is not used for steam raising primarily because it is difficult to burn due to the low volatile content, but also because it is scarce and is potentially more valuable for other applications.

FIGURE 1.4 Coal ranks. *(Adapted from www.geology.com).*

1.3.6 Graphite

Graphite, also called *meta-anthracite* is technically the highest form of coal. Graphite is produced by metamorphosized organic material originally deposited as sediment or mixed with sediment. Graphite is most often found as flakes or crystalline layers in metamorphic rocks such as marble, schists and gneisses. As organic material is metamorphozed, hydrogen and oxygen are driven off as water, leaving the carbon behind to form graphite (Figure 1.4f). Unlike coal, the processes which favour the formation of graphite can be simulated in industrial practice and synthetic graphite is now being produced commercially on a large scale. Graphite is structurally different from coal and has properties which distinguish it from coal. Unlike coal, it is soft and conductive, it has lubricating properties and it is not a good fuel because it is virtually devoid of oxygen and thus has poor burning characteristics. Graphite has been used for making pencils from early Egyptian times. It is an excellent solid lubricant especially for elevated temperature applications. It is used also as electrodes in metallurgy, batteries, fuel cells, etc.

Graphite is used also for auto brake linings. It is used as neutron moderator in nuclear reactors, the synthetic type is the base material for the production of carbon fibres, a very versatile material which finds applications from sports equipment, tyres, to aerospace materials. Many high-value industrial components including aerospace and aircraft components are fabricated from carbon fibre composites. For example, about 60% of the fuselage of the latest Boeing 787 airplane and Airbus 450 is made from carbon fibre composites.

1.4 AGE OF COAL DEPOSITS

The bulk of the world's known coal resources is believed to have been deposited in the Carboniferous Period between 250 and 350 million years ago when the Earth's surface had become covered with sufficient vegetation. However, there is ample evidence that some deposits had been formed much earlier in the Devonian Period (Figure 1.5).

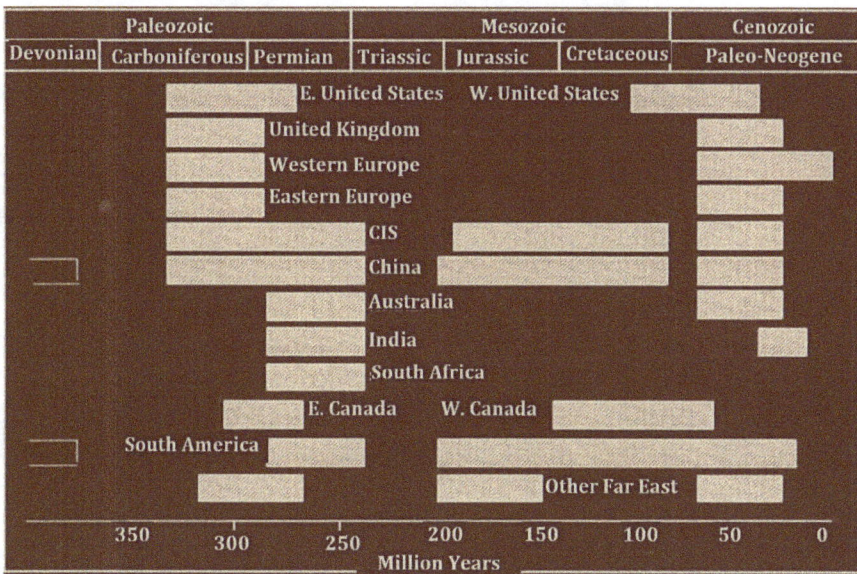

FIGURE 1.5 Geographical age of the world's major coal resources. *(Walker, 2000; Thomas, 2013).*

The process of coal formation is a continuous one and some coal deposits are less than 50 million years old. In effect, coal should be a continuously available energy resource. However the disruption of the natural ecosystem by human activities will inevitably impact negatively on future availability of coal.

1.5 COMPOSITION OF COAL

The principal characteristics of a coal seam are its thickness, lateral continuity, rank, micro-constituents and quality (Thomas, 2002). While rank is determined by burial environment and tectonic history, all the other properties are determined primarily by the prevailing factors at the time of burial. These factors include the type of mire, types of vegetation, growth rate, degree of humidification, base-level changes and rate of clastic sediment input (McCabe and Parrish, 1992).

Differences in the nature of the plant debris (mire) and the extent of its decomposition in the biochemical stage determine the chemical and micro-constituents of the ultimate coal (type) while temperature and pressure during the geochemical stage determine the rate of coalification and therefore the maturity (rank) of coal. These variables are also responsible for the heterogeneity of typical deposits, with different coal types within the same deposit and vertical and lateral variations within the same seam or deposit. Significant variations of physical and chemical properties within coal seams can also be caused by marine influxes and deposition of inorganic matter during the peat formation stage.

It is estimated that the process of transformation from peat to the most mature type of coal (anthracite) may take over three million years. Apart from *macerals* which are organic in nature, coal may contain inorganic mineral matter trapped from the environment and the ratio of the two constituents and chemical composition determine the type of coal formed (coal type) while the degree of coalification (metamorphism) determines the position of a particular coal in the lignite-to-anthracite ranking system (coal rank).

Carbon is the predominant constituent of coal, forming about 50-60% of the composition of lignite and over 96% of anthracite. Other elements present in significant proportions are oxygen, hydrogen and nitrogen. Carbon content increases as rank increases while hydrogen and oxygen contents decrease. Calorific value also increases with carbon content. Typical values of the elements in wood, peat, and various types of coal are given in Table 1.1.

TABLE 1.1 Typical chemical composition of wood, peat and various types of coals (dry, mineral matter-free basis). *(Pitt and Millward, 1979).*

COALIFICATION STAGE	% C	% H	% O	% N
Wood	50.0	6.3	42.7	1.0
Peat	57.0	5.2	36.8	1.0
Lignite	65.0	4.0	30.0	1.0
Sub-bituminous coal	79.0	5.5	14.0	1.5
Bituminous coal	88.0	5.3	5.0	1.7
Anthracite	94.0	2.9	1.9	1.2

Significant changes also take place as coalification advances (Figure 1.6). It is clear that, as coalification process progresses, the plant matter is carbonized and more aromatic structures are formed. Carbon content increases while both the hydrogen and oxygen contents decrease. Other elements which may be present include sulphur, calcium, chlorine, phosphorus and many others which are usually present in traces. Apart from chemical composition there are also differences in the nature and characteristics of the micro-constituents and their behavior on carbonization.

As discussed earlier, graphite is not classified with coals because it is not used as fuel. The biochemical history is similar to that of coal but the geochemical conditions of formation are significantly different. Natural graphite is usually impure and carbon content can vary from 90 to 95%. The morphologies of natural deposits also vary significantly and this influences the extent to which impurities penetrate the structure and the amenability of the ore to upgrading. However, purity as high as 99.96% carbon can be achieved by beneficiation. Apart from chemical composition there are also differences in the nature and characteristics of the micro-constituents of coal and their behavior on carbonization.

1.6 COAL PETROGRAPHY

Coal is made up mainly of microscopically recognizable, individual organic micro-constituents, with characteristic physical and chemical properties. These constituents known as macerals are coalified plant remains preserved in coal which change progressively, both chemically and physically as coal matures. Significant amounts of inorganic matter may also be present. Macerals can be classified into three major groups (Crelling and Dutcher, 1980):

(1) *Vitrinite* (*huminite*) which formed from woody tissue and cell walls
(2) *Liptinite* (*exinite*), a product of spores, plant resins and cuticles
(3) *Inertinite*, a group of plant material transformed by severe de-gradation during the peatification stage of the coalification process.

Rank		Low Rank		High Rank	
	Lignite	Sub-bituminous	Bituminous	Anthracite	
Age			Increases		
% Carbon	≈65-72	≈72-76	≈76-90	≈90-95	
% Hydrogen	≈5		decreases	≈2	
% Nitrogen			≈1-2		
% Oxygen	≈30		decreases	≈1	
% Sulphur	≈0	increases ≈4-15 decreases		≈0	
% Water	≈70-30	≈30-10	≈10-5	≈5	
Heating value (MJ/kg)	≈16	≈23	≈28-35	≥35	

FIGURE 1.6 Changes in coal chemistry with coalification.

Maceral recognition and distinction depend primarily on the morphology and optical properties. Reflectance is the major distinguishing feature of the maceral groups and is commonly used to rank coals. However, morphological differences are also critical factors in defining macerals. For example, most vitrinite macerals show botanical structures that clearly indicate the origin, and spores and resinous matter are identifiable on the basis of their morphologies (Figure 1.7).

The relative proportions of the maceral groups determine coal type while the vitrinite reflectance is taken as a measure of coal rank. Vitrinite has optical properties between those of liptinite (the darkest) and inertinite groups (the brightest) at low or medium rank maturation. At higher levels of organic metamorphism, the optical properties of vitrinite and liptinite converge and may be virtually indistinguishable (Figure 1.8). There are also significant differences in chemical structure and composition between macerals. In low rank coals, the carbon atoms in vitrinite are approximately divided evenly between aliphatic and aromatic bonding but hydrogen is held predominantly in aliphatic bonding. Aromaticity increases with increasing rank in all maceral groups and the chemical properties converge in higher rank coals (Figures 1.9a and 1.9b) (ICCP, 2011). A model of probable molecular changes with coalification is shown in Figure 1.10

The thermochemical behaviour of vitrinite is critical in assessing coal's utility in metallurgical coking. Vitrinite in some coals of bituminous rank softens at temperatures between 350 and 550°C or so. The degree of plasticity and extent of the temperature range largely determine the quality of the *semicoke* formed and the ultimate *coke* quality. Liptinite has even higher plastic properties but its presence relative to vitrinite in most coals is low. Only a relatively small group in the bituminous rank exhibits fluidity on carbonization, all other coals have little or no plastic properties.

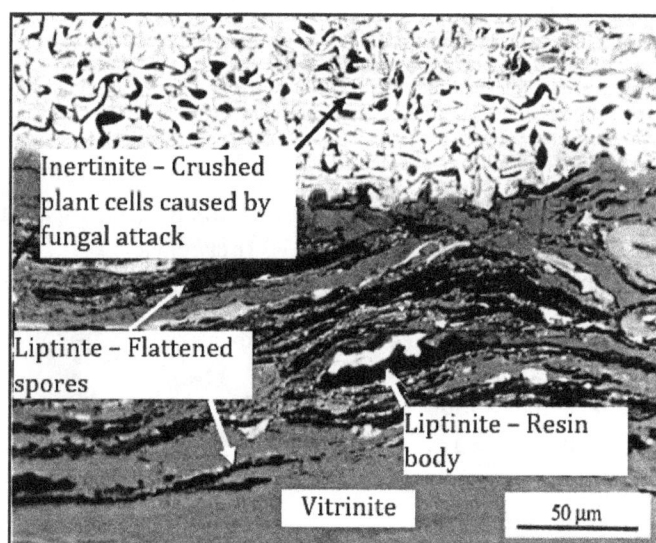

FIGURE 1.7 Main coal maceral groups. *(Taylor et al, 1998).*

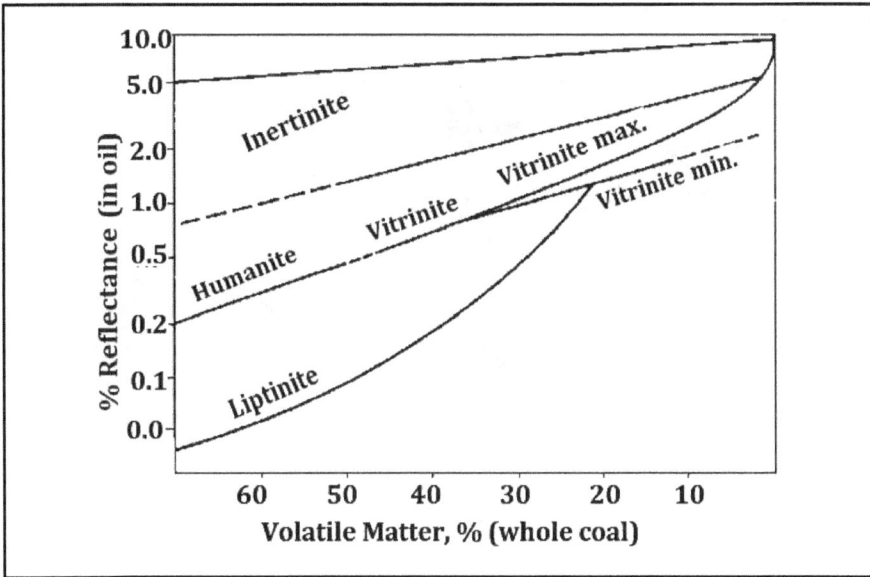

FIGURE 1.8 Relationship between reflectance of macerals and coal maturity. *(Alpern and Lemos de Sousa, 1970).*

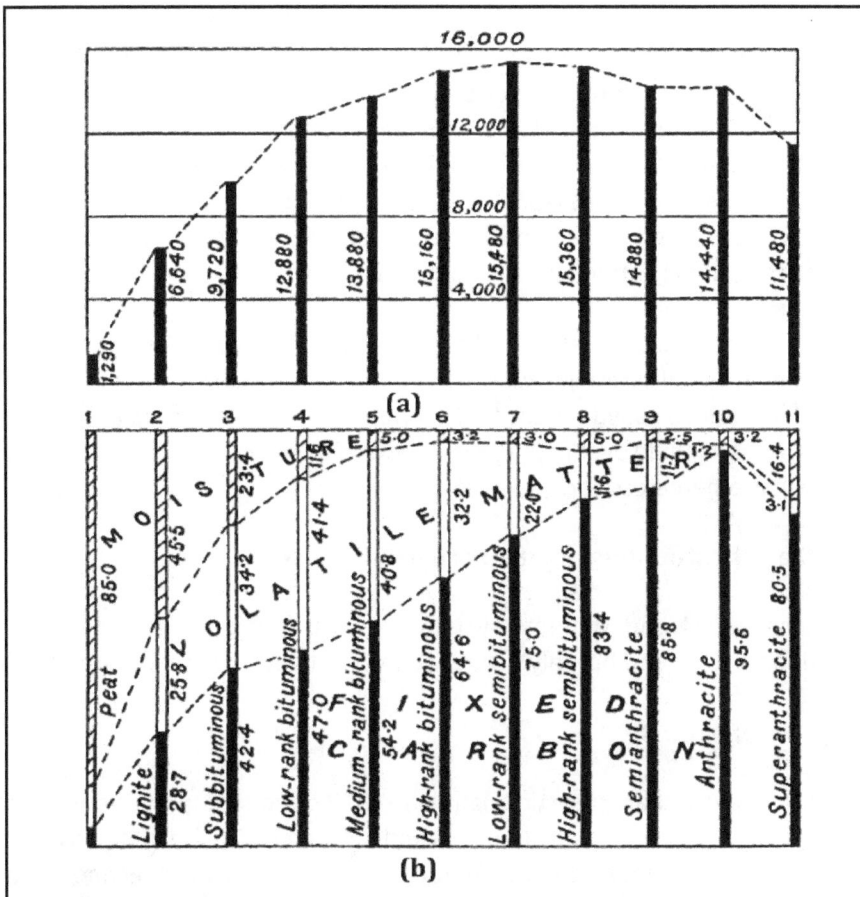

FIGURE 1.9 (a) Calorific values of different ranks of coal (Btu); (b) Coal rank from proximate analysis. *(Averitt, 1975, adapted from Campbell, 1926).*

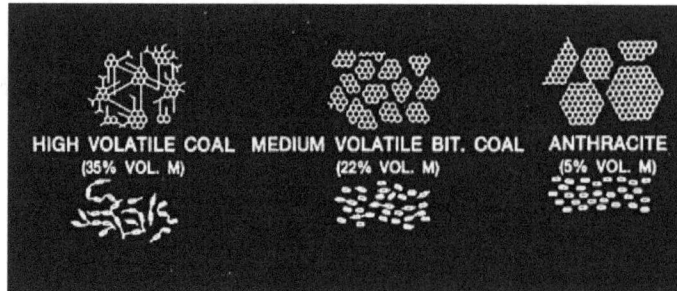

FIGURE 1.10 Molecular changes in vitrinite with maturity. *(Taylor et al., 1998).*

1.7 COAL CLASSIFICATION SYSTEMS

Coal is one of the most heterogeneous materials in existence. Composition and properties vary widely between coals, within the same coal deposit, and even within the same seam. The need for classification stems mainly from the desire to establish a basis for comparison of different coals and also for the determination of the suitability of a particular coal for a given application. Intercontinental trade in coal has always been strong and the end use determines the most appropriate classification system.

Coals destined for combustion are evaluated on the basis of calorific (heating) value, the amount of undesirable elements and compounds in the coal, moisture content, the ash fusion characteristics, etc. whereas coals meant for metallurgical use are specified on the basis of caking characteristics, sulphur and ash contents. These end-use specifications form the basis for most coal classification systems. Some systems use carbon/hydrogen, carbon/oxygen correlations to classify coals.

Many systems have been developed in various countries for the classification of coal. These range from very simple systems based on macroscopic and microscopic constituents to complex systems based on chemical composition, while others are based on utilization. Efforts have been made over many years to evolve a unified, internationally acceptable system but none has emerged so far.

1.7.1 Classification based on macroscopic constituents

One of the oldest and simplest systems of classification is based on the macroscopic composition of coal and broadly divides coals into two: *humic* coals and *non-humic* coals.

1.7.1.1 *Humic coals*

Humic coals have a characteristic bright, stratified structure. In the transverse section bright bands alternate with semi-lustrous bands and dull bands. The bright bands are fairly markedly fissured or cleated, a feature not common in the dull bands. The technological properties of humic coals vary with their petrographic composition, and also with the manner of distribution of mineral inclusions.

1.7.1.2 *Non-humic (Sapropelic) coals*

Non-humic (sapropelic) coals are coals in which the original plant material was more or less transformed by putrefaction. In contrast to humic coals, sapropelic coals are usually unstratified and show little or no development of cleat fractures. Complete seams of sapropelic coals are rare but layers or bands of varying thickness within seams are more frequent (IHCP, 1963). The coals have lower density than humic coals of comparable rank specifications and this makes them less easily cleanable by the jig and dense medium separation processes. The coking properties of sapropelic coals are poorer than those of humic coals of the same rank. However, the high hydrogen content makes the coals valuable as rich sources of gas and tar. Coals in this group are often sub-classified into *cannel* coals composed largely of spores or fine organic matter, and *boghead* coals composed of algeal matter. The two sub-groups of sapropelic coals are virtually indistinguishable macroscopically. However, cannel coals have characteristic microscopic structures which distinguish them from humic coals. The macerals are finer, more uniformly grained and more intimately mixed compared with humic coals. Also, cannel coals show a more uniform micro-stratification and a more homogeneous structure. The term, cannel coal has been used for coal burning with steady, luminous flame for centuries but is now confined to sapropelic coals containing spores.

Boghead coals are generally brown to black in colour, macroscopically unstratified but microscopic examination shows a significant amount of alginate and very finely dispersed vitrinite and inertinite. Some sapropelic coals show characteristics which fall between cannel and boghead coals since they contain both sporinite and alginate. These coals are classified as cannel-boghead or boghead-cannel coals in transition depending on the relative proportions of sporinite and alginate. Like cannel coals, pure boghead, cannel-boghead or boghead-cannel coal deposits are very rare.

1.7.2 Classification based on maturity (ranking)

In this system, coal is classified into five categories on the basis of the degree of coalification. The five ranks are lignite and sub-bituminous which are generally grouped as soft coals, and bituminous, semi-anthracite and anthracite, which are often called hard coals. Apart from chemical composition (carbon, volatile matter content), several other characteristics distinguish the various ranks of coal, for example, the reflectance of the vitrinite increases with rank (Figure 1.11). The vitrinite reflectance for low-grade coals is between 0.3 and 0.6. Hard steam and coking coals have values between 0.6 and 1.9. It should be noted however that the reflectance boundaries overlap in some coals. Volatile matter, in particular, oxygen and hydrogen content of coal decrease with increasing rank. The appearance of each of the three groups of macerals changes with advance in coal rank and the distinguishing characteristics of the group such as reflectance, refractive index and density converge as rank increases. They are hardly distinguishable in anthracite.

1.7.3 Classification based on plastic properties

The Hoffmann-Hoenne system of classification of coals is based on the plastic properties on heating. Coals are classified into four main categories depending on their behavior between

250°C and 550°C. The main groups are presented in Figure 1.12).

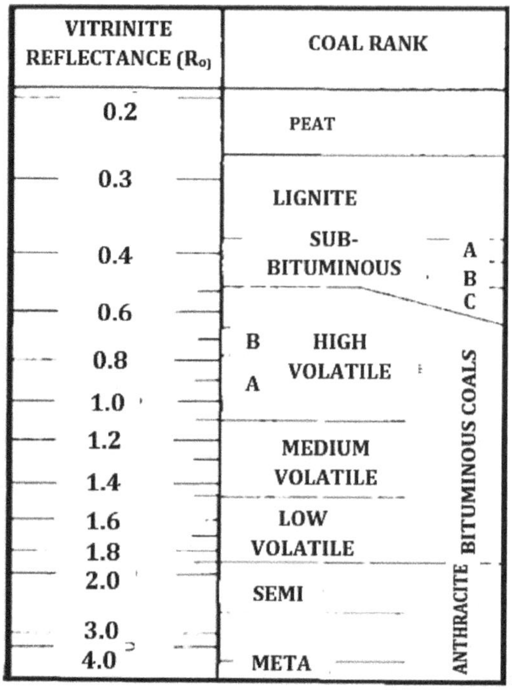

VITRINITE REFLECTANCE (R₀)	COAL RANK

FIGURE 1.11   A.S.T.M classification of coals by maceral optical properties.

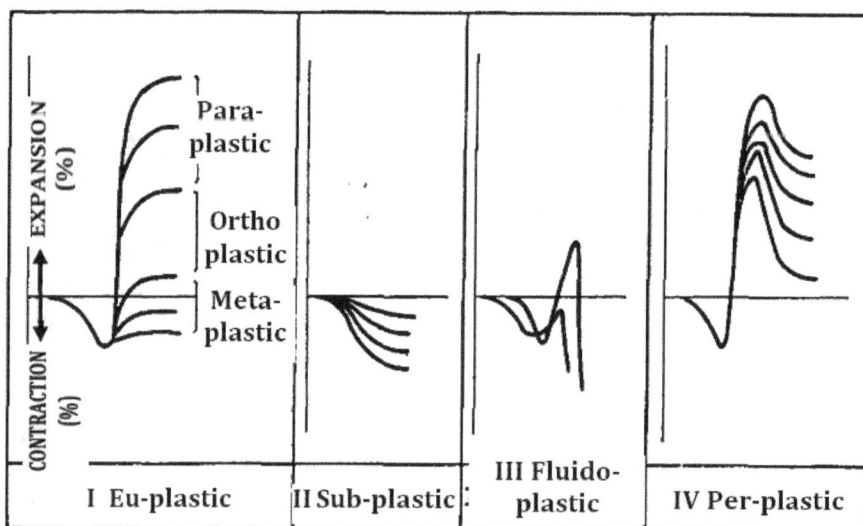

FIGURE 1.12   Hoffmann-Hoenne coal classification system based on thermoplastic properties.

1.7.3.1   Eu-plastic coals.

This group comprises coals which undergo an initial softening on carbonization, followed by contraction, dilation to an extent which depends on the relative proportions of fusible and infusible constituents. The coal re-solidifies and finally contracts to a small extent. The group may be sub-classified into ortho-plastic, para-plastic and meta-plastic coals, depending on the extent of dilation. Coals of the ortho-plastic type (50% to 140% dilation) have the best coking properties. Para-plastic and meta-plastic coals can be blended in appropriate proportions to obtain characteristics similar to the ortho-plastic type.

1.7.3.2   Sub-plastic coals.

These coals show an initial contraction but no subsequent dilation. This behaviour is characteristic of coals which contain high proportions of non thermoplastic matter, mainly inertinite and infusible vitrinite. This behaviour is typical of non-coking coals.

1.7.3.3   Fluido-plastic coals.

These coals behave like eu-plastic coals except that dilation is lower and post-solidification contraction is higher. The behaviour is typical of liptinite-rich coals which also contain appreciable amounts of fusible vitrinite. The plastic characteristics can be modified by blending with other coals.

1.7.3.4   Per-plastic coals.

Coals in this group behave like fluido-plastic coals but dilation is higher and the post-solidification contraction less for per-plastic coals. The behaviour is also typical of liptinite-rich coals which contain higher proportions of fusible vitrinite than fluido-plastic coals. Coke made from these coals is foamy, highly porous and unsuitable for metallurgical smelting. Again the plastic characteristics can be modified by blending with suitable coals.

1.7.4   Seyler's classification system

For some applications, classification of coal on the basis of rank or macroscopic properties is usually sufficient. However, classification based on chemical composition is usually necessary to determine the industrial applications for which a coal is best suited. One of the earliest systems was proposed by Seyler (1899, 1933) and is still in use in some parts of the world. Seyler's system is based on the elemental composition of coal which has been found to correlate well with technological properties. Coals are classified into four main groups according to carbon content and each group is divided into sub-groups according to hydrogen content (Figure 1.13). The Seyler system was designed specifically for British Carboniferous coals and has been found to be inaccurate when applied to low rank coals. Carbon content is the primary variable for ranking coals under the system. While this is largely true, many other coal properties have been found to be relevant. One important use of the Seyler charts is the calculation of calorific value of coal from carbon, hydrogen and oxygen contents which are much easier to determine.

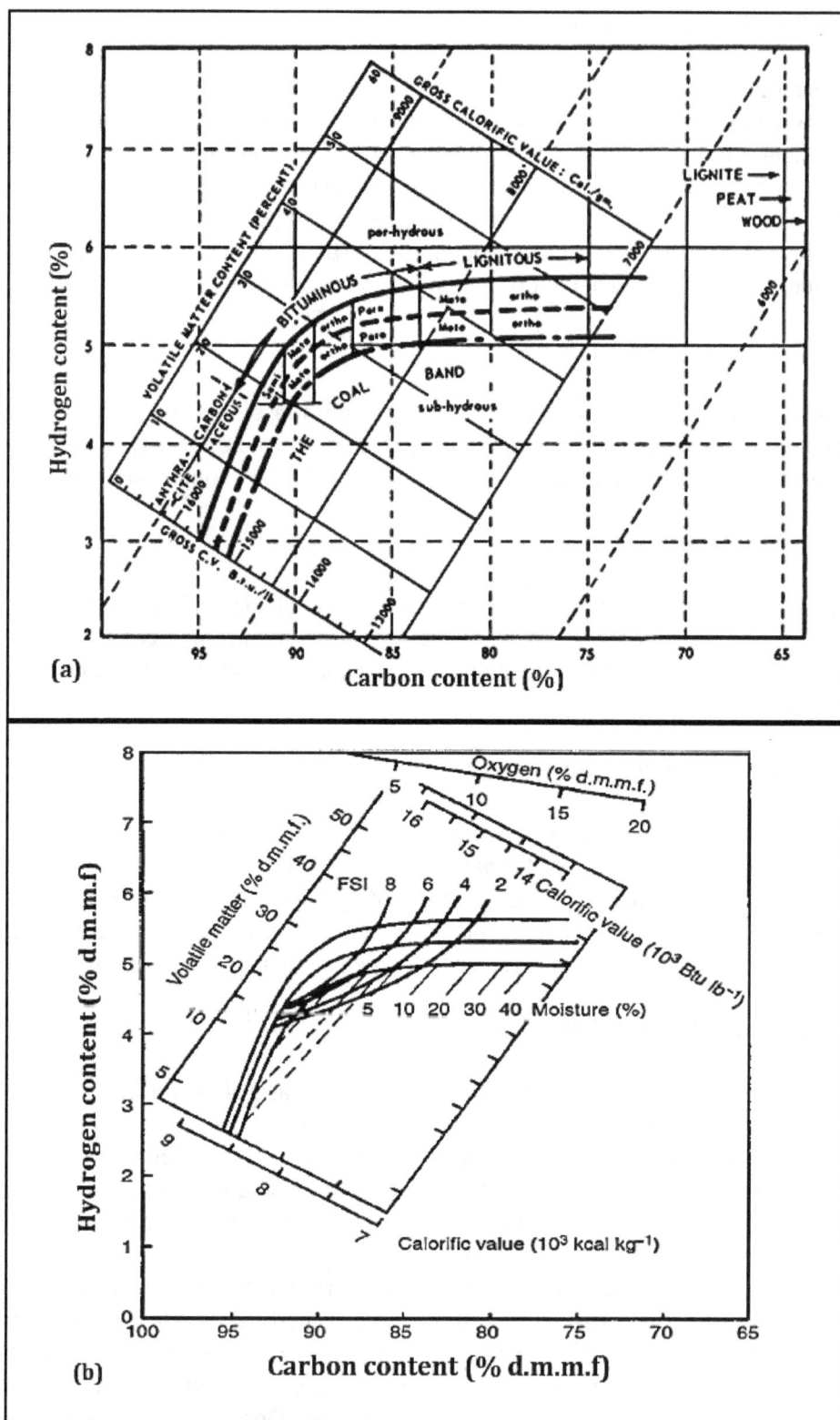

FIGURE 1.13 Seyler's coal chart. Version (b) shows relationships between elemental composition, volatile matter and moisture contents and caking properties. *(Berkowitz, 1979).*

Relationships between important coal variables have been derived for the Seyler's system as shown in Equations 1.1 to 1.3.

$$C = 0.59(Q/100 - 0.367VM) + 43.4 \qquad [1.1]$$

$$H = 0.069(Q/100 + VM) - 2.86 \qquad [1.2]$$

$$Q = 388.1 \times H + 123.9 \times C + 25 \times O_2 - 4269 \qquad [1.3]$$

Where

C = carbon content %w/w; H = hydrogen content (% w/w); VM = volatile matter content % (w/w), O_2 = oxygen content (% w/w), Q = calorific value (cal/g).

Although Seyler's system has a sound scientific basis, its complicated nature precludes its wide acceptance and application. Also, it is only applicable to coals between low-grade and the anthracite ranks.

1.7.5   National Coal Board classification system

A coal classification system developed in the United Kingdom in 1946 and modified several times since then, is based on two variables: the volatile matter content of coal on dry, ash-free basis and coking property as determined by the Gray-King coke type test. Coals are designated by a three-digit numerical code, the first two digits of which are based on the amount of volatile matter in the coal on a dry, mineral-matter-free basis and the third digit on the coking quality. The classification system is shown in Table 1.2. The Gray-King coke types given in the table for these coals merely indicate typical ranges normally found in practice. An abridged form of the British system which focuses on coal utility is shown in Table 1.3 and includes equivalent German classification system. If the ash content of a coal is over 10%, it must be washed by float and sink or similar processes prior to analysis. Also, coals having less than 19.6% volatile matter content are classified by this parameter alone. Coals that have been oxidized due to weathering have a suffix, 'W' and those that have been thermally altered by igneous intrusions are identified with a suffix, 'H'.

1.7.6   ASTM classification system

The American Society for Testing and Materials (ASTM D 388-99) system is based on a number of parameters obtained by prescribed tests. The variables are heating value, volatile matter, moisture, caking properties, ash and fixed carbon. The classification system is based primarily on the fixed carbon content (which is directly related to volatile matter content) when its value is 69% or more, (31% or less volatile matter) on a dry, mineral matter-free basis. This group comprises anthracite, semi-anthracite, low and medium volatile bituminous coals. For lower quality coals, classification is based on a combination of the calorific value and agglutinating properties of coals. The system specifically excludes non-humic coals and other coals with unusual properties. In general, the system is by rank but there are anomalies whereby a coal may be classified in a different rank because of its different thermo-chemical properties. The ASTM classification system is given in Table 1.4. Many countries that have substantial coal resources have their own classification systems - Russia, Germany, Canada, Australia, Japan, China, India, South Africa. Some are independent systems but many are modified versions of the ASTM or UK systems, or a combination of both. There is also a system developed by the international Maritime Organization (IMO) that focuses on variables which affect safe transportation of coal.

TABLE 1.2   National Coal Board (UK) coal classification system.

Group(s)	Class	Volatile Matter (dry, mineral matter-free %)	Gray-King Coke Type	General Description
100	101	>6.1	A	Anthracites
	102	6.1 – 9.0	A	
200		9.1 – 19.5	A – G8	Low volatile steam coals
	201	9.1 – 13.5	A - G	
	201a	9.1 – 11.5	A - B	Dry steam coals
	201b	11.6 – 13.5	B - C	
300		19.6 – 32.0	A – G9 and over	Medium-volatile coals
	301	19.6 – 32.0	G4 and over	
	301a	19.6 – 27.5	G4 and over	Prime coking coals
	301b	27.6 – 32.0	G4 and over	
	305	19.6 – 32.0	G – G3	Mainly heat-altered
	306	19.6 – 32.0	A - B	medium volatile coals
400 - 900		Over 32.0	A – G9 and over	High volatile coals
400	401	32.1 – 36.0	G9 and over	Very strong caking
	402			coals
500		Over 32.0	G5 – G8	
	501	32.1-36.0	G5 – G8	Strongly caking
	502	Over 36.0		coals
600		Over 32.0	G1 – G4	Medium caking
	601	32.1 – 36.0		coals
	602	Over 36.0		
700		Over 32.0	B - G	Weakly caking coals
	701	32.1 – 36.0		
	702	Over 36.0		
800		Over 32.0	C - D	Very weakly caking
	801	32.1 – 36.0		coals
	802	Over 36.0		
900		Over 32.0	A - B	Non caking coals
	901	Over 32.0		
	902	Over 36.0		

Note: Coals of groups 100 and 200 are classified by using the parameter of volatile matter alone. The Gray-King coke types quoted for these coals indicate the ranges found in practice and are not criteria for classification

TABLE 1.3 British and German coal classification systems based on coal utility.

German Classification	English Classification	Vol (%)	C (%)	H₂ (%)	O₂ (%)	S (%)	CV (KJ/kg)
Braunkohle	Lignite (brown coal)	45-65	60-75	6.0-5.8	43-17	≈1	<28,470
Flammkohle	Flame coal	40-45	75-82	6.0-5.8	>9.8	≈1	<32,870
Gasflammkohle	Gas flame coal	35-40	82-85	5.8-5.6	9.8-7.3	≈1	<33,910
Gaskohle	Gas coal	28-35	85-87.5	5.6-5.0	7.3-4.5	≈1	<34,960
Fettkohle	Fat coal	19-28	87.5-89.5	5.0-4.5	4.5-3.2	≈1	<35,380
Esskohle	Forge coal	14-19	89.5-90.5	4.5-4.0	3.2-2.8	≈1	<35,380
Magerkohle	Noncaking coal	10-14	90.5-91.5	4.0-3.75	3.8-2.5	≈1	35,380
Anthrazit	Anthracite	7-12	>91.5	<3.75	<2.5	≈1	<35,300
Note: Element content % is by weight.							

TABLE 1.4 A.S.T.M. coal classification of coal by Rank. *(ASTM D388-12).*

COAL RANK		FIXED CARBON LIMITS % dmmf	VOLATILE CONTENT % dmmf	GROSS CALORIFIC VALUE Btu/lb Moisture mmf	GROSS CALORIFIC VALUE MJ/kg Moisture mmf	AGGLOMERATING CHARACTERISTICS
Anthracite Class	Meta-Anthracite	≥98%	≤2%			
	Anthracite	92-98%	2-8%			
	Semi-Anthracite (Lean Coal)	86-92%	8-14%			
Bituminous	Low Volatile Bituminous	78-86%	14-22%			
	Medium Volatile Bituminous	69-78%	22-31%			
	High Volatile A Bituminous	≤69%	≥31%	≥14,000	≥32.557	Commonly Agglomerating
	High Volatile B Bituminous	≤69%	≥31%	13,000-14,000	30.232-32.557	
	High Volatile C Bituminous	≤69%	≥31%	11,500-13,000	26.743-30.232	
	High Volatile C Bituminous	≤69%	≥31%	10,500-11,500	24.418-26.743	Agglomerating
Sub-Bituminous	Sub Bituminous A			10,500-11,500	24.418-26.743	
	Sub Bituminous B			9,500-10,500	22.09-24.418	Non-agglomerating
	Sub Bituminous C			8,300-9,500	19.30-22.09	
Lignite	Lignite A			6,300-8,300	14.65-19.30	
	Lignite B			<6,300	<14.65	

1.7.7   International classification system

There are many other classification systems apart from the ones treated above and adoption of each system has tended to be localized. With the increasing growth of international trade in coal, it became necessary to evolve an internationally acceptable classification system. A system proposed in 1949 by a committee set up by the United Nations Economic Commission for Europe has now been adopted by the International Standards Organization.

The International Classification System is particularly suitable for classifying coals for industrial applications because it is based on the relevant physico-chemical properties of coal: volatile matter, swelling index, Gray-King coke type, moisture content, ash content and tar yield on carbonization at low temperature. Hard and soft coals are treated separately. Hard coals with calorific value (specific energy) more than 23.86MJ/kg on ash-free basis are specified by a three-digit system. When the volatile matter of the coal is less than 33%, the gross calorific value on a moist, ash-free basis is used instead. The first digit of the International Coal Classification system represents the class of the coal as determined by the volatile matter, the second digit based on swelling or Roga Index which indicates thermoplasticity specifies the group, and the last digit based on the Gray-King coke type or dilation value determines the sub-group.

Soft coals with calorific value less than 23.86MJ/kg, on a moist, ash-free basis are classified in a four-digit system. The first and second digits are based on the total moisture content of the ash-free coal and specify the class. The third and fourth digits based on the tar yield on dry, ash-free basis in low temperature carbonization, specify the group. Classification systems for hard and soft coals are presented in Tables 1.5(a) & 1.5(b). Another system developed by the same body and based primarily on reflectance values is shown in Figure 1.14.

1.7.8   International Energy Agency (IEA) System

All the systems reviewed above classify coal according to certain qualities or properties such as coal rank, chemical composition or utility. Some countries adapt established classification systems to suit their local requirements. China and India for example adapted the British system to meet domestic requirements, while Canada uses the ASTM classification system. A simple classification system developed by the International Energy Agency (IEA) and commonly used in coal trading is shown in Table 1.6. The IEA coal ranking system is based primarily on the age of the deposit. It is imprecise and inappropriate for classifying the wide range of coals that are available. Coals of the same geological age can have very different characteristics depending on the nature of the original plant debris and the prevailing environmental conditions that dominated the coalification process.

It is clear from the above that there is no universally accepted coal classification system. Coal deposits occur in many parts of the world and vary widely in rank and quality. An attempt to develop an internationally acceptable classification system has not been very successful. Many countries device classification systems that are based on the characteristics of domestic coals. For example, The ASTM system has been found applicable to many Chinese coals but is not suitable for classifying metallurgical coking coals, hence the British system has been adopted for this group of coals.

TABLE 1.5 (a) International classification system (high rank coals).
(Produced from data obtained from unece.org).

The first figure of the code number indicates the class of coal determined by volatile matter content up to 33% V.M and by calorific parameter above 33% V.M.
The second figure indicates the group of coal determined by caking properties.
The third figure indicates the sub-group, determined by coking properties.

GROUPS — determined by caking properties			CODE NUMBER										SUB-GROUPS — determined by coking properties		
Group No.	Swelling No.	Roger index	0	1	2	3	4	5	6	7	8	9	Sub-group No.	Dilatometer	Gray-King
3	>4	>45						435	535	635			5	>140	>G8
						334	434	534	634				4	>50-140	G5-G8
						333	433	533	633	733			3	>0-50	G1-G4
						332 A B	432	532	632	732	832		2	≤0	E-G
2	2½-4	>20-45				323	423	523	623	723	823		3	>0-50	G1-G4
						322	422	522	622	722	822		2	≤0	E-G
						321	421	521	621	721	821		1	Contraction only	B-D
1	1-2	>5-20			212	312	412	512	612	712	812		2	≤0	E-G
					211	311	411	511	611	711	811		1	Contraction only	B-D
0	0-½	0-5		100 A B	200	300	400	500	600	700	800	900	0	Non-softening	A
Class Number			0	1	2	3	4	5	6	7	8	9			
Class parameter	Volatile matter (d.a.f)		0-3	3-10 (<3-6.5, 6.5-10)	>10-14	>14.2 0	>20-28	>28-33	>33	>33	>33	>33			
	Calorific parameter[a]		·	·	·	·	·	·	>7750	>7200-7750	>6100-7200	>5700-6100			

As an indication, the following classes have approximate volatile matter content of:
Class 6.......33-41%
Class 7...... 22-44%
Class 8...... 35-50%
Class 9.......41-50%

NOTE: (i) Where the ash content of coal is too high to allow classification according to the present systems, it must be reduced by laboratory float-and-sink method (or any othe appropriate means). The specific gravity selected for floatation should allow a maximum yield of coal with 5 to 10% ash.
(ii)   332a.....> 14-16% V.M./332b.....>16-20% V.M
[a]   Gross calorific value on moist, ash-free basis (30°C, 96% humidity.

TABLE 1.5 (b) International (UNECE) classification system (low rank coals).
(Unece.org).

Coal Type	Gross CV (MJ/kg, m.a.f)	Moisture (%, a.r.)	R_o (%)
Low rank A	15.0	<75	-
Low rank B	15-20	-	
Low rank A	20-24	-	<0.6

RANK CATEGORY				DESCRIPTION	NEW BOUNDARY DEFINITION		OLD BOUNDARY DEFINITION	
					VITRINITE REFLECTANCE (MEAN RANDOM RoV%)	GCV MJ/KG m, af	% VOL. MATTER daf in VITRINITE (ca)	% C daf in VITRINITE (ca)
SOFT COALS	LOW RANK	LIGNITE SUB-BIT	A – B – C	LIGNITE		15		
				SUB-BIT		20		
HARD COALS	MEDIUM RANK	BITUMINOUS	D	VERY LOW RANK BITUMINOUS [VRLRB/SB]	0.4 ... 0.6	24	45	75
			C	LOW RANK BITUMINOUS [LRB]	0.6 ... 1.0		32	±82
			B	MEDIUM RANK BITUMINOUS [MRB]	1.0 ... 1.4		22	±88
			A	HIGH RANK BITUMINOUS [HRB]	1.4 ... 2.0		14	±90
			C	SEMI-ANTHRACITE	2.0 ... 3.0		8	±91
			B	ANTHRACITE	3.0 ... 4.0		2	
			A	META-ANTHRACITE	4.0 ... 6.0		0	100
GRAPHITE					10.0			

FIGURE 1.14 The new international classification of seam coal: categorization by rank.

TABLE 1.6 International Energy Agency (IEA) coal classification system. *(iea.org, 2010).*

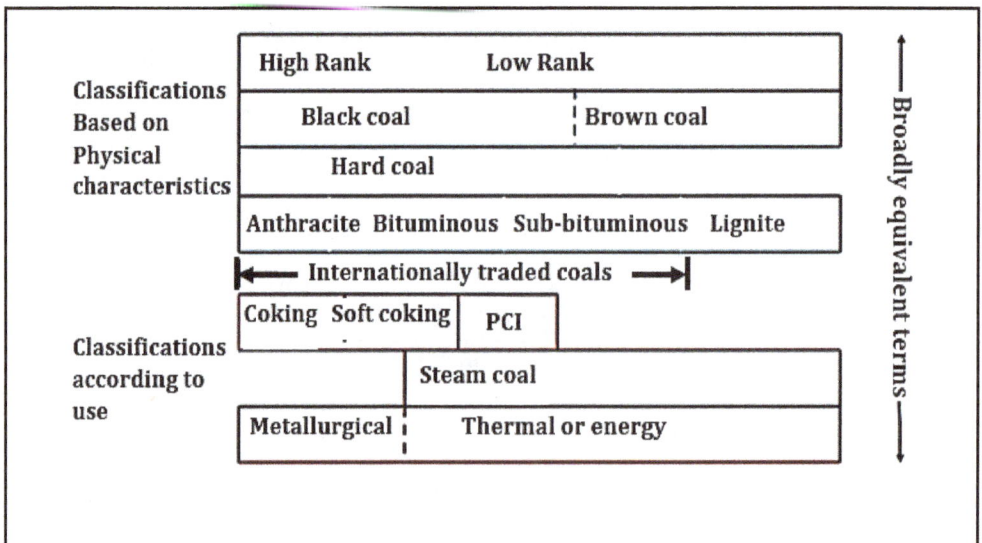

Classifications Based on Physical characteristics

High Rank	Low Rank
Black coal	Brown coal

Hard coal

Anthracite Bituminous Sub-bituminous Lignite

← Internationally traded coals →

Classifications according to use

Coking Soft coking	PCI

Steam coal

Metallurgical	Thermal or energy

← Broadly equivalent terms →

1.7.9   Suggate Classification System

The Suggate ranking system (Suggate, 1959) is based on the volatile matter content and calorific value. It classifies coals in 18 different ranks as shown in Figure 1.15. The system was developed specifically for New Zealand coals but has been found very useful in ranking a wide range of coals from different parts of the world. The Suggate classification system uses two primary coal variables: Volatile matter and gross calorific value, both expressed on dry, mineral matter-free basis.

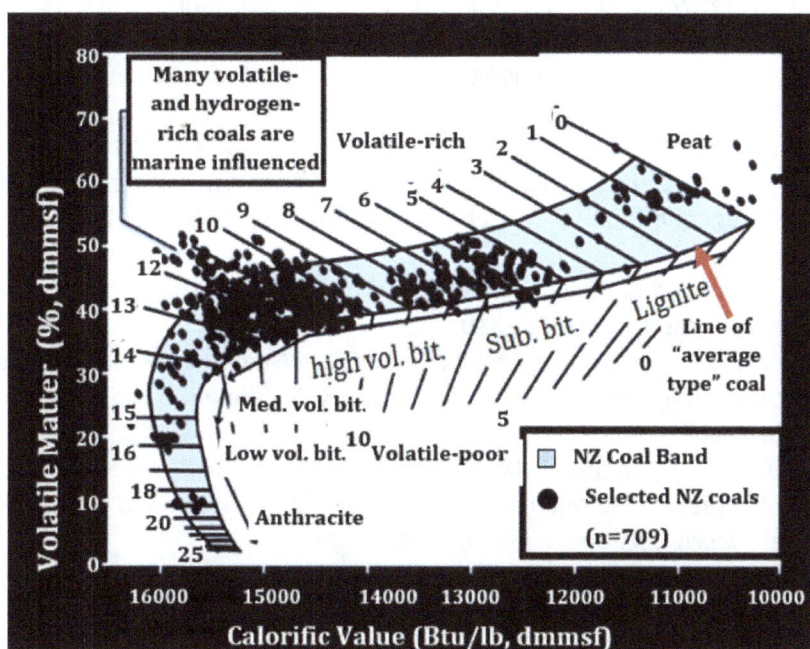

FIGURE 1.15 Suggate plots of New Zealand coals of different ranks. *(Suggate, 1959; 2000, 2002).*

REFERENCES

Alpern, B. and Lemos de Sousa (1970), in *Petroleum formation and occurrence*, 2nd ed. (eds Tissor, B. and D. H. Welte, 1984, p. 243. Springer-Verlag, New York, USA.

Averitt, P. (1975). "Coal resources of the United States." *U.S. Geological Survey Bulletin* 1412, 131p.

Bend, S.L. (1992). "The origin, formation and petrographic composition of coal." *Fuel,* vol. 71(8), August 1992, pp. 851-870.

Berkowitz, N. (1979). *An introduction to coal technology.* Academic Press.

Campbell, M. R. (1926)." Report on the operation of coal testing plant of the U.S.S.G.S, 1904 pp. 156, 173." *Internat. Conf. Bitum. Coal at Pittsburgh, 1926, vol. 1, p. 5.*

ASTM D388-99. *Standard Classification of coals by Rank.* American Society for Testing and Materials.

Crelling, J. C., and R. R. Dutcher (1980). *Principle and applications of coal petrology.* SEPM Short Course 8, 127 p.

Eberhard Lindner; Chemie für Ingenieure; Lindner Verlag Karlsruhe, S. 258.

Gastaldo, R. A. (2010). "Peat or no peat: why do the Ranjang and Mahakam deltas differ?." *Inter. J. Coal Geol.,* 83, 162-172.

EIA (2016). *International energy outlook, 2016.* Energy information Admin., eia.gov.

ICCP (2011). *International Committee for coal and organic petrology training manual.*

IEA (2010). *World Energy Outlook.* International Energy Agency. iea.org.

IHCP (1963). *International Handbook of Coal Petrography.* International Committee for Coal Petrology. Center National de la Recherche Scientifique, Paris, France. 2nd ed., 1963, n.p. Parts I and II in one volume. Part I- Alphabetical Lineup of Coal Nomenclature.

McCabe, P. J. and Parrish, J. T. (1992). "Tectonic and climatic controls on the distribution and quality of Cretaceous coals." In *Controls on the distribution and quality of Cretaceous coals* (ed. P. J. McCabe & J. T. Parrish. Geol. Soc. Am. Spec. Paper, 267, pp. 1-15.

Pitt, G. J. & Millward, G. R. (Eds.) (1979). *Coal and Modern Processing: An Introduction.* Academic Press, London. Pp. 210.

Schweinfurth, S. P. (2009). "An introduction to Coal Quality." *The National Coal Resource Assessment Overview.* Eds. Pierce, B. S. and K.O. Dennen. U.S. Geological Survey Professional Paper 1625-F.

Seyler, C. A. (1924). "Chemical classification of coal." *Fuel in Science and Practice,* 41, 79.

Seyler, C. A. (1933). "The relation between the volatile matter and elementary composition of coal." *Journ. Soc. Chem. Ind.,* I, II, 304- 306.

Singer, J. G. (1981). *Combustion – Fossil Power Systems – a Reference Book on Burning and Steam Generation.* Combustion Engineering Inc.

Suggate, R. P. (1959). " *New Zealand coals: their geological setting and its influence on their properties".* New Zealand Department of Scientific and Industrial Research, Bulletin 134.

Suggate, R. P. (2000). "The Rank (Sr). scale: its basis and its application as a maturity index for all coals." *New Zealand Journal of Geology and Geophysics,* 43:521-553.

Suggate, R. P. (2002). "Application of Rank (Sr), a maturity index based on chemical analyses of coals." *Marine and Petroleum Geology,* v. 19/8, pp. 929-950.

Taylor, G. H., M. Teichmuller, A. Davis, C. F. K. Diessel, R. Littke, and P. Robert (1998). *Organic petrology: Genruder Borntraeger,* Berlin and Stuttgart, Germany, 704 p.

Thomas, L. P. (2002). *Coal Geology.* John Wiley & Sons.

Thomas, L. P. (2013). *Coal Geology,* 2nd Edition. John Wiley & Sons.

UNECE (1998). *International Codification System for Medium and High Rank Coals.* United Nations Economic Commission for Europe, ECE/COAl/115; *Low Rank Coal Utilization – International Codification System,* ECE/ENERGY/50.

Walker, S. (2000). Major Coalfields of the World. IEA Coal Research, ccc/32, pp. 131. www.iea.org.

WCA (2004). *The Coal Resource.* World Coal Association. www.worldcoal.org.

WEC (1998). *World Energy Resources.* World Energy Council. www.worldenergy.org.

2 World energy resources

2.1 INTRODUCTION

Coal is perhaps the oldest of the fossil fuels known to mankind. There is evidence that coal collected mostly from outcrops has been used as fuel from the Stone Age. Similar archaeo- logical evidence has also been uncovered which proved that the Early Man made fires from gas leaks and oil seeps from underground. For hundreds of years coal was the most important energy resource in the world and, to a large extent made the Industrial Revolution possible.

Coal was the primary energy resource for the steam engines which supplied power to industry and transportation. It was (and still is) the primary energy source for the production of iron and steel, and one of the main fuels for the generation of electric power. For some countries like Australia, China and South Africa, coal is still the primary fuel for power generation. The dominance of coal as the primary energy resource worldwide began to decline with the development of other energy resources, in particular, oil and gas.

Oil and gas also have a very long historic background comparable to coal. The ancient Sumerians, Assyrians, Babylonians used crude oil and asphalt collected from large seeps from below the ground around the Euphrates river as mortar and for waterproofing some five to six thousand years ago, and the ancient Egyptians used liquid oil as fuel for lamps, medicine for treating wounds, and in embalming thousands of years ago. French explorers who arrived in America in early seventeenth century discovered that natives were igniting gases that were seeping into and around some lakes.

Lumps of asphalt formed from underwater seeps from the Dead Sea were in common use in the Middle East thousands of years ago and the Native Americans collected oil from the surface of streams and lakes for use as medicine, for waterproofing their canoes, and as adhesives and mortar binders. There is evidence that the Chinese were using drilling bits attached to bamboo poles to drill and recover oil from depths several hundred metres underground as early as 350 AD and the famous Italian explorer, Marco Polo observed oil seeps from the shores of the Caspian Sea as early as the 14th century. Natural gas seeps were first discovered in Iran and several other parts of the Middle East several thousand years ago and provided the fuel for the 'gods of fire' worshiped in ancient times in these areas.

In 1846, Canadian geologist Abraham Gesner distilled coal to obtain a clean-burning lamp fuel which he named kerosene. The fuel was cheaper and cleaner than the traditional whale oil fuel. A few years later, Lukasiewicz of Poland developed a process for obtaining kerosene from the more readily available crude oil, also known as rock oil. The first rock oil mine was built in Poland in the early 1850s, followed by a refinery in Baku, Russia a few years later which supplied most of the world's fuel requirements.

The first commercial oil and gas well in North America was drilled in Oil Springs, Ontario, Canada in 1858, followed by a major discovery in Pennsylvania in 1859 when considerable amount of oil seeped from a 20-metre deep dug water well (CEC, 2012). A method was devised to pump up the oil and this method is still the basis for modern crude oil drilling technology. The developments in the mid-nineteenth century are widely regarded as the humble beginnings of the modern history of crude oil drilling and petroleum (petro-leum means rock oil) refining industries. Gas is usually found in association with oil and most modern oil fields also yield gas as a by-product but there are many independent, mostly untapped dry gas fields in many parts of the world. Estimates indicate that the world gas resources are considerably larger that the oil reserves.

The discovery and recovery of oil in substantial quantities marked the turning point in the world energy scene and revolutionized transportation technology at the turn of the last century. Coal began to be replaced by oil in many applications because of the positive features of oil. It is abundant and, relative to coal, easy to produce and handle. Also, compared with coal, the negative effects of recovery and utilization on the environment are low. Gas is also becoming very prominent as an alternative energy resource in many applications, in particular, electric power generation. Several other developments have continued to change the world energy mix over the years, in particular, the development of hydro energy and nuclear energy. Unlike coal, oil and gas reserves are concentrated in only a few geographical zones and most needs worldwide are met through imports.

2.2   WORLD PRIMARY ENERGY RESOURCES

The world has enormous reserves of energy resources which may be classified broadly into two categories: *non-renewable* and *renewable* energy. Fossil fuels dominate the group of non-renewable energy resources, the only other major source being nuclear energy. Renewable energy resources include solar power, wind power, bio-energy, geothermal energy, marine and hydro-kinetic energy, and hydro-power (Figure 2.1). The world's proved *recoverable energy reserves* of primary energy as defined by the United Nations is "the proportion of total known *energy resources* that can be recovered in the future under present and expected local economic conditions with existing available technology." (Figure 2.2). One important implication of this definition is the recognition that reserves can continue to increase as new exploration and exploitation technologies become available in spite of an increasing rate of exploitation. Examples are the continuously increasing global reserves of oil and gas.

New oil drilling and coal mining technologies are coming on stream all the time and recent developments in hydraulic fracturing technology have made it possible to extract gas from the previously inaccessible but abundant resources of shale gas. It should be expected therefore that the life span of the fossil fuels will continue to be reviewed upwards in the foreseeable future. Resources are reserves that have been identified, are not fully developed, or cannot be recovered economically by currently available technologies. For example, most of the world's oil and gas reserves are off-shore and current drilling technologies are capable of no more than about 3km depth while half of the oceans which represent about a third of the planet are deeper than 3km. Also, commercial and economically viable technologies are not yet fully developed for the recovery of oil from the very large known reserves of shale and tar sands.

Furthermore, coal seams that are currently considered too thin, too sporadic or located in areas that are geologically unfavourable for underground mine development, can now be gasified in-situ and utilized for power generation. Again, developments of the appropriate technologies for this process are still evolving and some are at rudimentary stages. The known reserves and resources are believed to be no more than a fraction of the world's potential primary energy resources, many of which are presently unidentified due to severe limitations in current knowledge about the Earth's geology, and the investigative technologies that are currently available (Figure 2.3).

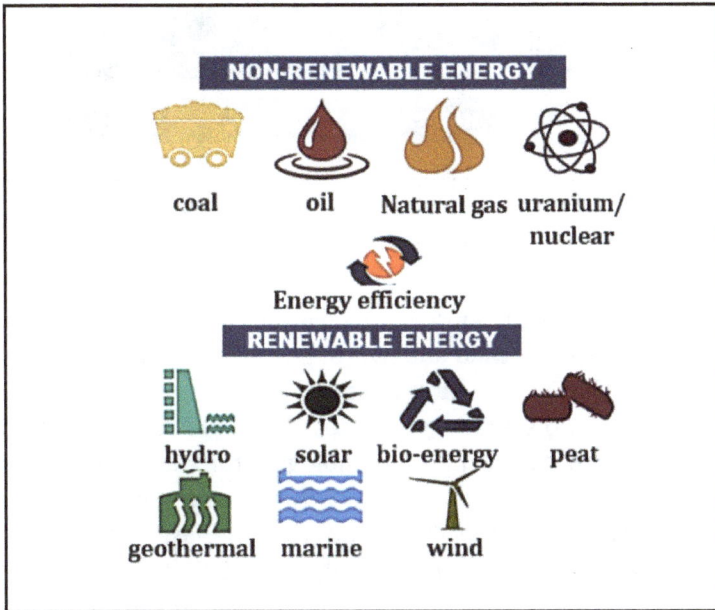

FIGURE 2.1 World's primary energy resources. *(Adapted from World Energy Resources WEC, 2013).*

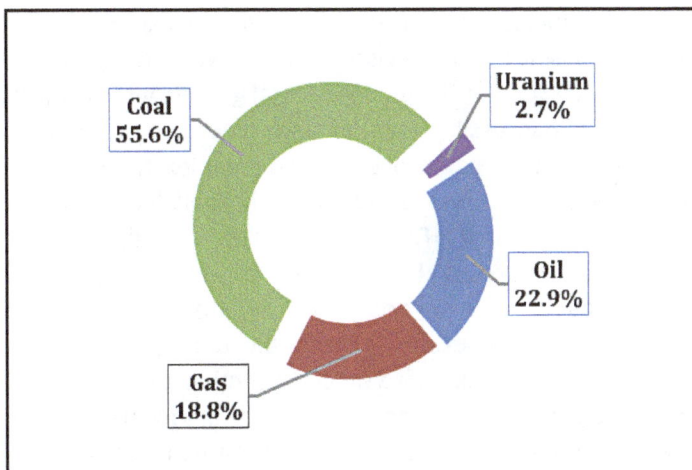

FIGURE 2.2 World's total primary energy reserves (in billion tonnes of coal equivalent). *(BP, 2011).*

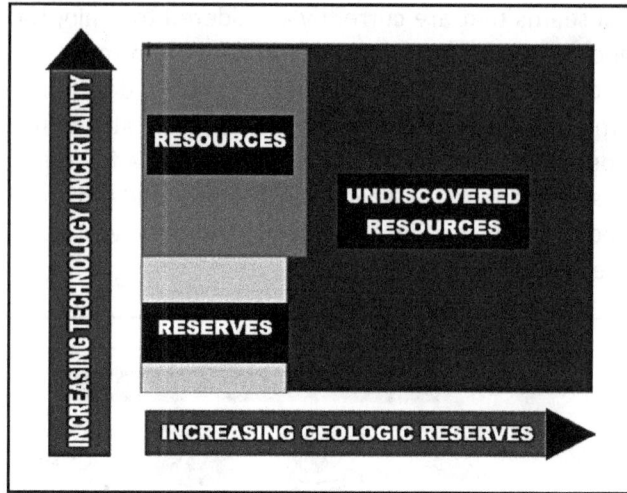

FIGURE 2.3 Schematic representation of world's primary energy resources and reserves.

Enormous potential primary resources are believed to exist on land and under the deep seas. The Sun's abundant energy, wind and hydro energy are largely untapped. The Earth's crust is also believed to hold abundant reserves of geothermal energy which could contribute significantly to the world's energy resources. Extensive research and development are in progress worldwide to evolve new technologies for retrieving energy sources that are considered unrecoverable presently. Due to impressive advances in technology, proven reserves of primary energy have increased over the last two decades or so despite the increasing exploitation rate. On average, only about 50% of the original oil-in-place volumes in reservoirs are recoverable using currently available advanced technologies (IEA, 2013). New enhanced oil recovery (EOR) drilling and seismic technologies, and CO_2 injection are raising recovery factors significantly in many oil fields around the world.

World reserves and resources of unconventional oil and gas are enormous and largely unexploited. These include high-viscosity oil and tar-sands, tight gas, shale gas, and coal-bed methane. New technologies are being developed for the exploitation of these vast resources, including steam processing to reduce viscosity and advanced separation processes to improve oil and gas extraction. Much of the global natural gas resources which had been considered uneconomical due to inaccessible locations are now being exploited using new, advanced drilling technologies, liquefaction processes and complex pipeline transportation.

The recoverable coal reserves using currently available advanced mining technologies are less than 5% of the resources (Table 2.1). New technologies are being developed to move more coal resources to reserves. These include advanced mining technologies to improve recovery factors of coal-in-place from operating and abandoned mines, and in-situ gasification technologies to recover and utilize inaccessible underground coal seams and deposits considered to be uneconomical, based on current mining techniques.

Over 80% of the global primary energy reserves are located in four regions, with Europe an Eurasia holding over 40% (Figure 2.4), although the geographical distribution varies significantly depending on the specific energy source. While only a few regions hold most of the oil and gas reserves, coal is much more widely distributed and is available in every continent of the world.

2.2.1   Fossil fuels

Fossil fuels (oil, gas and coal) which are currently the main sources of global primary energy supply are products of dead and decomposed plant and animal remains buried deep under the Earth's surface for millions of years (see Chapter 1). Most of the deposits being exploited today are around 360-350 million years old. The current estimates of world's total reserves are shown in Figures 2.5-2.7. Fossil fuels currently meet around 80% of the global energy demand and, in spite of pressures of increasingly stiff environmental pollution control regulations, projections indicate that the current dominant position of fossil fuels will prevail for decades. One major reason is the problem of lack of viable substitutes. Nuclear energy is crippled by safety and environmental problems and, with the slow pace of development and deployment of appropriate technologies for renewable energy exploitation, their role in the global energy equation will remain complementary in the foreseeable future.

2.2.1.1 *Oil and gas*

Five countries held about 63% of the global reserves of oil in 2011 (Table 2.2). However, many countries remain largely unexplored and may hold substantial reserves. Furthermore, there are large reserves of unconventional fuels - shale, oil sands, extra heavy oil and natural bitumen in many parts of the world.

TABLE 2.1   World's coal reserves and resources. (*IEA, 2013*).

COAL TYPE	RESERVES		RESOURCES	
	GIGA-TONNES (GT)	TRILLION BARRELS OF OIL EQUIVALENT (BOE)	GIGA-TONNES (GT)	TRILLION BARRELS OF OIL EQUIVALENT (BOE)
Hard coal	730	3.6	18,000	88.8
Lignite	280	0.7	4,000	19.7
Total	1,010	4.3	22,000	108.5

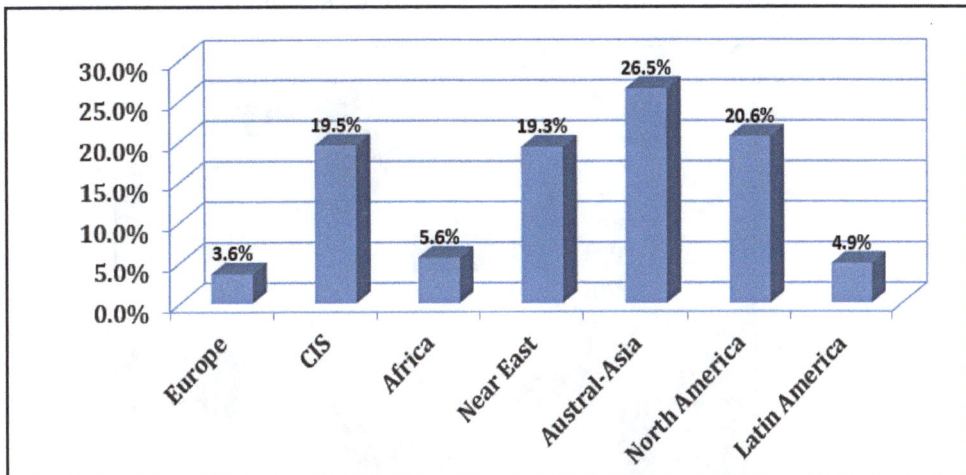

FIGURE 2.4   World's proven reserves of primary energy, 2011 (1,350 billion tonnes of coal equivalent). *(BP, 2011)*.

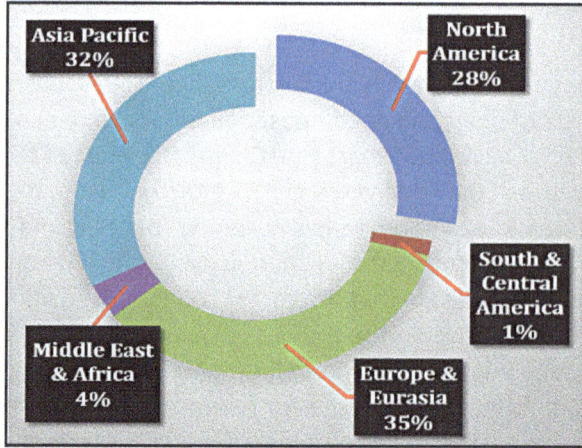

FIGURE 2.5   World's reserves of coal by region in 2013. (860,938 million tonnes). *(BP, 2014).*

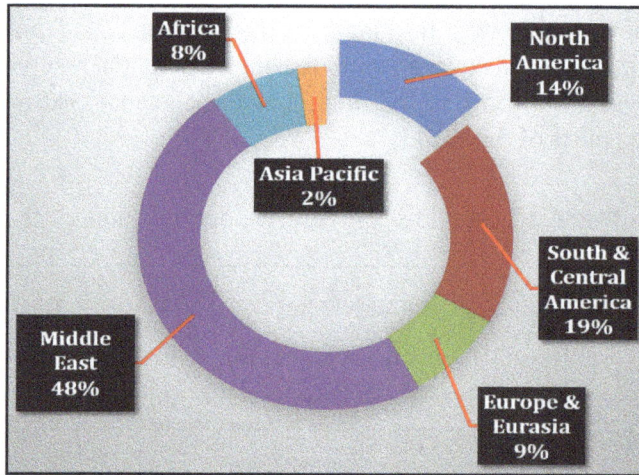

FIGURE 2.6   World's reserves of oil by region in 2013. (1687.9 thousand million barrels, 238.2 thousand million tonnes). *(BP, 2014).*

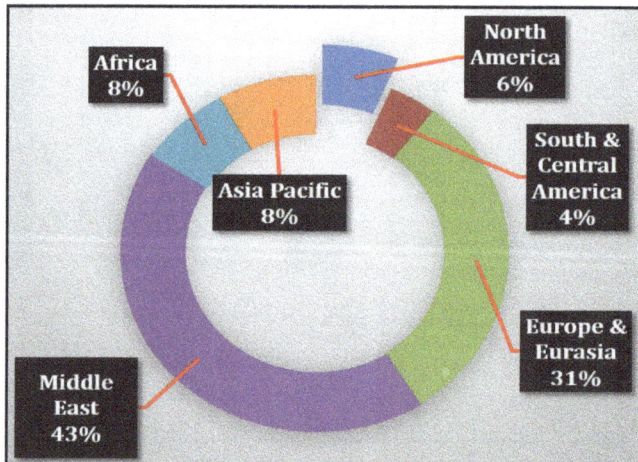

FIGURE 2.7   World's reserves of gas by region in 2013. (187.3 trillion cubic metres). *(BP, 2014).*

TABLE 2.2 Five countries which hold over 60% of global reserves of crude oil. *(WEC, 2013)*.

Country	Reserves (Mt) 2011	Reserves (Mt) 1993	Production (Mt) 2011	Production (Mt) 1993	R/P Years
Venezuela	40,050	9,842	155	129	>100
Saudi Arabia	36,500	35,620	526	422	69
Canada	23,598	758	170	91	>100
Iran	21,359	12,700	222	171	96
Iraq	19,300	13,417	134	29	>100
Rest of World	82,247	68,339	2,766	2,338	30
Global total	223,454	140,676	3,973	3,179	56
R/P is the estimated life of reserve at present rate of production					

All of the non-conventional fossil fuels (shale, heavy oil, bitumen, tar sand) are potential sources of oil and, if taken into account, total reserves could be four times larger than the current conventional reserves (WEC, 2013). Natural gas resources are estimated to be more than oil resources and, by current estimates, five countries hold about 67% of the total global reserves (Table 2.3). Interest in the development of natural gas resources has grown in recent years because of its increasing status as the most flexible and efficient fossil fuel for power generation. It also has the lowest carbon footprint of all the fossil fuels.

2.2.1.2 *Coal*

Investment in coal exploration and mine development has been low compared with oil and gas, despite the strong interest in the resource for power generation. In the last two decades, proven global reserves have declined by about 14% while production has gone up 68%. Coal deposits are abundant in every continent and around a hundred countries including many developing countries have proven reserves. Many more have potentially large but unexplored resources. The United States holds the world's largest coal resource base estimated at 28% of the global reserves and, with four other countries hold about 73% of the total world reserves (Table 2.4). China leads the other four countries as the world's largest producer of coal, and was also the world's largest consumer in 2012, accounting for nearly half of the global consumption.

TABLE 2.3 Five which hold over 60% of global reserves of natural gas. *(WEC, 2013)*.

Country	Reserves (bcm) 2011	Reserves (bcm) 1993	Production (bcm) 2011	Production (bcm) 1993	R/P Years
Russian Rederation	47,750	48,160	670	604	71
Iran	33,790	20,659	150	27	>100
Quatar	25,200	7,079	117	14	>100
Turkmenistan	25,213	2,860	75	57	>100
Saudi Arabia	8,028	5,260	99	36	81
Rest of World	69,761	57,317	2,407	1,438	22
Global Total	209,742	141,336	3,518	2,176	55

TABLE 2.4   Five countries which hold over 60% of global reserves of coal. *(WEC, 2013).*

Country	Reserves (Mt)		Production (Mt)		R/P
	2011	**1993**	**2011**	**1993**	**Years**
United States of America	237,295	168,391	1,092	858	>100
Russian Federation	157,010	168,700	327	304	>100
China	114,500	80,150	3,384	1,150	34
Australia	76,400	63,658	398	224	>100
India	60,600	48,963	516	263	>100
Rest of World	245,725	501,748	1,805	1,675	>100
Global total	891,530	1,031,610	7,520	4,474	>100

Coal will remain prominent in the global primary energy scenario for decades to come, primarily because there is no viable substitute for fossil fuels but also because, for many emerging countries, it is the only locally available primary energy source that can readily provide access to energy for their teaming populations.

2.2.2   Nuclear energy

Uranium constitutes less than 3% of the total world energy reserves (in tonnes of coal equivalent) (Figure 2.2). Over half (52.8%) of the reserves are located in Austral-Asia, 16.7% each in North America and CIS, 8.3% in Latin America, and 5.5% in Africa. There are no significant deposits in Europe. Current estimates put the global uranium reserves at 2ZJ and the resources at 17.2ZJ for once-through fuel cycles. More deployment of breeder reactors which are capable of producing more fissile fuel than they consume raises the resource estimate to 1000ZJ (IPCC, 2007). Equivalent estimates by the International Atomic Agency (IEA, 2005) are much higher (2500ZJ).

Advanced technologies for single-cycle and fast breeder reactors are available which could increase significantly the contribution of nuclear energy to the global primary energy supply. In spite of the low reserves of uranium compared with the other fuels, the fact that only small quantities are needed to generate enormous amounts of energy makes the fuel a very important primary energy source. However, due to safety, environmental and political concerns about proliferation of nuclear technology, nuclear power plants are being de-commissioned and plans to build new ones are being suspended. In 2012, nuclear reactors provided only 5.2% of the world's primary energy and 13% of the world's electricity (IEA, 2012). About a third of the reactors currently in operation are deployed to power naval vessels.

2.2.3   Renewable energy

Renewable energy is the group of energy sources that derive from resources which are naturally replenished. Technically, fossil fuels classified as non-renewable also fall in this category, the difference being the timescale. While it would take 350 to 400 million years to replenish the fossil fuel resources currently being exploited, energy is available on a continuous basis from the Sun, the wind, the tides, and a third of the Earth's surface is covered by water which could be harnessed for hydropower. Biomass resources can be widely regenerated on an annual timescale and geological activities in the Earth's crust which generate enormous geothermal energy occur on a continuous basis. Also, while fossil fuel resources are concentrated in relatively few geographical locations, renewable energy resources are available worldwide. The Sun's energy that falls on the Earth's surface daily is enormous and it is estimated that capturing less than 0.02% will provide enough energy to meet the world's total annual primary energy requirements (Tester et al., 2005). Despite the relative advantages of renewable energy over fossil fuels, only about 10% of the current global primary energy demand is filled from renewable resources, about two-thirds from biomass which is used mainly for heating. Many constraints limit fast deployment of renewable energy in spite of the fact that the technologies are well developed. Solar, wind, wave and tidal, hydro-power are all weather-sensitive and site-specific, solar cells are expensive, biomass cultivation is weather-dependent, and the usual inertia that characterizes transition from known to new technologies is also a constraint.

Energy efficiency is often classified with energy resources because most current primary renewable and non-renewable energy utilization technologies are grossly inefficient and even a 1% increase in efficiency can translate to a significant decrease in energy demand and extend the life of non-renewable resources.

2.2.4   New renewable energy technologies: Nuclear fission

The prime focus of worldwide research on energy is how to manage existing supplies effectively and create new sources. However, considerable research is also on-going on simulation and exploitation of the processes by which the Sun and the stars generate energy. In contrast with nuclear fission which involves splitting atoms to release enormous energy, the Sun generates energy by fusing two hydrogen atoms to form helium thereby releasing a large amount of binding energy. Hydrogen is the primary constituent of the Sun and the prevailing internal temperatures and pressures are sufficiently high to sustain the reactions on a continuous basis.

Nuclear fusion reactors which are currently under development derive hydrogen from sea water and other sources, and can produce enormous net energy that can be harnessed for electric power generation. The high temperatures and pressures required to sustain the reaction and availability of appropriate materials are major constraints and, although there are many research fusion reactors in the world, none has reached a commercial stage.

2.3   WORLD ENERGY SUPPLY

Energy is the prime mover of economic development and availability in sustainable and acceptable forms will to a large extent control the pace of world economic and human development in the foreseeable future. The world is endowed with abundant energy resources – fossil fuels (wood, coal, oil gas), renewable energy (solar, wind, water, biomass, geothermal), and nuclear energy. The problem is how to manage these resources to provide a balanced mix to satisfy the strongly growing world energy requirements without the exacerbation of the negative effects on the environment. The inequitable geographical distribution of fossil fuel resources has made energy supply, particularly oil a potent variable in geopolitics. Furthermore, it has made agreement on environmental issues associated with energy use very problematic, since coal, the most potent pollutant is also the most widely available, and often the only major source of primary energy in many developing countries.

The estimated world's total primary energy supply (TPES) in 2013 was 12730.4 million tonnes of oil equivalent (Mtoe) (BP, 2014). This did not include traded electricity between countries which is not regarded as primary energy because it is already accounted for by the primary energy demand. There has been a steady growth in demand over the last decade or so (Figure 2.8). Nearly all of the increase in demand over the same period has been driven by developing economies. In 2013, about 80% of the total increase in the global energy demand was in emerging countries.

2.3.1 Fossil fuels.

Fossil fuels (coal, oil and gas) have remained the dominant primary energy sources over the decade and accounted for about 87% of the total world energy demand despite the numerous diversification efforts by various countries (Figure 2.9). A projection by the World Energy Council (WEC, 2013) to 2020 is shown in Figure 2.10. Industry, transport, domestic energy, are all heavily dependent on fossil fuels (Figure 2.11). Many projections have been made in recent years by various bodies on the world's primary energy scenarios in 2035 and beyond and most predict a rise in demand of up to 40%, with fossil fuels still contributing about 75% (IEA, 2013). The bulk of the growth will be accounted for by the emerging and developing economies. The IEA projection is considerably lower than an earlier projection in 2003 which estimated a growth of about 65%. However, the latest projection was based on the assumption that all policy commitments and plans to reduce energy demand announced by all major consumers will be achieved by 2035.

It should be noted however that predictions made twenty years ago about global total primary energy supplies (TPES) today have proved to be gross underestimates. However, predictions on resources have also fallen far short of present estimates due to discovery of new resources and development of new advanced technologies for accessing resources that were considered unrecoverable twenty years ago.

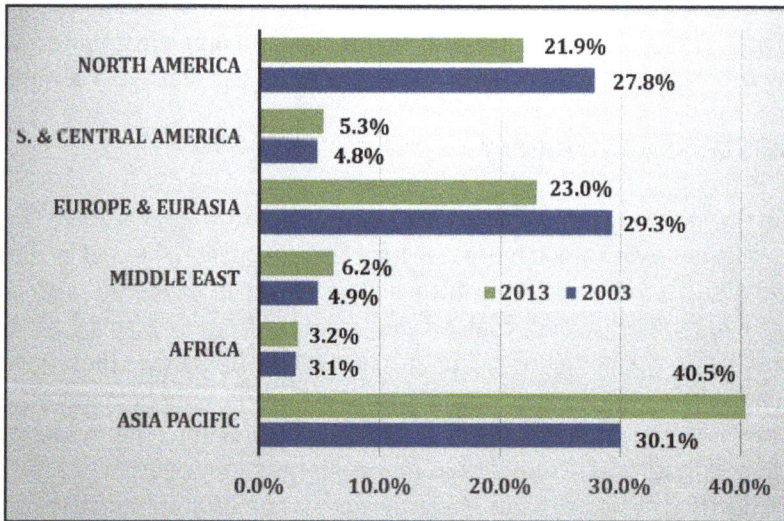

FIGURE 2.8 World's primary energy supply by region in 2003 and 2013. *(BP, 2014).*

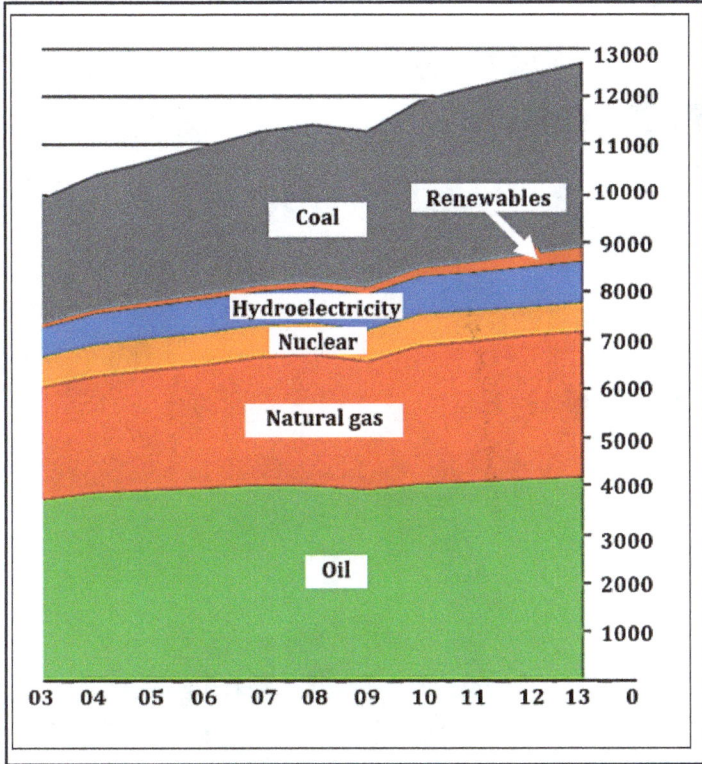

FIGURE 2.9 World's primary energy consumption, 2003-2013. (*BP, 2014*).

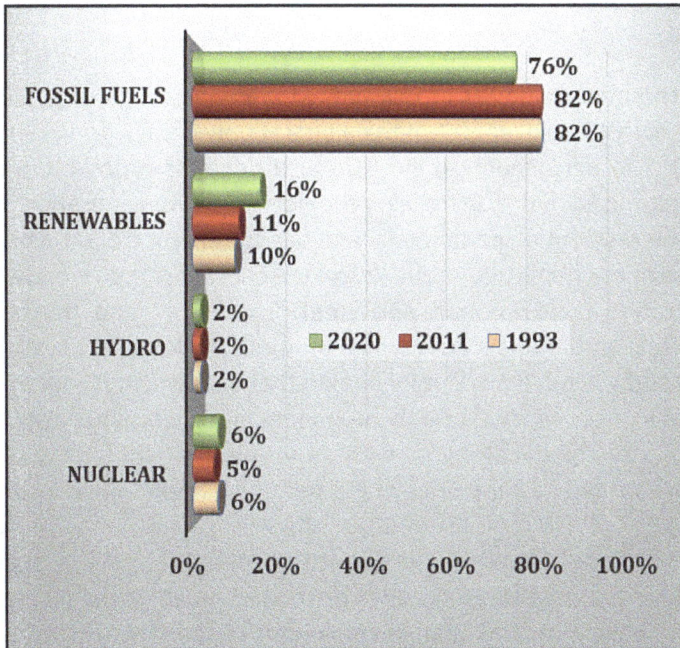

FIGURE 2.10 World's primary energy demand projection to 2020. *(WEC, 2013)*.

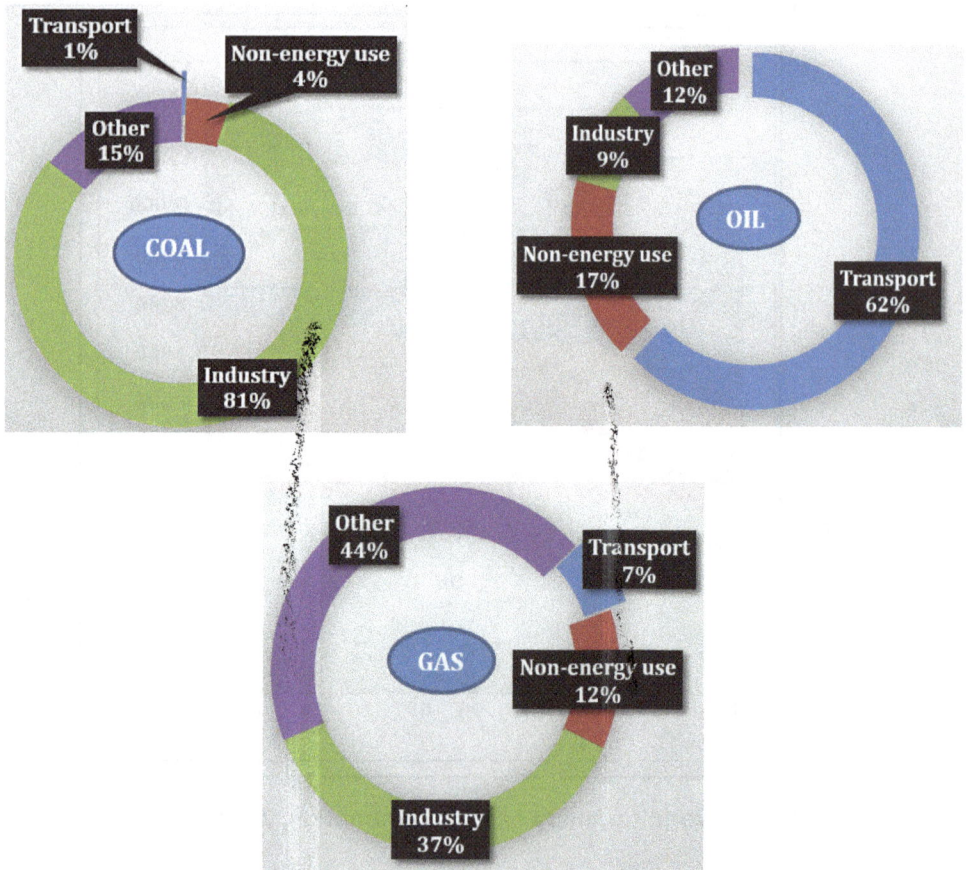

FIGURE 2.11 Share of world fossil fuel consumption in 2012. (*IEA, 2013*).

Oil remains the most important fossil fuel with a wide range of applications, in particular, transportation. In 2011, over 60% of the global supply of oil was used by the transport sector (Figure 2.11). The petrochemical industry also depends on by-products of oil refining for feedstock. Gas is used mainly in industry for electric power generation and industrial heating, and also in the residential, commercial and agricultural sectors. Gas is also a rich source of chemicals and petrochemicals. About 42% of the coal consumed worldwide is used for electric power generation, 14% for iron and steel production, and the balance for industrial, commercial, residential applications. The unpredictability and unreliability of supplies and pricing of oil and gas have made it inevitable that coal which is much more widely available at relatively low prices will remain very prominent in the world's energy mix in the foreseeable future. Coal is also cheap compared with other major fossil energy sources. For a country that has both coal and gas, the price of gas per unit calorific value is about fifteen times that of coal.

There is little doubt that the dominance of fossil fuels will be sustained at around the current level for the next three decades or so, and much of the increase in demand will be due to the virile pace of economic development of the emerging economies, in particular, China and India where significant diversification to other fuel sources is unlikely. World transportation is heavily carbonized and this will continue to be the case in the foreseeable future since there is still no mature alternative to fuel oil while transportation (air, sea, ground) which depends heavily on fossil fuels continues to rise internationally, at double digit rates in some emerging countries. Despite fluctuating oil prices in the last few years and instability in the oil supply chain, the forecast is that oil demand will continue to grow steadily in the foreseeable future (WEC, 2013). Also current efforts to reduce the inequitable worldwide accessibility to primary energy will contribute significantly to future power demand.

In 2004, only one in six people worldwide had access to energy required to provide the high standard of living enjoyed in the developed world. This small group in the developed world consumed about 50% of the world's primary energy supply while the lowest 10% of the world's population consumed only 4% (WBCSD, 2004). The global population is growing and this increases the pressure on social equity in the energy sector. About a third of the world's population could be excluded from the opportunities presented by access to energy. While the focus in most developed countries in recent years has been on the environmental damage caused by energy use, most developing countries have different priorities which focus primarily on access to energy in the first instance. About 70% of Sub-Saharan Africa's population (58% of the total African population), most of who live in rural areas, lack access to electricity.

Much of the investment in the global primary energy sector has been in the replacement of ageing plants and equipment in the developed economies, and installation of new traditional fossil-fueled plants to meet the booming demand for electricity in the developing economies (IEA, 2014). New coal-fired electric power plants are projected to dominate future investment in generation capacity in China, India, and many other countries of Asia, Latin America and Africa due to the relatively wide geographical spread and abundance of coal deposits. The relatively slow projected growth in the deployment of new coal-fired generating plants in developed countries has impacted negatively on investments in the improvement of efficiencies of existing technologies, or development of new alternative technologies. Much of the dynamism in the world's primary energy market is coming from the developing economies where investments in the primary energy sector are unpredictable because they are shaped by government policy measures and competing options for the commitment of funds.

2.3.2   Nuclear fuel

The world has substantial uranium resources, with a growth rate of over 10% in the last five years. Before 1954 when the first nuclear electricity plant was commissioned, uranium was used primarily for weaponry. Over the next fifty years or so, the number of installations grew very rapidly and nuclear plants were supplying nearly 20% of the global electricity in the late nineteen eighties. In some countries, nuclear power accounted for between 40 and 75% of electric power production in 2013 (Figure 2.12). However the share had declined to only 13.5% in 2012 due to a series of accidents: Three Mile Island in 1979, Chernobyl in 1986, and Fukushima in 2011. Japan has the world's third largest installed capacity of nuclear power (11.2% in 2013). In 2010, nuclear power provided about 26% of the country's primary energy supply (IEA, 2015b). However, the most recent Fukushima accident in 2011 caused Japan to shut down 52 of its 54 reactors and share of nuclear in the domestic electricity generation had dropped to less than 2% in 2013. On the contrary, China is accelerating the deployment of nuclear electricity plants. According to the International Atomic Energy Agency (IAEA, 2013),

28 of the 68 nuclear reactors which were under construction in 15 countries in 2012, were in the People's Republic of China.

2.3.3 Renewable energy

Unlike fossil and nuclear fuels which are exhaustible, renewable energy resources: water, solar, biomass, wind and geothermal, regenerate and can be sustained indefinitely. Solar and biomass energy resource utilization is as old as mankind but the discoveries of coal, oil and gas have stunted development of renewable energy which presently contributes only about 10% to the global primary energy supply. The regional share of renewable energy is shown in Figure 2.13.

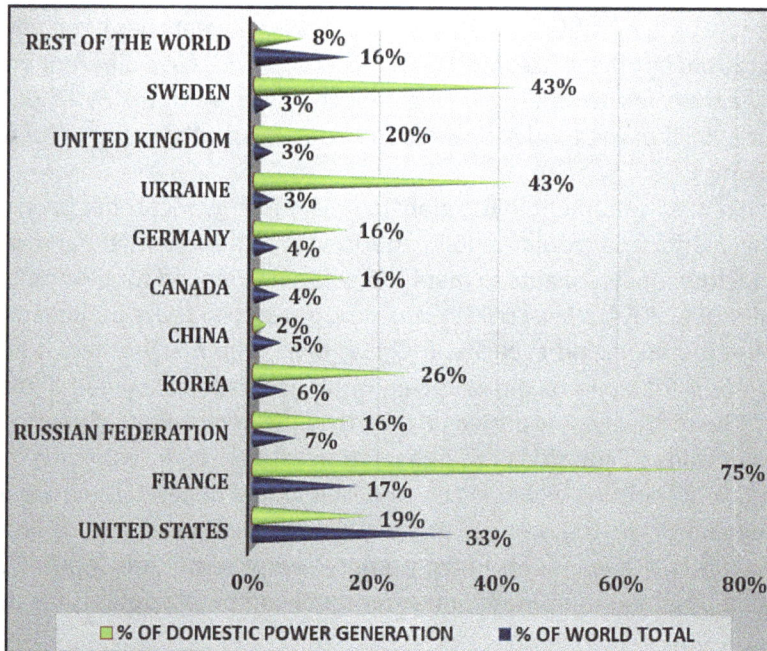

FIGURE 2.12 World's top ten producers of nuclear electricity in 2013. *(IEA, 2013b).*

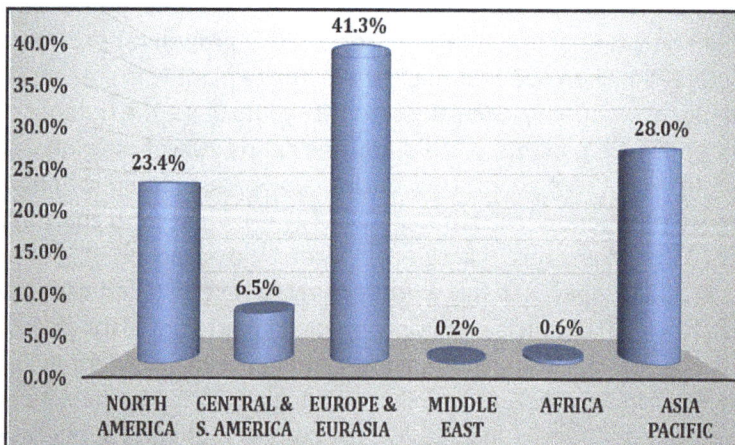

FIGURE 2.13 World renewable energy supply by region in 2013. *(BP, 2014).*

Europe and Eurasia lead the world with 41.3%, while the highest growth rates of around 24% between 2012 and 2013 had been in South and Central America and the Middle East. Global output of biofuels (ethanol and biodiesel) increased by about 6% in the last two years, with North America and South and Central America accounting for over 80% of the production in 2013.

The development of renewable energy, with the exception of hydro-energy has been slow despite the exponential growth of renewable resources, due to the intermittent nature of the resources and dependence on weather by solar, wind and water energy availability. Draught also affects water levels for hydropower generation and cultivation of crops for biomass energy. The bulk of renewable energy, in particular, biomass, solar and wind is used for electric power generation and the intermittent nature makes it difficult to integrate the power produced with national electricity grids. Furthermore, most of the private sector financial investments in the primary energy sector have been on fossil fuels and much of the development of renewable energy sector has been driven by public sector initiatives and incentives. However, this situation is about to change, considering the recent declaration by the world's major fossil energy producers to diversify and invest more in renewable energy.

In the last two decades or so, efforts have been made to develop bi-fuel hybrid power plants which combine wind, solar, biomass or tidal power with fossil fuel. Some hybrid solar-gas plants are already in commercial operation in some developed economies, in particular, the United States of America. Renewable energy is deployed whenever available and the intermittent gaps are filled by natural gas.

2.3.3.1 *Hydropower*

Hydro power contributed only 6.7% to the global primary energy mix in 2012 but its share of electricity generation capacity was significantly higher, at about 15%. Hydroelectricity is available in over 150 countries but only five countries accounted for about 55% of the total world's capacity in 2012 (Figure 2.14).

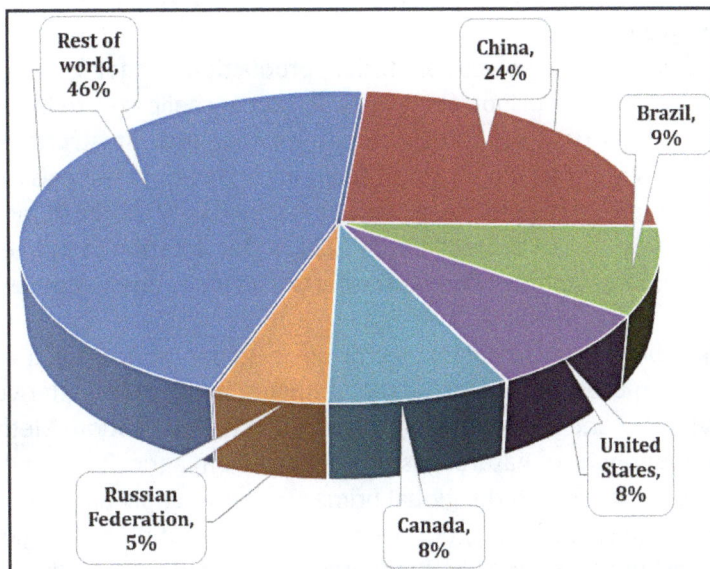

FIGURE 2.14   Countries with the highest hydro energy capacity in 2012.
(WEC, 2013).

The Asia-Pacific region led the world, generating 32% of the global hydro power in 2012. Brazil and Canada increased their shares to 10.2% and 10.6% respectively in 2013, China maintained the same share while the shares of the other three countries dropped slightly (BP, 2014).

In some countries, hydro power accounts for over 50% of all electricity generation. The potential for a rapid increase in global hydro power generation is very high and the United Nations is leading efforts to develop mini hydro dams in many developing countries. The growth of hydro power over the last two decades, estimated at about 55% has been facilitated largely by the relatively wide geographical availability. The increasing stringency of environmental pollution controls has also affected positively the deployment of hydro power.

Hydropower has many advantages over other fuels in generating electricity. The unit cost of the electricity produced is low especially when the dam is large, it responds flexibly to changing energy demands, it produces no direct waste, and the carbon footprint is relatively low. However, development is being curtailed by ecological problems, especially in the developed countries. Full utilization of installed capacities has also been affected by increasing shortages of water due to variations in climatic conditions.

2.3.3.2 *Bioenergy*

Bioenergy is a group of fuels of biological origin. The group includes wood which is perhaps the oldest source of energy used by mankind apart from solar energy. Other fuels in the group are charcoal, organic waste from forestry and agriculture, municipal solid waste, animal manure, human waste, and biofuels derived from crops including sugarcane and corn. Bioenergy is already contributing substantially to global primary energy supply. In 2013, bioenergy accounted for about 10% to the global primary energy supply, about two-thirds in traditional biomass use (space heating, etc) and the balance was used for heat and power generation or for conversion into road transportation fuels (IEA 2015a).

Projections indicate that 25% to 35% of the world's primary energy supply in 2050 will come from biofuels and the bulk of the increase will be in environment-friendly power generation and transportation fuel production (IEA, 2014; IPCC 2014; IEA, 2015b). This development would substantially reduce dependence on fossil fuels since potential areas of utilization include electric power generation, production of heat for industrial and domestic use, and production of transportation fuels. Potential challenges to bioenergy development include sustainability of raw material sources, logistics and infrastructure development, and environmental issues, transition from an economy driven by fossil fuels to a bioeconomy, and greater involvement of the private sector in a technology that has so far been driven largely by political strategies (IEA Biotechnology, 2016). The international Energy Agency (IEA) is coordinating research and development efforts that focus on the above issues in about fifty countries.

The bulk of bioenergy is used for electric power generation, production of heat and steam for industrial purposes, and space heating. Ethanol and biodiesel derived from sugarcane, corn, vegetable oils and animal fats are also used in transportation. Methane (biogas) which is the main component of natural gas is produced from agricultural and human waste. The contribution of bioenergy to the global primary energy supply has grown by 23% in the last twenty years or so (WEC, 2013), but the percent contribution to the global primary energy supply remains low. Traditionally, bioenergy has been produced by direct combustion of biomass. However, technologies have now become available for converting biomass to biofuels which are more efficient and can be utilized or transported more conveniently. The first generation of biofuels are made by fermentation of sugar derived from wheat, corn, sugar cane, sugar beets, molasses, etc., using microorganisms and enzymes. Starch (from corn, potato, fruit waste) and vegetable oils are also fermented to produce alcohols, mostly ethanol but propanol and butanol are also produced. While butanol can be used in gasoline engines directly as a replacement for gasoline, ethanol is blended with gasoline.

Brazil and the US lead the world in the production and utilization of biofuels from starch derived from corn, cassava, etc, while the main focus of Europe is on the production of biodiesel from oils and plant/animal fats (soy, rapeseed, palm oil, mustard, sunflower oils etc.) by catalyzed transesterification. The process of producing fuel from oils and animal fats involves reacting an alcohol with an ester to produce a different alcohol through an exchange of organic carbonyl carbon groups in the feedstock with organic groups of alcohol. Biodiesel produced by this process is more environment-friendly and produces 60% less pollution compared with fossil-sourced diesel oil.

2.3.3.3 *Solar Energy*

The Sun's radiation about 60% of which reaches the Earth is the most abundant primary energy source both in intensity and geographical availability. The annual potential of solar energy is several times larger than the total world primary energy consumption (UNDP, 2000). Solar energy is relatively simple to deploy, it is highly reliable when it is available, it can be quickly installed and dismantled, and it provides simple solution to power problems in remote areas and developing economies. Solar radiation is regarded as clean energy because it does not add directly to the global anthropogenic waste load. However, on a total life cycle basis, the carbon footprint is significant. There are formidable technical problems which restrict the proliferation of solar energy use. Most of the available solar energy capacity is committed to electric power generation and it accounted for only 2.8% of the world's total supply of about 5000GW in 2013. Only five countries produced over 90% of the total global power output (Figure 2.15).

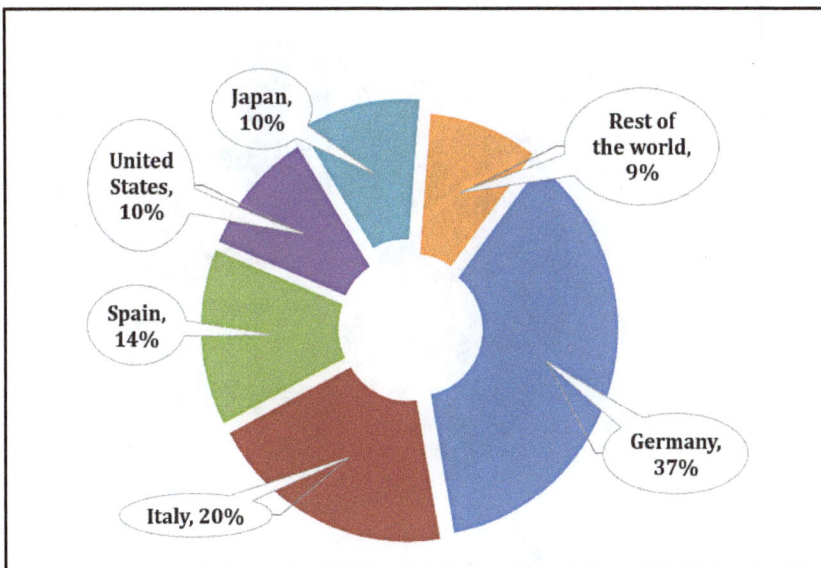

FIGURE 2.15 Global solar electric power generation in 2011. *(WEC, 2013).*

The intermittent nature of solar energy which is only available in the daytime, the variable geographical intensity, and the astronomical cost of solar photovoltaic (PV) panels had been major constraints to a widespread deployment of solar energy. However, the use of solar energy is now the fastest growing global source of energy with a growth rate of about 35% over the past few years due mainly to the decline in the cost of solar panels and development of parabolic collectors. The major investments in the past few years have been led by United States, China, Europe, Japan and India.

2.3.3.4 *Wind energy*

Like solar energy, wind power is available virtually everywhere on the Earth. In fact, wind energy derives from the uneven heating of the atmosphere by the Sun. The speed of movement of the wind is controlled by the Earth's terrain, the Earth's rotation, vegetation, oceans and seas, etc. Wind speed is usually strongest over open oceans which cover over 70% of the planet because of fewer obstructions. Strong wind passing through a propeller can generate considerable thrust, very much like in the aircraft, and can be used to drive a turbine to generate electricity. If it were possible to harness just 5% of the available potential wind energy, it would provide electric power equivalent to the current global total primary energy needs. World wind energy capacity has been doubling every few years, mostly used for electric power generation (Figure 2.16). Actual generation in 2011 was about 378,000GWh compared with only 53,000GWh for solar power.

Wind energy is very attractive especially on the relatively low environmental carbon footprint because no waste is produced. However, one major disadvantage of wind power is irregular and often unpredictable availability. In effect, wind power is usually fed into a grid whenever available or combined with another power source such as a gas-fired generating plant. Another potential drawback is the relatively high capital investment compared with fossil fueled power plants.

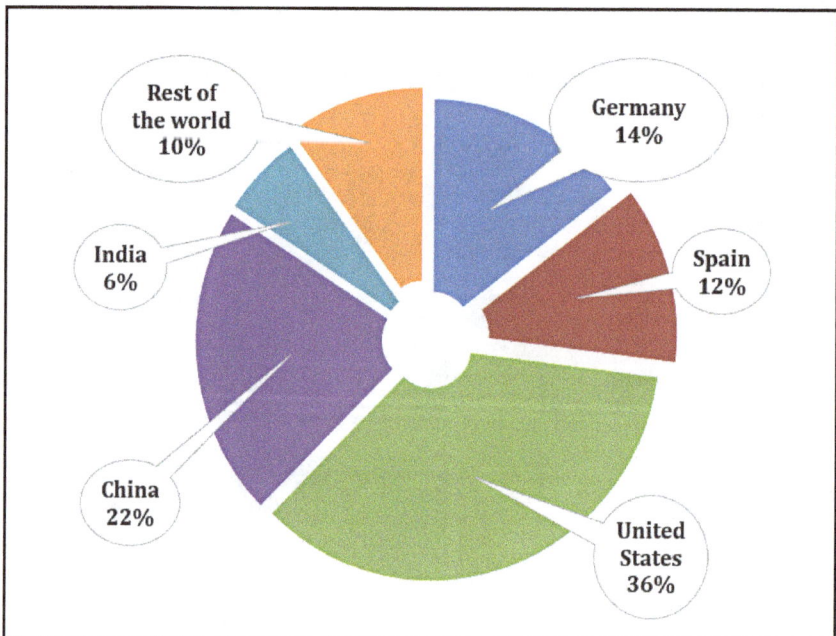

FIGURE 2.16   Global wind electric power generation in 2011. *(WEC, 2013).*

Wind power has several advantages over other primary power sources. Operating costs are much lower since there are no fuel costs, plant equipment is simple and maintenance costs are low. Hence the economics of a wind-powered power plant becomes competitive when compared with other power generators on a life-cycle cost basis. The fact that wind of the right quality is intermittent and cannot be made available when needed is a deterrent to faster deployment of wind-based power plants. However, developments in recent years have led to configuration of bi-fuel power plants featuring integrated wind power and a fossil fuel. The plants operate on wind whenever available and switch to a fossil fuel to fill gaps. Also, because the best wind farms are usually in remote areas with relatively few obstructions, many wind powered generating plants are deployed to serve remote communities.

2.3.3.5   *Geothermal energy*

The Earth's crust stores an enormous amount of energy, believed to have originated largely from radioactive decay of minerals but also during the Earth's formation. The core mantle boundary which is about ten kilometres deep is believed to be about 4000ºC, hot enough to melt rocks and set up a geothermal gradient between the core and the Earth's surface. High pressures force molten rocks towards the Earth's surface. Underground water that comes in contact with the heat becomes heated and penetrates to the Earth's surface as warm springs. In some locations the temperature could be as high as 380ºC, enough to produce superheated steam which can be captured for electric power generation. It is estimated that the upper ten kilometres of the Earth's crust contains 50,000 times as much energy as found in all the World's fossil fuels. Geothermal energy is abundant in areas located around the boundaries of tectonic plates which make up the Earth's outermost shell also known as the lithosphere. Countries which are rich in geothermal energy include North and Latin American countries, African countries, Japan, Australia and Russia.

Geothermal energy has been used since the ancient times as source of domestic hot water and space heating, and, currently, is being used extensively for space and industrial process heating, heating of pools and greenhouses, and water treatment. However, focus has shifted in the last decade or two to electric power generation and commercial power plants are operating in around 25 countries of the world including countries located along the Great Rift Valley of Africa (Kenya, Ethiopia, Uganda), and others located around the East Mediterranean. The world's largest geothermal plant (850 megawatts) is located in California, USA, and hundreds of new plants are being developed around the world. El Salvador and Iceland currently derive nearly 30% of their electricity from geothermal energy and Kenya plans to source over half of its electric power requirements from geothermal energy within the next decade (Brown, 2009).

New technologies are being developed which involve drilling to the hot rock layer, fracturing the rock and pumping water into the crack to generate superheated steam for driving steam turbines in power generating plants above ground. It is believed that when this new technology becomes fully developed, geothermal energy could supply the entire electricity requirements of many countries, including the United States.

2.3.4   Global primary energy outlook

There are so many variables involved in the global primary energy dynamics that make accurate prediction of future trends very difficult. Changes in the global oil market have a ripple effect on the prices of other fuels and new technologies can have unintended effects on projections. For example, introducing new technologies in oil and gas production in the United States has succeeded in transforming from an energy importer to a net energy exporter in less than a decade. However, this development has depressed the local energy prices to the extent that over 60% of oil and gas rigs have been shut down in the last two or three years.

Another unpredictable driver of global primary energy demand is the rate of growth in emerging countries many of which currently provide energy for less than a third of the population. India has a sixth of the world's population and, although it is the world's third largest economy, the country accounts for only 6% of the global primary energy use. Coal, the country's main primary resource will fuel the expected energy demand growth which is projected to be the most rapid in the world over the next two decades. Many other emerging countries in Asia and Africa are also expected to rely heavily on coal for the projected growth in primary energy production. All projections show that fossil fuels will continue to dominate the global primary energy supply for decades, accounting for nearly 80% by 2040 (EIA, 2016).

REFERENCES

BP (2014). *Statistical Review of World Energy.* www.bp.com

Brown, D. W. (2009). "Hot Dry Rock Geothermal Energy: Important Lessons from Fenton Hill." *Proc. Thirty-Fourth Workshop on Geothermal Reservoir Engineering.* Stanford University California, February 9-11, 2009. SGP-TR-187.

EIA (2016). "International Energy Outlook 2016." eia.gov.

IAEA (2013). "Gaseous emissions from coal stockpiles." Clean Coal Centre PF 13-1.

IEA (2005). *World Energy Outlook.* International Energy Agency.www.iea.org

IEA (2012). *World Energy Outlook.* International Energy Agency.www.iea.org

IEA (2013). *Resources to Reserves.* International Energy Agency. www.iea.org.

IEA (2014). *World Energy Outlook.* International Energy Agency. www.iea.org.

IEA (2015a). *Key World Energy Statistics.* International Energy Agency. www.iea.org

IEA (2015b). *Realizing The World's Sustainable Bioenergy Potential.* Proceedings, IEA Bioenergy Conference, Berlin. International Energy Agency. www.iea.org

IEA (2016). *Developing the Global Bioeconomy: Technical, Marker, and Environmental Lessons from Bioenergy.* Lamers et al., (Eds). Academic Press. International Energy Agency. www.iea.org

IPCC (2014). *Climate Change 2014: Mitigation of Climate Change.* Contribution of Working Group III to the Fifth Assessment Report of the Intergovernmental Panel on Climate Chang. Cambridge Inoversity Press.

Tester, J. W, Drake, E.M., Driscoll, M. J., Golay, M. W., and W. A. Peters (2005). *Sustainable Energy, Choosing Among Options.* MIT Press.

UNDP (2000). *World Energy Overview.* United Nations Development Program. www.undp.org.

WBCSD (2004). *The GHG Protocol: A corporate reporting and accounting standard (revised edition).* World Business Council for Sustainable Development. 116 pages.

WEC (2012). *World Energy Insight 2012. World Energy Council. www.worldenergy.org.*

3   Coal in the global energy mix

3.1 INTRODUCTION

Coal is one of the three major global primary energy resources, the oldest, and most widely used of fossil fuels despite its severe negative impact on the environment. It is the most widely available of the fossil fuels, and the least expensive. Coal has been in use from the ancient times and was the primary source of energy that powered the Industrial Revolution of the late nineteenth century. Despite earlier predictions that the use of coal will decline because of the harmful effects on the environment, the fuel has remained a vital source of energy. Most projections indicate that coal will maintain its prominence in the foreseeable future and may replace oil as the world's largest source of primary energy in the near future.

   Coal mining is a major source of direct and indirect employment, especially in developing countries and large scale coal mines are often the biggest source of employment for rural communities. In South Africa and Columbia, industry reform and stable economic conditions for export have attracted major foreign investment for the establishment of large coal mines which provide employment and export large proportions of their production.

3.2 COAL RESOURCES OF THE WORLD

Coal is widely distributed all over the world and major deposits have been found on every continent. About a hundred countries have coal deposits and many more remain unexplored. Estimates of the world's coal reserves range from 850 billion to one trillion tonnes, enough to last for over a hundred years at the current and projected rates of exploitation, substantially longer than oil and gas reserves. The regional distribution of the global coal reserves is presented in Figure 3.1. Six countries hold 76% of global reserves (Figure 3.2). The various published reserve estimates are considered low in view of the fact that coal exploration does not attract substantial investments compared with oil and gas and much of the world remains unexplored. Furthermore, many abandoned mines still contain valuable large amounts of coal which can be recovered using new technologies, or gasified in-situ to produce energy. Coal provides primary energy for a wide range of industrial processes including electric power generation, iron and steel production and steam raising.

   Coal's share of global energy mix has continued to rise in spite of the associated environmental problems, and coal accounted for about 30% of the world's total primary energy demand in 2013 (Figure 3.3). The Asia Pacific Region accounted for 70% of the total demand for coal in the same year (Figure 3.4).

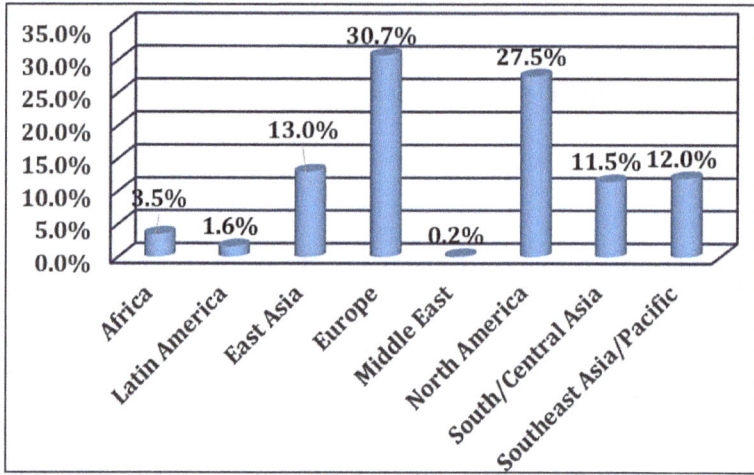

FIGURE 3.1 Coal reserves by region. *(Data from WEC, 2013).*

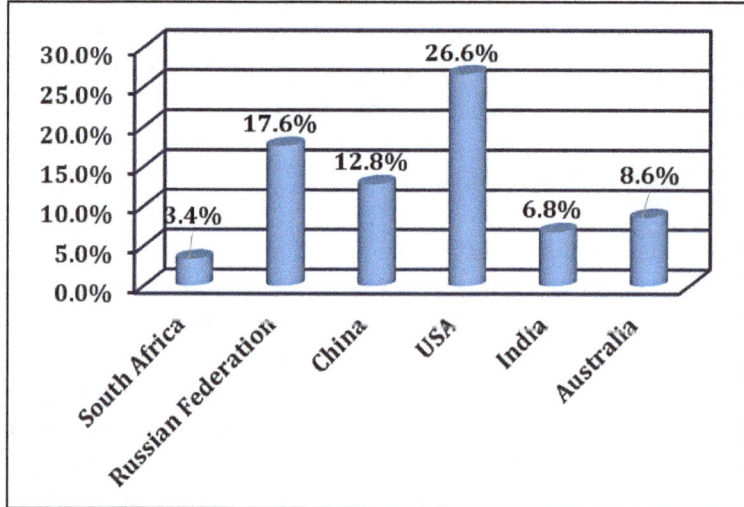

FIGURE 3.2 Six countries which hold 76% of the global coal resources. *(Data from WEC, 2013).*

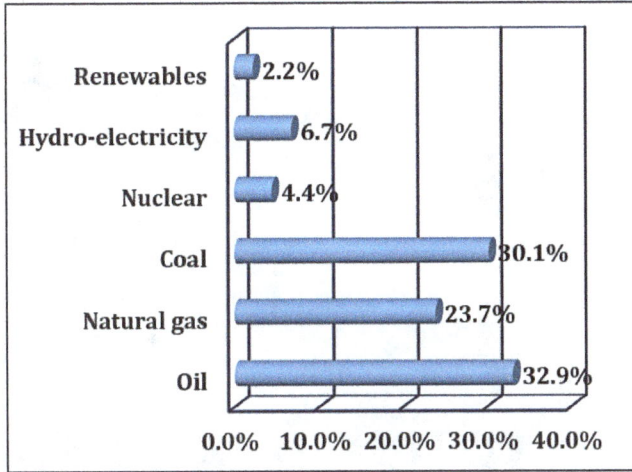

FIGURE 3.3 World primary energy supply mix in 2013. *(BP, 2014).*

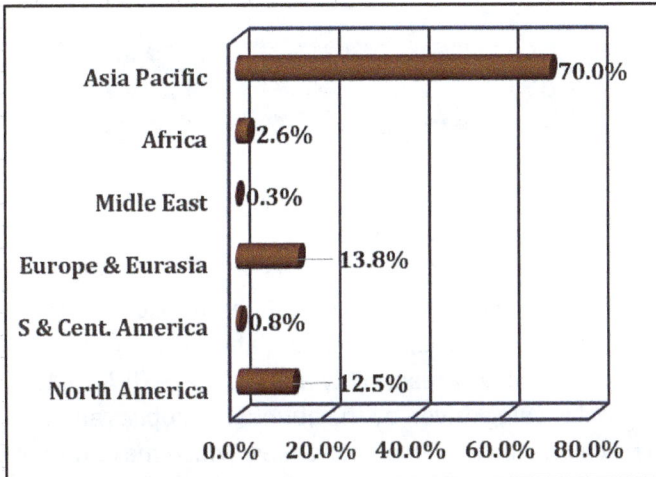

FIGURE 3.4 World coal demand in 2013. *(BP, 2014).*

Projections indicate that coal may surpass oil as the world's top primary energy source by the end of the decade and the demand will increase in every region with the exception of the United States, where coal is being gradually replaced by natural gas (IEA, 2013). Even in the United States, coal utilization is beginning to rise again due to the rising prices of natural gas and increasing opportunities for coal export. In a recent projection of world coal utilization (2012-2040), just three countries (The United States, China and India) will account for more than 70% of world coal use (EIA, 2016). The types of coals required for various applications are summarized in Figure 3.5.

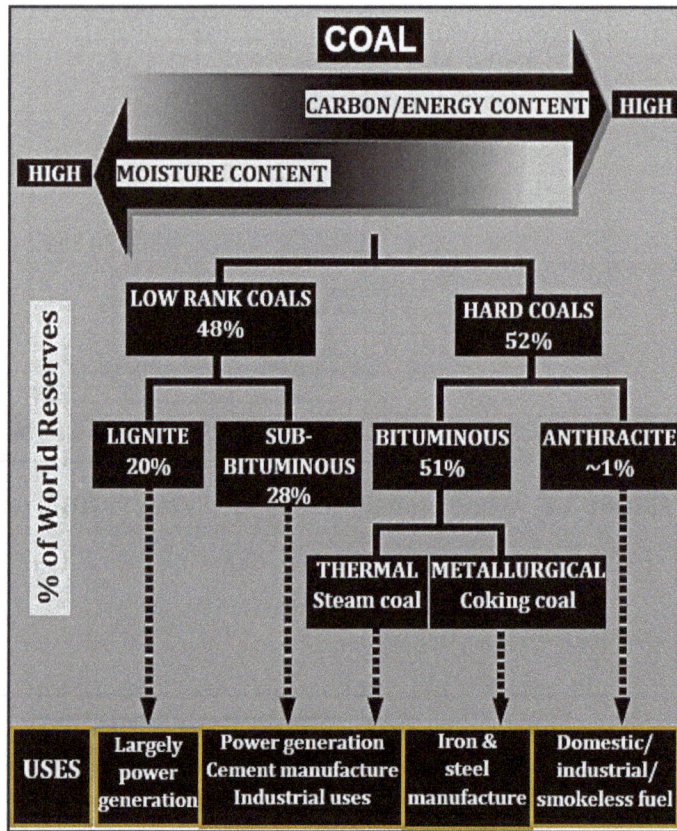

FIGURE 3.5 Types of coals required for various applications. *(worldcoal.org)*.

Coal's share of the global energy mix has continued to rise over the past decade or so and, although there has been a slight reduction in the rate in the last few years, projections indicate that it will catch and possibly surpass oil within a decade (IEA, 2013). The aggressive rise in the utilization of coal is being driven by the abundant supplies and wide geographical spread of resources, particularly in emerging economies, which make it relatively easy for them to meet the rapid growth in the demand for electric power. For many, it is the only locally available fossil fuel. The biggest demand for coal is in Asia which currently accounts for nearly 70% of global coal consumption and China and India are the highest consumers (WCA, 2012). China could surpass the rest of the world in coal production and consumption within the decade. By 2040, Asia is projected to account for 80% of coal consumed globally (IAEA, 2015c).

Although the demand for coal is decreasing in some countries, in particular, the United States due to an increasing use of natural gas, these countries are exporting more coal to other countries which have no access to cheap natural gas and have less stringent restrictions on carbon emission. Some countries such as Japan and South Korea which do not have sufficient energy resources to meet their requirements import significant quantities of steam coal for electricity generation and coking coal for steel production. Table 3.1 shows the major producers of coal in 2011.

TABLE 3.1 World's leading coal producers in 2011. *(IEA, 2012a).*

PRODUCERS	% OF WORLD TOTAL
China	45.9
United States	12.9
India	7.5
Australia	5.3
Indonesia	4.8
Russian Federation	4.3
South Africa	3.3
Germany	2.4
Poland	1.8
Kazakhstan	1.5
Rest of the world	10.3
World total	100.0

3.3   COAL FOR ELECTRIC POWER PRODUCTION

In 2012, 41% of the world's electric power generation was coal based (Figure 3.6) and coal was the fastest growing form of energy outside renewable energy, with an average annual growth rate of 4.4%. The dynamics of regional share of generation has also changed significantly in the last two decades (Figure 3.7), with Asia now the fastest growth region. Although global coal consumption in absolute terms is expected to continue to increase in the foreseeable future, its share of power generation is likely to reduce due to increasing availability of natural gas. However, the increasing prices of natural gas may slow down its use to some extent.

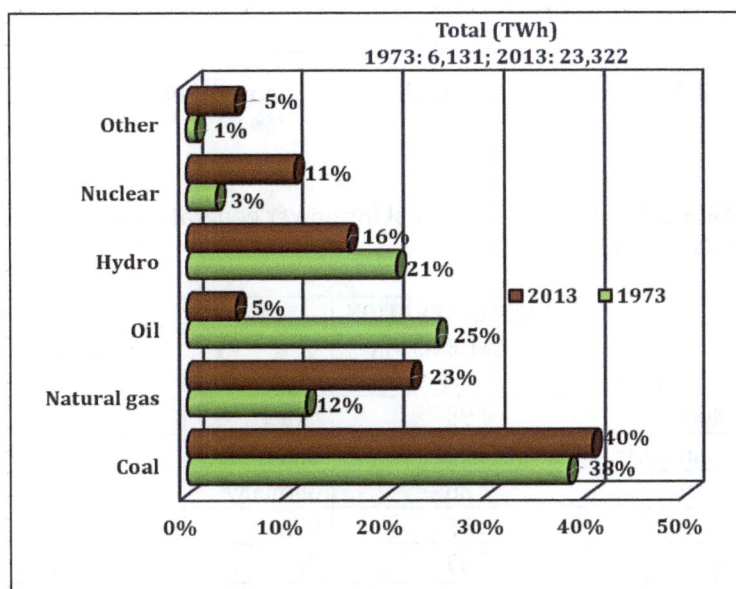

FIGURE 3.6   Fuel shares of electricity generation in 2013. *(IEA, 2015).*

Dependence on coal for power generation varies significantly between countries (Table 3.2). Mongolia and South Africa use coal for over 90% of electricity generation compared with only 43% by Germany. Emerging countries especially in Asia and Africa face significant challenges in meeting the demand for electricity supply and many are able to access readily available local, affordable, and secure coal resources to meet their goals. Already, China derives around 80% of its electricity production from coal-fired plants, India plans to increase its coal-based electricity generation capacity by over 50% by 2017, South Africa, Nigeria and several other countries are planning coal-fired generation plants. It is expected therefore that production of coal-fired electricity will grow rapidly in the near future, especially in the developing economies.

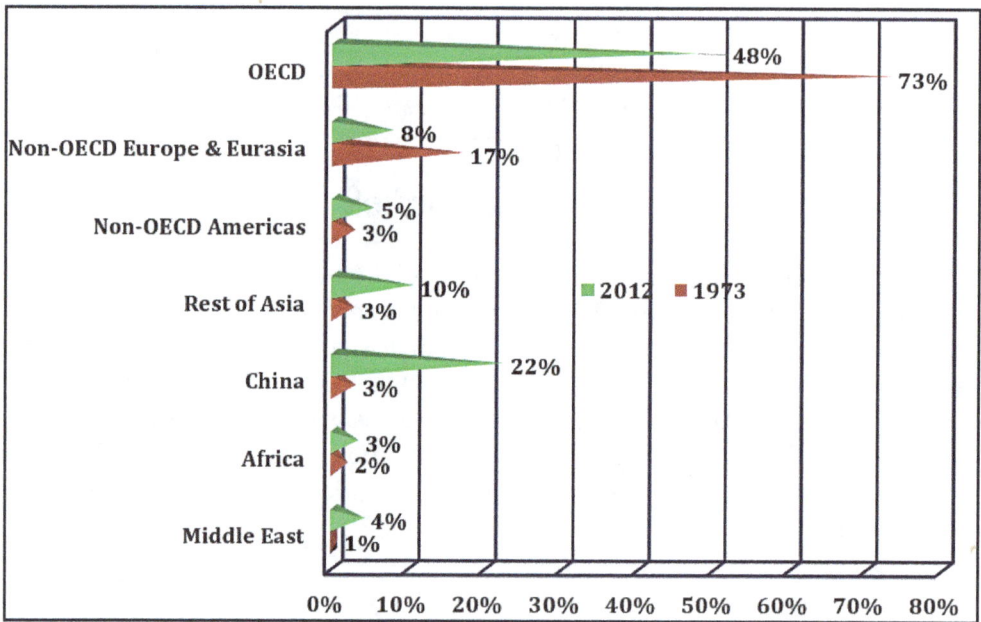

FIGURE 3.7 Regional share of global electric power generation in 2012. *(IEA, 2014).*

TABLE 3.2 Dependence on coal for power generation by countries.
IEA Coal Information, 2013).

COUNTRY	POWER GENERATION DEPENDENCE ON COAL (%)	COUNTRY	POWER GENERATION DEPENDENCE ON COAL (%)
Mongolia	98	Israel	59
South Africa	94	Indonesia	44
Poland	86	Germany	43
PR China	81	USA	43
Australia	69	UK	29
India	68	Japan	27

3.4   COAL FOR STEEL PRODUCTION

The steel industry is a prime mover of the world's economy. Steel is a strong building block and the primary material that stimulates industrial growth, and the fortunes of both have become interdependent. Strong population growth and rapid urbanization are driving demand for steel, with growth in demand for infrastructure development, energy supply from exploration to distribution, food processing, manufacture of consumer products, water, transport and communication projects. Shortage of steel supply impacts negatively on industry while any slow down in the world economy cuts the demand for steel significantly. Machine building, transportation, construction, housing, agriculture, water supply and infrastructure, energy generation industries all depend heavily on steel. About 200 billion steel cans are produced for the food canning industry each year. No other material offers the same unique combination of strength, formability, and versatility at affordable cost.

The total primary steel consumption worldwide in 2012 was about 1.6 billion tonnes, about 70% higher than in 2002, nearly 75% produced by just six countries: China, Japan, United States, India, Russia and South Korea (Figure 3.8). There has also been a pronounced divergence of coal production geographical pattern in the same period (Figure 3.9). South Korea leads the world in the consumption of primary steel and steel products (Figure 3.10) and China has more than doubled production in the last ten years, now accounting for nearly half of the world's total primary steel output (WCA, 2013). China and India together consumed over 50% of the total global crude steel in 2012 (WCA, 2013). Also, China alone consumed about 46% in comparison to only 23% in 2002. With China and India set to account for about 36% of the global population in 2025, the demand for steel is expected to grow in commensurate proportions.

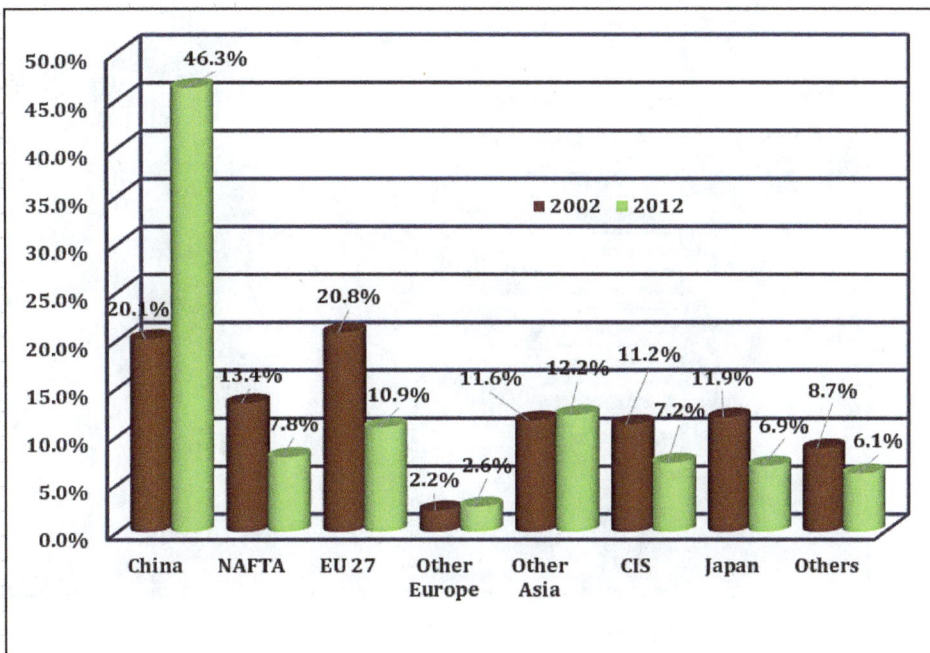

FIGURE 3.8 Regional share of global primary steel production in 2002 & 2012. *(WSA, 2013).*

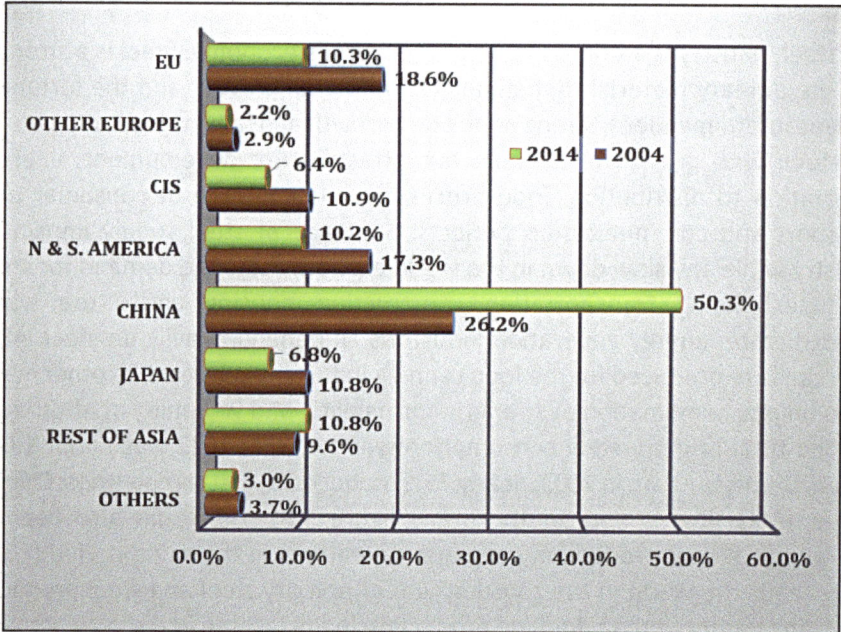

FIGURE 3.9 Regional share of global coal production in 2013. *(IEA, 2015).*

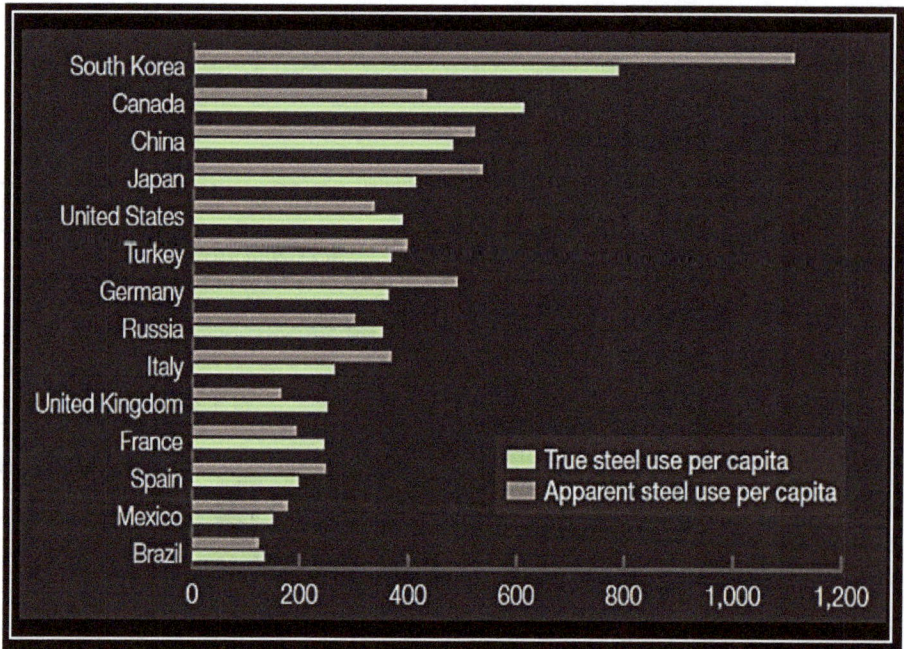

FIGURE 3.10 World steel consumption in kilograms per capita in 2014.
Note: True steel use is primary steel consumption, Apparent steel use includes use of steel products. (WCA, 2016).

Around 60% of the world's population are expected to be urban dwellers by 2030 and the need to develop infrastructure is expected to place even more pressure on the demand for steel. In 2013, about 72% of the global steel production was by the blast furnace-basic oxygen furnace route (BF-BOF) which depends primarily on coal (Figure 3.11). About 770kg of coal is required to produce 1 tonne of primary steel (WCA, 2014). Approximately 14% (around 1 billion tonnes) of hard coals produced worldwide in 2012 were used in producing steel by this process (WCA, 2014). Most of the balance of world steel production (around 28%) was by the electric arc furnace (EAF) route which also depends primarily of coal-based electric power supply. It is estimated that around 160kg of coal is required to produce the electric power needed to produce 1 tonne of steel by this route. The balance of world's steel production, around 3% is by the direct reduction-electric arc furnace (DR-EAF) route. Apart from the coal required for the electric power supply in this process, natural gas or gasified coal is required to produce the direct reduced iron that feeds the arc furnace.

3.5 OTHER USES OF COAL

Coal has many other uses apart from its use primarily for power generation and steelmaking. It provides energy for many industrial processes and is also a source of many industrial chemicals, petrochemicals and premium industrial materials. Furthermore, for many countries, affordable energy from coal is vital for building internationally competitive industries, and providing basic household energy for lighting, heating, cooking and refrigeration.

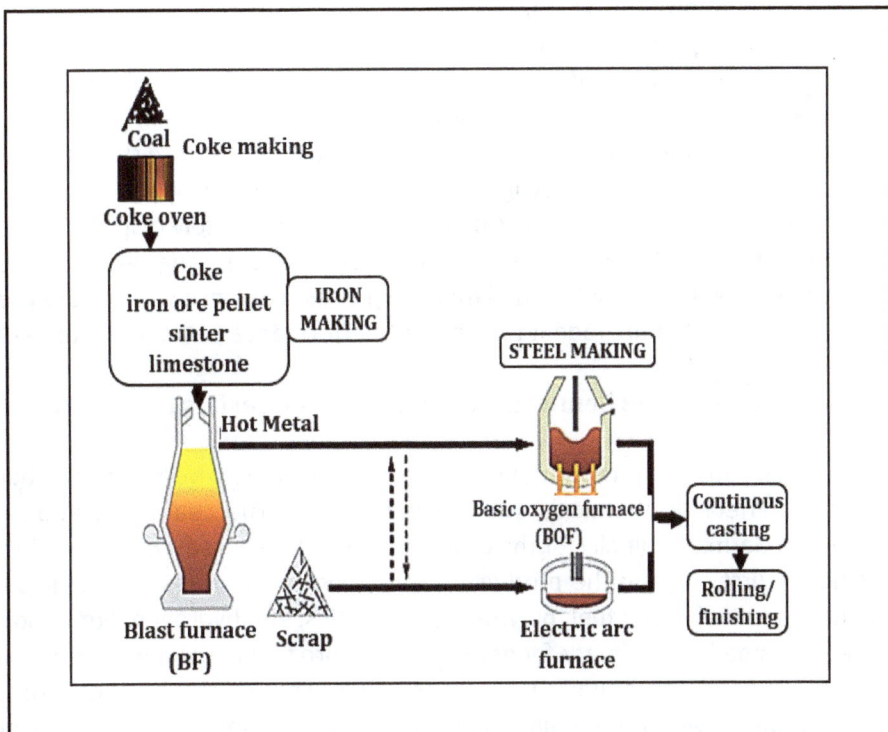

FIGURE 3.11 The blast furnace route for primary steel production. *(wci.org).*

3.5.1 Coal use for cement production, industrial and domestic heating

Coal is used extensively as an energy source in the manufacture of cement and about 220kg of coal is required to produce one tonne of cement. Coal combustion products such as fly ash are also mixed with clinker in the cement manufacturing process. This reduces basic raw material requirements and improves clinker quality because of its high alumina content and the presence of silica, iron and calcium all of which are important elements in cement production. Fly ash is also used to replace part of the cement in concrete up to about 50% to improve early strength, hardening rate and durability, particularly for hydraulic applications. Furthermore, sand aggregate in concrete can be replaced up to about 50% by fly ash to increase strength and elasticity.

Coal is used for the production of steam which is required for many industrial applications and many industrial ovens and furnaces are heated by coal. In many developing countries, coal is burned directly for cooking and home heating and coal that has been processed into smokeless briquettes is in common use in many developed countries for domestic heating, barbeques, etc.

3.5.2 Synthetic gas from coal

One of the major disadvantages of coal use is the obnoxious and harmful tarry, particle-laden smoke that evolves from coal combustion or carbonization. However, coal can be processed to obtain a synthetic gas known as *syngas* which is a mixture of hydrogen and carbon monoxide. Syngas is similar to natural gas but has lower heating value, and is used extensively for power generation and as domestic and industrial energy. The coal-to-syngas process which has been in use for many decades has now assumed a major status as an alternative to direct coal combustion because of its lower carbon footprint. Coal-based syngas is becoming the major fuel for power generation in many countries in preference to coal, and many coal-based power installations all over the world now feature a syngas unit as an integral part of the plant configuration. Syngas has several advantages over direct coal combustion, the most important of which is the ability to capture and safely dispose of the bulk of the pollutants from coal at the gasification stage. Furthermore, syngas-fired power plants are much more efficient than coal-fired plants and process control is much more precise. Syngas can also be enriched to upgrade it to a level of performance comparable to natural gas.

3.5.3 Chemicals, petrochemicals and premium materials from coal

Syngas contains mainly carbon monoxide and hydrogen, two very versatile building blocks for a wide variety of chemicals and petrochemicals. The coal gasification process was developed in Germany initially for the production of gasoline from coal in the nineteen thirties but is now a first stage in the production of a wide range or chemicals including ethanol, methanol, gasoline, diesel fuel, polymers and plastics, and hydrogen. For hundreds of years coal has remained the primary feedstock for the production of a wide range of chemicals including aromatic compounds used as intermediates in the synthesis of dyes, drugs, antiseptics and solvents. Premium chemicals including naphthalene, anthracene, phenanthrene, carbazole and heterocyclic hydrocarbons such as pyridines, and quinolines, and phenols are obtained from the distillation of coal tar, a by-product of coal carbonization, gasification or combustion. Most of the aromatic hydrocarbons such as benzene, toluene and xylene are now obtained mainly from petroleum but coal remains an important source of many organic chemicals. Anthracene, a solid polycyclic aromatic hydrocarbon used in the production of synthetic red dye alizarin, wood preservatives, and insecticides is still obtained from coal tar. Naphthalene a crystalline white solid hydrocarbon, also predominantly manufactured from coal tar is converted to phthalic anhydride for the manufacture of plastics, dyes (pigments) and solvents. The compound is also an intermediate for the production of chemicals in the leather industry and is used extensively as a building block to produce synthetic lubricants for auto engines, antiseptics, insecticide, and to enhance the flowability of concrete mixture which ultimately increases the strength of the concrete structure.

Phenanthrene is produced from coal tar and is used in the manufacture of wood preservatives and carbon black, an important constituent of auto tyres, paints and many other industrial products. Much of the phenol, cresols, benzene and xylenols required by the chemical industry are still obtained from coal tar. Acenaphthenes, fluorine, carbazoles, cresols, xylenols, phenols obtained from coal tar distillation are important feedstocks for the manufacture of dyes, explosives, drugs, and plastics. Pyridine obtained mainly from coal is a major feedstock in the production of medicines, pesticides, feeds, and other chemical products. Aspirin (acetylsalicylic acid), a versatile drug, anti-dandruff and anti-seborrheic preparations, pharmaceutical skin care preparations, shampoos, fabric dyes insecticides and disinfectants are also major products of coal tar distillation.

The decline in the use of coal as a source for chemicals in the last fifty years or so has reached a turning point and the future of coal as a major source of important feedstock for the chemicals industry is very bright. Important potentials include pyrolysis of coals to make a wide range of aromatic and phenolic chemical feedstocks, ammonia, ammonium sulphate. Coal tars produced from pyrolysis, gasification and carbonization are processed for the manufacture of aromatic and phenolic chemicals, naphthalene derivatives for future generation of new polymeric materials (IEA, 2005; Song et. al. 2005). Coal has several advantages as a feedstock for aromatic specialty chemicals and carbon materials. In most coals the major fraction of carbon is in aromatic structures which are dominated by polycyclic as well as monocyclic aromatic ring systems. Thus coal may be better for the production of aromatic and phenolic chemicals than alternative crude oil feedstocks in which much of the carbon is aliphatic, and especially because naphthalene derivatives have strong potential as monomers for the coming generation of new polymer materials.

Valuable liquid products including ultra-clean petroleum and diesel are now produced either by dissolving coal in a solvent at high temperature and pressure (direct liquefaction), or by gasifying the coal to form syngas which is then condensed over a catalyst (the Fischer-Tropsch or indirect process). The wide range of valuable products that can be produced from coal liquefaction processes includes heavy oil which closely resembles natural crude oil and can be processed by existing refineries. Other products are diesel oil, naphtha, waxes, lubricating oil, synthetic rubber, plastics, synthetic fibres, detergents, specialty waxes, ammonia, ammonium nitrate, urea, nitrogen, methanol, and many more. Hydrogen which is considered a major future clean fuel, and synthetic natural gas can also be produced from coal-derived syngas.

The solid products of coal tar distillation are the primary sources of feedstock for a wide variety of high-value industrial products including industrial pitch, graphite electrodes for electric aluminium and steel furnaces, mesocarbon microbeads, carbon fibres. Activated carbon, carbon molecular sieves, and specialty carbon materials such as industrial graphite, industrial diamond, fullerene, and graphene are produced from coal tar distillation. Other products include humic materials such as humic acids and calcium humates which are used as soil modifiers and fertilizers, and composite materials such as coal/polymer composites and coal/conducting polymer composites.

3.6   WORLD COAL TRADE

Coal is actively mined in over fifty countries and traded internationally. Steam coals made up nearly three quarters of the coals produced globally in 2011 and nearly half of the total production was by China. Much of the coal that is being produced around the world is for local consumption, primarily to fill local energy needs but also because transportation costs account for a large share of the total delivered price of coal. Furthermore, the heating value and other properties can deteriorate significantly during transportation by shipping and storage.

In the last decade or so, there has been a significant divergence of coal production patterns between the developed and emerging economies. While production has declined steadily in Europe, output from non-OECD countries has doubled over the same period. The bulk of the increase in global coal output over the period has been in high grade steam and coking coals (hard coals). Over 70% of coals mined in 2011 were steam coals (Figure 3.12). China was both the largest producer and importer of coal in 2011 while Indonesia was the leading exporter. Although over 11 million tonnes of coals comprising mainly steam and coking coals were traded internationally in 2011, it still only accounts for about 15% of the total coal consumed globally (WEC, 2013), which shows that coal is produced mainly for domestic use in most countries. The strong inter-regional and intra-regional export trade in coal is driven primarily by the demand for steam and coking coals which are relatively scarce and are used for power generation and steelmaking worldwide.

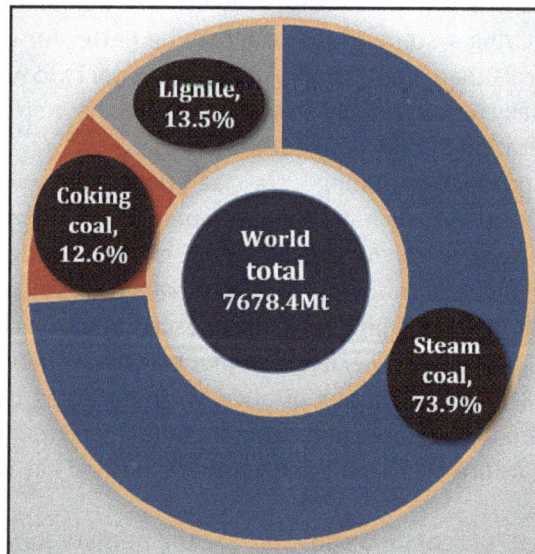

FIGURE 3.12   World production of coals by type in 2011. *(IEA Coal Information, 2012).*

A significant amount of low-grade and soft coals is also traded mainly within regions because they are more susceptible to quality deterioration over long transportation periods. Also, transportation of low-grade coals over long distances is uneconomical. The world's major producers, exporters and importers of coal in 2011 are presented in Table 3.3. Indonesia, Australia and Russian Federation top the list of exporters while PR China, Japan and South Korea are the main importers (Table 3.4). International trade in coal is mostly restricted to hard coal and, although lignite is used extensively for power generation, it is very rarely transported over long distances. The calorific value of lignite is low and the production and transportation costs (per unit calorific value) would be significantly higher than that of hard coal because of the high moisture content. Furthermore, a higher tonnage would be needed to produce the same amount of energy and special modifications may be required for boilers designed to burn higher grade coals to enable the use of low-grade coals. The common practice is to develop lignite mines and lignite-fired generating plants in close proximity as a single economic entity (embedded generation).

Although China is the largest producer of coal in the world, imports have increased very rapidly in recent years, over 30% between 2011 and 2012 (Carnot-Gandolphe, 2013). The country accounted for 23% of global imports in 2013. The main reasons are the increasing cost of domestic production and transportation from the coal deposits in the northern parts of the country to the south-eastern provinces where coal is needed. Also, increasing exports by the United States due to a decline in local demand caused by increasing replacement of coal by natural gas for local power generation, has depressed international prices of coal to a significant extent.

TABLE 3.3 Global production of coal by type in 2011. *(IEA, 2012).*

Total coal produced		Steam coals		Coking coals		Low-grade coals	
Country	% of total	Country	% of total	Country	% of total	Country	% of total
PR China	45.2	PR of China	49.9	PR China	52.1	Germany	17.0
USA	13.1	USA	15.0	Australia	15.1	PR China	13.0
India	7.6	India	9.0	USA	8.5	Russia	7.5
Australia	5.4	Indonesia	6.6	Russia	8.1	Turkey	7.2
Indonesia	4.9	S. Africa	4.4	India	3.7	USA	7.
Russia	4.4	Australia	3.5	Canada	3.1	Australia	6.6
S. Africa	3.3	Russian Fed.	3.1	Mongolia	2.1	Poland	6.0
Germany	2.4	Kazakhstan	1.7	Ukraine	2.1	Greece	5.6
Poland	1.8	Columbia	1.4	Kazakhstan	1.3	Czech Rep.	4.1
Kazakhstan	1.5	Poland	1.2	Poland	1.1	India	4.0
Columbia	1.1	Vietnam	0.8	Germany	0.6	Serbia	3.8
Turkey	1.0	Ukraine	0.7	Czech Rep.	0.5	Bulgaria	3.6
Canada	0.9	DPR Korea	0.6	Columbia	0.4	Romania	3.4
Ukranine	0.8	Canada	0.5	S. Africa	0.3	Thailand	2.2
Greece	0.8	UK	0.3	Mexico	0.3	Estoria	1.8
Czech Rep.	07	Mexico	0.2	Indonesia	0.2	Bosnia	1.0
Other	5.1	Other	1.1	Other	0.5	Other	6.1
Total (%)	100		100		100		100

TABLE 3.4 Global major exporters and importers of coal in 2011. *(IEA Coal Information, 2012).*

PRODUCERS	% OF WORLD TOTAL
China	45.9
United States	12.9
India	7.5
Australia	5.3
Indonesia	4.8
Russian Federation	4.3
South Africa	3.3
Germany	2.4
Poland	1.8
Kazakhstan	1.5
Rest of the world	10.3
World total	100.0

REFERENCES

BP (2014). *Statistical Review of World Energy.* www.bp.com

Carnot-Gandolphe, Sylvie. (2013). *Global Coal Trade, From Tightness to Oversupply.* Iffri.org

EIA (2016) *International Energy Outlook 2016.* U.S. Energy Information Administration. www.eia.gov/forecasts/ieo/coal.cfm.

IEA (2005*). Premium carbon products and organic chemicals from coal.* International Energy Agency, IEA Report PF 05-05. www.iea.org.

IEA (2012a). *World Energy Statistics 2012.* International Energy Agency. www.iea.org.

IEA (2012b). *Coal Information 2012.* International Energy Agency. www.iea.org.

IEA (2013a). *Coal Market Report 2012.* International Energy Agency. www.iea.org.

IEA (2013b) *Key World Energy Statistics 2012.* International Energy Agency. www. eia.org.

IEA (2013c) *Transition to Sustainable Buildings Strategies and Opportunities to 2050.* International Energy Agency. www. eia.org.

IEA (2014) *Key World Energy Statistics 2012.* International Energy Agency. www. eia.org.

IEA (2015) *Coal Information 2014.* International Energy Agency. www. eia.org.

IEA (2015b) *Key World Energy Statistics 2015.* International Energy Agency. www. eia.org.

IEA (2015c) *Key World Outlook 2015.* International Energy Agency. www. eia.org.

Song, C., Schobert, H. H. and J. M. Andersen (2005). *Premium carbon products and organic chemicals from coal.* IEA Clean Coal Centre, London, IEA CCC/98.

WCA (2012). *Coal and Steel Facts 2012.* World Coal Association. www.worldcoal.org.

WCA (2013). *Coal and Steel Facts.* World Coal Association. www.worldcoal.org.

WCA (2014). *Coal and Steel Facts.* World Coal Association. www.worldcoal.org.

WCA (2016). *World Steel in Figures.* World Coal Association. www.worldcoal.org.

WCI (2009). World Coal_Steel Report. World Coal Association. www.worldcoal.org.

WEC (2013) *World Energy Resources:Coal.* World Energy Council. www.wec.org

WSA (2013). World Steel in Figures. www.wsa.org.

4   Coal mining, beneficiation and transportation

4.1   INTRODUCTION

Coal has been a primary source of energy throughout history but the fuel was sourced from outcrops and large scale coal mining did not develop until the Industrial Revolution of the 18th century when coal became the main source of primary energy that fueled the industrial and transportation steam engine prime movers. Coal outcrops were no longer adequate and a systematic search for coal deposits began, nucleating a coal industry. The initial exploitations of coal were on a small scale and focused on deposits where the coal seams reached the surface, exposed by erosion, or were relatively shallow and the overburden could be removed easily. This method of mining known as *opencast mining* is still practiced in many parts of the world. Ultimately, underground mining methods were developed to reach deeper coal seams which are generally of higher quality.

There is ample evidence that coal was being traded in England as early as the beginning of 1200 AD. The first major coal mine opened in Scotland in 1575 and adopted an *underground mining* technique in which shafts were sunk to reach coal deposits under the sea. More coal mines developed in other parts of England, spreading to continental Europe and innovative mining techniques evolved rapidly. Coal was discovered in Eastern North America in the 18th century and mines quickly spread to other parts of the country.

The rapid growth of coal mining worldwide since the late 18th century has provided a strong impetus for the development of advanced opencast and underground coal mining techniques and equipment. Technologies were developed and have been modernized for efficient removal of coal from seams, evacuation to the surface, crushing to desired sizes, washing, and transportation to customers. Mine safety issues also developed rapidly and led to rapid deployment of safety measures (See Riazi and Gupta, 2016). Gasification of coal in-situ is becoming increasingly common. Residual coal in abandoned coal mines and coal deposits which were considered unmineable because of depth of deposit, low seam thickness or geological issues are now being converted in-situ to valuable syngas.

4.2   COAL MINING

Coal seams are found at varying depths in the Earth's crust, from depths of a few hundred to several thousand metres. Massive earth movements can cause coal which was formed at great depths to move closer to the Earth's surface and some seams are less than 50 metres deep. Mining technologies that have evolved and developed over the last three centuries have centered on two techniques: opencast (surface) mining and underground mining. The choice between the two options depends primarily on depth of burial, density of the overburden and thickness of the coal seam. Opencast mining is significantly safer and more economical than underground mining and the yield of coal is higher. Also, surface mined coal is usually relatively clean because it is easy to avoid mining shale and overburden along with the coal. Furthermore all coal seams are exploited and about 90% or more of the coal deposit can be recovered. Comparative figures for underground mining could be less than 50%. Coals recovered by both techniques usually require washing to remove as much of the associated debris and mineral matter as possible but many surface mined coals are not washed. Figure 4.1 shows a summary of mining techniques currently in use.

4.2.1 Surface (open cast) mining

Surface mining is believed to have been practiced on a small scale since around 200 AD. However, commercial surface mining is believed to have grown and spread in the nineteenth century. Basically it involves the removal of the overburden which could include small mountains to expose the coal surface which is then broken up by drilling, hammering and explosive blasting. The coal is loaded onto conveyor belts or trucks for transportation to the coal preparation plant. Seams that are close to the surface, at depths less than about 50 metres are usually mined by open cast techniques (Figure 4.2). When coal seams are thick, of the order of 20-30 metres (often numerous seams which lie parallel to each other), opencast mining technology may be used for depths up to about 450 metres below ground level. Opencast mining produces about 40% of the annual world output of coal but the intensity of use varies among countries. In Australia it accounts for about 80% of production, in the US, about 70% (WCA, 2011). Opencast mining raises serious environmental issues including the disruption of ecosystems. Apart from the particulate pollution during breakup and transportation, a typical opencast mine covers areas of many square kilometres and leaves deep pits which need to be rehabilitated when mining has been completed. In developed countries there are usually strict laws enforcing post mining rehabilitation of land but in many emerging nations wide areas of land around abandoned mines have been destroyed.

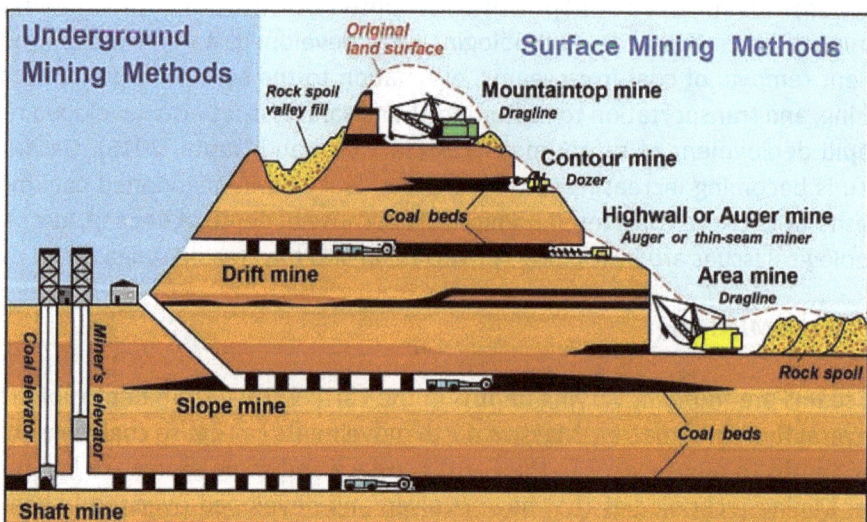

FIGURE 4.1 Coal mining techniques. *(uky.edu).*

4.2.2   Underground mining

Underground mining involves tunneling through the ground to reach the coal seam. Shafts are sunk and work areas cut into the coal beds, with areas of the coal bed left for roof support as pillars, supplemented by timber, thus creating room for miners to work on the coal seams. The early underground miners used pickles and hydraulic drill/hammers to work the coal face. Many miners today in some parts of the world still use this crude method which is often referred to as *room-and-pillar mining* (Figure 4.3).

FIGURE 4.2 An opencast coal mine. *(www.tea.com).*

FIGURE 4.3 An underground coal mine (a) manual technique (b) mechanized technique. *[Sources: (a) Obaje, 2009; mining-technology.com; (b) maheshwaree.com.]*

Only between 30% and 60% of a seam can be recovered by the room and pillar technique since the coal pillars have to be left behind. Inevitably, coal recovery by manual techniques is slow and dangerous, and the yield is low. However, many modern mines are now mechanized and use remote controlled equipment to provide large hydraulic mobile roof supports and mechanized equipment to break up the coal bed. Also, the processes of loading the mined coal onto conveyor belts or shuttle cars and evacuation to the surface are now partially or fully automated. One of the most important developments in mining technology is the introduction of the *longwall mining* technique which facilitates the full extraction of coal from a section of the coal seam, often as long as 100-350 metres, using mechanical shearers (Figure 4.4). Mobile hydraulic roof supports are deployed during mining, then withdrawn when mining is completed, allowing the roof to collapse. This is the most expensive underground mining technique but the coal yield is high and operations are relatively safe.

FIGURE 4.4 (a) A long wall coal mining machine (b) Roof supports.
[(a) maheshwaree.com); (b) worldcoal.org].

Around 70% – 80% of the coal in a seam which can be as long as 3km, can be recovered by the long wall technique. Also, the introduction of long wall mechanization has dramatically improved efficiency and tonnage of coal produced per worker and significantly reduced the level of skills required of miners. Furthermore coal mine accidents are relatively low. Many modern mines are now mechanized. They use remote controlled equipment to provide large hydraulic mobile roof supports and mechanized equipment to break up the coal bed. The coal is then transported to the surface on conveyor belts. Abandoned mines are now being re-opened and more coal is being recovered by supplementary mining which is also called *retreat mining.* The technique involves removal of the coal pillars as miners retreat. The roof is then allowed to collapse and the mine is abandoned. New developments in mining technology also allow unmineable coal to be gasified in-situ and the product gas which is a mixture of hydrogen, carbon monoxide and methane is used in nearby electric power generating plants or by local communities as energy source. In spite of the obvious advantages of this technology, proliferation has been slow due to various technical problems. Furthermore, not all coal mines are suitable for in-situ gasification.

4.2.3   Mine safety

Coal mining, especially underground mining has been and still is one of the most hazardous professions. Roof collapses, explosions, fires, flooding, are regular occurrences in coal mines and frequently lead to loss of lives, often on a large scale. Mine dust and methane released into the atmosphere in the mining process are also major fire and health hazards. Mechanization has greatly reduced mine accidents and human fatalities but has exacerbated the level of coal mine pollution emissions. Most mining deaths in recent times have occurred in developing countries due apparently to the relatively slow rate of infusion of modern mining and safety technologies into mining systems.

All types of coal mining carry safety risks but underground coal mining involves a higher safety risk than opencast mining due primarily to the potential for mine collapse and problems associated with mine ventilation and emergency evacuation of miners. Coal mining, whether by underground or opencast technique, involves the use of heavy machinery for excavation and evacuation of coal which poses considerable danger to workers and has caused many mine accidents. Mining explosions are particularly common due to the release of methane from coal seams. Methane previously trapped in the coal bed during coal formation processes is released as a direct result of the physical process of coal extraction, whether by opencast or underground mining technique.

For safety reasons, coal bed methane (also known as coal mine methane, CMM or coal seam gas, CSG) is usually vented into the atmosphere through boreholes prior to mining. However, the gas is a potentially valuable fuel and exploitation technologies have been developed, led by the US since the 1970s. Other countries actively utilizing this fuel resource include Australia and Canada. Methane emissions also arise from the collapse of the surrounding rock and strata after a section of the coal seam has been mined and the roof supports are removed as mining progresses to another section of the seam. Apart from the fact that methane inhalation could be deadly, concentrations in air as low as 5-15% can be ignited by sparks from any source including moving machinery (WCA, 2014). The resulting explosion often causes fatalities as well as collapse of mine roof supports, trapping miners.

Most modern coal mines have rigorous safety and health standards and procedures, and extensive continuous training programmes for workers. This has led to a significant reduction of coal mine accidents and fatalities. However, there are private, unregulated or poorly monitored coal mines in many developing countries and most of the mine accidents with heavy casualties in recent times have occurred in such mines.

4.3   COAL BENEFICIATION TECHNOLOGIES

Coal seams are always interspersed with discrete layers and intrusions of shale, minerals and rock. In opencast mining the overburden is removed and coal recovered is relatively clean, usually not requiring cleaning. The manual system of underground mining also produces relatively clean coal but modern mechanized methods produce dirty coal mixed with shale, minerals and rock often referred to as *gangue,* and cleaning is normally required to remove inorganic, non-combustible matter such as rock, shale, clay, and mineral matter. Treatment upgrades the value of coal, reduces transportation costs, and meets customer quality and size specifications which are becoming increasingly stringent.

A coal preparation plant (CPP) is designed to remove undesirable inclusions in mined coal, also known as run-of-mine (ROM) coal. It is usually an integral part of a modern coal mine and mined coal is transported from the mine to the washery by conveyor belt systems to avoid higher cost of transportation by any other means. Portable, semi-portable coal washing plants with capacities of a few hundred tonnes an hour are also used in small coal mines. Mined coal is mostly consigned for utilization in one of three areas: carbonization for ironmaking; heating and power generation; or conversion to syngas, petroleum fuels, or solid smokeless fuel. The specifications for each application differ but require that most run-of-mine coals be washed to reduce impurities and moisture. Around 13% of the annual coal consumption worldwide is converted into coke for blast furnace iron and steelmaking. Coal serves as the heat source and reducing agent for converting iron ore to pig iron. Ash which results from the combustion of inorganic content of coal must be as low as possible because some of the coal energy and flux are wasted in fusing and flushing it out in form of slag. Sulphur is also undesirable in coal because part of it forms anthropogenic sulphur oxides while the balance passes into the molten metal. Sulphur in molten metal is difficult and expensive to remove at the steelmaking stage and ends up in steel, causing embrittlement. For these reasons, extensive cleaning is usually required for coking coals.

Coals for heat raising, power generation and conversion usually are not subjected to intensive cleaning compared with coking coals but there are compelling reasons for beneficiating coals destined for combustion as well. These include increasing global efforts to reduce environmental pollution from coal utilization through stringent regulations, and the increasing demand by end users for high heating value, low impurity content, and closely-sized coal. Modern coal washeries achieve 75-80% reduction in ash, 85-90% heating value recovery, 30-80% reduction in pyrites and 20-80% reduction in trace elements (with the exception of mercury and selenium due probably to the fact that these elements are intimately associated with the organic matrix (Rosendale et al, 1993). However, coal cleaning processes are expensive: a typical coal preparation plant may account for up to 15% of the capital cost of a coal mine and 20-25% of the cost of mining coal. Coal preparation raises the cost of clean coal by around 10% and the cost per tonne of cleaning fines is 3 to 4 times higher than the cost of cleaning larger size coal. If dewatering and thermal drying are required the cost of preparation increases further. Coal cleaning processes are categorized as either physical or chemical cleaning. Physical cleaning is by far the most widely applied and involves mechanical separation of coal from the impurities using differences in density, in processes known as *gravity separation*. Modern coal cleaning plants combine an array of solid-solid and solid-liquid separation processes, the configuration depending on the product size range desired. Typical processes adopted are crushing, grinding, screening, classification, dense media separation, gravity concentration, froth flotation, magnetic separation, centrifugal separation, filtration, thickening and dewatering. The design of a flowsheet for a coal starts with a determination of the amenability of coal to cleaning, also known as *washability characteristics* which vary from coal to coal. The response of coal to a particular cleaning process depends on the physical and chemical nature, and the morphology of the associated gangue. The maceral composition of coal also determines the density which is an important variable in determining the cleaning process parameters. At various stages of the cleaning process, samples are taken and analyzed for chemical composition and washability characteristics.

Coal coming from the same mine can have widely varying properties due to variations in the mining conditions, the seam coal quality, and dilution by inorganic seam intrusions. A typical flowsheet comprises a series of unit operations from grinding to dewatering, arranged sequentially, with duplication of some units to achieve desired cleaning efficiencies. Coal cleaning schemes vary among plants but most feature four basic stages: initial preparation, coarse coal processing, fine coal processing, and final preparation. A typical flow diagram is shown in Figure 4.5. After washing, both the clean product and the refuse must be processed further to provide a product with the right moisture content and size to meet customer specifications, and a refuse that can be disposed of in an environmentally acceptable manner.

4.3.1 Coal sorting, comminution and screening

Coal delivered from mines to washeries often contains a considerable amount of shale and rock. The first stage of coal cleaning involves deshaling the coal by removing bulk non-organic matter. Some old washeries use hand picking to remove lump shale, a grossly inefficient and labour-intensive method, others use rotary breakers and oscillating sieving tables. *Coal comminution* is a system of processes used to reduce run-of-mine coal to desired size classes and usually involves processing in a series of size reduction units. Coal delivered to the washery may contain lumps above 70mm and is broken down and crushed to the size range 75 x 6mm which is the desirable size range in many coal washing plants.

A wide range of equipment is available for comminution operations, including jaw crushers, cone crushers, rotary crushers, hammer mills, impactors, and roll crushers (Bise, 2013), (Figure 4.6). If size range below 6mm is required, further grinding is carried out in rod or ball mills. Some modern mines use movable screen jigs for separating large stones from coal in the size range 400 x 40mm while others use dense-medium separation which is the most expensive of them all. Run-of-mine coal is sorted into narrow particle size classes using vibrating screens for coarser particles and cyclones for the finer size range. Crushing in jaw or cone crushers and further grinding in rod or ball mills may be required depending on the desired size classes. Crushing is normally done on as-received basis but grinding is done wet to reduce dust. Each comminution unit has an integrated sieving unit which may be made of wire mesh or crossed bars. Screens may be arranged in series or as multi-deck systems and

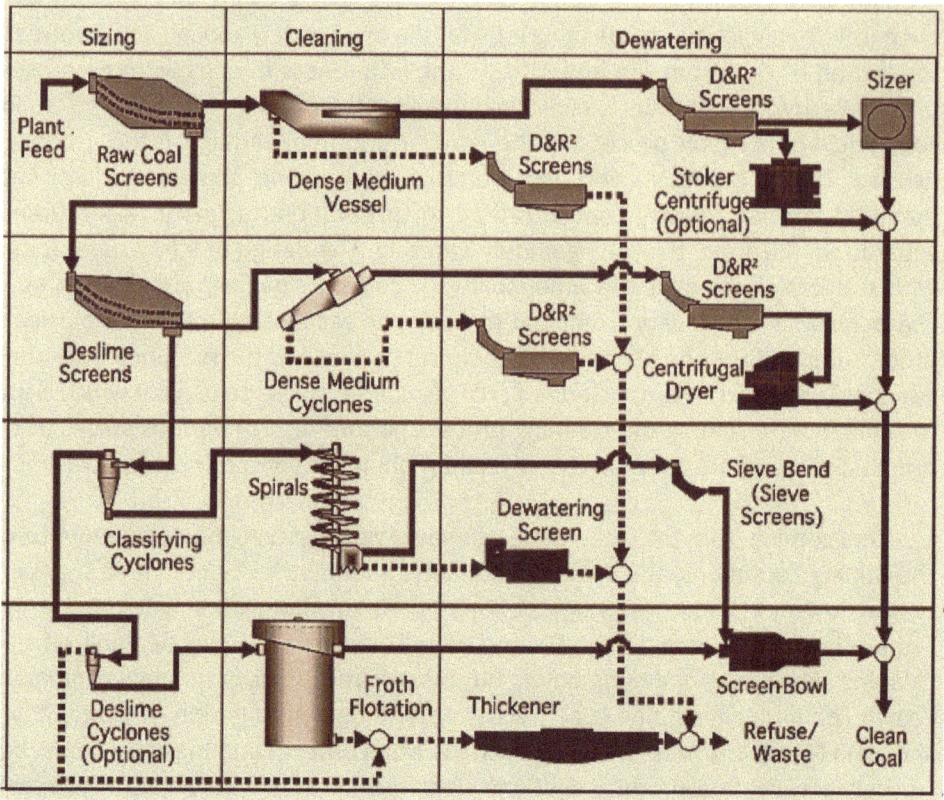

FIGURE 4.5   A simplified flowsheet for a modern coal cleaning plant. *(www.energy.vt.edu).*

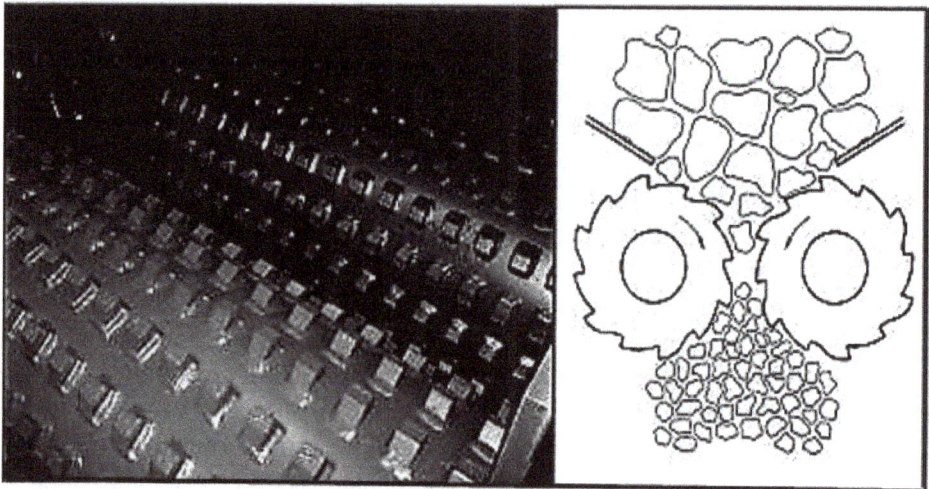

FIGURE 4.6   A double roll crusher. *(Bise, 2013).*

may be of vibrating (for coarse coal) or cyclone type (for fine coal). The crushing programme may be designed to liberate as much as possible of the amount of mineral matter locked in the coal particles but over-crushing may produce excessive fines. Crushed coal (-150mm) is commonly split into four size categories to meet the feed specifications of the various washing processes:

Coarse	(-150+13mm)
Intermediate	(-13+1mm)
Fine	(-1+0.15mm)
Ultrafine	(-0.15mm)

4.3.2   Coal concentration processes

Because the densities of the mineral matter impurities are higher than those of the coal, separations are primarily done on the basis of density differences. However, for ultrafine particles (e.g. < 0.15mm) separations are made primarily on the basis of surface differences. Coarse coal is separated from undesired burden by taking advantage of differences in the specific gravities compared with coal, a process known as sink and float or gravity concentration. The coal size fractions from the comminution unit are conveyed to their respective cleaning units. Coarse coal is cleaned in dense media vessels (DMVs), intermediate size range is treated in dense media cyclones (DMCs), spiral concentrators are used for treating the fine material, while ultrafine coal is cleaned in froth flotation cells (Bethell and Barbee, 2007).

Figure 4.7 shows the types of processes that are suitable for treating different particle size ranges. Figure 4.8 also shows typical flow charts for different coal particle size ranges. Material feed for flotation cells is usually deslimed (- 0.044mm slime removed) to improve cell performance and reduce handling problems. Since most separation (concentration) processes are carried out wet, dewatering is necessary to reduce the product moisture to an acceptable level as required by the consumer, and to reduce transportation costs. Dewatering is done using screens, centrifuges and/or filters depending on the size range involved. The coarse, intermediate and fine refuse fractions are dewatered by screening. The ultrafine refuse is thickened by sedimentation and possibly dewatered further by filtration while the process water is recycled.

4.3.2.1   *Dense Media Vessels (DMV)*

Coals of size range 150 + 6mm are cleaned in static or oscillating dense media or fluidized-bed vessels. The efficiency drops significantly at the lowest end of the particle size range and some plants process coal below 13mm in dense medium cyclones. Although magnetite or ferro-silicon mixed with water is the most common medium used to control density, some cleaning plants use finely sized sand with reasonably good results. A dense media separator is shown in Figure 4.9. The fluidizing unit is a cylindrical tank with a perforated disk bottom above which is a bed of coarse coal. Raw coal is fed into the vessel while a strong jet of a liquid is injected from below into the coal bed. The lighter coal particles rise and are removed from the top of the bed while the heavier portions are removed from the bottom.

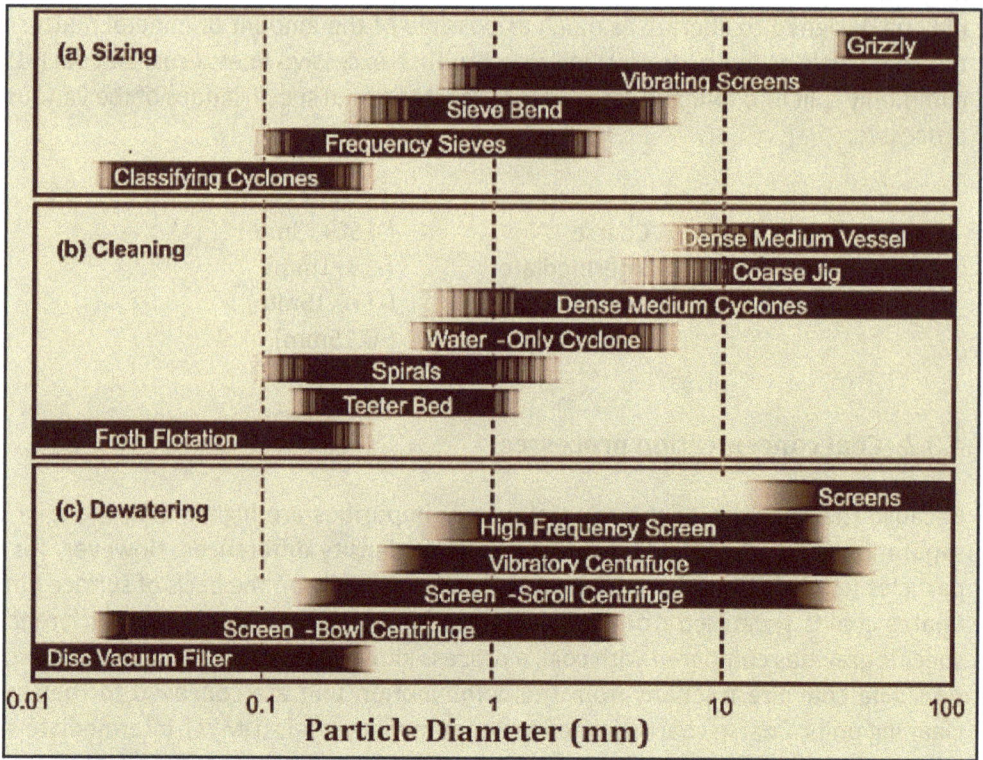

FIGURE 4.7 Effective range of particle sizes treated by various coal preparation processes. *(www.energy.vt.edu).*

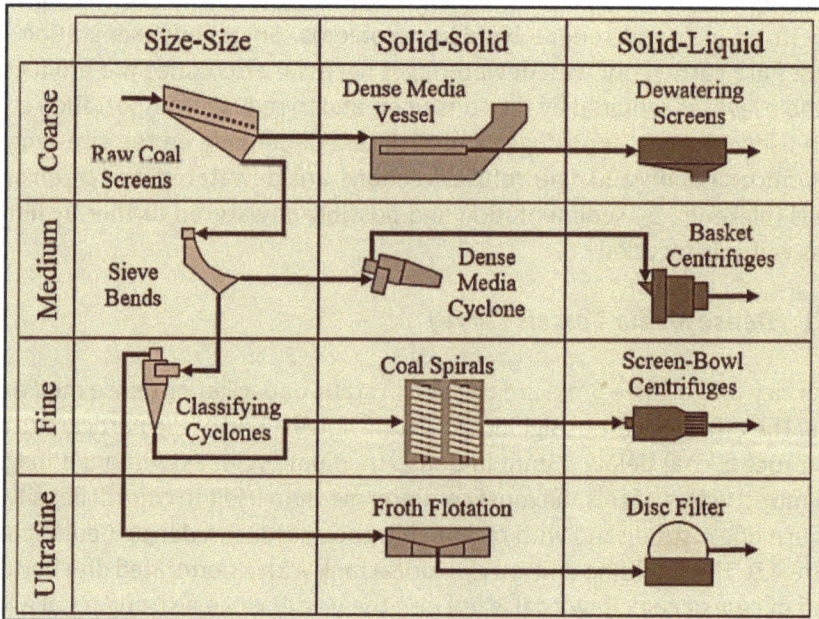

FIGURE 4.8 Typical cleaning process routes for different coal particle sizes. *(www.castconsort.org).*

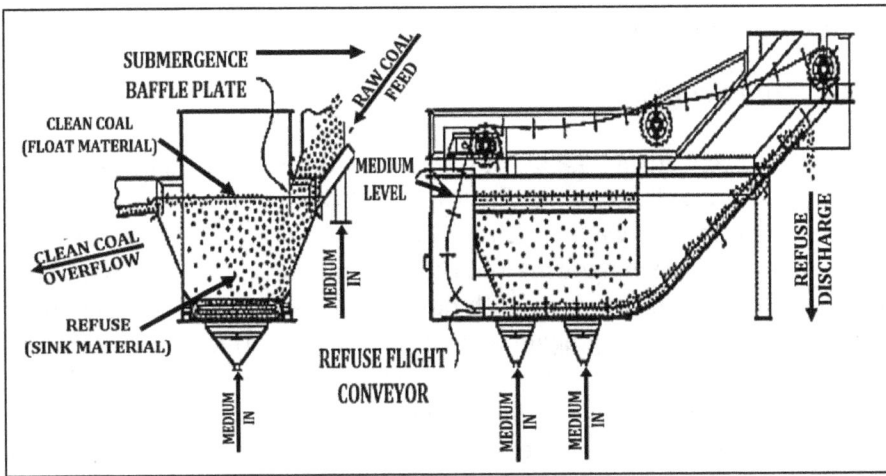

FIGURE 4.9 A dense medium vessel (DMV). *(Bise, 2013).*

Water is the usual medium for the coarse fractions but the specific gravity may be adjusted by mixing fine magnetite with the water. Magnetite is particularly favoured because of its high specific gravity (5.0-5.2), and ease of recovery using low-intensity magnetic separator. Another advantage is that it allows for the control of the specific gravity of the mixture with great precision and particles with a wide range of densities can be separated. Water systems have a major disadvantage in that they increase the moisture content of the coal thereby reducing quality and requiring expensive dewatering and drying. This has led to the development of pneumatic systems which use high-pressure air as the fluidant. In some cases media is added in the form of magnetite and crushed limestone. However, with the mechanization of mining which has greatly increased the inorganic and fine content of run-of-mine coal, pneumatic systems have been found to be inefficient and use has declined in recent times.

4.3.2.2 *Dense medium cyclone*

One major limitation of dense-medium vessels is the significant drop in cleaning efficiency with decrease in size of coal fed to the washer hence coal of size range -13+1mm is washed in a dense-medium cyclone washer. However, some modern jigs with feldspar ragging, for example Baum jigs are also capable of cleaning particle sizes below 5mm. Coal suspended in a dense medium, usually a mixture of water and fine magnetite, is injected tangentially and at high pressure into the cyclone (Figure 4.10). The fast rotation of the suspension drives heavy minerals to the conical wall, where they sink to the bottom and are extracted, while float waste minerals are sucked from the center of the vortex.

The earliest dense-medium cyclones known as cylindroconical cyclones can process coal sizes up to 50 x 0.5mm but the more modern cylindrical systems can handle sizes up to 10 x 0.5mm. Some more recent spiral concentrators known as hydrocyclones or autogenous cyclones which feature vortex finders and ceramic liners use water only as the medium, thus eliminating the need for expensive magnetite and desliming (Figure 4.11). Several concentrators may be nested to increase capacity. These cyclones can clean coal in the particle size range -1+0.15mm and are particularly good at processing fine, oxidized coal which cannot be cleaned by froth flotation.

4.3.2.3   *Froth flotation*

Froth flotation takes advantage of differences in the surface properties of fine materials. Some materials are difficult to wet (hydrophobic) while others have a tendency to interact with or be dissolved by water (hydrophilic). The process is very effective in separating carbonaceous coal from mineral matter. Ultrafine raw coal (-0.15mm) is mixed with water to form a slurry in a vessel known as froth flotation cell and the coal is rendered hydrophobic by the addition of a frothing agent known as frother, surfactant or a chemical known as collector. Air (or nitrogen) is passed through the slurry to produce bubbles to which the hydrophobic particles attach, forming a concentrate froth at the surface of the slurry from which they can be removed (Figure 4.12). A wide range of frothers is used to preferentially coat the fine coal particles to improve hydrophobicity. These include alcohols, polyglycols and mineral oil, kerosene, diesel oil.

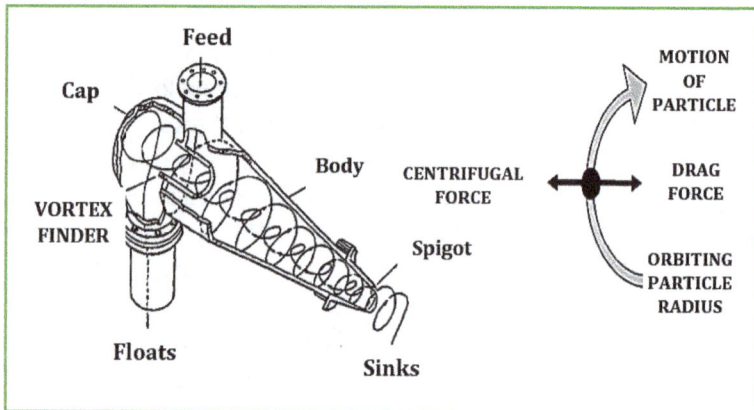

FIGURE 4.10 Coal dense media cleaning cyclone (DMC).

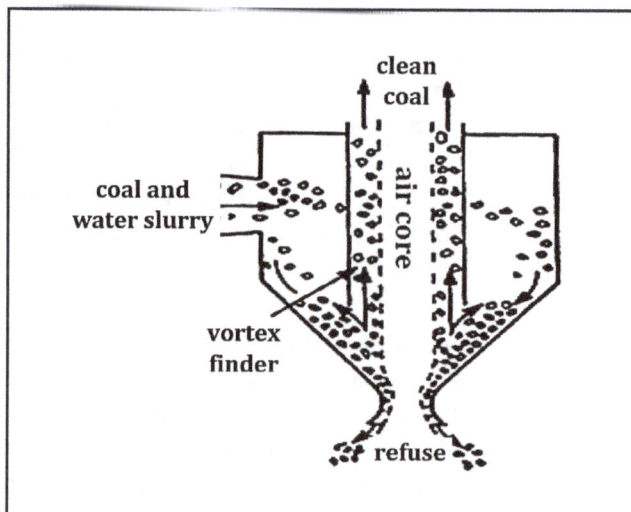

FIGURE 4.11 Hydrocyclone. *(Couch, 1991).*

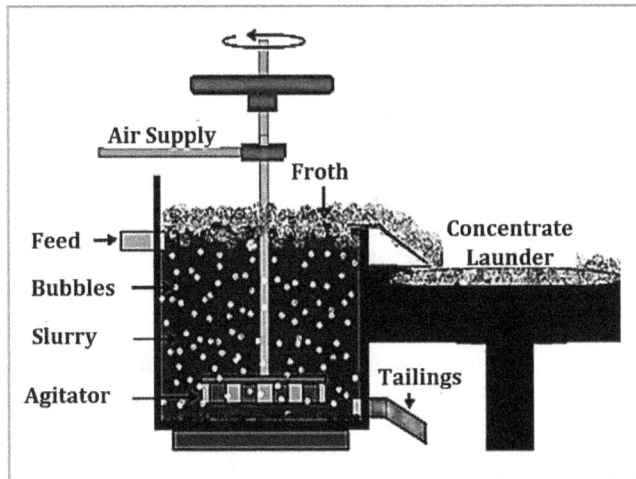

FIGURE 4.12 Coal froth flotation cell. *(www.911metallurgist.co).*

4.3.2.4 *Dewatering*

Coal that has been wet-washed needs to be dewatered in order to improve the value of the product, reduce transportation costs, meet customer specifications, and recover as much of fine coal as possible. Many washeries use centrifugal dryers for dewatering the coarser sizes up to 35 x 0.5mm. Finer sizes below 0.5mm are treated in disc, drum or vacuum filters and pressure filter presses.

4.3.3 Washability characteristics of coal

Coal washing processes utilize three primary physical properties of coal: hardness, density and surface properties. Hardness determines the energy required to break down coal to the desired particle size range while density and surface properties determine the behavior of coal when immersed in a fluid. These properties vary for coals of different ranks, for coals of the same rank, and even for coals mined from the same deposit. Furthermore, weathering and oxidation can alter all the three variables significantly. Raw coal is a mixture of pure coal which has a specific gravity in the range 1.25 to 1.29 mixed with shale, clay and other minerals of specific gravity of 2.70 and above. The response of such coals to cleaning depends on the morphology and distribution of inorganic matter in the coal. Because of the potential variability in the properties of coal mined even from the same seam it is vital to characterize coal to determine washability characteristics on which the design of a cleaning flowsheet is based. This is done in laboratory tests to determine grindability and float and sink characteristics of coal. Tests are carried out periodically to determine the need for adjustments of process parameters.

4.3.3.1 *Sampling*

In view of the variability in properties of coal arising from inhomogeneity of coal seams, variations in mining conditions, and coal particle size degradation during handling and transportation to the washery, coal properties from the same mine can vary by as much as 20% (Hern, 1994). As a result, coal preparation plants must be capable of processing feed material with widely varying properties. It is vital also that samples taken from a coal batch for laboratory tests and analysis must be representative. There are established standards for washability tests and methods of taking representative samples (Nicol and Swansen, 1997).

4.3.3.2 *Grindability*

The ease with which coal can be ground is assessed by determination of a grindability index. There are several methods but the Hardgrove Grindability Index (HGI) is the most common. It measures coal's resistance to crushing and grinding, and allows for the evaluation of coal performance in the crushing and grinding mills, including estimation of grinding power requirements, grinding and throughput capacities. The test is discussed more extensively in Chapter 6.

4.3.3.3 *Washability curves*

Inorganic matter in coal may arise from several sources: some elements, mainly sulphur and nitrogen may be present in the original coal-forming vegetation or produced in the course of peatification, mainly by bacteria in the plant matter decay and peat accumulation stages. Such elements are usually part of the predominantly chemically-bonded elements in the coal microstructure. This type of matter is classified as *syngenetic* or fixed ash and can be removed only by chemical or thermal processes. Inorganic clay and quartz may migrate into the peat swamp and eventually form finely disseminated unbounded deposits in the coal microstructure. This type of impurity is known as *epiclastic* and can be partially removed by washing if the coal is finely ground. A third group of mineral matter – kaolin, pyrite, calcareous and carbonaceous shale, iron oxides – migrates from inorganic matter in the seam overburden and intrusions into coal fissures and fractures possibly through diffusion or underground flooding. This group is classified as *epigenetic* or free ash and tends to occur in discrete lumps or flakes in coal cracks.

Epigenetic ash can be freed by crushing and grinding, and removed by washing relatively easily. Depending on the degree of mechanization, considerable inorganic matter may be mined with coal but can easily be removed by beneficiation. Syngenetic inorganic matter in coal is usually small, often less than 1% compared with epiclastic and epigenetic matter and modern coal cleaning processes which often include fine grinding can remove nearly all mineral matter (Figure 4.13).

In summary, the washability characteristics of run-of-mine coal must be determined prior to the design of a flowsheet for beneficiation. This is done by carrying out specific gravity fractionation known as float and sink analysis on representative laboratory samples in a laboratory simulation of the dense medium vessel (DMV) process. Float-sink laboratory tests rely on differences in specific gravities of the different components of run-of-mine coal to separate coal from non-carbonaceous material (Table 4.1). Several baths of progressively denser liquids are prepared using mixtures of chemicals, typically mixtures of aqueous salt solutions (e.g. zinc chloride, sodium metatungstate) or organic liquids (e.g. toluene, perchloroethylene).

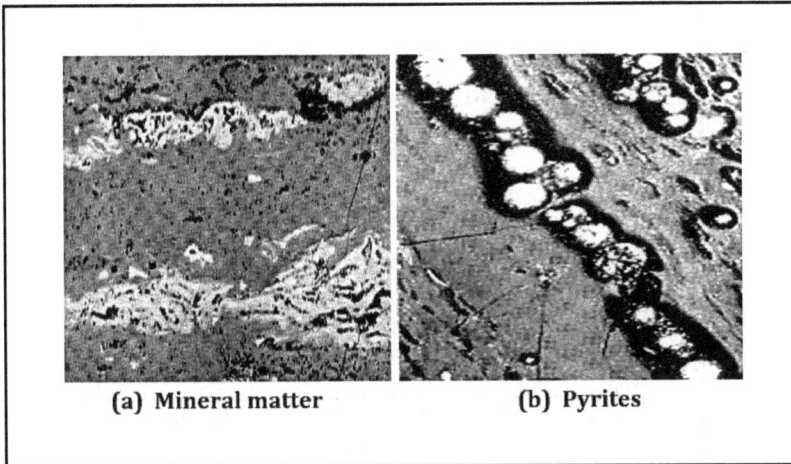

(a) Mineral matter   (b) Pyrites

FIGURE 4.13 Epiclastic and epigenetic inorganic matter. *(ICCP, 1963).*

TABLE 4.1 Specific gravities of constituents of typical run-of-mine coal.

Coal	1.23-1.40
High-ash coal	1.6-2.0
Carbonaceous shale	2.0-2.6
Gypsum	2.3
Shale	2.6
Clay	2.6
Sandstone	2.6
Calcite	2.7
Pyrite	5.0

A coal sample of known particle size range is placed into the bath containing the lowest specific gravity liquid. The material floating is recovered, dried and weighed, while the sink material is removed and placed into the next higher specific gravity bath. This procedure continues until the desired number of fractions is obtained, typically four to six. The amount of recovered float and sink portions for each range of specific gravities is determined. The resulting samples are analyzed according to a standard procedure (e.g. ASTM D 5142) to determine the appropriate attribute values such as ash, total sulphur and calorific content. Data obtained is used to generate washability curves, comprising yield curve, cumulative float curve, cumulative sink curve, partition (Tromp) curve. Washability curves show the theoretical (maximum) percentage of clean coal that can be produced for a specified attribute value and at what specific gravity the separation should be made. A typical set of washability curves is shown in Figure 4.14.

Most coal consumers have detailed specifications on the undesirable mineral contents, moisture content, calorific value, size mix, etc. Suppliers place considerable importance on coal cleaning, starting with washability tests which provide data for an appropriate flowsheet design. In fact, coal washeries are located as part of many coal mining complexes.

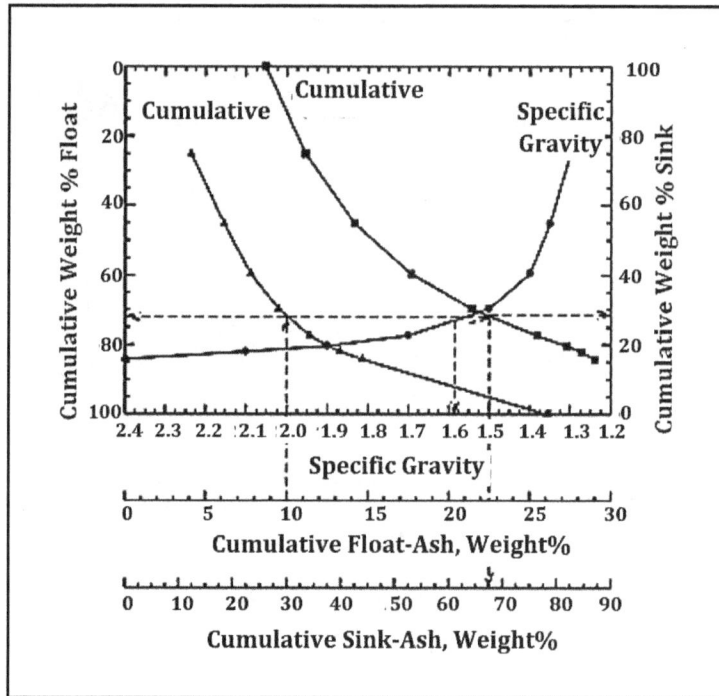

FIGURE 4.14   A set of washability curves. *(SMME, 2013).*

4.3.4   Coal desulphurization

The sulphur content of coal, although usually lower than other mineral contents, is regarded as the greatest obstacle to coal utilization because of the potential damage of the environment by post-combustion emissions of sulphur oxides. Also, sulphur is generally undesirable in coals destined for use in the metallurgical industry. Specifications for most coal applications require maximum sulphur content of the order of 0.5-0.7%. Sulphur is present in most coals in quantities ranging from around 0.5% to over 12% hence substantial removal by cleaning is almost always necessary. Low-rank coals usually have low sulphur content, less than 1% while many coals of all ranks can contain high sulphur depending on the geological history of the deposit, the chemistry, geo-stratigraphy and morphology of the overburden, and interspersed inorganic intrusions in the coal seam.

Much of the sulphur in low sulphur coals derives from the sulphur content of the material constituents of the original mire/peat. Higher sulphur contents usually derive from the depositional environment. For example, sea water and brackish water in coal beds contain sulphates which can be converted by bacteria to hydrogen sulphide. The gas reacts with iron in the water and other inorganic deposits in the mire to form pyrites, or with the organic material to form organic sulphur structures. Sulphur may be in organic and inorganic forms. Organic sulphur is strongly bonded to other elements, typically carbon or iron in the coal structure. Organic sulphur within the coal structure is still a subject of intensive research and many models have been proposed, some of which are shown in Figure 4.15. Most coals contain only small quantities of organic sulphur in the range 0.5% to 2% w/w. Inorganic sulphur may also occur in coal as gypsum, marcasite, sulphate or elemental sulphur in relatively low quantities.

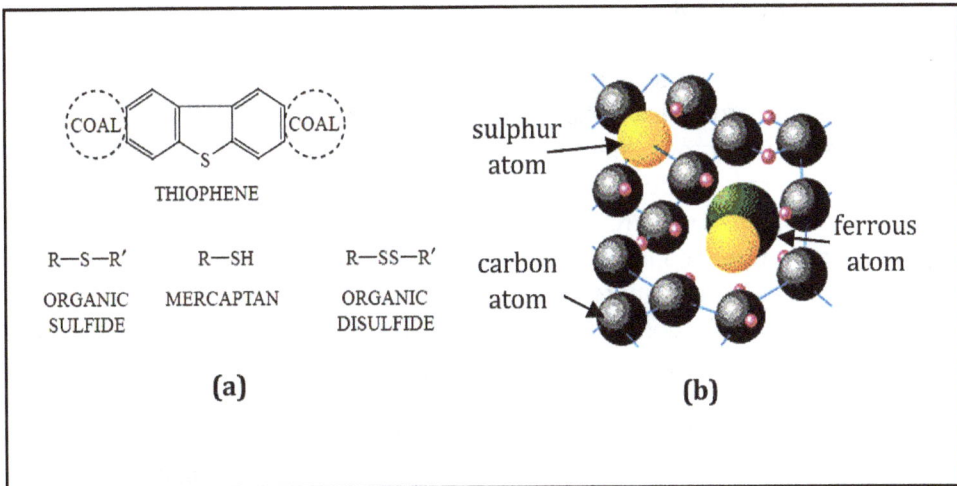

FIGURE 4.15 Forms of syngenetic sulphur in coal (a) organic sulphur in the coal structure (b) organic sulphur atom, carbon atom, and ferrous atom. *(energy.us.gov).*

The presence of elemental or sulphate sulphur in coal is believed to be as a result of chemical reactions which occur during coal weathering. Around 50% or higher of inorganic sulphur in coal can be removed by washing depending on the forms present and how finely the coal is ground. Microcrystalline pyrite and organic sulphur cannot be removed by physical cleaning. Removal requires heat or chemical degradation. Pyrites of epigenetic origin is easily released during coal fracturing that occurs during mining and processing and is easy to remove during cleaning. The pyrites that is present due to bacterial action during deposition of the coal-forming material is either an integral part of the bonds in the coal structure (Figure 4.15) or finely disseminated in the coal matrix and intimately associated with the coal structure. Size reduction to below 0.075mm is usually required to liberate the fine pyrite.

Only pyritic and sulphate sulphur can be removed by physical cleaning since organic sulphur is spread throughout the structure of the coal and cannot be liberated by crushing and grinding. Physical separation methods are effective when the inorganic sulphur is present in a form that can be liberated from the coal by crushing and grinding. In order to liberate fine inorganic sulphur compounds the coal must be crushed to a particle size that is smaller than the particle size of the inorganic sulphur inclusions. Even then, a significant amount of pyrite particles may remain locked with carbon and behave as if they are coal particles. This makes them very difficult to remove by physical cleaning (Kawatra and Eisele, 2001). Furthermore, cleaning becomes very difficult if the coal has to be ground smaller than around 10 microns.

Many advanced physical processes capable of removing liberated sulphur from coal are now available. The most effective processes combine several physical processes including gravity separation, flotation, magnetic separation. Microbial leaching can also remove pyritic sulphur from coal. Thermal and chemical processes available for removal of organic sulphur include molten caustic or base aqueous leaching, microwave desulphurization and oxy-desulphurization. However, most thermal and some chemical processes destroy the caking properties of coal. For coking coals this is undesirable but it enhances quality if the coal is processed for combustion.

4.4   COAL TRANSPORTATION

Much of coal used worldwide is consumed locally, but there is also a strong global trade in coal, particularly hard coal. It is expensive to transport coal, often more expensive than mining the coal, hence many consumers, in particular coal-fired electric power plants are near coal mines to lower transportation costs. Intrastate movement of coal is mainly by conveyor over short distances, or by rail or road to longer destinations, but some developed countries now transport coal suspended in water or oil by truck, rail, barge on rivers and lakes, by ship or through pipelines to consumers.

Coal for the international market is normally transported by ship. Slurry pipelines have been in use for over a century, used to transport solid material including coal, limestone, copper and iron ore over long distances of up to 2000km. There are two main types: coal-water slurries (CWS or CWSF) or coal-fuel slurry (CFS). The main advantage of pipeline transportation is the significant reduction of pollution which usually occurs during loading and transfer between different forms of transportation. Coal-fuel slurry systems are particularly attractive because the slurry can be fed directly to combustors in electric power generating plants, thereby cutting pre-combustion treatment and handling costs considerably. Also weathering and loss of heating value can be reduced considerably when transporting low-grade coals.

Coal for pipeline transportation must be ground to desirable particle size range, depending on customer specifications or the adopted slurry medium (Table 4.2). The ground coal is then distributed in a liquid medium, typically water, methanol, crude oil, by agitation and fed into the pipeline. Coal:medium ratios for optimal transportation are usually in the range 60:40 and 70:30. There are four main systems: conventional fine coal; conventional coarse coal; stabilized flow; and coal-water mixture (Lee et. al., 2007). The coal is separated from the liquid at the destination for subsequent use in combustion and the liquid is either discarded, reused as fuel, or recycled in a two-way slurry pipeline to the loading point. Coal transported in water medium has to be de-watered and dried before use as combustion fuel. This constitutes a disadvantage compared with methanol or crude oil media since coal-fuel slurries can be fed straight into combustors (if the coal is ground to 20 microns or below). On the other hand, large and reliable source of crude oil and methanol is required for fuel-coal slurry transportation. New co-systems in which methanol or synthetic oil is produced from coal at the feed end of the pipeline are being developed. Also, technologies for direct combustion of coal-water slurries and pipeline transportation of coal/liquefied carbon dioxide slurries are under development.

TABLE 4.2 Coal-water slurry pipeline systems. *(Hsu et. Al., 1988).*

Coal-water slurry systems	Particle sizes
Conventional fine coal	Lee than 1 mm
Conventional coarse coal	50-150 mm
Coal-water mixtures	-30 and -150 μm
Stabilized flow coal	Less than 0.2 mm (fine) and less than 50 mm (coarse)

Coal-water slurry systems are becoming increasingly popular because cost per unit energy output can be significantly lower than for other fuel sources. Depending on geographical location, the price per unit energy of CWS may be 30% to 70% lower than the equivalent amount of oil or gas. However, each supplier-consumer system must be evaluated from the technological, geographical and economic perspectives to determine feasibility and potential economic advantage over other available fuels.

Shipment of coal to international destinations is restricted mainly to hard coals for two reasons: shipping is expensive and the economics is based on the heat value/cost ratio (HV/Cost). Since low grade coals usually have low calorific values and high moisture content, shipping over long distances is not considered economically feasible because of the low heat value to cost ratio. Another reason why low grade coals are not traded internationally is the potential for weathering, oxidation and spontaneous self-ignition of coal piles during storage or transportation over long distances. Low-grade coals are much more susceptible to weathering which leads to deterioration in coal quality. Also, cases of spontaneous self-ignition during storage and transportation of low-grade coals are very common. However, some trading in low-grade coals occurs between neighbouring countries, although special precautions are usually necessary to minimize the risk of spontaneous self-ignition.

REFERENCES

Bethell, P. J. and Barbee, c. J. (2007). "Today' coal preparation plant: A global perspective." In *Designing the Coal Preparation Plant of the Future.* Edited by B. J. Arnold, M. S. Klima, and P. Bethell. Littleton, CO:SME. Pp. 9-20.

Bise, C. J. (2013). *Modern American Coal Mining: Methods and Applications.* Society for Mining, Metallurgy and Exploration.

Couch, G. (1991). *Advanced coal cleaning technology.* IEA Coal Research, International Energy Agency. London, U.K.

Hern, B. (1994). "Automatic On-line Process Controls at Bearco." *Coal Magazine,* Sept., Pp. 32-34.

Hsu, B. D., Leonard, G. L., and Johnson, R. N. (1988). "Progress on the investigation of coal-water slurry fuel combustion in a medium speed diesel engine: Part 3 – Accumulator injector performance." *J. Eng. Gas Turbines Power.* 110, 516-520.

ICCP (1963). *International Handbook of Coal Petrography.* International Committee on Coal Petrology, Second Edition, 1963.

Kawatra, S. K. and T. C. Eisele (2001) (Eds). *Coal desulphurization.* Taylor and Francis Inc., New York.

Lee, S., Speight, J., and S. K. Loyalka (2007). *Handbook of Alternative Fuel Technologies.* CRC Press.

Nicol, S. K. And A. R. Swanson (1997). *Principles of Coal Preparation.* Australian Coal Preparation Society, 370p.

Obaje, N. G. (2009). "Geology and Mineral Resources of Nigeria." Springer Dordrecht.

Rakes, P. H. (2012). *Coal Mine Mechanization.* e-WV: the west Virginia encyclopedia, 19 June, 2012. Web 24 September, 2014.

Riazi, M. R. and R. Gupta (2016) (Eds.). *Coal Production and Processing Technology.* CRC Press.

Rosendale, L. W., Devito, M. S., Conrad, V. B. and G. F. Meenan (1993). "The effects of Coal Cleaning on Trace Element Concentration." Air and Waste Management Association. Annual Meeting, Denver, June.

SMME (2013). *Modern American Coal Mining: Methods and Applications.* Society for Mining, Metallurgy & Exploration.

WCA (2011). *Coal Facts.* World Coal Association. www.worldcoal.org.

WCA (2014). *Coal Facts.* World Coal Association. www.worldcoal.org.

5 Structure of coal

5.1 INTRODUCTION

The structure of coal is to a large extent determined by the nature of the original plant debris from which it was formed, the swampy conditions under which debris was immersed, the nature and extent of the overburden, seam intrusions, marine and flood incursions, and the age and geological history of the deposit. All these variables also have a major influence on the physical, chemical and thermo-chemical properties, and utilization of coal (Suarez-Ruiz, 2008, 2012).

5.2 MACROSTRUCTURE OF COAL

One of the first attempts in describing the macrostructure of coal was by Stopes (1919) who specified four different types of macroscopic constituents which are visually observable in banded lump coal known as humic coal. These are vitrain, clarain, durain and fusain, usually called lithotypes (Schapiro and Gray, 1960). The characteristics peculiar to each constituent are given in Table 5.1. In some coals, the visibly stratified structure characteristic of humic coals is absent. Such coals known as sapropelic coals derive from a homogeneous mixture of specific kinds of fine grained organic matter including plant spores and algae. Humic coals on the other hand derive from a heterogeneous mixture of a wide range of macroscopic plant debris and have a banded structure. Humic coals have a very high percentage of spores while sapropelic coals are composed of a high proportion of algal remains. The bituminous coal shown in Figure 5.1 exhibits clear horizontal bands. The bright bands are well preserved woody material such as stems and tree branches. The dull bands contain mineral matter deposited by flood intrusions during peatification, weathered, carbonized, degraded or oxidized plant matter.

5.3 MICROSTRUCTURE OF COAL

An examination of a polished coal section under the microscope in reflected light usually reveals a number of micro constituents, which are easily distinguishable on the basis of reflectance measurements, morphology and botanical affinities in origin, and the mode of preservation within the sediment, or similarities in properties. These coal constituents are known as macerals commonly classified into three groups: vitrinite, liptinite (exinite) and inertinite, the sub-groups of which are given in Table 5.2.

TABLE 5.1 Macroscopic constituents of coal.

LITHOTYPES	CHARACTERISTICS
Vitrain	Has a bright, coherent and uniform structure with brilliant, glossy, vitrous texture. It tends to be more brittle than other macroscopic coal constituents, often breaking with a conchoidal fracture.
Clarain	Similar to vitrain but has a definite and smooth surface when broken at right angles to the bedding plane and these faces have a pronounced gloss or shine, with pronounced and inherently banded structure. It commonly contains fine vitrain bands alternating layers of duller layers
Durain	Has a dull appearance with a hard, close and firm texture which appears rather granular to the naked eye; A broken face is never truly smooth but always has a fine lumpy or matt surface. It is relatively harder than other lithotypes and tends to break into large, blocky fragment.
Fusain	Occurs chiefly as patches or wedges; it consists of powdery, readily detachable somewhat fibrous strands; it is soft and friable and loosely resembles charcoal from which it takes its name.

FIGURE 5.1 Bituminous coal with banded structure. *(geology.com).*

TABLE 5.2 Microscopic maceral groups and sub-groups of coals.
(ICCP, 2001; www.intechopen.com).

VITRINITE GROUP	LIPTINITE GROUP	HUMANITE GROUP	INERTINITE GROUP
Telinite	Sporinite	Textinite	Fusinite
Collotelinite	Cutinite	Ulminite	Semifusinite
Vitrodetrinite	Resinite	Attrinite	Funginite
Collodetrinite	Alginite	Densinite	Secretinite
Corpogelinite	Suberinite	Corpohumanite	Macrinite
Gelinite	Chlorophyllinite	Gelinite	Micrinite
Telovitrinite	Fluorinite	Telohumanite	Inertodetrinite
Detrovitrinite	Bituminite	Detrohumanite	
Gelovitrinite	Exudatinite	Gelohumanite	
	Liptodetrinite		

The maceral groups differ in chemical composition, in particular, the relative distribution of the total carbon, volatile matter, hydrogen and oxygen content of coal. Carbon is highest in vitrinite, volatile matter and hydrogen in liptinite, and oxygen in inertinite. They also differ in the relative distribution of aromatic and aliphatic bonds involving carbon and hydrogen, particularly in low grade coals. However, aromaticity increases in all macerals with increasing rank and this results in convergence of features and properties in high rank coals. Macerals behave differently in coals of different types and ranks on carbonization. All macerals in low rank coals are unreactive and do not undergo any significant thermoplastic phase, most of micrinite and fusinite and part of semifusinite in most coals are also unreactive while vitrinite, resinite and part of semi-fusinite in some coals soften in the temperature range 350 to 550°C (the plastic or fluidity range, also called the mesophase). Above the mesophase temperature range, the coal residue reconsolidates to form a strong mass often referred to as semicoke which transforms into coke on further carbonization. The ability to go through a plastic phase on carbonization is the main property that distinguishes caking and non-caking coals.

The macerals also differ in physical properties, in particular, reflectance. Inertinite has the highest reflectance and vitrinite has values between inertinite and liptinite which has the lowest reflectance. Again, as coalification advances, there is a convergence in reflectance values. Micrographs of some of the macerals and coal ranks are shown in Figures 5.2(a) - (j). Remnants of plant matter are evident in young coals and become minimized as coalification progresses. The structure becomes largely amorphous in anthracite, the highest coal rank, and plant matter remains are hardly found (See Figures 5.2(a) and 5.2(b)). The vitrinite reflectance of a coal is a common basis for classifying coals into ranks while the relative proportions of macerals determine coal type. A comprehensive atlas of maceral micrographs has been published by USGS (2013). See also SIU, 2015; Scott (2002).

5.3.1 The Vitrinite Group

Vitrinite is the principal coal maceral group, the primary constituent of most bright coals. Macerals in the group derive from the humification of wood and leaf tissues and other parts of plants which are chemically composed of the polymers, cellulose and lignin. Vitrinite in low grade coals often shows prominent signs of incomplete humification typified by remnants of plant tissue, and is known as humanite. This maceral transforms into largely amorphous vitrinite in mature coals (Figure 5.2(a).

The reflectance of humanite is low and vitrinite with a reflectance value of less than 0.5 is classified as humanite. Vitrinite is the predominant group and makes up 50 to 90% of the total micro-constituents of most coals. However, there are also many coals which are low in vitrinite but have relatively high inertinite content, for example, some Canadian, Indian and South African coals. The transformation of humanite to vitrinite is accompanied by the polymerization of biodegraded humic substances to form condensed aromatic and hydro aromatic ring structures which are connected by covalent cross-links. As coalification progresses, cross-links become degraded, non-aromatic matter content is reduced and tightly packed aromatic structures are formed.

Vitrinite comprises several sub-macerals, the most prominent of which are shown in Table 5.2. The varieties describe the nature of the plant tissues from which they were formed. When vitrinite has a structure of woody tissue it is called telinite and the structureless type is called collinite while varieties featuring extensive comminution are classified as vitrodetrinite. The mean maximum reflectance of vitrinite increases with maturity of coal, from about 0.5% for sub-bituminous coal to about 6.5% for anthracite.

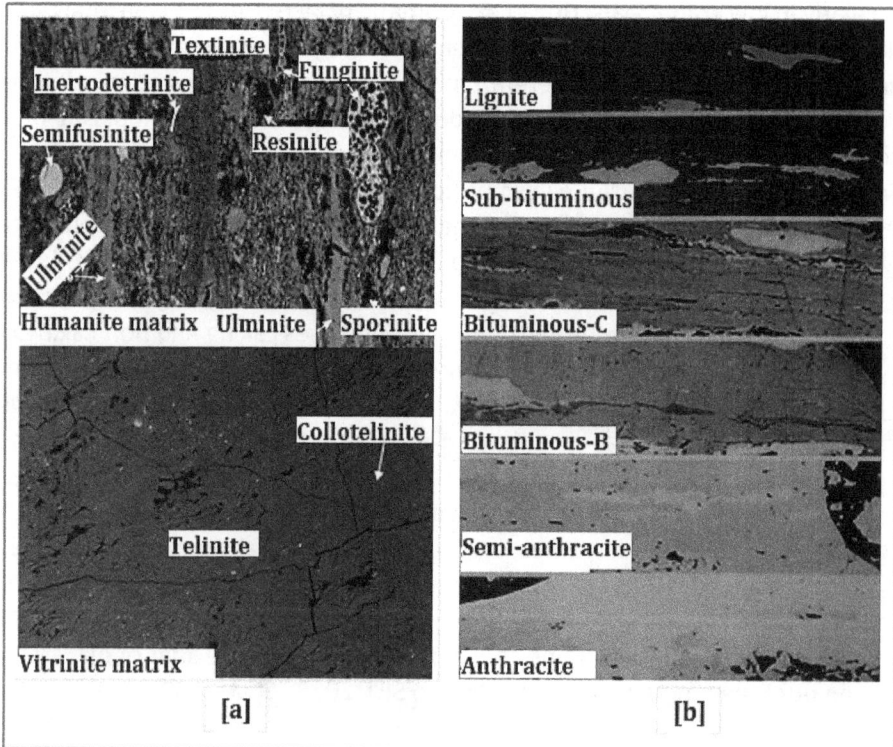

FIGURE 5.2 Microstructures of different ranks and types of coal showing the main macerals. *(www.energy.usgs.gov; www.siu.edu).*

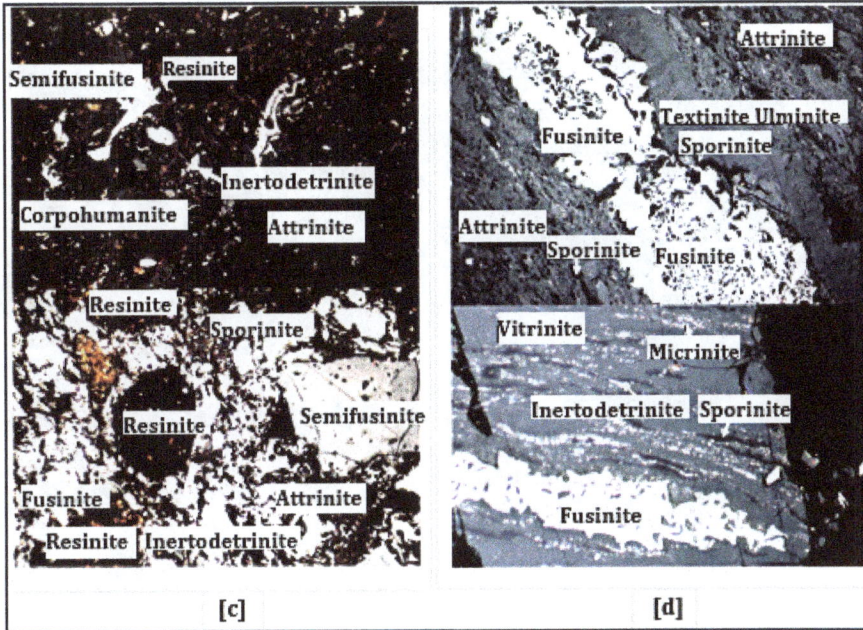

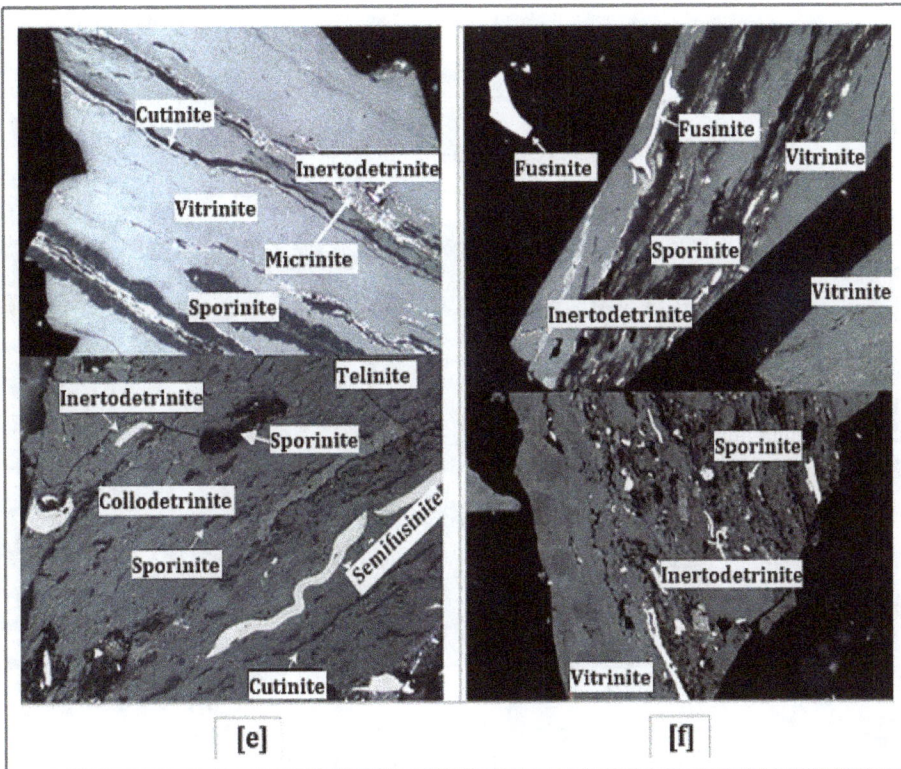

FIGURE 5.2 continued.

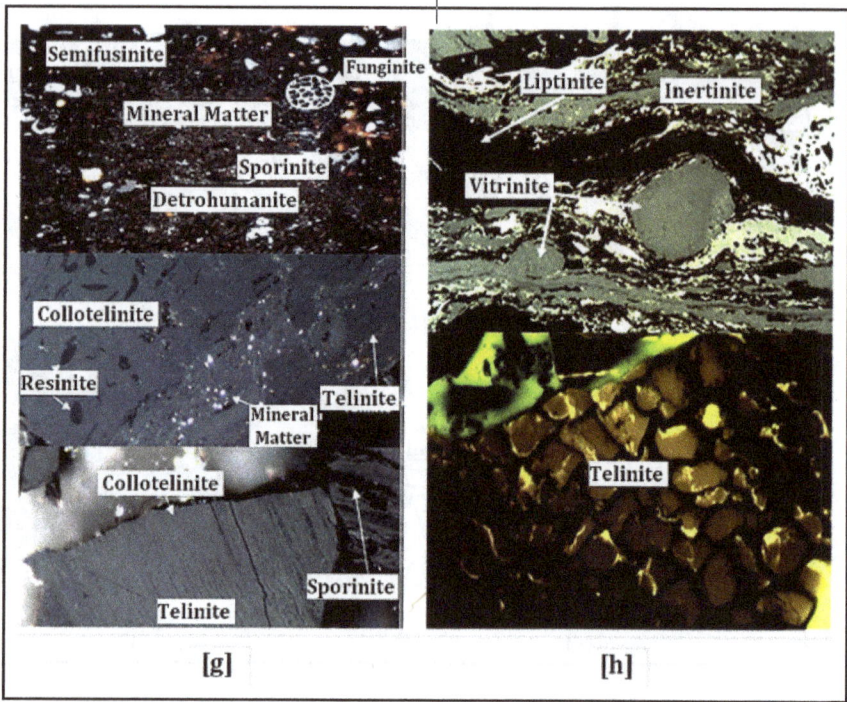

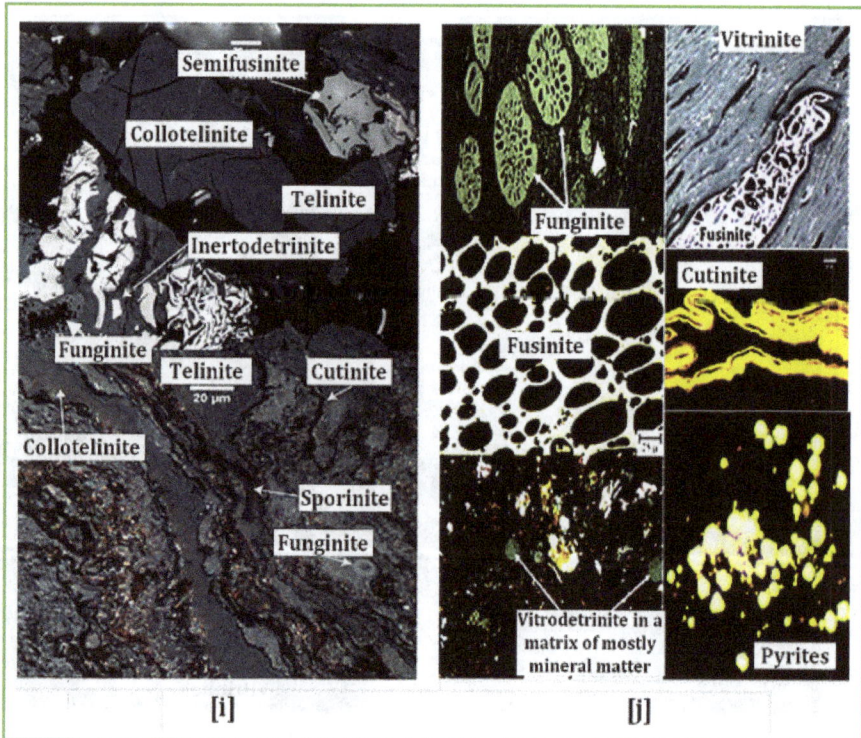

FIGURE 5.2 continued

5.3.1.1 *Collinite*

Collinite is the variety of vitrinite that displays an absence of cell structure under normal microscopic conditions. Morphologically, collinite has four sub-macerals whose structure ranges from homogeneous (corpocollinite) through colloidal gel (gelocollinite), completely gelified (desmocollinite) to the telocollinite variety which occurs in wide layers whose margins are visible.

5.3.1.2 *Tellinite*

Tellinite is a maceral in the vitrinite group which shows distinct though deformed cell structure of the initial plant maceral precursors such as wood. The cell cavities may be completely compressed to lines or may be round or oval, filled with other substances such as collinite, resinite, micrinite, clay minerals etc. Within the vitrinite group, tellinite usually has the highest reflectance.

5.3.2 The Liptinite (Exinite) Group

Exinite was first used by Stopes (1935) to describe the micro-constituent of coal that originated from exines of spores and contains high volatile matter and hydrogen relative to other micro-constituents. Seyler (1933) described the now more widely accepted liptinite group which includes exinite and other micro-constituents that do not necessarily derive from exines but appear to have similar technical properties. These include macerals which originated from spores, pollen, dinoflagellate cysts, leaf cuticles, and plant resins and waxes which are resistant to weathering and diagenesis (Taylor et al., 1998). Macerals in the liptinite group tend to retain their original plant structure in form of plant fossils and phyterals. Differences in these features are used to sub-classify the maceral group.

Liptinite content of many coals is usually low, between 5 and 15% but the maceral dominates some coals. In boghead coals, alginate is abundant and sporinite dominates cannel coals. The main feature of the liptinite group is the lower reflectance than that of vitrinite in the same coal. Liptinite may be completely absent in some coals and, when present, is usually small compared with the other main macerals. The sub-macerals in the liptinite group tend to retain their original plant-fossil structure but start to disappear as coalification advances and are often absent in medium and low volatile coals.

5.3.2.1 *Sporinite*

Sporinite belongs to the liptinite sub-group. It comprises distinct micro and megaspores and pollen grains whose colour ranges from rusty brown, dark gray or nearly black in low-rank coals, to lighter gray with increasing rank, depending on the starting materials, that is, the plants from which the spores originated. The spores are usually embedded in collinite. Reflectivity (Rm, oil) ranges from 0.10 to 1.5%. The amount of sporinite in coals varies from 10 to about 30% and the maceral is an important tar and gas producer in low-temperature carbonization because of its richness in hydrogen. As coalification proceeds, sporinite displays properties very similar to vitrinite and may no longer be distinguishable.

5.3.2.2   *Cutinite*

Cutinite derives mainly from the cuticles of leaves and stems. Depending upon the origin of the cuticles and relation of the microscopic section to the bedding structure, cutinite may display serrated bands of varying thickness. The colour in reflected light, (bright field, oil immersion) varies from black to dark gray. The maceral is hydrogen-rich (7.5–12%) and yields 50-60% tar and 22–27% gas on low-temperature carbonization.

5.3.2.3   *Resinite*

Resinite in coals is attributed to plant resins and essential oils in leaves which are converted during coalification to spherical or oval shaped bodies and occasionally lenses or thin bands of fluorescing material (Stach et al., 1982). Resinite consists mainly of small resin bodies and microspores embedded in collinite. The shapes may be spheroidal, lenticular or angular and the size may be so large that it is visible to the unaided eye. The colour of resinite may vary from matt in high-rank coals to brown or brownish-black in low-rank coals. It is rich in hydrogen (8 – 11%) and softens readily on carbonization. It yields up to 90% tar on low-temperature carbonization, the highest being from brown coals. Most of the balance is gas and liquor.

5.3.2.4   *Liptodetrinite*

Liptodetrinite comprises macerals of the liptinite (exinite) group consisting of extremely small particles. The maceral refers to constituents in the group which because of the extreme fineness cannot be assigned with certainty to any other maceral in the group. It may consist of fragments or relics of sporinite, cutinite, resinite, suberinite or alginate. The colour in reflected light varies from black to dark gray or dark brown. The technological properties have not been determined but are believed to be similar to those of the other members of the group.

5.3.3   The Inertinite Group

Inertinite is a group of macerals which derive from plant material that has been strongly altered and degraded by oxidation at the peat forming stage of coal formation due to weathering or pyrolysis in minimum air. For example, fossil charcoal is the inertinite maceral known as fusinite. Possible causes of degradation include charring of plant matter by peat fires, moldering, fungal attack, biochemical gelification, heating by igneous intrusions (volcanic lava). Inertinite is formed from the same plant tissues that form humanite/vitrinite when not subjected to degradation. The inertinite macerals have the highest reflectance of all the macerals in a coal. The group includes fusinite, semi-fusinite, sclerotinite, macrinite, micrinite, funginite and inertodetrinite (Table 5.2).

The various macerals in the inertinite group are distinguished by their relative reflectances and structures, and concentration in coals varies widely. Inertinite content of coal can be as low as 5% in some American coals or as high as 70 - 88% in some western Canadian, South African and Indian coals. Macerals in the inertinite group undergo very little change on carbonization and coals with high inertinite content are generally unreactive. Also, they are characterized by a relatively high carbon content and strong aromatic cross linkage structures which make them less combustible than other coal macerals, resulting in high unburnt carbon in the ash. However, the heat value of inertinite is higher because of the high carbon content and the maceral burns readily at higher temperatures, in oxygen-rich atmospheres, and at longer residence times. Furthermore, some recent studies have shown that inertinite in low rank coals may have a reflectance below 1.30% and could be more reactive than the vitrinite in the same coal (Choudhury et al., 2007). The macerals in the inertinite group are undesirable in coal carbonization (coking) because they behave as unreactive diluents and additional heat and flux are needed to remove them.

5.3.3.1 *Sclerotinite*

Sclerotinite is a prominent sub-group of the inertinite group which derives its name mainly from its behaviour on heating. It neither softens nor becomes plastic. The group is more important for its undesirability in coal than for any useful properties because of its relatively inert nature (chemical and thermal) compared with vitrinite and exinite groups. Sclerotinite may occur in coal in round or oval bodies or as interlaced fungal remains. The colour varies from light gray to white but it can be deep black in some soft brown coals. It is fairly widely distributed in coals but usually in small quantities.

5.3.3.2 *Fusinite/Semi-fusinite*

This sub-maceral group shows well defined cellular structure of wood or sclerenchyma. The cell cavities may vary in size and shape from round to oval to elongated. The colour is yellowish white in reflected light. It is common in coals and, because it is friable, it is usually in small dispersed fragments. Semi-fusinite has similar structure as fusinite but the cell cavities are generally smaller and sometimes less well defined. Also the technological properties lie between those of fusinite and vitrinite, hence the sub-maceral is considered as a transition material between vitrinite and fusinite.

5.3.3.3 *Micrinite and Macrinite*

Micrinite and macrinite are members of the inertinite group and occur in coal either in isolated discrete form or as a groundmass. They are generally non-angular, exhibiting no relic plant cell wall structure. They are usually only a few percent of the total inertinite in coal. The colour of micrinite/macrinite varies from light gray to white in reflected light. When the size is less than 10 μm, the sub-maceral is classified as micrinite while coarser particles are classified as macrinite. Very little is known about the technological behaviour of micrinite/macrinite except that it is chemically resistant. In any case, the concentration in many coals is usually negligible. This also explains why not much research has focused on the technological properties of the sub-macerals.

5.3.3.4 *Inertodetrinite*

All microstructures in coal which display a higher reflectivity than the corresponding vitrinite and which cannot be classified as micrinite, macrinite, semifusinite or sclerotinite are grouped under the term inertodetrinite. The colour varies from light gray to white in reflected light and it has no cell structure. Inertodetrinite may be present in significant quantities in some coals, it is believed to raise the strength and reduce fissuring of microlithotypes, and it is chemically and thermally inert.

5.3.4   The Mineral Matter Group

Mineral matter is not a maceral but a group of inorganic minerals which have become associated with coal by various processes during the various stages of coalification. These include inorganic constituents of the original plant material (plant ash). Mineral matter may be intimately intergrown with organic matter in coal. This depends on the conditions in the peat swamp (depth of coverage by water, water movement, chemical composition and acidity of the swamp, etc). Typical intergrown minerals include finely intergrown pyrites and siderite, fine-grained quartz and intercarlations of clay minerals (Figure 5.3).

Some elements in the mineral matter, for example sulphur may be covalently bonded to carbon in the organic structure of coal. Mineral matter can also be deposited in discrete forms in seam cleats and   fissures, the extent depending on activities in the overburden and underburden of the seam. Many of these minerals form solutions which may ascend or descend and penetrate into coal seams. Mineral matter in coal is important because its presence is significant in determining the economic value and market acceptability of coal. Also, the quantities and characteristics of mineral matter in coal determine whether or not the coal can be cleaned economically. A wide range of mineral matter may be present in coal either intimately intergrown with the coal or as discrete matter deposited in cracks and fissures in the coal. The principal minerals are clays, carbonates, iron minerals, sulphides, oxides, quartz and salts (chlorides, sulphates, etc) (Table 5.3).

Mineral matter in coal is classified on the basis of its origin into two categories: *inherent mineral matter* and *adventitious mineral matter*. Inherent mineral matter derived from the organic and inorganic constituents that previously formed part of the tissues in the peatification stage. Most of the mineral matter is chemically and colloidally combined with organic coal matter. It is not removable by washing, only by thermal or chemical degradation of the coal aromatic structure.

Inherent mineral matter may also derive from minerals transported into the swamp or bog by air or water (detrital minerals). Minerals transported by flood water may be deposited as mineral-rich partings in coal beds and some may diffuse into the decaying organic matter, forming intimate bonds with carbon. Mineral matter may be deposited into bogs by wind, especially when the bog is located in close proximity to active volcanic regions (Thomas, 2002). Adventitious (authigenic) mineral matter derived from both pre and post-peatification deposition of mineral matter by oceanic transgression, land slides, etc. These events may occur during or subsequent to coalification. Depending on the geological history of the deposit, adventitious mineral matter deposited during peatification may be present in chemical or colloidal combination with the coal substance. This type of mineral matter cannot be removed by cleaning. Adventitious mineral matter may be in discrete aggregates finely disseminated throughout the coal structure, or may be present in coal seam cracks, fissures and open voids, seam intrusions, seam overburden, etc. While lump mineral matter can be easily removed from coal by cleaning, finely disseminated mineral matter is difficult to remove and requires that the coal is finely ground.

In addition to the elements in the major mineral matter in coal a number of elements are often present in very small amounts and may be present as either inherent or adventitious matter, or in both forms. This group of elements known as trace elements includes mercury, zirconium, zinc, uranium, tin, vanadium, lead, molybdenum, nickel, silver, strontium, germanium, copper, cobalt, bismuth chromium, gallium, beryllium, antimony. Mineral matter is not combustible and ends up as ash at the bottom of combustion chambers or in coke. The chemical composition of mineral matter is important in estimating loss of heat value of coal due to endothermic decomposition of many minerals, and its effect on the fusibility and removal of the ash formed during coal carbonization or combustion. In effect the presence of mineral matter in coal is inevitable but undesirable, and the quantity, morphology and chemical nature can degrade severely the value of coal. Also, removal prior to or after carbonization/combustion could be expensive and many of the constituents, in particular, sulphur and trace metals are anthropogenic.

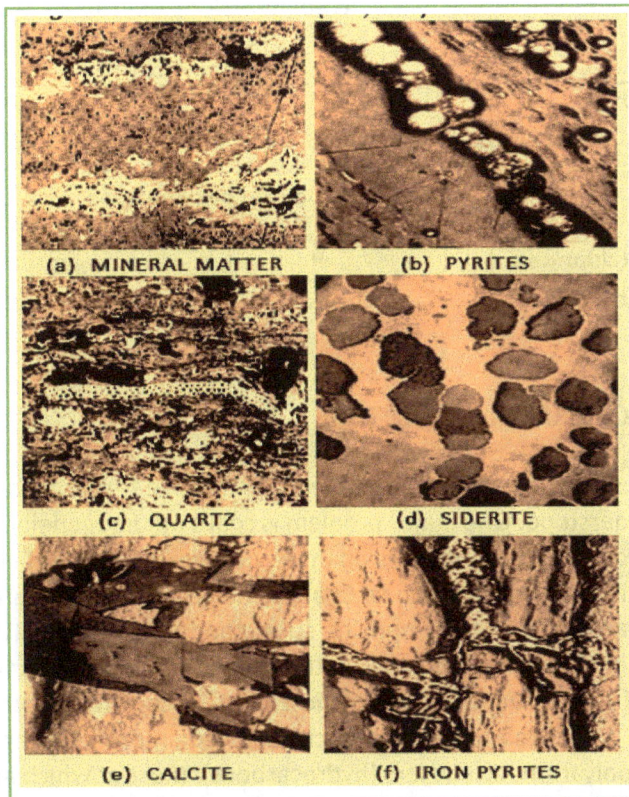

FIGURE 5.3 Mineral matter in coal. (a) shows a deposit of complex mineral matter; (b) intimate intergrowth of pyrites and coal, the most common being iron pyrites; (c) shows intimate intergrowth of quartz and coal; (d) shows intimate intergrowth of siderite accretions and coal; (e) shows calcite deposits in cleats and fissures; and (f) shows iron pyrites deposits in cleats and fissures. *(ICCP, 1963).*

TABLE 5.3. Main mineral matter inclusions in coals. *(ICCP, 1963).*

Type of mineral	Deposited by water or wind	Originally formed in the peat	Deposited in cleats and other fissures (coarsely inter-grown)
Clay minerals	Illite,sericite, kaolinite, leverrierite, montmorillonite, etc.		
Carbonate minerals (spars)		Accretions of siderite, dolomite, (ankerite), and calcite, $FeCO_3$ and $CaCO_3$ in fusite.	Calcite, Ankerite.
Sulphide minerals		Accretions of FeS_2, Accretions of FeS_2-$CuFeS_2$-ZnS, Melnikovite.	Pyrites, Marcasite, Zinc blend, Copper pyrites, Galena.
Oxides		Limonite, Hematite.	Gothite, (needle iron ore).
Quartz	Granular quartz.	Chalcedony and quartz produced by deposition of aluminium silicates.	Quartz Chalcedony
Salts, (chlorides, sulphates, etc)	Rock salt, thenardite, gypsum		

5.4 COAL MOLECULAR STRUCTURE

Coal molecular structure has been a subject of intensive research for over a century, but is still not well understood. One major problem is coal's inhomogeneity. Another problem is its insolubility in common solvents. It is opaque and non-crystalline, hence difficult to analyze by chemical or instrumental methods. Coal has no repetitive structure and is believed to be a mixture of macromolecules comprising strong and weak bonds. The molecular structure varies not only between ranks but also with each seam and each sample (Neavel, 1981).

A model of coal structure is shown in Figure 5.4 but many other models have been proposed. Coal is believed to be a highly cross-linked amorphous copolymer and comprises stable 2-4 ring polyaromatic clusters (hydrocarbons), most of which are connected together by hydroaromatic, etheric or aliphatic linkages while others are relatively mobile and unconnected (Mathews and Chaffee, 2012). The degree of carbon aromaticity varies between about 60% in the youngest coal (lignite) to around 90% in anthracite. Most of the hydrogen is in aliphatic groups and the oxygen in phenolic units.

The molecular structure of coal is unimportant if it is for carbonization, combustion or gasification because the heat is sufficient to break down the aromatic bonds. However, it has a major effect on coal conversion processes including liquefaction, catalysis and conversion to premium carbon materials.

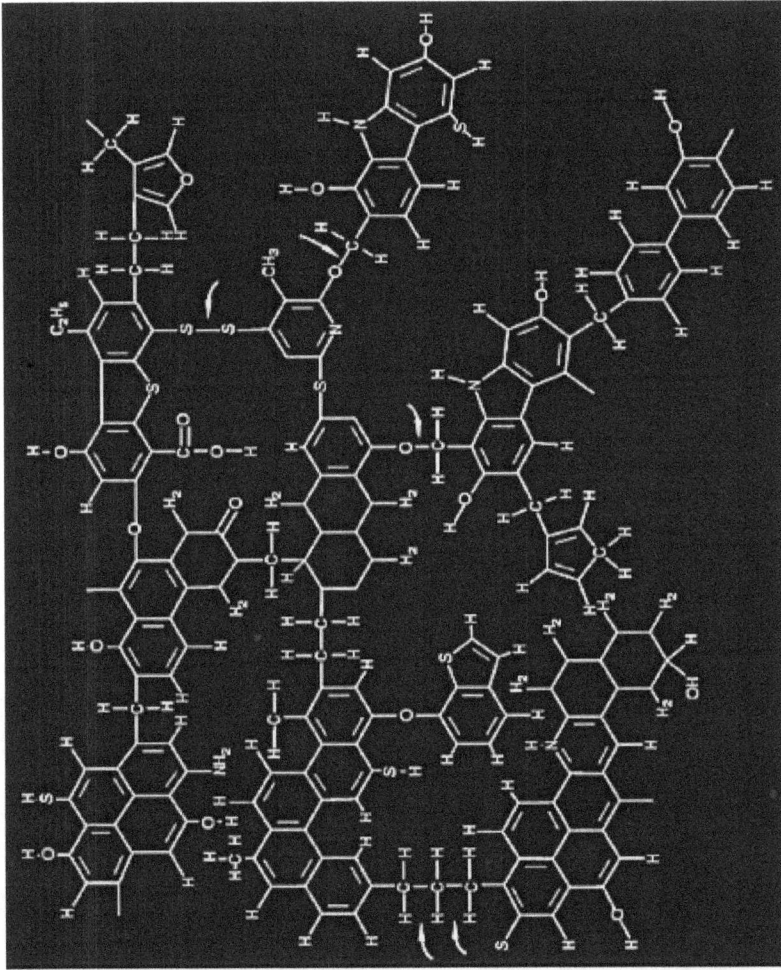

FIGURE 5.4 Wiser's molecular model for coal. *(Whitehurst, 1978).*

5.5 RELATIONSHIP BETWEEN STRUCTURE
AND CHEMICAL PROPERTIES OF COAL

There have been many studies on the relationship between structure and chemical properties of coal. Dormons et al., (1957) measured the reflectance of a range of coals of different ranks and related the variable to carbon content of vitrinite (Figure 5.5). van Krevelen (1961) defined relationships between hydrogen:carbon and oxygen:carbon ratios (Figure 5.6). Bustin et al (1985) studied a wide range of coals and related vitrinite reflectance to volatile matter content of vitrinite. Numerous studies more recently have confirmed the strong correlation between vitrinite reflectance and coal maturity Figure 5.7).

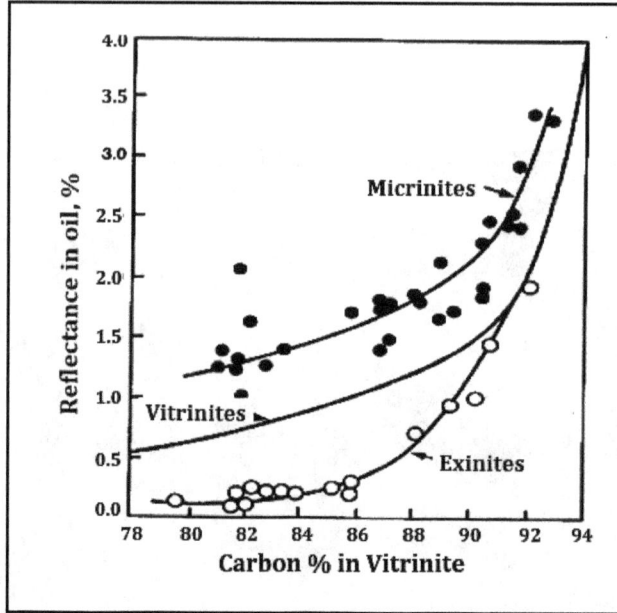

FIGURE 5.5 The reflectance of three macerals at different levels of rank. *(Dormons et al., 1957).*

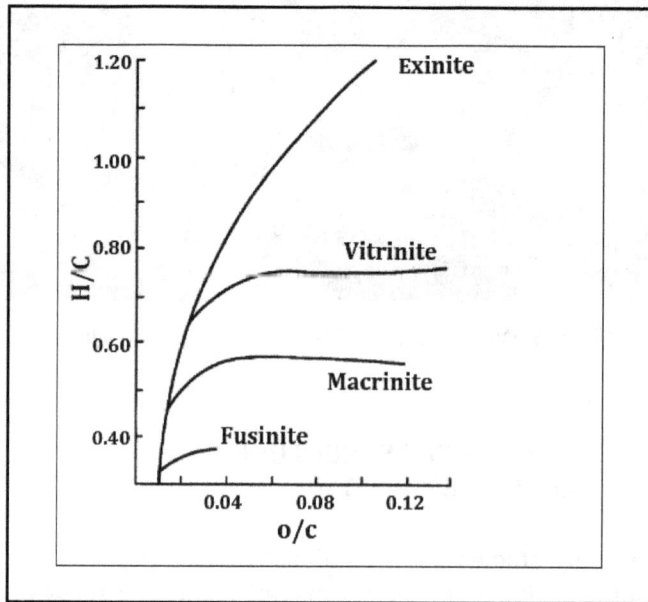

FIGURE 5.6 Coalification tracks of different macerals based on H/C:O/C atomic ratios. *(van Krevelen, 1961).*

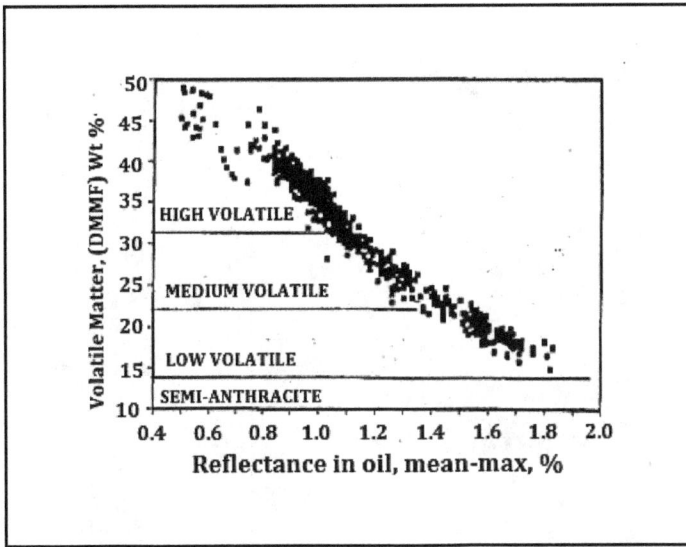

FIGURE 5.7 Relationship between reflectance of vitrinite and volatile matter of US coals. *(ASTM D388-98a).*

Vitrinite reflectance has also been used for ranking coal (Table 5.4), and for grouping coals for utilization. Correlation with coal fluidity during carbonization has also been established by many studies (Figure 5.8). Petrographic analysis of coal is a major tool for predicting coke quality obtainable from a coal or coking blend. The mean maximum reflectance of vitrinite in coking coals varies between 0.85% and 1.35% and the minimum vitrinite content is about 50%, although there are coals with lower content. Also, the quantity and chemical nature of other macerals and mineral matter in the coal are important. Coal reactivity is largely determined by chemical properties and relative proportions of the maceral constituents while the chemical nature and quantity of ash formed in combustion or carbonization depend on the mineral matter content of coal.

TABLE 5.4 Rank categories of coal based on reflectance.

Reflectance	Rank
≤ 0.050	Low-rank (lignite/sub-bituminous
0.50 – 1.12	High-volatile bituminous
1.12 – 1.51	Medium volatile bituminous
1.51 – 1.92	Low-volatile bituminous
1.92 – 2.50	Semi-anthracite
> 2.50	Anthracite

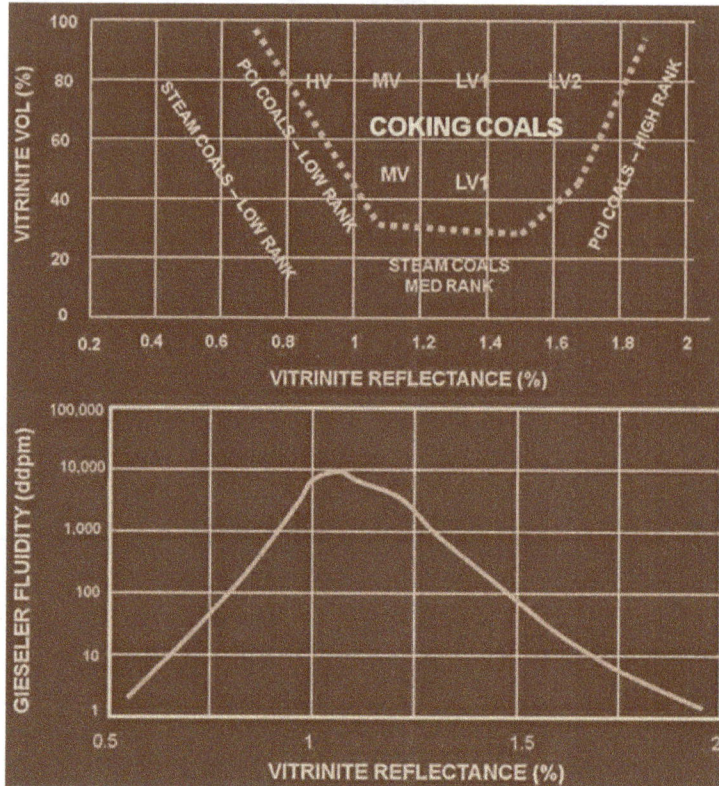

FIGURE 5.8   Relationship between vitrinite reflectance and coal plasticity and utilization. *(Cook, 2008, www.coalanalysislaboratory.com).*

5.6 STRUCTURE OF COAL, GRAPHITE AND DIAMOND

Coal, graphite and diamond are three naturally occurring allotropes (forms) of carbon. Several synthesized allotropes include fullerenes and carbon nanotubes. Coal is predominantly amorphous, graphite is partially crystalline and diamond is fully crystalline (Figure 5.9). All three allotropes of carbon (coal, graphite and diamond) were formed from a carbon source in the great depths of the Earth's crust under conditions of high temperature and pressure. Coal is formed from plant debris buried in swamp and covered with overburden over time estimated to be between 250 and 350 million years. Typical coal deposits are no more than two or three kilometers deep in the Earth's crust. Graphite is believed to metamorphosize from coal or other organic material that is trapped in sedimentary rock which exerts much higher temperatures and pressures than are required for coalification.

Graphite most often occurs in metamorphic rocks formed from regional metamorphism or contact metamorphism of organic-rich sedimentary rocks, such as organic-rich marble, quartzite, schist, gneiss, and metamorphosized coal. The favorable conditions promote the almost total expulsion of hydrogen and oxygen as water, leaving carbon-rich graphite. Depending on the nature of the deposit environment, largely amorphous flake graphite or semi-crystalline lumps may form. However, most graphite deposits are infiltrated by inorganic matter and graphite deposits purer than around 95% carbon are hardly found.

Diamond is believed to have been formed from carbon sources under high temperatures and pressures deep in the Earth's mantle beneath the stable interiors of continental plates around 150-250 kilometers from the surface. Temperatures in parts of the mantle could reach or exceed 1000-1400°C (Mitchel, 1995, Erlich and Dan Hausel, 2002; Shirey, 2013). Such temperatures combined with the very high pressures in the sub-continental lithospheric mantle area are thought to promote the formation of diamond from suitable carbon sources.

The prevailing conditions at great depths of the Earth are also conditions under which diamond is known to be stable. At shallower depths, where temperatures and pressures are lower, diamonds are not stable and any carbon present will occur as graphite. Due to extensive residence in this domain of stability, formed diamond goes through processes of recrystallization, refinement and growth. Formed diamonds are believed to be transported to the Earth's surface during volcanic eruptions through tunnels known as kimberlite and lamproite pipes. Diamonds weathered and eroded from these eruptive deposits are usually found in the sedimentary deposits of streams and coastlines (AMNH, 1998).

Coal and graphite are possible carbon sources for the diamond-forming process. However, there is a growing body of evidence that the formation of virtually all the world's diamonds predated the appearance of vegetation on the Earth's surface by about 100 million years. Most diamonds that have been dated were formed in the Precambrian Period, between the Earth's formation about 4,600 million years ago and the Cambrian Period about 542 million years ago whereas land plants did not appear on Earth until about 450 million years ago. More probable carbon sources for diamond formation are from the subduction of oceanic plates which are rich in carbon sources such as carbonate rocks (limestone, marble, dolomite) and potassic carbonatitic fluids (Gurney et al., 2010), Figure 5.9(d).

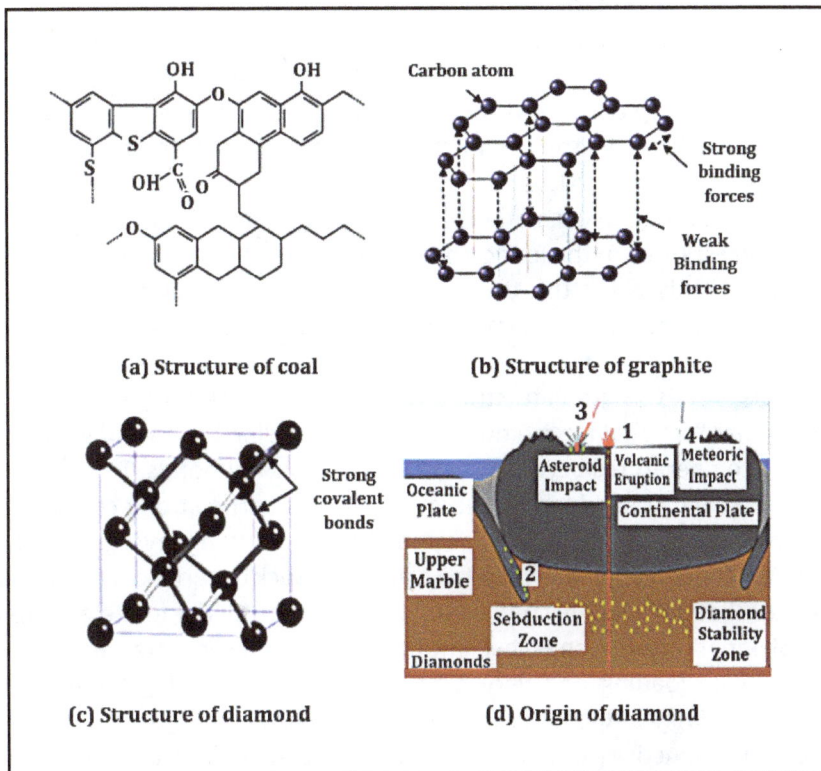

(a) Structure of coal

(b) Structure of graphite

(c) Structure of diamond

(d) Origin of diamond

FIGURE 5.9 Structures of coal, graphite and diamond.

Another possible source of carbon is the ingress of carbon-rich mobile fluids and melts and C-O-H bearing fluids to the continental lithosphere (Shirey et al, 2013). Some researchers believe that diamonds form from carbon in methane or other hydrocarbon gases that ascend through the upper mantle from greater depths of the Earth (Mitchel, 1991). These fluids undergo simple redox reactions to produce carbon, for example, the reactions shown in Equations 5.1 and 5.2.

$$CO_2 = C + O_2 \hspace{6cm} [5.1]$$

$$CH_4 + O_2 = C + 2H_2O \hspace{5cm} [5.2]$$

There are other theories of diamond formation from meteorites and large asteroids hitting the Earth's surface at very hyper-velocities and generating more than enough energy to metamorphosize carbon sources such as graphite and coal on or beneath the Earth's surface.

Although coal, graphite and diamond are chemically similar allotropes of carbon, their different crystal structures give them very different physical and technological characteristics [Figures 5.9 a-c)]. For example, coal is largely amorphous, graphite is partially amorphous while diamond is fully crystalline. Coal and graphite are opaque and black while diamond is highly transparent. Coal and graphite are soft while diamond is the hardest naturally occurring material known. Coal and graphite have low thermal conductivities while diamond has the highest thermal conductivity of all known naturally occurring materials. Coal and diamond are non conductors of electricity and good insulators while graphite is a good conductor in the planar direction. Both graphite and diamond have extremely high melting points, (3927°C and 3550°C respectively) while coal, like pure carbon has no melting point at atmospheric pressure. Also, because of the largely amorphous nature and the numerous impurity contents, coal ignites at relatively low temperatures. Diamond is very hard and brittle while graphite is one of the best known high-temperature solid lubricants.

The structure of coal Figure 5.8(a) shows carbon bonded to oxygen atoms (O) and hydroxyl groups (OH) in an irregular manner. The structure of graphite [Figure 5.9(b)] consists of flat, planar layers. In each layer the carbon atoms are arranged in a regular hexagonal array and covalently bonded to each other while the planes are loosely held together by weak van der Waals forces. Three of the four carbon electrons in graphite carbon are involved in forming the covalent bonds while the fourth forms a delocalized electron cloud which spreads uniformly over all the carbon atoms. For this reason graphite conducts electricity parallel to its plane.

Also, the planes in graphite structure can glide on each other under low loading (for example planar frictional forces) due to the weak binding forces and yet remain stable even at very high temperatures because of the planar covalent bonds. This explains why graphite is such a good and stable high-temperature lubricant. The strong covalent bonds between the carbon atoms at the planar level of graphite stay stable at very high temperatures because of the ultra-high melting point while the planes can stick to surfaces and glide over each other easily under shearing load. In comparison with graphite all the four electrons of each carbon atom in diamond form strong, covalent bonds between carbon atoms which remain stable under very high loading and very high temperatures, making diamond neither a good conductor nor a lubricant, but the hardest known substance [Figure 5.9.(c)]. The exceptionally high thermal conductivity of diamond is due to the ability of the very stable carbon-carbon bonds to transfer heat through strong vibration.

REFERENCES

AMNH (1998). *The Nature of Diamonds.* American Museum of Natural History.

ASTM D388-12. *Standard Classification of coals by Rank.* American Society for Testing and Materials.

Bustin, R. M. A. R Cameron, D. A. Grieve, and W. D. Kalkreuth (1985). *Coal Petrology, its principles, methods and* ICCP (2001). *International Committee for coal and organic petrology training manual.* International Committee on Coal and Organic Petrology.

Mathews, J. P. and A. L. Chaffee, (2012). "The molecular representations of coal – A review." *Fuel*, Volume 96, June 2012, pp. 1.14.

Mitchel, R. H. (1991). " Kimberlites and lamproites: Primary sources of diamond." *Geoscience Canada,* Volume 18, Number 1, pp. 1-16.

Mitchel, R. H. (1995). "Kimberlites, orangeites and related rocks." Plenum Press, New York.

Neavel, R. C. (1981). "Origin, Petrography and Classification of coal." *Chemistry of Coal Utilization, Ch. 3.* M. A. Elliott, (Ed.). Wiley, New York.

Schapiro, N., and R. J. Gray (1960). "Petrographic classification applicable to coals of all ranks." *Proceed. Ill. Min. Inst., p*p. 83-97.

Scott, A. C. (2002). "Coal petrology and the origin of coal macerals: a way ahead." *Int. Jnl. Coal. Petrology, Vo*l. 50, pp. 119-134.

Seyler, C. A. (1933). "The relation between the volatile matter and elementary composition of coal." *Journ. Soc. Chem. Ind.,* I,II, pp. 304 T 306 T.

Shirey, S. (2013). "Diamonds and the Geology of Mantle Carbon." *Reviews in Mineralogy & Geochemistry*, Vol. 75, pp. 355-421, 2013.

Shirey, S. B. and J. E. Shigley (2013). "Recent Advances in Understanding the Geology of Diamonds." Gems & Geology, Winter 2013, Vol. 49, No. 4.

SIU (2015). *Crelling's Petrographic Atlas of Coals and Carbons.* coalandcarbonatlas.siu.edu. Solomon, 1981).

Stach, E. M. Mackowsky, M. Teichmuller, G. H. Taylor, D. Chandra and R. Teichmuller (1982). *Stach's textbook of coal petrology.* Stuttgart¨ Gebruder Born-traeger.

Stopes, M. C. (1919). "On the Four Visible Ingredients in Banded Bituminous Coal: Studies in the Composition of Coal." *Proc. Royal Society B,* Vol. 90, Issue 633.

Stopes, M. C. (1935). "On the petrology of banded bituminous coal." *Fuel,* Vol. 14, Pp. 4-13.

Suarez-Ruiz, I and J. Crelling (2008). *Applied Coal Petrology. The Role of Petrology in Coal Utilization.* Elsevier.

Suarez-Ruiz, I. (2012). "Organic Petrology: An Overview. In Petrology – New Perspectives and Applications." *Ed. Ali Ismail Al-Juboury. Intech.*

Taylor, G. H., M. Teichmuller, A. Davis, C. F. K. Diessel, R. Littke, and P. Robert (1998). *Organic petrology.* Genruder Borntraeger, Berlin and Stuttgart, Germany, 704p.

Thomas, L. P. (2002). *Coal Geology.* John Wiley & Sons.

USGSS (2013). *Photomicrograph Atlas of Coal.* energy.usgs.gov/Coal/OrganicPetrology/PhotomicrographAtlas.

van Krevelen, D. W. (1961). *Coal.* New York, Elsevier.

Whitehurst, D. D. (1978). "Chemistry and constitution of coal." In J. W. Larsen (Ed.). *Organic Chemistry of Coal.* ACS Symposium Series 71, pp. 1-35.

Billion tonnes oil equivalent

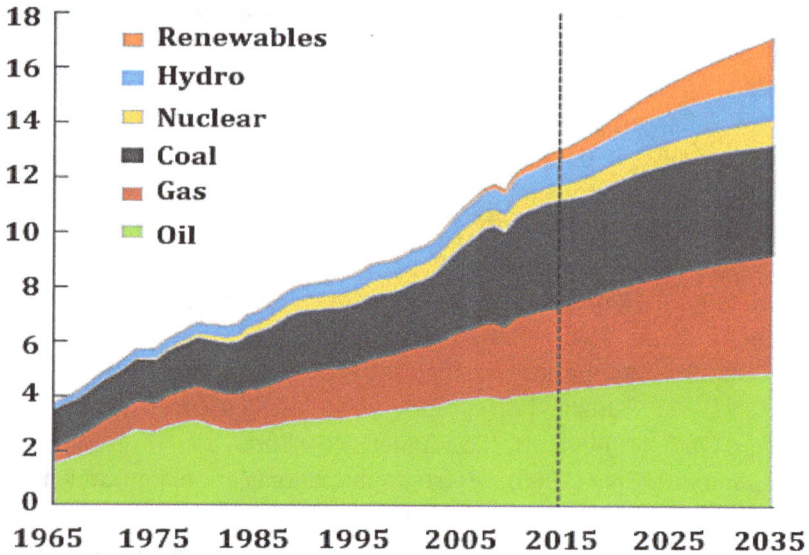

Billion tonnes oil equivalent

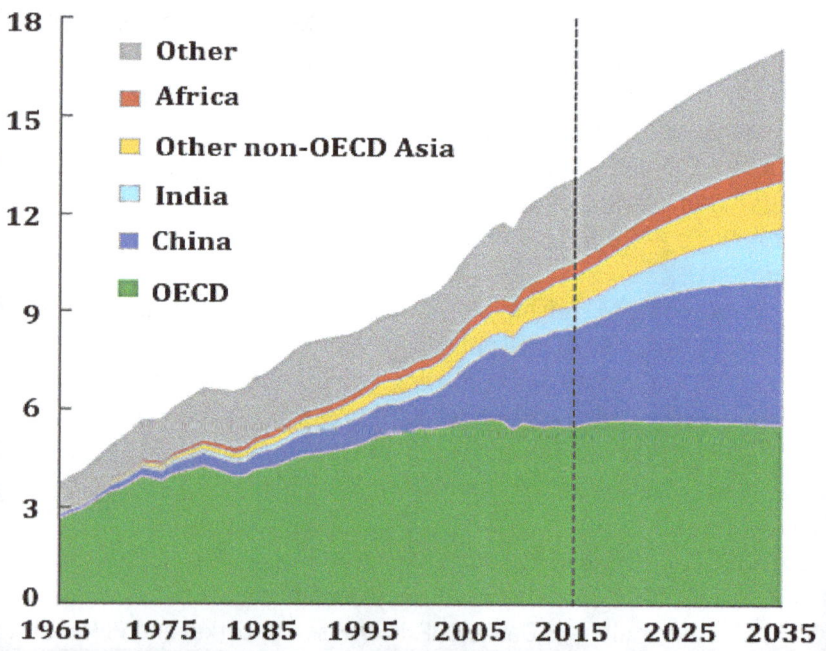

Global primary energy demand projection

(BP 2017 Energy Outlook)

6   Technical properties of coal

6.1 INTRODUCTION

The physical, chemical and thermochemical properties of coal largely determine its industrial properties and utilization potentials. These include density, optical properties, electrical properties, magnetic properties, elastic, and thermo-rheological (agglomerating) properties. The topic has been treated exhaustively in van Krevelen (1961). Only those associated with coal preparation and behaviour on combustion or carbonization will be treated in this chapter. There is also extensive literature on coal analysis and testing (see Speight, 2015) hence only summaries will be presented in this chapter.

6.2 DENSITY OF COAL

The density (weight per unit volume) of coal depends on its rank, type and inherent mineral matter content. It is also an important parameter for the determination of the degree of coalification. The density of coal decreases gradually with rank and passes through a minimum at about 85% carbon. Beyond this, the density rises steeply and continues smoothly towards a highest value for graphite. The initial drop in density is believed to be due to losses in the relatively heavy element, oxygen, whereas the hydrogen content remains nearly constant. The density therefore tends to fall off. When nearly all the oxygen has been liberated, the very light hydrogen starts to disappear and the coal becomes increasingly richer in carbon with a consequent steep rise in density (van Krevelen, 1961).

Apparent or relative density (density with reference to water at 4°C) has a significant influence on the response of coal to cleaning methods. Bulk density (the way bulk coal packs into a confined space) expressed as kg/m^3 is an important variable in coal carbonization. This variable is influenced by particle shape and size distribution, particle density, particle surface properties, and moisture content. High bulk density is known to improve coke quality. The bulk density of coal is always higher than the sum of the densities of the size fractions packed separately due to the increased intergranular spaces, the wider the spectrum of sizes of particles, the higher the bulk density of coal. Moisture content also has a significant influence on bulk density of coal, in particular, high rank coals.

The hydrophobic nature of higher rank coals retards moisture from penetrating into the coal. Most of the moisture is retained on the particle surface and increases the particle volume, thereby reducing bulk density. This variable is an important parameter in coal combustion and carbonization. High bulk density has a positive influence on the productivity of furnaces and also on metallurgical coke quality. It is common practice therefore to pre-process coal by drying, mixing with hydrocarbon oil, or small additions of a chemical reagent to modify the surface characteristics of the coal particles. Coke ovens are often stamp-charged to increase bulk density.

6.3 STRENGTH OF COAL

The strength of coal is of vital importance in mining, handling, transportation and grinding of coal. It is dependent on the rank of coal and the physico-chemical condition, in particular, the extent of oxidation. Compression strength decreases steeply with increasing rank, reaching a minimum corresponding to 20 to 25% volatile matter, and then rises steeply as coalification proceeds further. The throughput and energy required for grinding coal depend on its hardness. Determination of grindability is performed in a Hardgrove (1932) laboratory grinder (Figure 6.1) in accordance with specific standards (for example ASTM D 409, 1991; BS 1016-112, 1995, ISO 5074, 1994). A charge of 50g ± 0.01g of coal is removed from analysis sample and poured on the grinding track. Grinding ball is inserted, the upper rotating part is loaded and the grinder drive is started up. After 60 rotations the grinder stops and the ground coal is screened for 10 minutes on the screen with square openings with eye length of 0.075mm. The grains caught at the bottom of the screen are removed by means of a fine brush. The coal screening and cleaning are performed twice again after 5 minutes. After the completion of screening the grains obtained by cleaning of screening surface are added to the undersize. The undersize and oversize are weighed with accuracy of 0.01g. The total weight of screening products can differ from the original weight at most by 0.75g. The HGI grindability value is calculated from the following formula:

$$HGI = 13 + 6.93\,w \qquad\qquad [6.1]$$

Where w = weight of the test sample passing through 75μ sieve and retained on 300μ sieve after grinding.

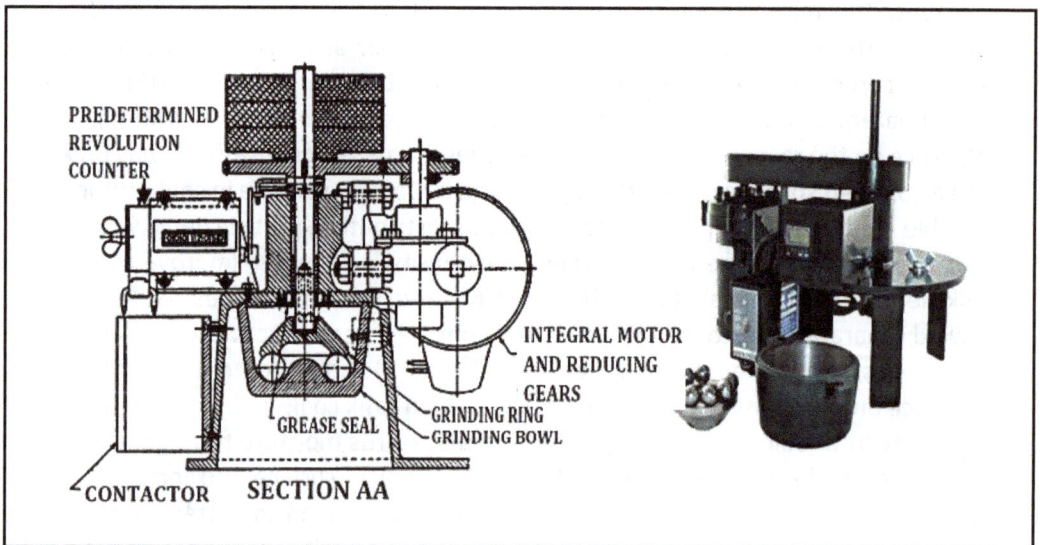

FIGURE 6.1   A Hardgrove laboratory grindability machine. *(www.preiser.com).*

The HGI machine is calibrated using a reference coal which is assigned an index of 100 and all other measurements have values relative to this standard coal. Coals with high HGI (> 50) are relatively soft and easy to grind while coals with lower index are hard and difficult to grind or pulverize. Grindability increases steeply with rank, reaching a maximum corresponding to minimum coal density (Figure 6.2). The pattern of this behaviuor closely follows that of density discussed above. Grindability and pulverization of coal depend on the rank and type of coal. In general, lignites and anthracites are more resistant to grinding than bituminous coals as shown in Figure 6.2. This is due to the changing hardness and relative proportions of coal microlithotypes as rank increases.

The Hardgrove grindability index (HGI) value is influenced by the chemical and petrographic compositions of coal, levels of content of volatile matter, carbon, hydrogen and mineral matter, and physico-chemical nature of mineral matter content (Fitton et al., 1957). An increase in content of volatile combustible matter up to approximately 30% improves the grindability, beyond which the grindability deteriorates. Similarly the HGI value increases with increase in carbon content. The grindability then drops rapidly with the contents of carbon exceeding approximately 92%.

Petrographic composition of coal is believed to have a significant influence on coal grindability. Vitrinite enhances grindability while coals with high micrinite and liptinite contents decrease HGI values (Hower, 1987; Hober, 1988). Grindability is also influenced significantly by the physical and chemical nature of the ash-forming contents of coal (Wang, 1996). In effect, coals of the same rank can have very different grindability depending on the coal type and water content. While high strength of coal is desirable to minimize degradation during handling and transportation, low strength facilitates mining and grinding.

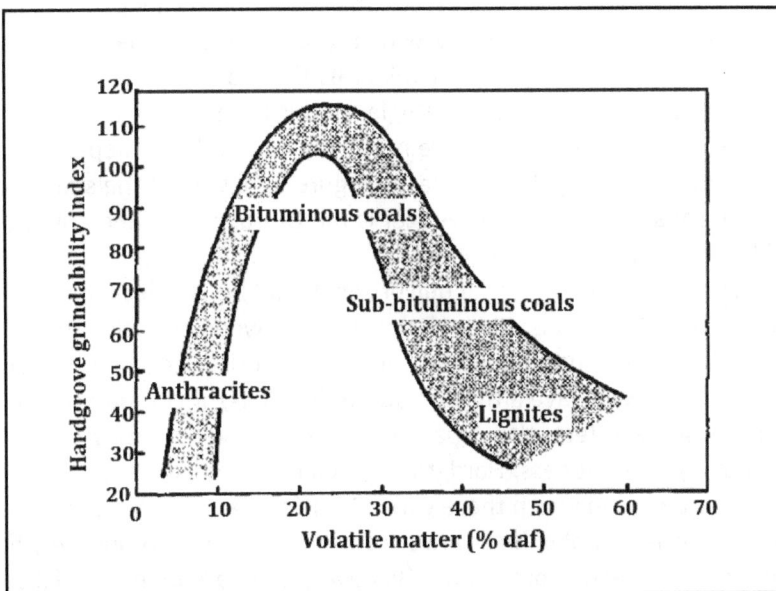

FIGURE 6.2 Schematic variation of Hardgrove Grindability Index with coal rank. *(van Krevelen, 1961).*

In many commercial operations, coals are required to be crushed to a desirable size mix or pulverized. The relative ease with which either process can be carried out depends on the strength of the coal and determines the strain on the grinding machinery, the wear of components, and the energy requirements. Two variables have been standardized for comparing the relative grinding (comminution) characteristics of coal: grindability index and abrasion index. Grindability is defined in terms of Hardgrove Grindability Index (HGI) which defines the relative ease with which a coal can be ground compared with a standard coal. It also determines the specific energy intensity which is the major process cost variable. The abrasive index determines the wear of equipment which is also a factor in evaluating the cost of grinding. The third variable is the particle size distribution which is important if the coal is to be used as a coal-water slurry boiler fuel.

Although some coal-fired boilers in power generation plants still burn lump coal, most modern boilers use coal pulverized so that 70-80% particles are below 75µm. This requires coarse grinding in a roll grinder or hammer mill and pulverization, most commonly in a ball mill. The energy requirement for both processes increases with increase in coal hardness and coals with HGI below 50 are considered hard and difficult to grind. It is estimated that mill throughput decreases by 1% for every unit decrease in HGI (that is, increase in coal hardness). Austin et al. (1995) found that the capacity of a roll grinder doubled as HGI increased from 44 to 106. Others have estimated that a 53 HGI coal requires 2.5 times the number of revolutions of the pulverizer as a 110 HGI coal to produce a fineness required for pulverized-fuel combustion (Fitton et al. ,1957; Hower, 2008; Suarez-Ruiz, 2008). Furthermore, pulverizing a blend of coals of different HGIs could lead to undesirable segregation (Milenkova et. al. 2003; Hills, 2007).

Most coals contain minerals, in particular quartz and pyrite which can drastically alter the strength and abrasion characteristics. Excluded minerals (those minerals in cracks and fissures that are liberated from the carbon matrix during grinding) are particularly abrasive. Included mineral matter associated with the coal organic matter can also be abrasive depending on the morphology. Other minerals that can cause abrasive wear are kyanite, alumina and topaz (Wells et al. 2004; Wigley and Williams, 2005). Moisture content of coal affects the HGI. The optimum moisture is between 10 and 25% depending on the coal type and rank. Outside this range HGI increases (Figure 6.3). One of the standard parameters for specifying the capacity of a coal pulverizer is the throughput when grinding coal with a HGI of 50 and less than 10% moisture.

The particle size output of a coal pulverizer should be 70% less than 75 microns and 1 or 2% greater than 300 microns. The relationships between particle size, HGI and pulverizer capacity are shown in Figure 6.4. For example, a pulverizer grinding coal of HGI = 50 to 70% less than 75 microns will have a throughput of 56,000kg/hr compared with 46,000 kg/hr for coal of HGI = 40. In effect, the capacity of a coal pulverizer may be defined as the coal throughput at a given fineness, Hardgrove grindability, moisture content and coal feed size and the different ways in which these variables affect capacity are shown in Figure 6.5.

Abrasive wear is complex and varies from one situation to another, hence there is no universally accepted testing procedure. However, there are many specialized abrasive wear testers in use in different parts of the world. One of the common methods involves grinding coal of specific size consist in a standard grinder with four metal blades for a specified duration. The wear of the blades is measured in milligrams and an Abrasion Index (AI) is expressed as mg/kg of coal processed.

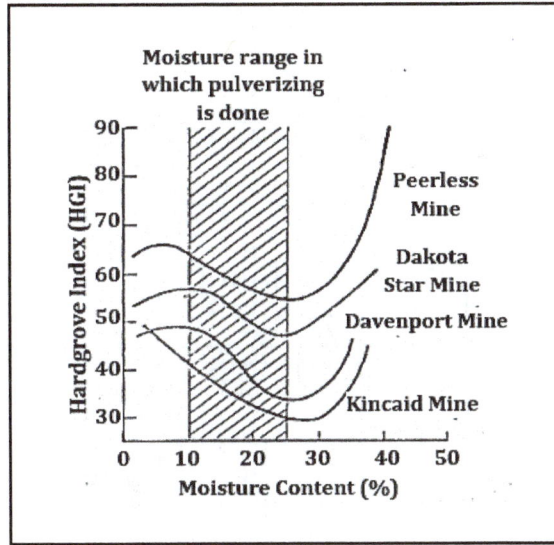

FIGURE 6.3 Variation of grindability index with moisture content of four different types of coals. *(Ellmann & and Belter, 1955).*

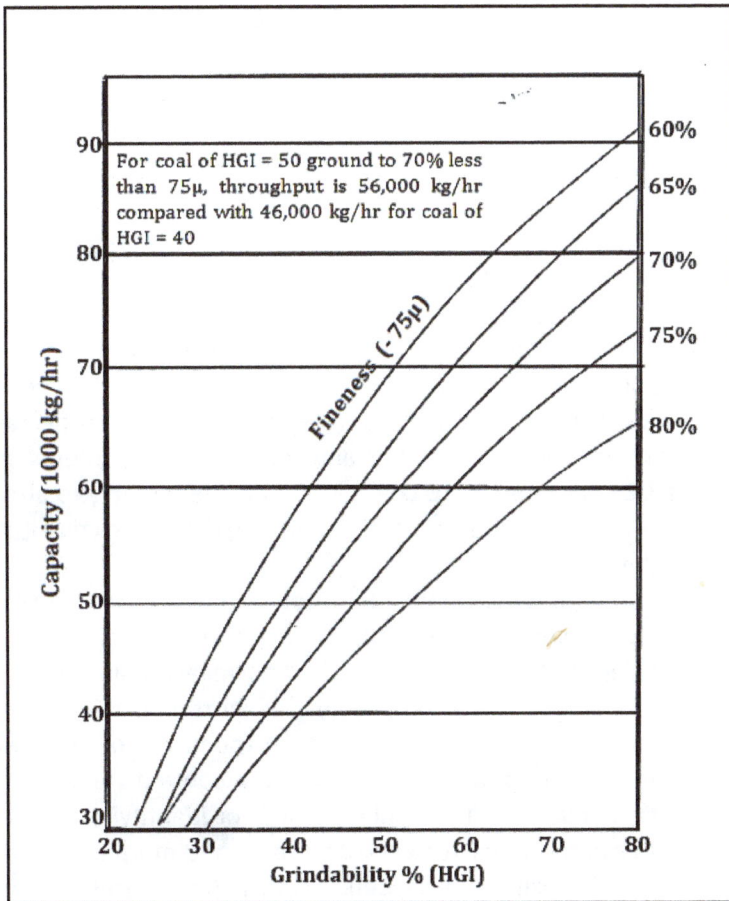

FIGURE 6.4 Variation of pulverizer capacity with Hardgrove grindability index for various particle size ranges. *(Adapted from Storm, 2015).*

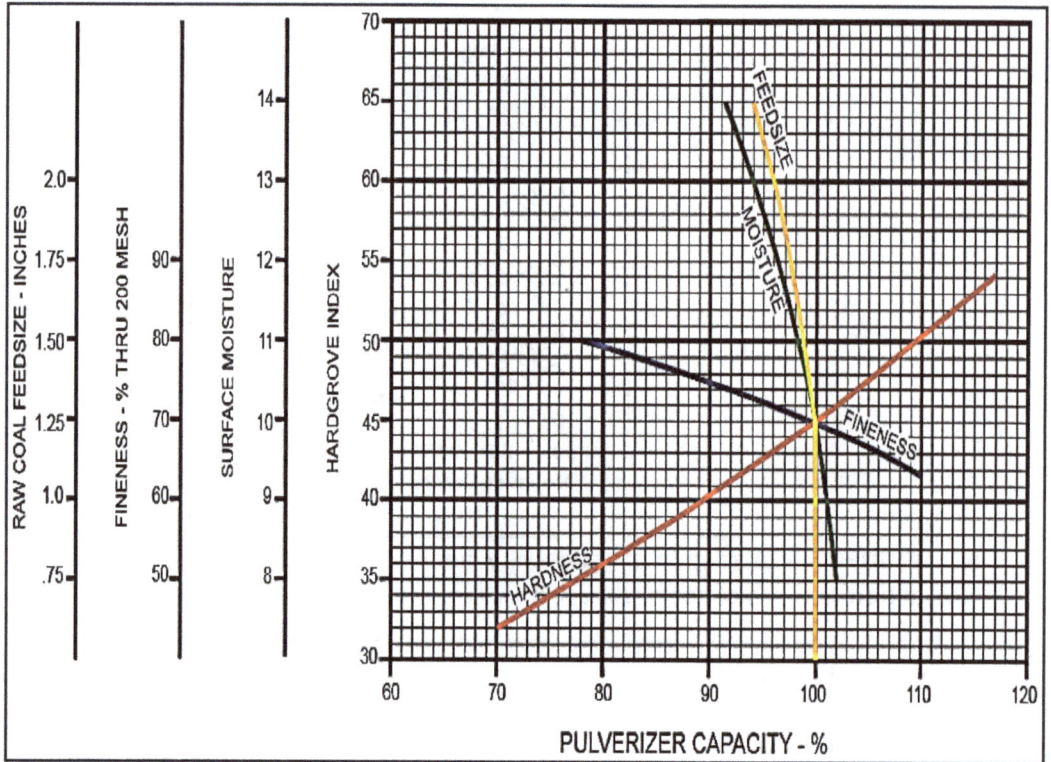

FIGURE 6.5 Coal pulverizer capacity specification factors. *(Fitton et al., 1957).*

6.4　OPTICAL PROPERTIES OF COAL

The reflectance of coal is the most important optical property from the point of view of industrial utilization. There is a correlation between coal rank and the reflectance of vitrinite, the most important maceral group in coal. The mean maximum reflectance in oil increases gradually with rank as indicated by carbon and volatile matter contents, from about 0.3% for lignite to about 1.2% for bituminous coals, and then rises steeply to about 6.5% for anthracite. (Figures 6.6 and 6.7). The determination of this variable also constitutes an easy method of distinguishing between maceral groups.

Reflectance is of importance in the utilization of coal because it is the only parameter of rank which can be measured on individual particles of coal and can therefore be used to characterize coal blends. Reflectance is also the only measurable parameter of rank which is not influenced by changes in the proportions of inertinite, mineral matter and maceral concentrations. Reflectance data have also been found to correlate well with important variables such as carbon content, volatile matter content, and calorific value of coal. Measurement of reflectance is a valuable method for identifying the different macerals in coal and the relative proportions. Also, the degree of metamorphism (and therefore the rank of coal), and potential thermochemical behaviour can be determined from reflectance data. Reflectance can give very useful information on the quality and yield of coke that can be obtained from a coal on carbonization. It is also very useful in the design of coking blends.

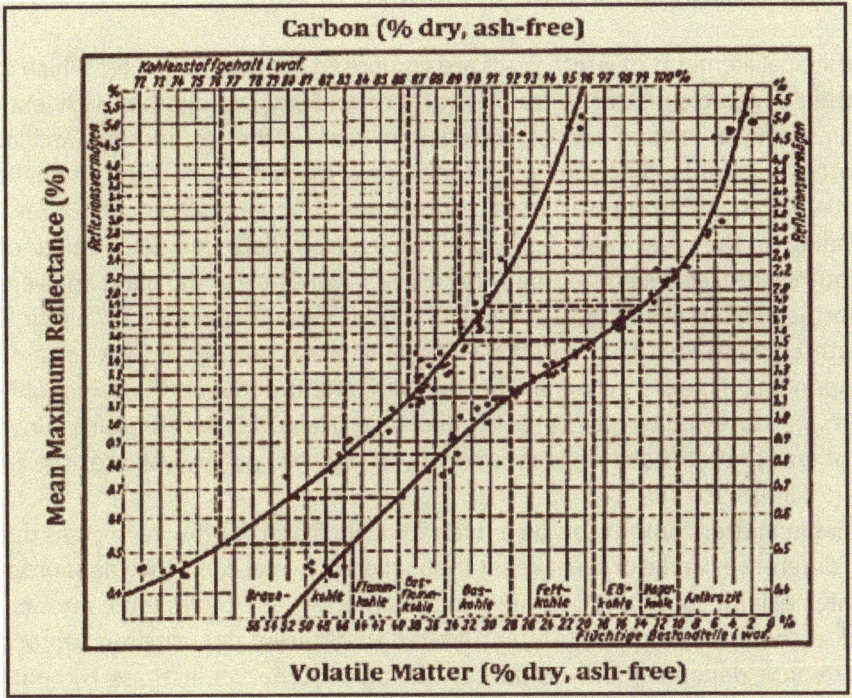

FIGURE 6.6 Mean Maximum Reflectance of coal as a function of carbon and volatile matter contents. *(van Krevelen, 1963).*

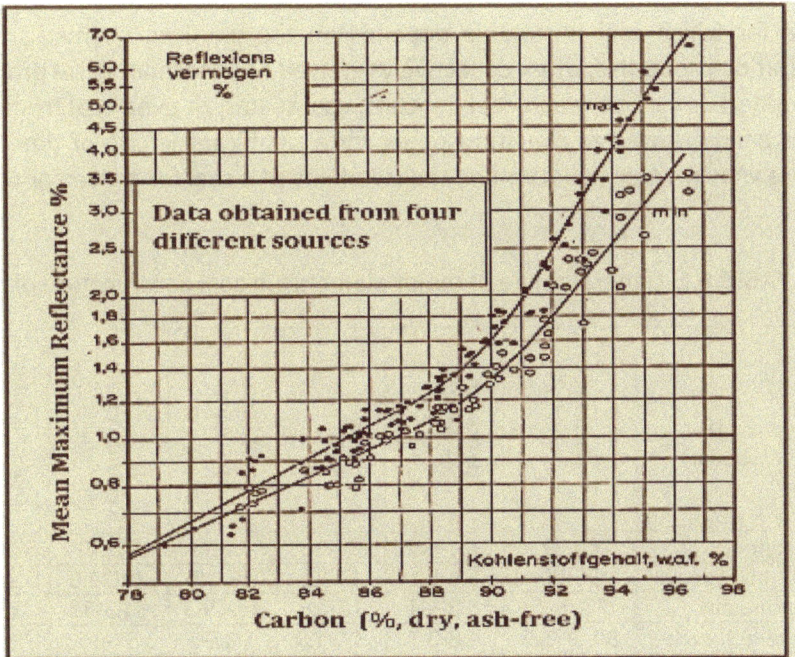

FIGURE 6.7 Mean Maximum Reflectance of coal as a function of carbon content. *(ICCP, 1963).*

6.5 CHEMICAL COMPOSITION OF COAL

The principal elements of which coals are composed are the same ones which make up wood and other vegetable matter - carbon, hydrogen and oxygen together with lesser amounts of sulphur mainly in form of pyrites, nitrogen, gypsum. Other compounds include calcite, clay and other elements characteristic of inorganic matter. Some sixty elements have been identified in coal. Carbon is the main constituent of coal making up between 60 and 96% depending on the rank. Table 6.1 was compiled from data on a wide range of coals world-wide but the ranges are not mutually exclusive. The nature of the plant and vegetable matter that formed coal as well as the geological conditions and age have a major influence on the ultimate composition. As the process of coalification raises the rank of a coal deposit, a corresponding increase occurs in the carbon content and calorific value while the volatile matter, moisture and oxygen contents decrease. These relationships are fairly valid through most stages of coalification or metamorphosis but may not be valid for the final anthracite stage (Gilchrist, 1974).

Volatile matter content can be a misleading indicator for low rank coals due to compositional differences in the original vegetation, and abnormal coals such as sapropels and those with high exinite contents have high volatile matter and hydrogen contents. Variations in chemical composition of coal are not limited to different deposits but can occur within the same deposit depending on the depth. Carbon content can increase by some 10% over a depth of one thousand metres, volatile matter can decrease by as much as 25% over the same depth while calorific value can increase by some 50% over a depth of 2000 metres (Gilchrist, 1974).

6.6 COAL ANALYSIS

The use for which coal is suitable depends on the physical, chemical, combustion and carbonization properties, often collectively referred to as technological properties. Many of these properties vary between coal deposits, coal seams, or even coal from the same seam that has been exposed to the atmosphere for a significant length of time. Coal analysis is therefore vital in order to determine the suitability of a coal for a given application.

TABLE 6.1 General range of major elements in coal as a function of rank.

Coal (% dmmf)	Carbon (% dmmf)	Hydrogen (% dmmf)	Oxygen (% dmmf)	Volatile matter (% dmmf)	Calorific Value (Mj/kg dmmf)	Inherent moisture (% wet basis)
Lignite	60-77	6.4-4.9	33-15	65-38	22-30	20-15
Sub-bituminous	75-80	5.6-5.1	17-12	47-38	29-32.5	16-12
Bituminous	78-91.5	5.8-4.4	15-2.5	47-19.5	31.5-37	15-0.4
Semi-anthracite	90.5-93.5	4.8-3.7	3.5-1.9	19.5-9	37-36	0.5-2.0
Anthracite	92-96	3.9-1.9	2.8-1.8	9-2	37-34.5	0.9-2.9

Many methods of determining the technological properties of coal have been standardized (British Standards, American Society for Testing and Materials, Japanese Industrial Standards, International Standards Organization, etc.) and there are many excellent books and standards on coal analysis (Bureau of Mines, 1967; Speight, 2015; ISO, 1974-2013). It is important however to emphasize that the reliability of coal analysis data depends on two key factors: sampling and specification of the basis of analysis. Coal is a heterogeneous material and run-of-mine coal is always associated with non-carbonaceous inherent and exogenous material. All coal standards specify methods of taking representative samples from bulk coal and the guidelines must be strictly followed to obtain any useful results.

It should be noted also that laboratory tests on coal and coke only provide useful information on the potential performance of coal in metallurgical and thermal applications and may not correlate with actual end use performance because actual industrial conditions are difficult to simulate in the laboratory. In effect, there is no substitute for pilot plant tests. The results of coal analysis can be expressed in many different ways depending on the treatment of the as-received coal. For example, carbon content can be expressed as "as received", "dry", "dry, ash-free" or "dry, mineral matter-free". Both proximate and ultimate analyses can be expressed in any of these and values can be significantly different depending on the basis of analysis (WCA, 2011). The basis for the expression of test results can make significant differences in value (Table 6.2). For example, fixed carbon expressed on dry, ash-free basis could be higher than the as-received value by over 30%. It is vital therefore to state the basis of analysis because it can make significant differences to values. This is also necessary in order to be able to classify a coal, compare coals, or relate data on the same coal obtained from various sources. The important laboratory tests on coal are outlined in the following sections.

6.6.1 Proximate analysis of coal

Proximate analysis gives the relative amounts of light organic compounds (volatile matter), and non-volatile organic compounds (fixed carbon) as well as the amount of the moisture content in the coal, and of the inorganic components left as residues when coal is burned (ash). If corrections are applied to compensate for the inorganic impurities, proximate analysis enables the organic components of one coal to be compared with those of another and hence provides much of the basic data for systematic classification of coal. It also gives useful information on the rank and potential heat value of a coal (Figure 6.8).

TABLE 6.2 Analysis of coal showing significant differences in values for the same coal depending on the basis of analysis. *(worldcoal.org).*

Coal content	As-received (ar)	Air-dried (ad)	Dry basis (db)	Dry, ash-free (daf)
Total moisture	11.0			-
Inherent moisture	2.0	2.0		-
Ash	12.0	13.2	13.5	-
Volatile matter	30.0	33.0	33.7	39.0
Fixed carbon	47.0	51.8	52.8	61.0
Sulphur	1.0	1.1	1.12	-

6.6.1.1 *Moisture content of coal*

Determination of moisture content of coal is based on the loss in mass between a sealed sample as received in the laboratory and that of the same sample fully dried at 110°C. The loss in mass due to elimination of moisture content is expressed as a percentage of the original sample.

6.6.1.2 *Volatile matter in coal*

Volatile matter content of coal is determined by heating a specific amount of coal in a standard crucible, out of contact with air at 900°C for a specified number of minutes. The volatile matter is calculated from the decrease in weight of the sample. A deduction is made for the decrease in weight due to moisture.

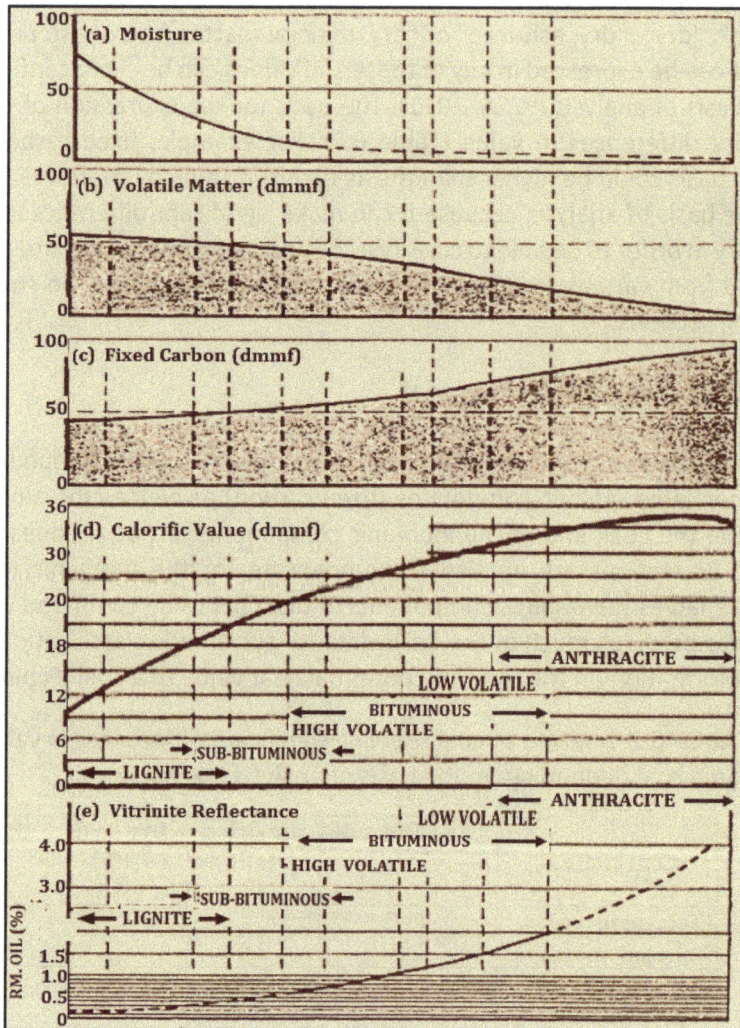

FIGURE 6.8 Variation in coal parameters with rank. *(AISI, 2015).*

6.6.1.3 *Ash in coal*

The ash in coal is the non-combustible inorganic residue that remains when coal is burned in air. It is determined by heating a specific amount of coal in air at a specified temperature in the range 815 to 850°C to constant weight. The percentage ash remaining is calculated from weight of ash residue after incineration.

6.6.1.4 *Fixed carbon in coal*

The fixed carbon content of coal represents the decomposition residues of the coal's organic components. It carries with it small amounts of nitrogen, sulphur, hydrogen and possibly oxygen as absorbed or chemically combined material (Rees 1966). The fixed carbon content is not determined directly, but derived by subtracting the sum of moisture, volatile matter and ash in air-dried coal from 100%. Appropriate corrections are made if the analysis is to be reported on any other basis apart from dry-ash-free basis.

6.6.2 Ultimate analysis of coal

The organic components of coal consist essentially of chemical compounds made from carbon, hydrogen, nitrogen, oxygen and sulphur and ultimate analysis involves the determination of the percentage of each of these elements in a coal sample. With the exception of nitrogen, these elements are also found in many of the mineral species which occur in coals. Carbon, hydrogen and oxygen are of great commercial significance in assessing the coking, gasification and liquefaction properties of coal, while nitrogen and sulphur represent possible sources of pollution in coal to be used in carbonization and combustion applications. The percentages of carbon, hydrogen and oxygen in organic fraction can also be used as indices of rank and parameters for classification. Another important application of ultimate analysis data is its use in combustion calculations to determine the theoretical air required for combustion, mainly for conversion of carbon, hydrogen and sulphur to carbon dioxide, water and sulphur dioxide respectively. The principles and procedures specified in detail in the various standards are again similar and will be outlined below.

6.6.2.1 *Carbon and Hydrogen in coal*

Carbon and hydrogen occur mainly as complex hydrocarbon compounds in coal. They are liberated as carbon dioxide and water when coal is burned and as a result, are most readily determined together. Techniques for determination of both elements are based on heating air-dried coal in a stream of dry oxygen and collecting the carbon dioxide and water liberated in a series of absorption tubes.

6.6.2.2 *Nitrogen in coal*

The nitrogen found in coals appears to be mainly contained in the organic compounds present (Francis, 1965). Techniques used in determining nitrogen content of coal involve converting the nitrogen to ammonium sulphate by catalytic digestion in sulphuric acid. The amount of the ammonium sulphate formed is then determined by titration.

6.6.2.3 *Sulphur in coal*

Sulphur is present in coal as organic sulphur, sulphide sulphur, sulphate sulphur and, in very rare cases, as elemental sulphur. The total sulphur content of coal is determined by converting it to sulphate for wet chemical analysis.

6.6.2.4 *Oxygen in coal*

Oxygen is a component of many organic compounds in coals. It is also found in the moisture of the many mineral species present in coal such as clays and carbonates. The oxygen content of coal is not determined directly but it is derived by subtracting the amount of other chemical components (C, H, N and S) from 100%. Oxygen is highest in low grade coals, as high as 40% in lignites, and reduces with the degree of coalification to around 1-2% in anthracite.

6.6.3 Coal ash analysis

Coal ash consists almost entirely of decomposed residues of silicates, carbonates, sulphides and other minerals. The composition of ash gives a useful guide to the type of minerals present in the coal, and the fusion characteristics are important in coal burning processes. Such data are also of commercial significance in caking coals, as the ash plays a part in the formation of blast furnace slag. Ash has a negative impact on energy consumption since energy is required to fuse and flush it out. The ash in coal may be dissolved by various means, treated with appropriate reagents and the resulting solutions analyzed using either an optical absorption spectrophotometer or flame photometer. Alternatively, techniques such as atomic absorption spectrometry, emission spectroscopy or x-ray fluorescence may be used for the analysis (Lowry, 1968).

6.6.4 Forms of Sulphur in coal

Although the total sulphur contents of coal often provides sufficient data for most commercial applications, a knowledge of the relative amounts of the three forms of sulphur is useful in determining the extent to which the element is removable by coal cleaning processes and other chemical or thermal treatments. It is possible to reduce pyritic and sulphate sulphur by coal beneficiation processes depending on their morphologies, but the removal of organic sulphur requires thermal or chemical treatment. Determination of the amount of sulphate sulphur is based on solution of sulphates in hydrochloric acid and their precipitation as barium sulphate for gravimetric determination. The total iron content of the sulphate fraction is also determined for use in the analysis of pyritic sulphur. The amount of pyritic sulphur is determined by dissolving both sulphates and sulphides in nitric acid and determining the total iron content in solution. The amount of iron in the sulphide fraction is then determined by subtraction of the amount of iron liberated in the analysis of sulphur content. There is no reliable method of determination of organic sulphur and it is presently determined by calculating the difference between the total sulphur and the sum of the pyritic and sulphate sulphur contents.

6.6.5   Coal mineral analysis

Various standard methods of chemical analysis are applied to determine the minerals in coal in form of oxides. The major proportion of the mineral matter in coal is usually a form of shale mixed with varying proportions of free silica, silicates and other compounds of aluminium, iron, calcium, magnesium, sodium, potassium, sulphur, phosphorus, titanium. Silica is usually the most prominent and makes up 50 – 90% of all the minerals in coal ash.

On combustion, some of the original inorganic compounds in coal are decomposed and the residual materials may re-combine or interact with other derivatives of the coal ash. For example, the sulphides, sulphates, and carbonates usually decompose, or oxidize, leaving their basic radicals free to combine with any excess of silica, and shales lose water of hydration. The ash left on combustion is therefore not a true measure of the nature or amount of mineral matter originally present (Francis 1965). Coal ash when heated does not melt sharply at any definite temperature but starts to soften at a substantially lower temperature than that at which it becomes molten. The major standards prescribe specific methods for the determination of mineral matter in coal and the ash fusion characteristics.

Although a knowledge of mineral matter in coal is important and valuable in calculating other important variables on a mineral matter-free basis, it is the composition of the ash that is of greater practical significance since it affects furnace operation. Ash fusibility is affected by the composition. For example, the higher the alumino-silicate content, the greater the refractoriness of the ash. On the other hand, lime, magnesia, ferric oxide, sulphur, oxides of sodium and potassium all lower the ash softening point. Ferrous oxide produced in the decomposition of ferrous carbonate and sulphate or the partial oxidation of iron sulphides also has a marked capacity for lowering the ash fusion point.

The fusion and flow temperatures of ash to a large extent determine its behaviour in combustion processes. Low fusion point ash tends to form clinker in boiler furnace grates. Clinker is a hard mass of refractory particles of ash, fine coal, and fused ash, that clogs a furnace grate, reducing the flow of air. In severe cases, it may cause a shut-down of the furnace or boiler. The formation of clinker is a complex process and the major causes or accelerators have been treated exhaustively by Francis (1965).

The mineral matter content and ash fusion characteristics also significantly influence the viscosity and chemical nature of slag in the blast furnace and other thermal metal extraction processes. Both are vital variables in thermal metallurgical extraction processes. Slag plays a primary role of flushing out impurities contained in metal ores and is removable from the furnace relatively easily when the viscosity is low relative to the metal. Also, acid or basic slag may be required depending on the chemical nature of the impurities to be flushed out of the molten metal, often requiring special furnace lining.

6.6.6   Calorific Value of coal

Calorific value is determined by completely burning a known weight of coal in excess oxygen in a bomb calorimeter placed in a tank of water at a pressure of 20-30 atmospheres. The heat produced during combustion is measured by recording the increase in temperature of the surrounding water. The products of the combustion are ash, sulphuric acid, carbon dioxide, nitric acid, nitrogen and liquid water. The heat generated is then calculated and expressed in Joules/kg (or Btu/lb).

The water produced is condensed and its latent heat recovered, hence the calorific value obtained is higher than the one obtained in actual burning of coal and is termed *gross calorific value.* In actual practice the water produced escapes as steam and its latent heat is not utilized. A correction is made for this to obtain a *net calorific value* which is of greater practical importance as it represents the actual useful heat that will be available for utilization. The two values are related as shown in Equation 6.2 (WCI, 2007).

$$Q_{net} = Q_{gross} - 50.6H - 5.85M - 0.191O \tag{6.2}$$

Where Q = calorific value, H = hydrogen%, Moisture % and O = oxygen% (All data on dry basis).

Calorific value is one of the most important parameters for evaluating the potential use of coal for combustion or carbonization. It is a measure of how much useful energy can be obtained from a given quantity of coal. It is also a useful variable for defining coal rank for much of the rank range (Figure 6.9). Calorific value is the leading the primary specification in coal trade.

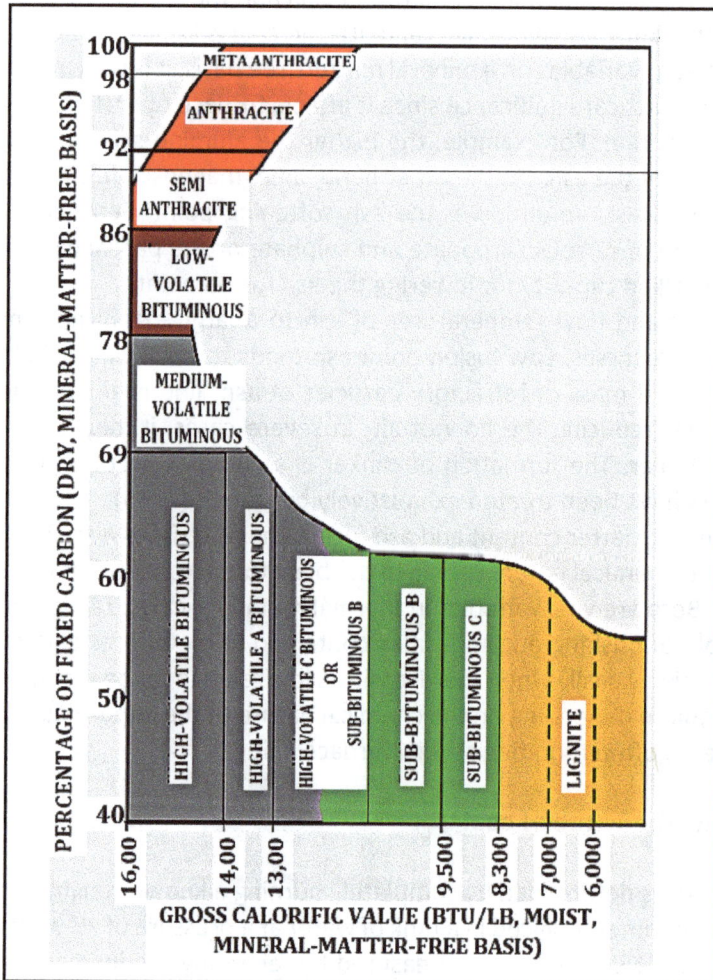

FIGURE 6.9 Gross calorific value with coal rank. *(Schweinfurth, 2009).*

6.6.7 Correlations between ultimate analysis and calorific value of coal.

There are many formulas dating back to Dulong formula (1880, see Mott et al., 1940) for calculating the gross calorific value of coal from the chemical composition. The most widely used have been reviewed recently by Mathews et al. (2014), and are summarized in Equations 6.3 - 6.6 (van Krevelen, 1993).

Dulong (1820) Q_{gross} = (80.8 x C) + (344.6 x H) - (43.1 x O) + (25 x S)   [6.3]
Seyler (1938) Q_{gross} = (123.9 x C) + (388.1 x H) + (25 x O) - 4269   [6.4]
Neavel (1986) Q_{gross} = (81.05 x C) + (316.4 x H) - (29.9 x O) + (23.9 x S) - 3.5 x ASH)   [6.5]
Given (1986) Q_{gross} = (78.3 x C) + (339.1 x H) - (33.0 x O) + (22.1 x S) + 152   [6.6]

Where Q_{gross} = Gross calorific value, C = carbon, H = hydrogen, O = oxygen, S = sulphur,
[All data on a dry basis (db) expressed as % weight].

All the above equations are empirical, derived from extensive coal data bases. Equations 6.4 - 6.6 are modifications of the original Dulong equation, based on comprehensive experimental data. The Neavel and Given equations are considered the most reliable because they were derived from large data bases comprising wide varieties of coals. Dulong's formula is one of the best known and most extensively tested methods for predicting the calorific value of a coal from its chemical composition. A popular version which is still used extensively is shown in Equation 6.7.

GCV = 14554 x C + 62028 x [H - O/8] + 4050 x S   [6.7]

Where:
C % = Fuel carbon content on a dry matter free mass basis
H % = Hydrogen content on a dry matter free mass basis
O % = Oxygen content on a dry matter free mass basis
S % = Sulphur carbon content on a dry matter free mass basis

Sulphur in coals can range from around 0.05% to 13%. Reactions between the element and oxygen are exothermic hence can contribute significantly to heating value of coal. On the other hand, reactions involving mineral matter are largely endothermic and could reduce calorific value significantly. It should be noted also that all the above equations apply only to single coals and are not valid for coal blends.

6.6.8 Correlation of proximate analysis with calorific value of coal.

Seyler (1948) also developed equations relating proximate analysis and calorific value as follows:
For coal below the rank of anthracite:

VM = 10.61H - 1.24C + 84.15   [6.8]

For anthracite:

Log VM = 0.23364 H - 0.02706 C + 2.579   [6.9]

For coals below the anthracite rank:

H   = 0.069 (Q_{gross}/100 + VM) - 2.86   [6.10]
C   = 0.59 (Q_{gross}/100 - 0.367VM) + 43.4   [6.11]

Where VM = volatile matter, H = hydrogen and C = carbon.

Equations 6.8 to 6.11 were combined to plot the graph shown in Figure 6.10. From the above equations, it is possible to calculate the carbon and hydrogen contents of coal from the values of volatile matter and calorific value which are relatively easy to determine empirically. The Seyler Chart has been found very useful in studies of a wide variety of coals.

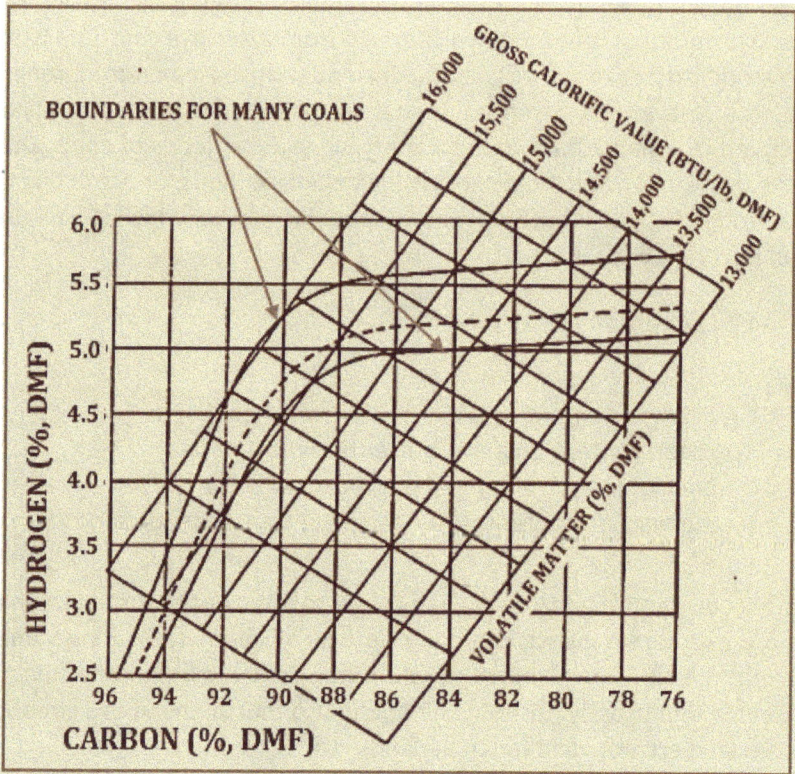

FIGURE 6.10 Seyler's Coal Chart showing boundaries for different ranks and types of coals from many parts of the world. (All variables are expressed on dry, mineral matter-free (dmmf) basis. (See BCURA, 2002).

6.7   THERMOCHEMICAL PROPERTIES OF COAL

Within a limited range of rank, coals soften on heating and resolidify in the temperature range of around 350°C to 550°C. Coals that are capable of thermoplastic behavior are known as caking coals, (or coking coals when the thermo-rheological (agglomerating) properties are within the range considered suitable for metallurgical coking). Concurrently, thermal decomposition and chemical reactions occur and gases evolve in the plastic phase. The complex physical and chemical changes which coal undergoes on heating are still not clearly understood. However, the results of many studies indicate that this unstable phase also called metaplast or mesophase, is formed after the moisture is driven from the coal. As heating proceeds bonds are ruptured and the molecular structure of the coal loosens. Molecules become mobile and are trapped in condensing tar to form a thin, viscous layer which envelopes the coal particles and thickens as heating progresses. On further heating a cracking process takes place in which the vaporized and non-aromatic groups are split off. This cracking process is accompanied by condensation and formation of semicoke. This material is porous due to the exit of gases from within the coal particle through the surface plastic layer. Virtually all the tar is expelled at this stage but the material still contains a considerable amount of volatile matter.

At higher temperatures the semicoke undergoes further devolatilization and contracts, causing fissures in the final coke. Relatively few coals in the bituminous rank (caking/coking coals) exhibit thermoplastic property which is critical in determining the suitability of a coal for many applications including metallurgical coke production for iron and steelmaking, coal gasification, combustion, hydrogenation (Figure 6.11). For example, prime metallurgical coking coals must have optimum thermo-fluidity and coals which cake or swell cause problems in combustion, carbonization, gasification and hydrogenation. The behavior of coal in the thermoplastic temperature range has been studied extensively for many years and many theories have been proposed. Perhaps the most prominent is the plastic caking mechanism which was developed by Van Krevelen (1961).

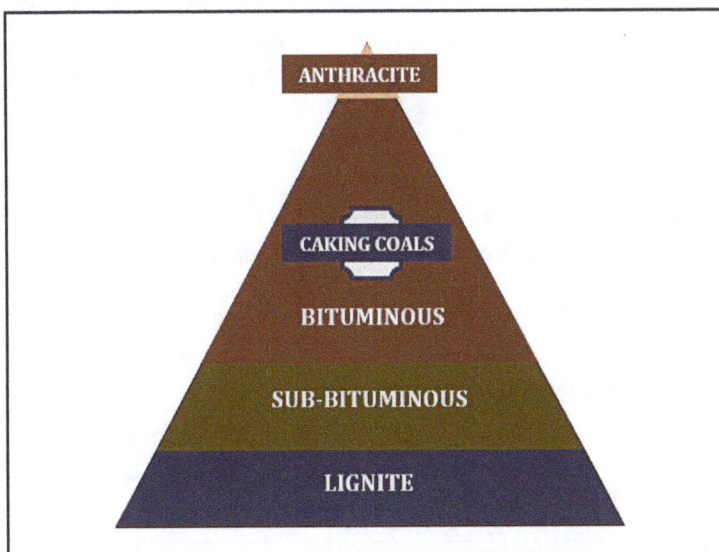

FIGURE 6.11 World coal resources showing relative abundance of different ranks and coking coals.

The softening process of coal under heating starts at around 620K with the decomposition of the organic matter content of the coal. Gases, (predominantly tar) start to evolve, part of which condense to bind the softened coal particles, forming a plastic mass. Between 720K and 820K the mass reconsolidates, forming a solid, porous mass called semicoke. Above 820K further decomposition of the organic matter continues, methane and hydrogen are evolved, and contraction of the semicoke occurs, resulting in cracks and fissures. A second theory proposed by Brooks and Taylor (1968) and several other researchers identified a mesophase between the formation of the plastic mass and the formation of solid semicoke.

The mesophase is polymeric and comprises lamelar liquid crystals which may coalesce to form a highly viscous mass if the fluidity of the coal in this temperature range is sufficiently high. Further heating consolidates the lamelar into a solid, porous semicoke with anisotropic structure. No correlation between coal rank and coal thermoplasticity has been established. In fact, two coals of the same rank and similar chemical properties can have very different thermoplastic properties. For example, consider two coals A and B (Table 6.3) which are similar, based on proximate analysis. Coal A has the highest Swelling Index (FSI) value of 9 compared with coal B with FSI value of 1.5. The two coals are of the same rank (a measure of degree of metamorphism) but are of different types, based on the petrographic composition. (See O'keefe et. al., 2013).

Vitrinite, the most prominent maceral in coal may soften (in some coals in the bituminous rank) while the maceral in younger or older coals is unreactive in the thermoplastic temperature range. Also, liptinite which is present in most coals in varying proportions may be reactive or unreactive. Inertinite and mineral matter are unreactive in the plastic range. In effect, thermo-fluidity properties depend not only on coal rank but also on type.

6.7.1 Measurement of Thermoplastic Properties of coal

There are four basic methods of measuring the thermoplastic properties of coal, although many variants have been developed in many parts of the world. All the methods involve the heating of a sample of coal under suitable conditions and measuring directly or indirectly, quantitatively or qualitatively, the properties of the plastic phase. These methods are discussed briefly below.

TABLE 6.3 Specifications for two coals of the same rank but different types, based on thermoplasitc properties. *(Falcon, 2013).*

TECHNICAL PROPERTY	COAL A	COAL B
	Rank	
	Bituminous	Bituminous
Inherent moisture (%)	2.5	2.6
Volatile matter (% air-dried)	24.0	24.5
Ash content (% air-dried)	10.0	9.5
Fixed carbon (%)	63.5	63.4
Free Swelling Index (FSI)	9	1.5

6.7.1.1 *Swelling Characteristics of coal (crucible swelling test)*

One gram of the analysis sample of crushed coal is placed and leveled in a translucent silica crucible with a prescribed size and shape. The crucible is heated rapidly in a special furnace to 800°C ± 5°C in 2½ minutes during which all the volatile matter would have been expelled. The profile of the coke button formed is then compared with a set of standard profiles (Figure 6.12). Plasticity varies from a minimum of free swelling index (FSI), (or crucible swelling number) of 0 to a maximum of 9. The FSI test is qualitative and is designed to measure the viscosity of the plastic layer formed during carbonization as well as the resistance it provides to gases which are being expelled. If the layer is highly viscous, the button will swell heavily due to the internal pressure build-up (FSI = 6-9). On the other hand, a non-plastic coal will not swell at all (FSI = 0-1).

A popular variant of the Free Swelling Index test developed in Europe is the Roga Index test. Rather than observe a change in coherence, the mechanical strength of the residue is determined. An intimate mixture of 1g of coal and 5g of standard anthracite is pressed into a button and heated to 850°C under a specified load. The cohesion of the coke produced is determined by subjecting it to a drum test in a strictly prescribed manner and the Roga Index is calculated from the results (ISO 335:1974).

6.7.1.2 *Gray/King/Fischer Carbonization Assay.*

A method known as the Gray-King Test for determining coal plasticity and evaluating the by–products was developed in the United Kingdom. The test involves heating a crushed sample of coal in a sealed cylinder laid horizontally in a special furnace from 300 to 600°C under standard conditions (ISO 502:1982). The mass of coke formed is compared with standard residue profiles to determine the coke type (see Figure 6.13).

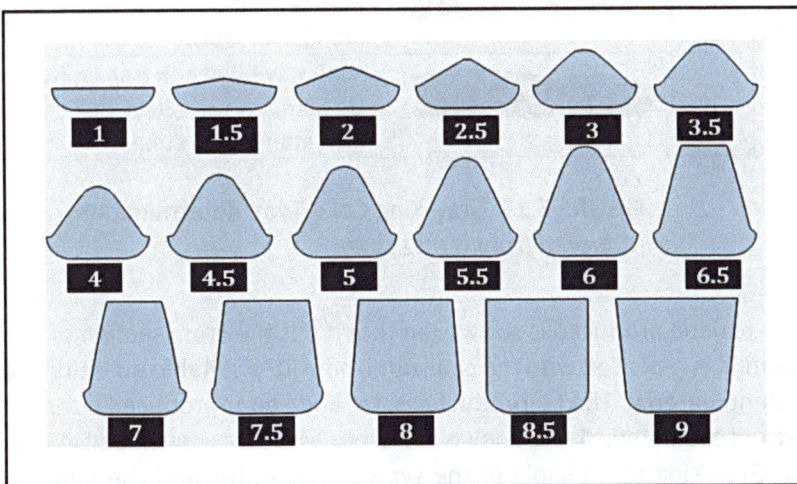

FIGURE 6.12 Free Swelling Index samples. *(ISO 501, 2012).*

Type	Carbonized coal	Description
A		Initial volume is retained. Non-coherent.
B		Initial volume is retained. Barely-coherent.
C		Initial volume is retained. Coherent but friable.
D		Slightly shrunken. Only moderately hard. Surface easily scratched with finger nail. Seen to be barely fused on examining cross section.
E		Very shrunken and hard, with several fissures. Seen to be fused on examining cross section.
F		Shrunken, hard and strong. Seen to be fused on cross examination.
G		Initial volume is retained. Fused and hard.
G1		Slightly swollen. Needs to be blended with electrode carbon to give standard G coke.
G2		Swollen. Needs to be blended with electrode carbon to give standard G coke.
G8		Swollen and fragile. Needs to be blended with electrode carbon to give standard G coke.
G9		Highly swollen and fragile. Needs to be blended with electrode carbon to give standard G coke.

FIGURE 6.13 Gray-King Coke Type determination.
(Adapted from IS 1353,1993).

If the coke residue produced is so swollen that it fills the cross section of the retort tube, the determination is repeated with the coal admixed with a suitable quantity of electrode carbon or equivalent material. The by-product gas, tar and liquor produced during carbonization are collected and evaluated. A test known as Fischer assay was developed in Germany and differs from the Gray-King test mainly in the type of apparatus and the heating rate employed. Although the conditions for the Gray-King coke type and Roga Index tests differ significantly from those which obtain in the crucible swelling test, there is fair correspondence between the values obtained from the three tests as shown in Table 6.4.

6.7.1.3 *Coal dilatometry*

In the Dilatometer test, a sample of crushed coal is pressed into a pencil in a special apparatus and heated at the rate of 3°C/minute from 300 to 600°C in a dilatometer. A typical curve is shown in Figure 6.14. The initial contraction and subsequent expansion of the pencil are measured as a function of temperature and the data presented as a function of heating rate and coal rank (Figure 6.15). There are many variants of the coal dilatometer, the most popular being the Ruhr and Audibert–Arnu dilatometers for low-temperature characterization and the Chevenard-Joumier dilatometer for high-temperature testing.

TABLE 6.4 Approximate comparison between Crucible Swelling Number and other carbonization tests.

Crucible Swelling Number	Gray-King Coke Type
0 – ½	A – B
1 – 4	C – G2
4½ - 6	F – G4
6½ - 8	G3 – G9
8½ - 9	G7 and above
Crucible Swelling Number	**Roga Index**
0 – ½	0 – 5
1 – 2	5 – 20
2½ - 4	20 – 45
> 4	> 45

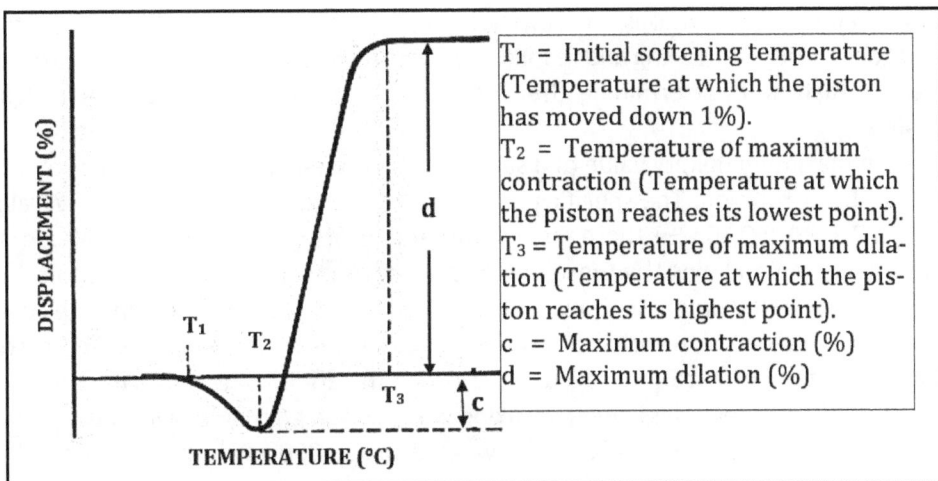

T_1 = Initial softening temperature (Temperature at which the piston has moved down 1%).
T_2 = Temperature of maximum contraction (Temperature at which the piston reaches its lowest point).
T_3 = Temperature of maximum dilation (Temperature at which the piston reaches its highest point).
c = Maximum contraction (%)
d = Maximum dilation (%)

FIGURE 6.14 Characteristic Ruhr coal dilation curve.

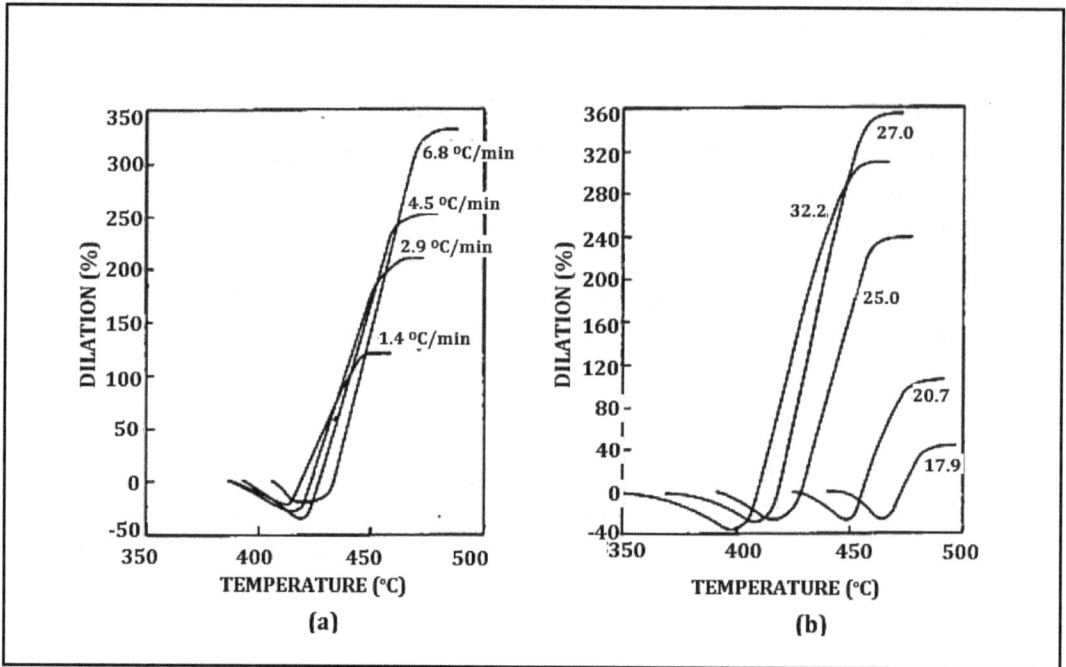

FIGURE 6.15 Effect of heating rate and coal rank on coal dilation (a) variable heating rate (b) variable volatile matter content of coal. *(van Krevelen, 1963).*

6.7.1.4   *Coal plastometry*

The coal plastometer is a type of viscometer that monitors the fluidity of crushed coal over a range of temperatures, usually 300 to 600°C at the rate of 3°C/minute. There are two basic types of equipment: constant torque and variable torque plastometers. In either type, the motion of a stirrer around which the crushed coal is compressed, is measured on a dial drum graduated into 100 divisions, (dial divisions per minute) over a range of temperatures as the coal goes through softening and resolidification. The plastic range in °C and maximum fluidity are key factors in determining the optimum blends of coking coals for a desired coke quality.

The maximum fluidity value of a sample is expressed in dial divisions per minute (ddpm) of the stirrer rotation. The specific conditions for the test are detailed in several standards (ASTM D 2639; ISO 10329).There are several variants of plastometers in use all over the world, the most common being the Gieseler (constant-torque) plastometer. An automated version of the Gieseler equipment is shown in Figure 6.16. The crucible containing the coal is heated in a molten solder bath furnace and the stirrer movement is displayed and recorded as a function of temperature. Typical curves showing the effect of heating rate are also shown in Figure 6.17. In general, high volatile caking coals have high rheological values between 5000 and 30,000 ddpm, medium volatiles between 0 and 20,000 and low volatiles between 20 and 1,000 ddpm. A variant of the plastometer is the penetrometer. Instead of the rotating paddle in the plastometer, a needle under load penetrates a coal sample heated over a range of temperatures. The rate of penetration of the needle as a function of temperature is a measure of the plasticity of the coal.

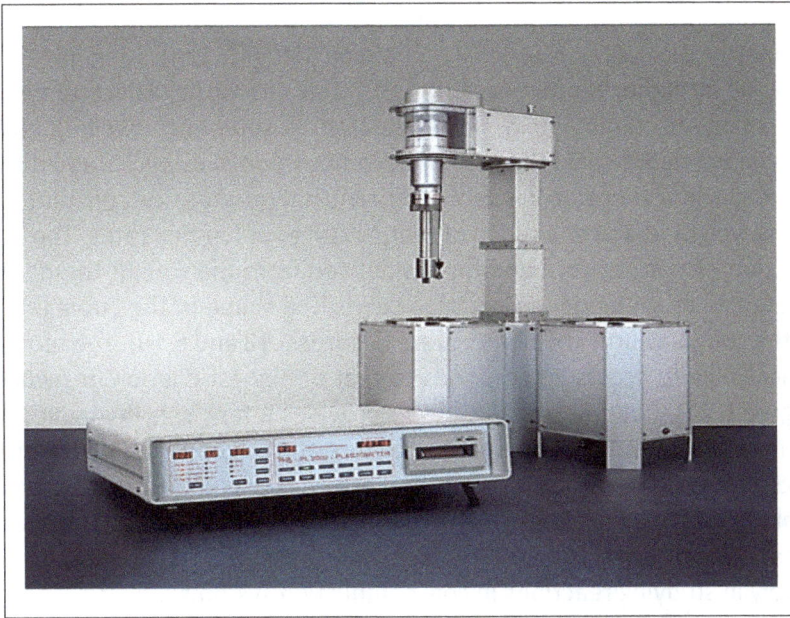

FIGURE 6.16 An automated Gieseler Plastometer.
(R. B. Authomazione http://www.rb-autom.com).

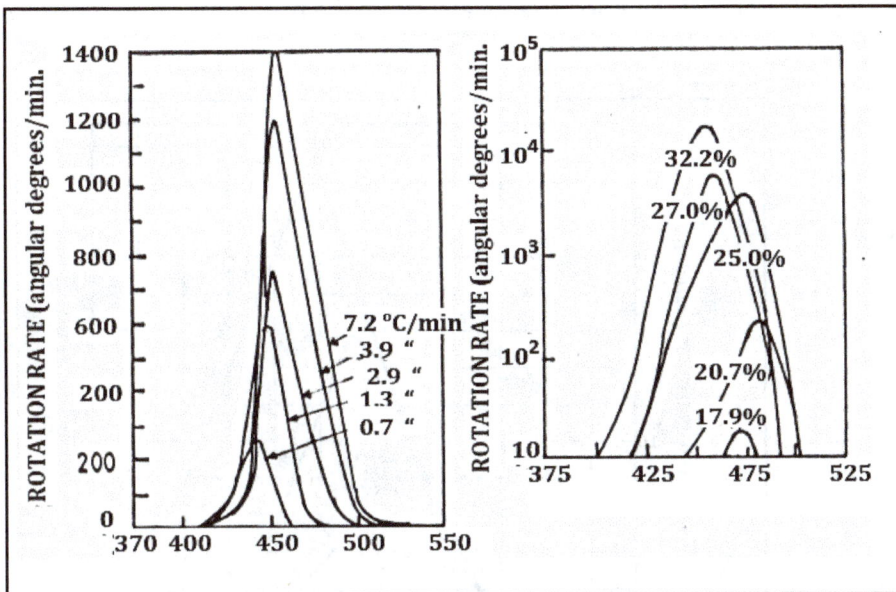

FIGURE 6.17 Effect of (a) heating rate (b) coal rank (volatile matter) on fluidity curves.
(van Krevelen, 1963).

6.7.1.5   Coal thermogravimetry

Volatile matter is expelled from coal on heating, initially comprising mainly of moisture and tar, followed at higher temperatures by gas. The rate of degasification is indicative of the decomposition reactions and is closely related to softening, swelling and resolidification phenomena in heated coal. The rate of decomposition is usually studied by heating coal of known weight in a thermobalance at constant temperature or constant heating rate, and measuring weight loss over a range of temperatures or heating rates. The rate of weight loss as a function of temperature and time is derived from the weight-temperature curve commonly referred to as the devolatilization curve. The shape of the curve is dependent on the rank of the coal as well as the heating rate (Figures 6.18 and 6.19). The temperature at which the highest fluidity occurs corresponds to that of highest dilation or highest degasification rate (Figure 6.20). Coal thermogravimetry provides a fast, sensitive and precise method of determining coal rank, volatile matter, moisture and ash contents with a precision comparable to BS and ASTM methods. The method provides useful information on the physical and chemical reactions taking place in coal on heating or liquefaction. The onset of the various reactions taking place can be determined with precision. Coal thermogravimetry is also used extensively in studying reactions in coal combustion systems.

Over the last thirty years or so, attempts have been made to use thermogravimetry to determine the proximate analysis of coal, with little success. However, a Simultaneous Thermal Analyzer (STA) developed recently by PerkinElmer is capable of using the method to determine the proximate analysis of coal and coke with good results compared with ASTM tests. The method is relatively fast and gives reproducible results (Cassel et al., 2012). The calorific value of coal can also be calculated from the proximate analysis results.

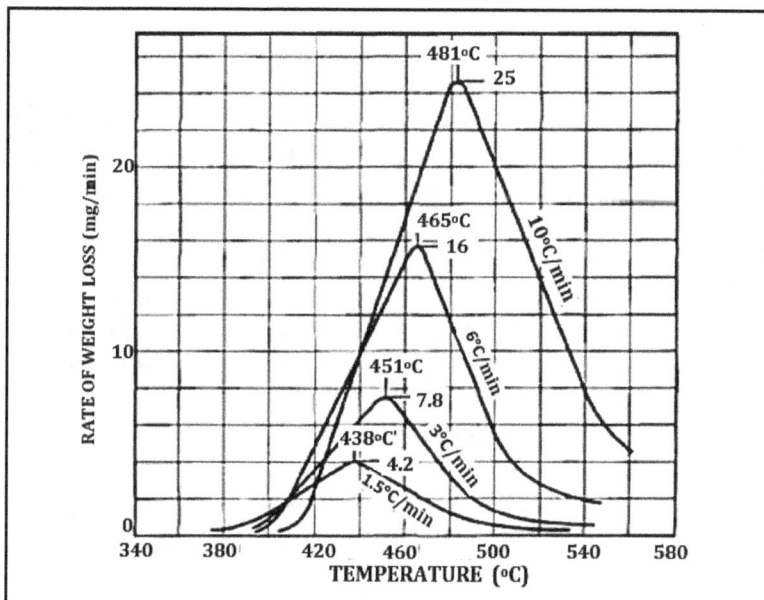

FIGURE 6.18 Effect of heating rate on coal fluidity curves. *(van Krevelen, 1963).*

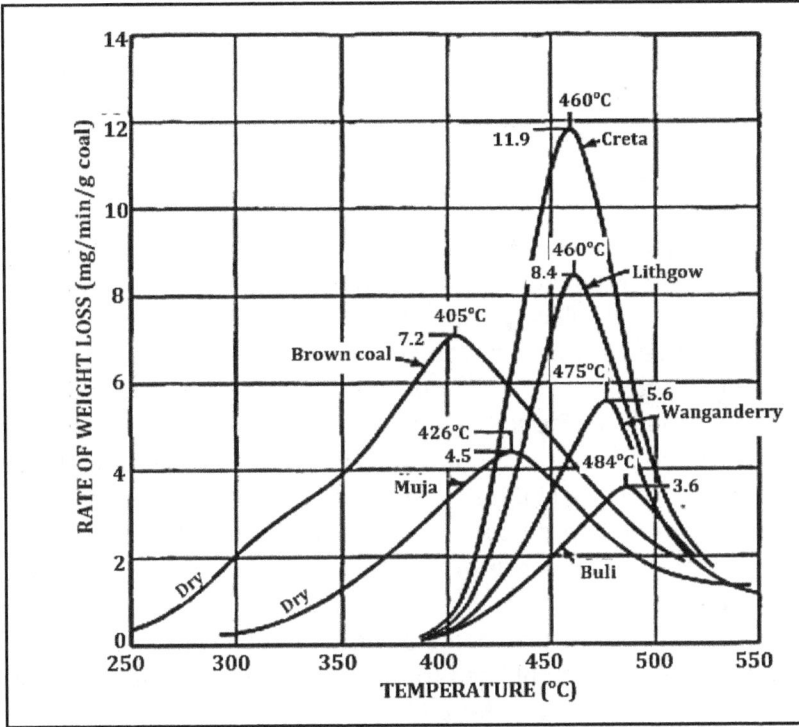

FIGURE 6.19 Effect of coal rank on coal devolatilization curves.
(van Krevelen, 1963).

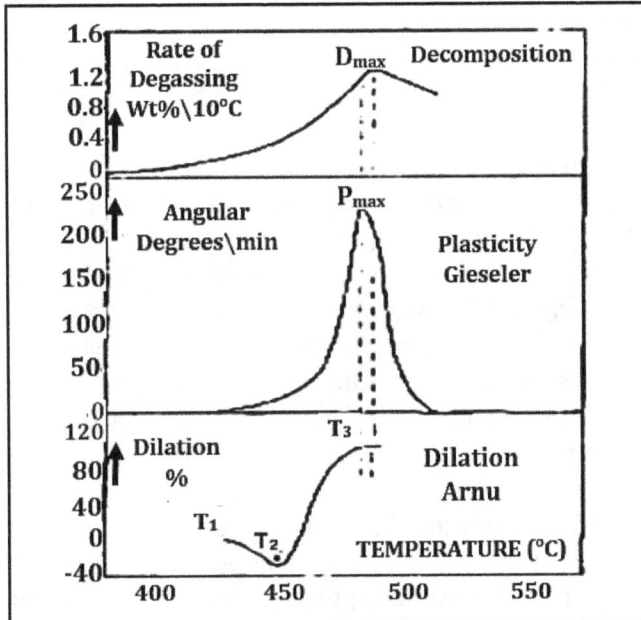

FIGURE 6.20 Correspondence between coal dilation, fluidity
and devolatilization curves. *(van Krevelen, 1963).*

6.7.1.6   *Coal petrography*

Coal is formed from accumulated vegetable matter moved from neighbouring high land by flowing water and deposited in elongated depressions in the Earth's surface. The presence of water and bacteria will cause the matter to decay and the pressure may cause subsidence of the base of the depression and the decaying matter starts to sink. Matter deposited in the depressions is not only organic in nature but also includes inorganic matter, animal remains, fossils, etc. The deposition process may occur over a very long period and both the physical and chemical conditions of the environment and the botanical nature of the plant debris being deposited will vary over time.

The plant matter that survives complete decay may eventually form peat swamp if the environmental conditions are favourable and increasing pressure on the peat due to in-creasing overburden may facilitate progressive coalification. The different types of chemical substances present in the deposited matter have differing degrees of resistance to decay. These include cellulose, lignin, resins, waxes, tannins all of which are present in vegetable matter. A typical coal deposit comprises a series of vertical layers of coal separated by a regular and repeating sequence of inorganic sediments. Many coal seams have visible banded structure which reflects the nature of the original deposits that formed the coal. Coal petrography involves macroscopic and microscopic examination, and evaluation of coal. Valuable data can be obtained from which the rank, physical, chemical and thermo-plastic behavior of coal can be inferred.

Early petrographic classification systems identified four groups of concentrates (macer-als) in the visible band of coal: vitrinite formed from woody tissue; liptinite/exinite formed from resinous and waxy materials of plants (spores, seed coatings, cuticles, etc); fusinite, believed to have been formed from woody tissue that had been subjected to excessive temperature and pressure and transformed into natural charcoal; and micrinite which represents matter of unknown source possibly inorganic matter (van Krevelen, 1961, 1993). Subsequent and extensive microscopic investigations have resulted in numerous new groups and sub-groups (see Chapter 5).

The physical and chemical nature of macerals have a profound influence on the bulk physical, chemical and thermal behavior of coal and applications for which a particular coal is potentially suitable. Vitrinite is the most predominant maceral in coal, accounting for 70 to 80% in most coals. It is also the most important in determining the age and rank of coal. Liptinite (exinite) largely determines the ability of coal to form gas, tar and coherent coke on carbonization while fusinite, micrinite (and other sub-divisions now collectively known as inertinite indicate the amount of infusible matter in coal which will eventually form ash on combustion. Fusinite contains the highest amount of carbon of all mecerals but in general the proportion of the maceral relative to vitrinite is low and constitutes only a few percent.

The reflectance of vitrinite in reflected light has been shown to be a good indicator of the coal rank. To determine this important parameter, a representative sample of coal for petrographic examination is crushed, ground to minus 0.850mm and embedded in an epoxy mounting medium in a cylindrical mould. A surface is ground, polished and examined at 500x magnification in oil immersion, in reflected light on a microscope (see ASTM Standard D-2798-91). The various macerals and sub-macerals can be identified by colour, physical structure and reflectance in oil, and the amount present quantified by point counting technique. Reflectance (the proportion of incident light that is reflected by a plane) is a good indicator of the degree of metamorphism of coal and values for vitrinite, being the most abundant, most homogeneous, are the most important. Reflectance measurements are made at many points (typically 100) on the sample surface and the results are plotted in a histogram to determine the mean maximum reflectance (%R_o max) as shown in Figure 6.21. Typical vitrinite values for coals of different ranks are shown in Table 6.5.

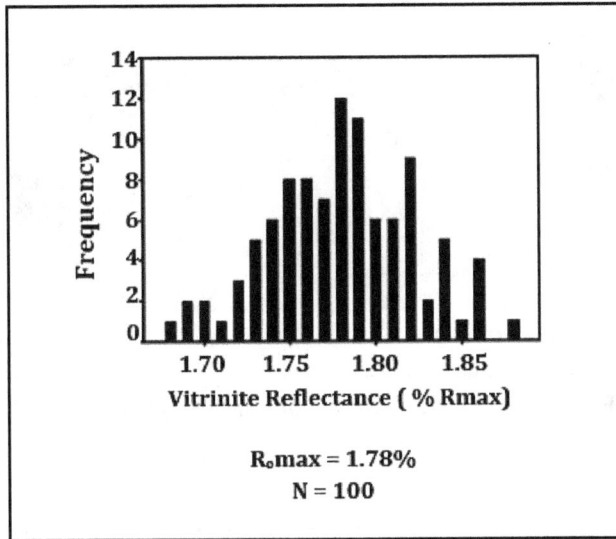

FIGURE 6.21 Vitrinite reflectance histogram. *(Cardott, 2012).*

TABLE 6.5 Variation of vitrinite reflectance values for coals of different ranks. *(Davis (1978, Stach et al., 1982).*

Coal rank	Reflectance (% maximum)
Liginte	<0.27 – 0.38
Sub-bituminous C	0.38 – 0.43
Sub-bituminous B	0.43 – 0.48
Sub-bituminous A	0.48 – 0.67
High-volatile bituminous C	0.47 – 0.57
High-volatile bituminous B	0.57 – 0.71
High-volatile bituminous A	0.71 – 1.10
Medium volatile bituminous	1.10 – 1.50
Low volatile bituminous	1.50 – 2.05
Semi-anthracite	2.05 – 3.0 (Approx)
Anthracite	>3.0

The relative contents of carbon, hydrogen and oxygen in coal macerals vary. Relative to other macerals in a coal, vitrinite contains more oxygen, exinite more hydrogen, and inertinite more carbon. Fusinite (inertinite) has the highest reflectance and exinite the lowest while vitrinite reflectance is intermediate between them (Figure 6.22). A correlation between mean maximum reflectance of vitrinite and other indicators of coal rank has also been established by many authors (Thompson and Benedict, 1974; Bustin et al., 1985). Some of the important results are presented in Figures 6.23 and 6.24. Coal petrography has become a very useful method of assessing the utilization potential of a coal deposit (Suarez-Ruiz and Crelling, 2008).

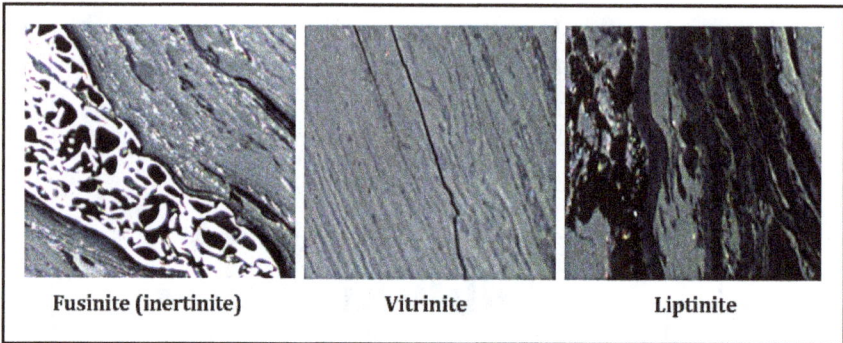

FIGURE 6.22   Petrographs of fusinite, vitrinite and liptinite.

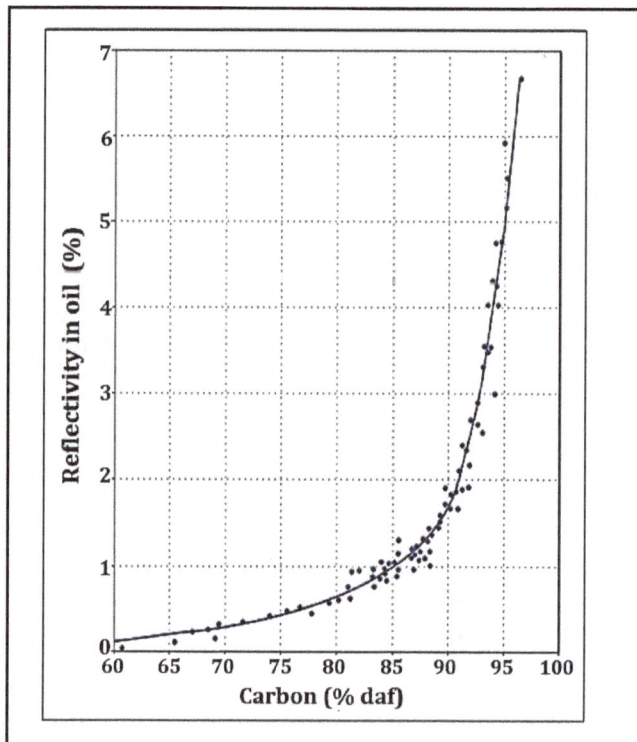

FIGURE 6.23   Vitrinite reflectance by rank. *(Berkowitz, 1979).*

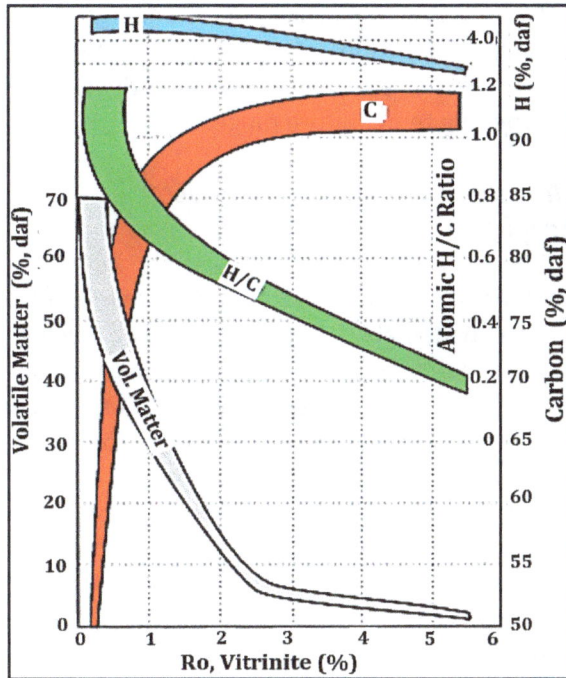

FIGURE 6.24 Measures of coal rank. *(Stach et al., 1982).*

6.7.2 Factors which influence thermoplastic behaviour of coal

Less than 5% of world reserves of coal have any appreciable thermoplastic (caking) properties and they all belong to the bituminous class. Fluidity varies from very low to very high and the range of temperatures from softening to reconsolidation also varies between caking coals. Several factors determine the thermoplastic behaviour of coal, the most important of which are discussed briefly below.

6.7.2.1 *Nature of the original plant debris*

The nature of the various components in the original organic matter from which coal is formed has a profound effect on different petrographic macerals in coal, and their thermoplastic properties. The organic, and cellulose materials have different resistance to bacterial attack and greatly influence the nature of the peat formed which in turn determines the ultimate properties of coal. Vitrinite in some types of bituminous coals (caking/coking coals) is thermally reactive but inert in most other coals, especially low-rank coals. Exinite is almost always thermally reactive and produces gas and tar on heating but then its presence in coal is usually low (0 – 15%). Most of the inertinite in all coals is thermally unreactive (a proportion of semi-fusinite may be reactive). Vitrinite and resinite in high rank coals are the most important macerals that influence coal fluidity. Vitrinites in low rank and the highest rank coals show no plastic behaviour at all if the carbon content is lower than 80.5% or higher than 91%.

Caking coals which have volatile matter content in the range 20-32% become plastic before active decomposition and degasification occurs. The residual coke is usually strong and abrasive-resistant. Outside this range caking coals undergo decomposition both before and during the plastic temperature zone. Usually, the coke formed is weak and fragile due to forceful and violent expulsion of a relatively large quantity of volatile matter through the thin plastic layer in the softening temperature range.

6.7.2.2   Environmental conditions and degree of coalification (rank)

A swampy environment is a prerequisite for the formation of peat, the first stage in the process of coalification. Bacterial activity helps to break down the plant material and has a significant influence on coal type, but the degree of degradation depends on oxygen availability and acidity of the environment. It is widely believed that microbiological decomposition of the plant material is only significant at the peat formation (humification) stage. Peat fires , hot volcanic intrusions, etc can carbonize part of the peat deposit, forming natural charcoal and other inert material which constitute thermally unreactive inertinite. Biochemical activity is greatly retarded as the depth of burial of the peat increases and biochemical factors have little or no influence on the conversion of peat to lignite. This process and further coalification are believed to be influenced only by geophysical factors of which temperature and pressure are the most significant.

6.7.2.3   Rate of carbonization

The plastic behaviour of coal on heating to a very great extent depends on the heating rate. High heating rates increase fluidity and swelling, and extend the temperature range over which the coal is plastic. Some coals which show no softening when heated slowly may soften and swell when heated more rapidly. On the other hand, all coals, irrespective of their rank and plastic properties can be fully carbonized without showing any fluidity or swelling provided the heating rate is sufficiently slow. Heating rate influences the kinetics of degasification and other chemical reactions which occur in the plastic zone.

6.7.2.4   Degree of oxidation or hydrogenation

The plastic properties of coals and indeed many other coal properties are affected markedly by even mild oxidation. In general, oxidation decreases fluidity and swelling, and contracts the temperature range over which coal is plastic. The effect is influenced significantly by the rank of the coal, with reactivity to oxygen falling with increasing rank. Guyot and Pollard (1974) carried out oxidation tests at 80°C under controlled conditions and found that the rate of loss fluidity of a high volatile (low rank) coal was twice as fast as that of a low volatile (high rank) coal. Mild hydrogenation has the opposite effect on plastic properties: fluidity, swelling and plastic range are increased.

REFERENCES

AISI (2015) *Coal Utilization in the Steel Industry.* American Iron and Steel Institute. www.steel.org/learning/howmade/coal.htm.

ASTM D 2639 Standard Test Method for Plastic Properties of Coal by the Constant-Torque Gieseler Plastometer. www.astm.org.

ASTM D 409. (1991). *Standard Test Method for Grindability of Coal by the Hardgrove-Machine Method.* American Society for Testing and Materials. www.astm.org.

ASTM D 2798-91 (1991). *Standard Test Method for Microscopical Determination of the Vitrinite Reflectance of Coal.* www.astm.org.

Austin, I. G., Trubeljal, M. P., and H. M. Seeback (1995). "Capacity of High Pressure Grinding Rolls." *Mine and Metallurgical Processing* May, p. 65.

BCURA (2002). "British coals plotted on Seyler's Coal Chart." The BCURA Coal Sample Bank: A Users handbook. British Coal Research Association. www.esg.co.uk

Berkowitz, N. (1979). *Introduction to Coal Technology.* Academic Press, New York (1979), 345p.

Brooks, J, D. and G. H. Taylor (1968). "The formation of some graphitizing carbons." In P. L. Walker (Ed.), *Chemistry and Physics of Carbon*, Vol. 4, Marcel Dekker.

BS 1016-112 (1995). Methods for analysis and testing of coal and coke - Determination of Hardgrove Grindability Index of hard coal.

Bureau of Mines, *(1967). Methods of Analyzing and Testing Coal and Coke.* Bureau of Mines Bulletin 638.

Bustin, R. M., A. R Cameron, D. A. Grieve, and W. D. Kalkreuth (1985). *Coal Petrology, its principles, methods and applications. 2nd edition:* Geological Association of Canada Short Course Notes 3, 230p.

Cardott, B. J. (2012). *Introduction to Vitrinite Reflectance as a Thermal Maturity Indicator.* www.searchanddiscovery.co; www.scribd.com

Cassel, B., Menard, K., And Earnest, c. (2012). "Proximate Analysis of Coal and Coke using the STA 8000 Simultaneous Thermal Analyzer." www.perkinelmer.com. (Accessed 12/2016).

Davis, A. (1978). *The reflectance of coal. In Analytical methods for coal and coal products,* Vol. 1, ed. C. Karr, Jr.; New York: Academic Press. pp. 27-81.

Ellmann, R. C. and Belter, J. W., (1955). "Grindability Testing of Lignites." *U. S. Bureau of Mines,* R1 5167, 38p.

Falcon, R. (2013). "Coal Geology, Types, Ranks and Grades." www.fossilfuel.co.za

Fitton, A., Hughes, T. H., Hurley, T. F. (1957). "Grindability of British Coals – a Laboratory Examination." *J. Inst. Fuel,* Vol. 30, No. 193, pp. 54-65.

Francis, W. (1965). *Fuels and Fuel Technology, Volume One.* Pergamon Press. London.

Gilchrist, R. (1974). "Principles of coal systematics." *Symposium on Coal Utilization.* Alberta Research Council, October, 1974.

Guyot, F and F. Pollard (1974). "Effects of oxidation on Coal Properties Related to Industrial Use." Volume 74, Issue 18 of P.R. (Australian Coal Industry Research Laboratories).

Hardgrove, R. M. (1932). "Grindability of coal." *Trans. Amer. Soc. Mech. Eng.,* Vol. 54, pp. 37-46.

Hills, L. M. (2007). *Solid Fuel Grindability: Literature Review.* PCA R & D Serial No. 2986. www.cement.org

Mott, R. A. and Spooner, C. E. (1940). "The Calorific Value f Carbon in coal: The Dulong Relationship." *Fuel* 19, 226-231, pp. 242-251.

Hober, J. C. and Wild, G. D. (1988). "Relationship between Hardgrove Grindability Index and Petrographic Composition of High-Volatile Bituminous coal from Kentucky. " *Journal of Coal Quality,* Vol. 7, pp. 122-126.

Hower, J. C. (2008). "Maceral/Microlithotype Partitioning with Particle Size of Pulverized Coal: Examples from Power Plants Burning Central Appalachian and Illinois Basin Coals." *International Journal of Coal Geology,* V. 73, Pp. 213-218.

ICCP (1963). *International Committee for coal petrology.* International Handbook of Coal Petrography. Center National de la Recherche Scientifique, Paris, France. 2nd ed., 1963

IS 13531(993*). Methods of test for coal carbonization – caking index.* https://archive.org/details/gov.in.is.1353.1993

ISO 335 (1974*). Hard coal – Determination of Hardgrove Grindability Index.* International Standards organization.

ISO 502 (1982). *Determination of caking power – Gray-King Coke Test.* International Standards Organization. www.iso.org

ISO 5074 (1994*). Hard coal – Determination of caking power – Roga test.* International Standards Organization.

ISO 10329 (2009*). Hard coal – Determination of plastic properties – Constant-torque Gieseler plastometer method.* International Standards Organization.

ISO 501 (2012). *Determination of the crucible swelling index of coal.* International Standards Organization. www.iso.org

Lowry, H. H. (Ed.) (1968). *Chemistry of Coal Utilization, Supplementary Volume.* John Wiley.

Mathews, J. P., Krishnamoorthy, V., Louw, E., Tchapda, A. H. N., Castro-Marcano, F., Karri, V.,Alexis, D. A., and G. D. Mitchell (2014). "A review of correlations of coal properties with elemental composition." *Fuel Processing Technology,* 121(2014) 104-113. Elsevier.

Milenkova, K. S., Boreggo, A. G., Alvarez, D., Xiberta, J., and R. Menendez (2003). "Tracing the origin of unburned carbon in fly ashes from coal blends". *Energy and Fuels,* V. 17, no. 5. Oct. 2003.

O'Keefe, J. et al. (2013). "On the fundamental difference between coal rank and type." *International Journal of Coal Geology.* 118 (2013) 58, 87.

Rees, O. (1966) *Chemistry, Uses and Limitations of Coal Analyses.* Illinois State Geological Survey Report of Investigations, 220, 55p.

Schweinfurth, S. P. (2009). "An introduction to Coal Quality. The National Coal Resource Assessment Overview." Eds Pierce, B. S. and K. O. Dennen. U.S. Geological Survey Professional Paper 1625-F.

Sehgal, R. S. and B. Wong (1976). "Significance of calorific value, ultimate analysis, ash fusion and mineral analysis of ash." *Symposium on coal evaluation,* Fryer et al (Eds.), Alberta Research Council, Canada.

Seyler, C. A. (1948). "The past and future of coal – the contribution of petrology." *Proc. S.W. Inst. Eng.,* 63, pp. 213-243.

SGS (2016). *Coal Calculations.* www.sgs.com.

Speight, J. G. (2015). *Handbook of Coal Analysis,* John Wiley. (Second Edition, 2015).

Stach, E. et al.(1982). *Textbook of Coal Petrology. Third edition.* Bomtraeger, Atutgart and Berlin, Germany, 42.

Storm (2015). *Pulverizer Capacity.* www.stormeng.com.

Suarez-Ruiz, I and J. Crelling (2008). *Applied Coal Petrology. The Role of Petrology in Coal Utilization.* Elsevier.

Thompson, R. R. and Benedict, I. G. (1974)." Vitrinite reflectance as an indicator of coal metamorphism for cokemaking." Geol. Soc. Am. Spec. Paper No. 153, pp 95-108.

Van Krevelen, D. W. (1961, 1963). *Coal.* New York, Elsevier.

van Krevelen, D. (1993). *Coal, Typology – Physics – Chemistry – Constitution.* 3rd Edition. Elsevier Science.

Wang, X. H., Guo, Q., Yingling, J, C., and B. K. Parekh (1996). "Improving pyrite liberation and grinding efficiency in fine coal comminution by swelling pretreatment." *Coal Preparation,* V. 17, pp 185-198.

WCA (2011). *The Coal Resource.* World Coal Association. www.worldcoal.org.

WCA (2011). *Coal Facts,* World Coal Association. www.worldcoal.org.

WCI (2007). *Coal Conversion Facts.* World Coal Institute. www.worldcoal.org

Wells, J. J., Wigley, F., Foster, D. J., Gibb, W. H., and J, Williamson (2004). "The relationship between excluded mineral matter and abrasion index of a coal." Fuel, 83, no. 3, pp. 359-364.

Wigley, F., and J. Williamson (2005). *Coal Mineral Transformations – Effects on boiler Ash Behaviour.* Report No. COAL. R278 DTI/Pub URN 05/659.

Opencast coal mining
(teara.govt.nz)

Underground longwall coal mining
(wagnerequipment.com)

7   Coal oxidation and weathering

7.1   INTRODUCTION

All coals exposed to air have the propensity for oxidation and weathering although the rate depends on coal rank and environmental conditions. Coal properties could deteriorate slowly or rapidly due to oxidation, even at ambient temperatures. Important properties such as volatile matter and heating value can depreciate by as much as 20% after exposure to air for a few days. Weathering alters the coal surface chemistry, the size consist, and volatile matter content. Other negative effects on properties include lowering of tar yield, changes in combustion behaviour, lowering of caking properties, and tendency for spontaneous combustion (Speight, 2015).

The effect of weathering on coal depends significantly on reactivity to oxygen which is a function of coal rank since affinity for oxygen falls with increasing rank. Both the organic and inorganic constituents of coal are affected by weathering. Although oxidation reactions involving both occur simultaneously, they are generally regarded as separate processes. Coal is often held in large stockpiles at coal mines and ports prior to transportation to consumers and prior to use by them (Figure 7.1). With increasing world trade in coal, large stockpiles are shipped over long distances and oxidation can cause significant disparities between the specifications quoted by the supplier and those that apply to the delivered coal.

FIGURE 7.1   A coal stockpile. *(fossilfuelfoundation.org).*

Coal weathering has been a subject of intensive study for over a hundred years and literature on the subject is extensive (Parr and Wheeler, 1908; Parr and Kressman, 1911; Wheeler, 1918; Francis and Wheeler, 1926; Francis and Morris, 1931; Wheeler and Woodhouse, 1932; Huggins et al, 1987, 1989; Stracher, 2007, 2011; TSOP, 2013). In-situ underground coal is in an environment which is water-saturated and oxygen-free. Any change in the environment including temperature, moisture or oxygen partial pressure will affect the coal's stability. Various types of degradation may occur depending on the coal constituents and the nature of the environment.

7.2 MECHANISM OF COAL OXIDATION AND WEATHERING

The changes in properties that occur during oxidation have been a subject of extensive and intensive research and are well documented in literature (van Krevelen and Schuyer, 1957; Dryden, 1963; Guyot and Pollard, 1974; Afonja, 1973; Wender et al., 1981; van Krevelen, 1981; Bend et al., 1989; TSOP, 2013). In spite of intensive research over decades, the exact mechanisms of oxidation, significant parameters, and the physical, chemical and structural changes that occur in the coal during oxidation are still unclear because coal is a physically heterogeneous and chemically complex substance. Literature is also extensive on the possible mechanisms that are prominent in coal weathering and oxidation (Banerjee, 1985; Larsen et al., 1986; Nelson, 1989; Clemens et al., 1991; Arisoy et al., 2006; Ulanovski, 2012; Mirosh-nichenko et al., 2012; Butakova et al., 2013). Coals contain a variety of functional groups involving carbon, oxygen, nitrogen and sulphur, the most important of which are the phenolic (-OH) and carboxylic (-COOH) groups. On exposure to air, even at very low temperatures, oxygen is physically adsorbed on the exposed coal surface, although the process is not uniform throughout the coal surface. From ambient temperatures the adsorption rate increases rapidly, especially in the first few days after which the rate starts to slow down due to resistance posed by the initially formed oxidized surfaces.

When the coal is stored in large stockpiles, moisture and oxygen start to diffuse inwards. At temperatures above about 80°C, adsorption changes to chemisorption and weak oxy-coal-water complexes form. Above 100°C, moisture is released and temperature starts to rise due to oxidation reactions between the carbonaceous material of the coal and oxygen, with the release of carbon monoxide and carbon dioxide. During the process of moisture release the temperature may fall slightly because the heat required for the moisture evaporation process is higher than the heat generated by the oxidation process. As temperature rises, the oxy-coal complexes become more stable. If ventilation is poor and the heat produced by exothermic reactions is not dissipated, temperatures within the stockpile could rise to a point at which spontaneous (self) ignition may occur.

The threshold temperature for self ignition varies with coal rank and ranges from about 200°C for lignite to 400°C or above for anthracite. However, actual ignition temperatures could be significantly higher or lower depending on coal rank, coal type, coal pile geometry and prevailing environmental conditions. The dependence of coal oxidation on rank is believed to be due to variations in the quantities and chemical nature of some major petrographic components of coal, particularly vitrinite. Also, low grade coals tend to be more friable than higher grade coals, hence they fragment relatively easily and expose new surfaces for oxidation. Furthermore, low grade coals have greater affinity for moisture and oxygen, both of which promote oxidation.

Hydrocarbons in coal may be divided into two broad groups. The aromatic group (benzene and similar chemical compounds) is very stable and requires high temperature aqueous solutions as well as catalyzation to decompose. On the contrary, the aliphatic group (alkanes, etc) decompose readily at low temperatures liberating heat which is believed to be the main source of self-heating in coal piles. Also, peroxides which are oxidation products of some ethers decompose readily at low temperatures, liberating heat. Hydroxyls, carboxyls, carbonyls and inorganic carbonates decompose at higher temperatures and are not believed to contribute significantly to self-heating (Moroeng, 2015).

Other coal constituents such as ash and mineral matter may accelerate or moderate the heat release processes leading to spontaneous ignition. For example, pyrites may accelerate the heat build up during the incubation period through exothermic oxidation while other minerals such as alumina and silica may have a retarding effect. The weathering process is not restricted to mined coal but can occur also in-situ in seams of coal mines, depending on the oxygen partial pressure and humidity of the coal bed environment. Also, a number of other factors can play an important role in the oxidation processes that lead to mine fires. These include the thickness of the overburden, seam thickness, ignetious intrusions into the seams, presence of faults and zones of impuissance around faults which enhance the penetration of oxygen into the coal seams, shale bands which separate coal seams, underground mine design (pillar design, mine ventilation, method of working).

The aerial oxidation process is a result of gas-solid reactions between the atmospheric oxygen and functional groups which cause changes to the coal surface. The oxidation processes are believed to proceed in three stages: the chemisorption of oxygen on to the surface of the coal to form weak oxygen complexes with the carbonaceous constituents of the coal; the decomposition of the coal-oxygen complexes; and the formation of oxy-coal (Speight, 2015). In the first step, oxygen molecules get physically adsorbed to the coal surface and move into the pores by diffusion. The oxygen reacts with the coal to form coal-oxygen complexes. This process which may start at a temperature as low as -80°C becomes significant at around -5°C after which adsorption becomes quite fast and the oxidized surface layer builds up. However, the oxide layer formed initially now acts as a partial barrier to the diffusion of oxygen into the pores.

The adsorption process slows down and becomes insignificant above around 50°C. (Munzner and Peters, 1965, 1966). It is estimated that a heat of 2 to 4 calories is evolved for every 1ml oxygen adsorbed. The heat starts to build up and temperatures in the pores start to rise unless the heat is dissipated. The coal may gain some weight initially due to the absorption of oxygen while the coal constituents remain intact. It is probable also that the chemisorption of oxygen on to the coal surface promotes reactions between carbon, oxygen and moisture to form a transient hydroperoxide and/or a peroxide, also commonly referred to as peroxygen complex, in accordance with the following equations (Given, 1984).

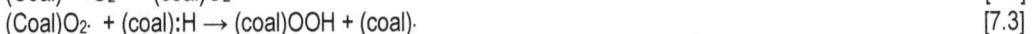

$$(Coal){:}H + O_2 \rightarrow (coal){\cdot} + HO_2{\cdot} \qquad [7.1]$$
$$(Coal){\cdot} + O_2 \rightarrow (coal)O_2{\cdot} \qquad [7.2]$$
$$(Coal)O_2{\cdot} + (coal){:}H \rightarrow (coal)OOH + (coal){\cdot} \qquad [7.3]$$

The formation of these transient products may contribute to the initial weight gain at the onset of oxidation. Studies have shown that up to 50% of the oxygen absorbed is retained by the hydroperoxides/peroxides.

In one study involving the low-temperature oxidation of coal (25-140°C) it was found that only a small proportion of the oxygen absorbed during oxidation ends up in the formation of carbon monoxide, carbon dioxide and water, the bulk being used in the formation of oxy-coal, especially at low temperatures, in accordance with the oxidation model below (Itay et al., 1989).

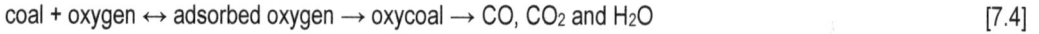

$$\text{coal + oxygen} \leftrightarrow \text{adsorbed oxygen} \rightarrow \text{oxycoal} \rightarrow CO, CO_2 \text{ and } H_2O \qquad\qquad [7.4]$$

In the second stage the coal-oxygen complexes decompose, yielding CO_2 and H_2O, forming carboxyl (COOH), carbonyl (C=O) and phenolic (OH) groups, and generating more heat. At higher temperatures, around 100°C, the groups formed in the second stage decompose releasing CO, CO_2, H_2, H_2O, methane, ethylene, propylene, and more heat. As the temperature rises further, the aliphatic structure breaks down and more carbon monoxide, carbon dioxide and water vapour are released (Bhowmick et al., 1959; Swann and Evans, 1979). However, as the weathering process advances, the carbon and hydrogen contents of the coal decrease markedly (Figure 7.2). Clearly there are limiting values for the two elements, depending on the coal rank. The effect of oxidative weathering on aliphatic structure of coal has been studied by Joseph and Mahajan (1991), using acid-catalyzed transalkylation of two coals with phenol. It was found that oxidative weathering reduces concentrations of several types of straight chain and branched chain aliphatic crosslinks, due apparently to their conversion to carbonyl groups.

In summary, research has established that low temperature aerial oxidation of coal results in the formation of oxygen complexes and peroxides, saturated structures (CH_3 and CH_2 groups) are reduced, moisture and oxygen play critical roles in the oxidation process, but the mechanisms involved are still subjects of intensive research.

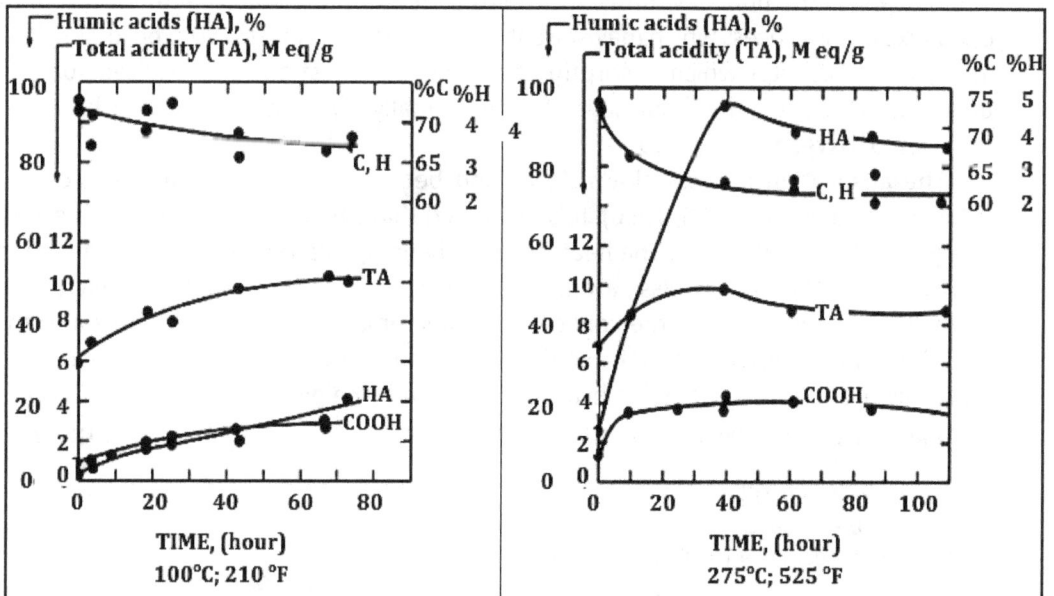

FIGURE 7.2 Aerial oxidation of a sub-bituminous coal at different temperatures. *(Jensen et al (1966); Speight (2013).*

7.3 FACTORS WHICH INFLUENCE COAL OXIDATION AND WEATHERING

Studies on oxidation kinetics have identified several factors which may contribute to coal weathering. These include intrinsic factors such as coal type, rank and chemical composition, inherent moisture content, the types, quantities and distribution of mineral matter in the carbonaceous matrix, while such extrinsic factors as prevailing environmental conditions, in particular humidity and ventilation, the particle size distribution and mix of coal piles, high surface polarity leading to high heats of wetting, coal pile geometry, igneous intrusions into coal seams, are also important.

7.3.1 Effect of coal moisture content

In-situ coal is normally saturated with moisture. Free water in coal pores will significantly inhibit the rate of diffusion of oxygen to the internal surfaces of coal and thereby slow down the rate of aeration. On mining and exposure to the atmosphere coal loses water and forms equilibrium with the ambient atmosphere. Inherent, chemically bound water remains in the coal and activates oxidation. The coal begins to lose free moisture, a process which causes increased porosity in the coal structure and facilitates the penetration of aerial oxygen. Environmental changes during storage such as increase in moisture due to rain will cause the coal to absorb moisture, the rate depending on the temperature, the initial moisture content, and the coal rank. The processes of desorption and re-absorption of moisture cause volume changes which occur differentially from the surface to the core, hence setting up internal stress gradients which weaken the coal structure and cause physical degradation. Decrepitation exposes new surfaces and facilitates more oxidation.

The initial moisture content of coal influences the rate of oxygen absorption. High moisture content accelerates oxidation because the loss of the moisture leaves a relatively large number of pores and the increased internal surface area of the coal facilitates the absorption of oxygen. On the other hand, extensive drying of coal adversely affects the pore structure and the consequent reduction in the internal surface area results in a marked reduction in the oxidation rate. Coal that has been thoroughly dried is very difficult to oxidize (Ogunsola and Mikula, 1992). On the other hand, exposure of dried coal to moisture can accelerate oxidation due to the high heat of wetting and the absorption of oxygen at an increased rate.

7.3.2 Effect of coal oxygen concentration

The rate of oxygen absorption by coal is a function of the oxygen partial pressure of the environment to which it is exposed. The rate of oxidation will slow down with time and the oxygen absorption rate is estimated to be inversely proportional to the $1/4^{th}$ root of the amount of oxygen absorbed (La Grange, 1950). The oxygen concentration also affects oxygen absorption rate. However, there is a wide variation in the rate of absorption of oxygen by different coals. For many high rank coals the rate is very slow but many low rank coals have high oxygen absorptive capacity. For example, the oxygen content of freshly mined lignite can increase several percent within a few weeks of exposure (Kimpel, 1992). The effect of oxygen concentration on the chemical processes and products during weathering is still not well understood but it seems to differ for different ranks of coal (Green et al., 2012).

7.3.3 Effect of temperature

The rate of most chemical reactions is temperature-dependent and in many cases the rate approximately doubles for every 10°C increase in temperature. This had been assumed to be approximately true for coal oxidation (Yohe, 1958). However more recent studies have shown that reaction rate for coal oxidation decreases with increasing temperature, due apparently to the loss of both the equilibrium moisture and oxygen-containing functional groups from the coal (Ogunsola and Mikula, 1992).

Severe drying of the coals promotes a marked reduction in the reaction rate due to a reduction in the formation of pores and therefore a reduction in new internal surface area for potential oxidation. In fact oxidation virtually stops after exposure for an extensive length of time. It appears however that different oxidation mechanisms are dominant below and above 70-80°C. (Jones and Townsend, 1949; Axelson et al., 1987; Green et al., 2012). The peroxygen complexes are formed mainly below this critical temperature range, accompanied by the production of small amounts of carbon monoxide and even smaller quantities of carbon dioxide. However, at higher temperatures these complexes become unstable and break down, there is a major increase in the oxygen content of the coal, and in the yield of gaseous oxidation products, carbon dioxide being the predominant gas.

The oxidation rate of coal below 70-80°C increases markedly with temperature and the porosity of the coal, although there may be a decrease in the oxidation rate with increasing particle size or with increasing rank. The presence of moisture is critical for reactions which occur in the temperature range and this has been confirmed by the results of Ogunsola and Mikula (1992). In a recent study by Green et al., (2012), coals of different ranks ranging from lignite to bituminous coal were isothermally heated at 95°C in air atmosphere for a period of up to six months. Their results supported the suggestion of several previous authors that different weathering mechanisms predominate at temperatures below and above about 100°C. There were also distinct differences between lignite and higher rank (sub-bituminous and bituminous) coals. Lignite exhibited only carbon centered radicals with an adjacent oxygen atom compared with only carbon centered radicals in the higher rank coals. Also, there was no significant change in radical concentration due to long term oxidation treatment while the higher rank coals showed a distinct increase in the radical concentration.

7.3.4 Effect of rank

Chemical changes associated with the coalification process and the physical consolidation of the coal deposit result in the gradual expulsion of water. Low rank coals contain a number of oxygen functional groups which make them highly hydrophilic, hence the high water content of up to 65%. Coals in this group also have high affinity for oxygen. As coal matures, it becomes increasingly hydrophobic (Yu et al., 2013). High rank coals generally have low affinity for oxygen content and low moisture holding capacity , and therefore do not oxidize as rapidly as low rank coals. The difference in response to oxidation is due to the coal structure. High grade coals have more condensed aromatic chains which are more resistant to oxidation while lower grade coal structures contain more aliphatic chains which have greater affinity for oxygen.

The presence of hydroxyl groups in the aromatic structure of lower grade coals also enhances oxidation. In effect, the rate of low-temperature oxidation of coal varies from very low for anthracite to very high for lignite. In fact, lignite often cannot be stored for any length of time because of its propensity for rapid oxidation and possible spontaneous ignition. Generally, the higher the rank, the less susceptible the coal will be to oxidation. There is a close relationship between coal rank and moisture holding capacity which in turn depends on the internal pore structure and heat of wetting characteristics. It should be noted however that all coals have a tendency to ignite spontaneously if the intrinsic and extrinsic conditions are favourable. Anthracite, the highest coal rank has been known to self-ignite in storage (Midgalski, 1956). The high tendency of lignite and sub-bituminous coals to self-ignite can preclude their widespread application due to transportation and storage problems. Such coals are rarely stored and tend to be used in the locality of the coal mines.

7.3.5 Effect of coal petrographic composition

Most coals are made up of three major micro-constituent groups, namely, vitrinite, liptinite and inertinite. Vitrinite is usually the most predominant, forming between 15% and 95% of coal macerals. Liptinite is rarely more than 20%, while inertinite can be over 80%. The processes that control coalification are neither temporal nor uniform. Variables that control coal formation and maturity include variations in precursor material, palaeogeography, original depositional environment, hydrology, changes in geographical and geological location of deposits over time due to earth movement, oceanic incursion, etc. Inevitably, these factors cause significant variations in the chemistry, petrology and utilization behavior of coal, even within the same seam or geographical location (Bend, 1991).

   The major constituents of coal have different affinities for oxygen and therefore have different weathering characteristics. Vitrinite is the most reactive and various studies have identified significant changes in the mean maximum reflectance of the maceral due to oxidation (Avila, 2012, Figure 7.3). This is contrary to previous views that vitrinite reflectance is not affected by oxidation and weathering. Liptinite is also reactive but, since it is usually present in coals in small proportions relative to vitrinite, the overall effect on coal oxidation is assumed to be small. Inertinite is unreactive and since it can be present in coal in high proportions, it can also have a damping effect on coal oxidation (Beamish and Arisoy, 2008).

7.3.6 Effect of coal particle size

A coal pile comprising large particles in a narrow size range presents a high external surface area for oxidative reactions and therefore enhances oxidation rate. It is estimated that oxygen absorption varies directly with the 0.4 power of the surface area (as determined by screen analysis), and slightly less than the first power of the oxygen concentration. It is probable however that the internal pore surface area also contributes to the oxidation rate in view of reports that the relationship between particle diameter and oxygen intake capacity is not linear (La Grange, 1950, Akgun and Arisoy, 1994; de Korte, 2001; Zuo et al., 2012). Coal-oxygen reactions in a coal pile are controlled primarily by the rate of internal gaseous diffusion. Large particle size mix allows free circulation of air in a coal pile and fast dissipation of heat generated by oxidation. Small coal particles or mixture of a wide spectrum of sizes promotes greater compaction and lower porosity and enhances heat build-up within the coal pile.

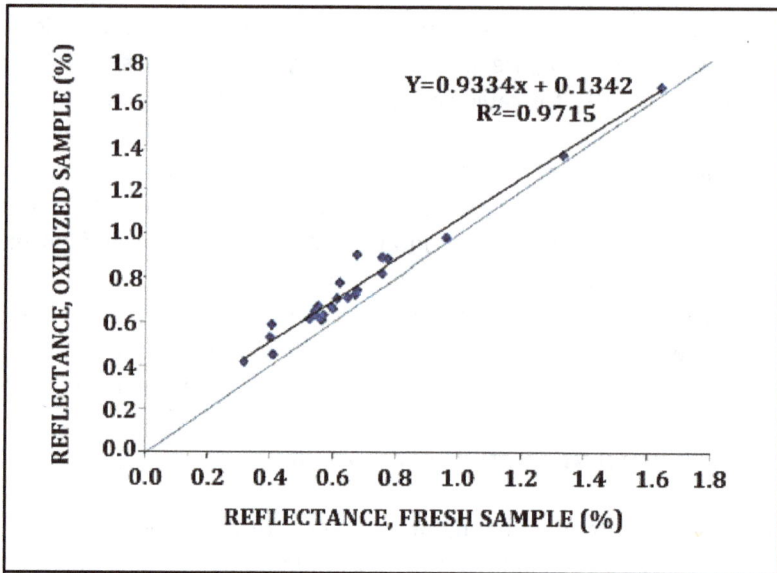

FIGURE 7.3 Comparison of reflectance between fresh and oxidized coal. *(Avila, 2012).*

It should be noted however that some high rank, low volatile coals are friable but do not oxidize readily despite the excessive fines they produce on handling and the corresponding increase in surface area. On the other hand, many low rank coals are not friable and do not degrade readily on handling, yet they oxidize easily because of their affinity for moisture. As oxidation proceeds, particle breakdown occurs due to physical and chemical processes and new surfaces are exposed. If the prevailing physical and environmental conditions restrict the dissipation of the heats of reaction, spontaneous combustion could occur. Although the heat generated in a weathering coal pile depends primarily on the coal reactivity to oxygen, particle size mix determines pile porosity and largely controls the rate of heat dissipation by convective and conductive heat transfer to the pile surface.

7.3.7   Effect of exposure time

The rate of absorption of oxygen is highest at the initial stages of exposure and slows down as the amount of oxygen uptake increases. Low grade coals tend to weather very rapidly in the first few weeks of exposure after which the oxidation rate slows down significantly (La Grange, 1951). However, oxidation rate can rise again if there is considerable size degradation and more particle surface area is exposed.

7.3.8   Effect of mineral matter content

Coal containing a significant amount of shale tends to be friable because the inorganic mineral matter are often present in the coal structure as discrete, brittle compounds which create fissures and render the coal brittle. Weathering of the inorganic compounds may result in the formation of new brittle compounds which further weaken the coal structure and increase the tendency to degrade and form fines. However, as mentioned earlier, friability by itself does not necessarily enhance oxidation, it depends on the coal rank and type. The chemical nature of the inorganic compounds is more important in determining the effect of their presence on coal oxidation. Generally, high mineral matter content inhibits oxidation unless the mineral matter contains a high percentage of iron compounds, lime, soda, iron pyrites, and some alkalis which tend to accelerate oxidation because they react with some of the oxygen to form new compounds, releasing heat in the process.

The pyrite in coal is oxidized initially to iron sulphates, which are then transformed to iron oxy-hydroxide (Huggins et al., 1983). Pyrite oxidation as expected is accelerated by increasing humidity, temperature, and oxygen partial pressure. Other minerals usually present in coal such as alumina, silica borates, calcium chloride remain unchanged by weathering and act as diluents, with a slowing down effect on oxidation (Huggins et al., 1980; Huggins et al., 1982; Huggins et al., 1983). However, recent work by Beamish and Arisoy (2008) involving bituminous coals indicated that mineral matter which had hitherto been assumed to be unaffected by weathering do not only act as heat sink but can also be involved in physico-chemical reactions which may retard the oxidation processes.

7.3.9 Effect of bio-oxidation (microbial metabolism)

Coal that is exposed to weathering will lose most of the sulphate sulphur content due to leaching. The pyritic sulphur content could reduce by 30-50% within months due to bacterial activity but the organic sulphur content remains approximately the same because it is largely unaffected by leaching or bio-oxidation (Savage, 1951). There is ample evidence in literature about the capability of certain micro-organisms to degrade coal. Freshly mined coal is usually resistant to microbial degradation. However, when the conditions are favourable, (high humidity, presence of minerals, abundant carbon source), certain fungi and microbes can grow on coal. These microorganisms are capable of modifying the coal structure significantly.

The intensity of activity of microbes varies with coal rank. Hard coals are more resistant to microbial degradation compared with brown coals and lignites. Intensive research in the last two decades or so has established the biotechnological significance of microbes in coal desulphurization, biological conversion of coal into useful products such as chemicals and fuels, and in the reclamation and rehabilitation of disused coal mining areas (Hofrichter and Fakoussa, 2001). Many types of bacteria have been investigated for coal demineralization and depyritization, with varying degrees of success. The mesoacidophylic bacteria have been identified as the most effective. *Thiobacillus ferooxidans* has been found effective in oxidizing sulphur and iron, *Thiobacillus thiooxidans* oxidizes sulphur, and *Leptospirillum ferooxidans* is an iron oxidizer (Prayuenyong, 2002).

Coal oxidation by *Thiobacillus ferooxidans* has been investigated extensively and two possible mechanisms have been identified: the direct process in which the bacteria cell attaches itself to the sulfide crystal surface in the presence of oxygen and moisture. Oxidation occurs in a thin film located in the interspace between the bacterial outer membrane and the sulphide surface. The products are ferrous sulphate and sulphuric acid (Equation 7.5).

$$\overset{\textit{T. ferrooxidans}}{2\,FeS_2 + 7\,O_2 + 2\,H_2O \quad \rightarrow \quad 2\,FeSO_4 + 2\,H_2SO_4} \qquad [7.5]$$

This mechanism is considered possible only in certain coals since many coals have fine pores and the bacteria are too large to penetrate. The indirect process is considered more feasible and involves the oxidation of ferrous iron to ferritic iron in accordance with Equations 7.6 and 7.7. The ferritic iron then reacts with the iron pyrite.

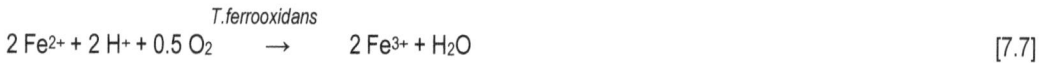

$$FeS_2 + 14\ Fe^{3+} + 8\ H_2O \xrightarrow{\text{T.ferrooxidans}} 15\ Fe^{2+} + 16\ H^+ + 2\ SO_4^{2-} \tag{7.6}$$

$$2\ Fe^{2+} + 2\ H^+ + 0.5\ O_2 \xrightarrow{\text{T.ferrooxidans}} 2\ Fe^{3+} + H_2O \tag{7.7}$$

It is possible also that the ferritic iron oxidizes the ferrous iron in the pyrite, to yield elemental sulfur (Equation 7.8). The elemental sulphur is then oxidized to hydrogen sulphate as shown in Equation 7.9.

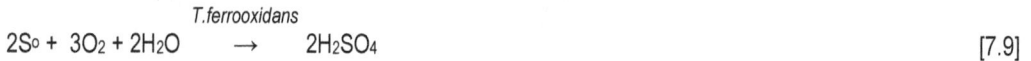

$$FeS_2 + 2Fe^{3+} \rightarrow 3Fe^{2+} + 2S^o \tag{7.8}$$

$$2S^o + 3O_2 + 2H_2O \xrightarrow{\text{T.ferrooxidans}} 2H_2SO_4 \tag{7.9}$$

It is still not very clear which environmental conditions favour any or all of the reactions in the above equations. For example, the reduction of elemental sulphur to sulphuric acid may be incomplete and this may explain why the element which is not known to exist in natural coal is often identified in weathered coals (de Korte, 2001). Other complex compounds including white gypsum may also be formed.

7.4 EFFECT OF WEATHERING ON TECHNOLOGICAL PROPERTIES OF COAL

There is general agreement that weathering (also frequently referred to as sub-critical heating or pre-oxidation) can cause severe deterioration in the physical and chemical properties of coal, with undesirable consequences that make the coal less suitable for technological utilization. Weathering has a deleterious effect on the way coal behaves In beneficiation, combustion, carbonization, gasification, liquefaction processes.

Weathering renders coal less responsive to some cleaning processes, in particular froth flotation, and reduces yield in liquefaction and solvent extraction processes. The rheological properties of coals are also diminished and this makes them less suitable for metallurgical coking. It should be noted however that, although weathering negatively affects many technological properties of coal, it can also have some positive effects. For example it has been established that weathering can positively moderate the fluidity of some highly plastic coals and make them more suitable for metallurgical coking. Weathering can remove the fluidity of coals meant for combustion thus reducing agglomeration problems. Also, It has been reported (Nelson, 1989) that pre-oxidation of coal may result in improved gasification reactivity and may improve the surface area of char produced from such coal.

There is considerable controversy on the mechanisms involved and how various coal properties respond to weathering. One main problem is the heterogeneity of coal in terms of contents and structure which means that coal from the same deposit or even the same seam can exhibit significant variations in properties. Another problem is the fact that most investigators have used laboratory-scale artificial oxidation to simulate industrial conditions. While this technique is relatively easy and convenient, some investigators have found that there could be big differences between the results from such tests and industrial realities (Cimadevilla et al., 2005a&b). A study by Lee (2012) which was carried out on industrial scale coal piles also supports this conclusion.

7.4.1 Effect of weathering on coal strength

The physical effects of weathering cause enlarged fissures in the coal structure and irregular micro-fissures in the vitrinite (Figure 7.4) (de de Korte, 2001). Coal containing a significant amount of mineral matter (depending on the nature, extent and variety) will usually show greater physical deterioration than clean coal due to fissure caused by brittle gypsum, sulphur etc formed during oxidation. The Hardgrove Grindability Index decreases as weathering progresses and weathered coal is more difficult to grind (Beafore et al., 1984).

The differential gradients of moisture, temperature and oxygen partial pressure between the surface and internal areas of a coal lump or coal pile may induce internal stress gradients which may cause fissuring and production of fines. This results in the formation of a large number of stress boundaries which weaken the coal structure, a phenomenon often referred to as slackening. Slackening can accelerate oxidation by increasing the surface area of the coal exposed to aerial oxidation (Fryer, 1973; Gray et al., 1976). Lee (2012) identified significant cracks in the vitrinite structure in industrial coal weathering experiments extending over a period of over 200 days. Another contributor to slackening of coal is the stress system caused by cyclic sorption and desorption of moisture which produce fissures and cracks that mechanically weaken and decrepitate the coal. (Schmidt, 1945; Fryer, 1973; Gray et al., 1976;). The thermal cycling that results from the exothermic and endothermic reactions also contributes to slackening.

FIGURE 7.4 Micro-fissures in weathered vitrinite. (*de Korte, 2001*).

The reaction of iron pyrites with oxygen and moisture which yields products of larger volume than the original pyrite can cause fissures in the coal structure and enhance the penetration of oxygen. Slackening is much more rapid and extensive in lignites and sub-bituminous coals than in higher rank coals (Fryer and Szladow, 1973; Gray et al., 1976). Slackening can also accelerate oxidation by increasing the exposed surface area of the coal to air. The oxidation of coal's organic constituents could participate in slackening through the creation of new hydrophilic sites which would promote moisture sorption.

7.4.2 Effect of weathering on coal cleanability

The purpose of beneficiation is to upgrade the quality of coal so that it can meet specific end-use requirements by reducing the mineral matter and clay content of the coal. Weathering changes the electro-kinetic properties and surface characteristics of coal. Fresh coal normally exhibits hydrophobicity and tends to repel water. The characteristic tendency of fresh coals to repel moisture is important in coal cleaning by froth flotation, a process which utilizes this surface properties to separate the associated mineral matter on the basis of their hydrophilic properties and tendency to wet readily (Hamza et al., 1983; Labuschange and Van der Watt, 1983; and Gala et al., 1989).

Weathering coal absorbs oxygen and acidic oxy-coal complexes are formed on the coal surface. Further oxidation of these compounds results in the formation of acid-like humic compounds or hydroxy-carboxylic acids. The acids degrade further to form water-soluble acids (Sadowski et al., 1988). The various oxidation reactions deplete the hydrocarbon content of the coal, hence the coal surface becomes less hydrophobic and more hydrophilic. The new oxygen-containing functional groups that are created when the coal is oxidized also increase its adsorptive affinity for polar molecules such as water as well as its water solubility (Murata, 1981; Kelekek et al., 1982; Taweel et al., 1982). Subsequent oil agglomeration also becomes more difficult because the organic contents of weathered coal are more hydrophilic and the inherent moisture is increased.

Dewatering and drying become more energy intensive and less effective with coal oxidation (de Korte, 2001). The physico-chemical changes which occur on the coal surface and structure as a result of weathering make dewatering and drying of coal more difficult and less efficient as the coal's degree of oxidation increases, and significantly higher energy is required per unit weight of coal. The stability and viscosity of coal-water and coal-oil slurries could also be affected by the more polar surface properties of particles of oxidized coal. Additionally, the peroxides produced during coal oxidation could potentially have a detrimental effect on the stability and viscosity of coal-oil slurries. The effect of surface oxidation on the hydrophobicity of coal and its response to cleaning appears to be a function of rank. It has been found that low-rank coals may exhibit an improved response to flotation after a low temperature thermal oxidation treatment whereas the opposite is true for coals of higher rank (Savage, 1951; Crelling et al., 1979).

In summary, weathering reduces hydrophobicity of coal and this makes it more difficult to selectively separate mineral matter particles and therefore reduces the coal's response to froth flotation. The efficiency of beneficiation by flotation, agglomeration and flocculation processes is also reduced and more energy is required for processing oxidized coal compared with fresh coal.

7.4.3 Effect of weathering on calorific value

The effect of weathering on the calorific value of coal depends on the rank, type and the storage conditions. While many high rank coals largely retain the original calorific values despite extensive exposure, the effect on low-rank coals can be quite severe. British high-rank coals have been found to lose only 5% of their original calorific value after 25 years of storage while low-rank sub-bituminous coals can lose over 50% within weeks (La Grange, 1950). The difference in response of various ranks of coal to weathering is due apparently to the relative differences in the affinity for oxygen. According to Fryer (1973), Nelson (1989), high rank coals lose approximately 156 kcal/kg for every 1% increase in oxygen content compared with 106 cal/kg for low rank coals. In spite of the higher rate of heat loss, the low affinity of high rank coals for oxygen makes the total loss in heating value insignificant even after exposure for many years.

On the other hand, the oxygen content of freshly mined lignite coals can increase several percent in a matter of weeks when the coal is stored in air at ambient temperature. Such a rapid oxidation can lead to a rapid loss in calorific value and an exponential increase in temperature if the coal is held in a stockpile in moist, aerated environment. As a result, spontaneous self-ignition can occur. The results of recent work by Lee (2012) appear to contradict the widely held view that higher rank coals lose heating value more rapidly than lower rank coals during the initial stages of weathering. Lee found that loss of heat value for every 1% increase in oxygen content was higher for the lower rank coals.

The disagreement between Lee's results and those of several other authors on the effect of weathering on coal calorific value may be due to the fact that Lee's work was on large, 20-ton piles in natural environment compared with laboratory-scale experiments on small samples in simulated environment by most previous workers. It is also possible that the heterogeneity of coal structure and variations in microconstituents, even within the same rank, make it difficult to generalize on weathering behaviour.

The work of Lee supports the conclusion of many previous workers on the insignificant deterioration of the heating value of high rank coals even after extensive exposure to an environment which enhances oxidation. Four coking coals showed no significant deterioration of calorific value after exposure to oxidation enhancing environment for over 200 days. In contrast the oxygen content of freshly mined and crushed sub-bituminous coal and lignite can increase several percent in just a few weeks of exposure to air at ambient temperature and the drop in heating value due to weathering can be very high (Fryer, 1973; Herring, 1983; Kimpel, 1992).

7.4.4 Effect of weathering on volatile matter and carbon content

The effect of weathering on volatile matter content of coal is unclear (Bend et al., 1989). Some authors have reported an increase (Fredericks et al, 1983; Pisupati et al., 1993a, 1993b, 1999) while others reported a decrease (Kalemma and Guvalas, 1987; Gerus-Piasecka and Jasienko, 1989; Wang et al., 2007). Clearly the gain or loss of volatile matter due to weathering depends not only on the coal rank but on the form in which oxygen is incorporated in the coal structure which may vary from coal to coal (Pisupati et al., 1993; Beier, 1994). Recent work by Avila (2012) appears to have explained the differences in conclusions reached by various authors on the effect of oxidation on volatile matter of coal.

In a comprehensive study involving many coals of different ranks and reactivity using thermogravimetry, Avila found that low grade coals have high affinity for moisture and tend to lose weight over a wide temperature range compared with higher rank coals due to evaporation. The relatively slow process of moisture evaporation in low grade coals delays the chemical reactions which are propelled by the adsorption of oxygen. On the contrary, moisture in high grade coals is expelled over a relatively short temperature range and chemical reactions lead to the expulsion of some of the volatile matter, notably carbon monoxide and carbon dioxide (Figure 7.5). Furthermore, the type and chemical nature of inorganic constituents of coal can have a significant effect on volatile matter content. In effect, change in volatile matter content as a result of weathering would depend on both the rank and type of coal.

Most authors who have studied the effect of weathering on coal have reported decrease in carbon and hydrogen content and increase in oxygen content. However, some authors have reported little or no change in the content of the elements, even after nearly two years of exposure in spite of considerable deterioration in plastic properties (Cimadevilla et al., 2003), again emphasizing the significance of rank and possibly coal type.

7.4.5   Effect of weathering on thermal processing

Oxidation can have positive effects on the response of coal to thermal processing (combustion, carbonization, gasification, liquefaction), and the technology for pre-oxidation of coal meant for thermal processing is quite well established. The effects of oxidation on coal's behaviour during thermal processing can be predicted to be profound but there is still considerable controversy on the mechanisms. The fact that coal loses much or all of its agglomerating properties on exposure to oxygen is well established. While this is a positive asset for coal meant for combustion, it greatly devalues coking coals.

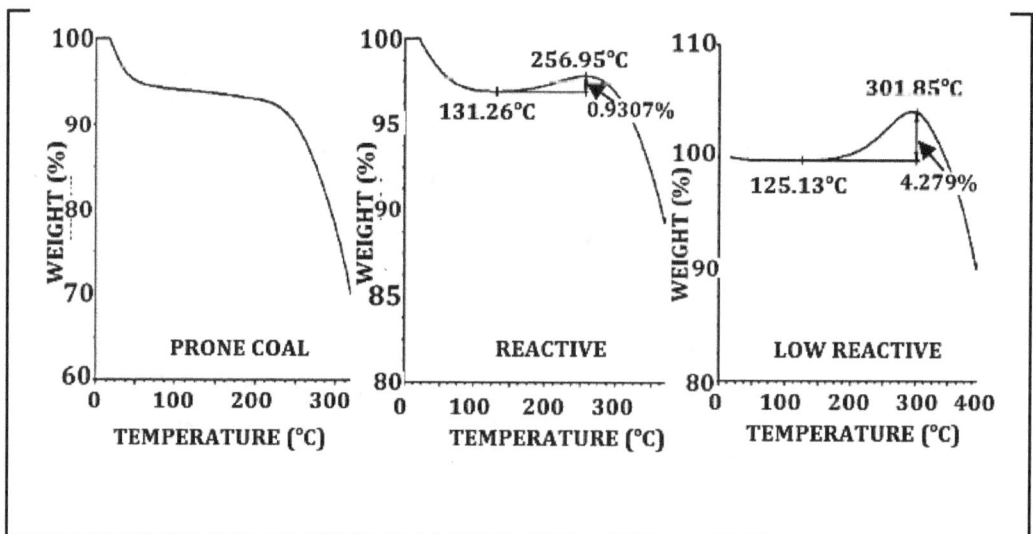

FIGURE 7.5 Thermograms of coals of different ranks. *(Avila, 2012).*

There is general agreement that oxy-coal functions are formed during oxidation and this leads to increased yields of carbon monoxide and carbon dioxide at the expense of tar and liquid yields during gasification, liquefaction, pyrolysis and thermal de-polymerization in non-hydrogen-donating solvents. Oxidation promotes increased cross-linking of the macromolecular structure of coal thereby reducing swelling during heating and increasing the molecular weight distribution of the extractible material and low-temperature tars. The surface area of the char produced is enhanced by a factor of up to 10 and gasification reactivity by a factor of up to 40. However, extractibility by organic solvents is reduced   (Schmidt et al., 1940; Schmidt, 1945; Brooks, 1957; Ignasiak et al., 1972; Wachowska and Pawlak, 1977; Crelling et al., 1979; Mahajan et al., 1980; Neavel, 1982; Furimsky et al., 1983).

7.4.6   Effect of weathering on coal thermoplasticity.

Some coals in the bituminous rank go through a plastic phase on heating in minimum air, usually between 300°C and 550°C after which the coal re-consolidates to form semicoke. This rheological behaviour is a primary requirement for coals destined for use in metallurgical processes and is believed to be due to the cracking of weak aliphatic carbon chains in the coal such as CH, CH_2 and CH_3 and the release of gases. It is generally agreed that weathering reduces the fluidity of thermoplastic coals due to chemical reactions which alter the nature of certain thermo-reactive components of coal. However, the reaction mechanisms involved, identification of compounds which undergo chemical changes and the new compounds formed are still subjects of intensive research. Exposure of coking coal to aerial oxidation results in a pre-carbonization cracking of the weak aliphatic chains thereby reducing or completely eliminating the fluidity of the coal which largely determines the quality of coke that can be produced from a coking coal or coking blend. Furthermore, the relatively stable complexes that are formed as a result of weathering are non-fusible and further reduce the agglomerating properties of coal.

The detrimental effects of weathering on coking properties of coal have been investigated extensively (La Grange, 1950; Savage, 1951; Gray et al., 1976; Crelin et al., 1979 ; Lowenhaupt et al., 1980, 1982; Huffman et al., 1985; Nelson, 1989; Alvarez et al., 2003; Desna, 2011). Winschel and Burke, (1988) suggested that the reduction in agglomerating properties of weathered coals is caused by both the increase in the stable oxygen crosslinks such as ether and ester and a decrease in the weak aliphatic CH contents. Other investigators of the effect of weathering on coal fluidity have identified shortened and decreased aliphatic chains and increased stable oxygen complexes as causes of loss of rheological properties of coal caused by weathering (Alvarez et al., 1998; Ota et al., 2003, Keating et al., 2004, Cimadevilla et al., (2005b) .

Exposure of coal to mild temperatures for short periods of no more than ten days or so has no significant effect on plastic properties. However, at temperatures above 75°C, progressive reduction of agglomerating properties occurs with exposure time. The calorific value, carbon and hydrogen contents also reduce while the oxygen content increases. The coal also becomes more friable and size degradation increases. There are reductions in Gieseler fluidity, swelling and dilation properties of coal and a significant narrowing of the plastic temperature range. The process can be quite rapid depending on the type of coal and the prevailing environmental conditions.

Guyot and Pollard (1974) carried out oxidation tests at 80°C under controlled conditions and found that the rate of loss fluidity of a high volatile (low rank) coal was twice as fast as that of a low volatile (high rank) coal. According to La Grange (1950), Leeder et al., (2011), the thermoplastic properties of coking coal can be destroyed completely in one to three years of storage.

A number of tests are widely applied to determine the effect of weathering on the agglomerating properties of coal. Maximum fluidity as determined by Gieseler plastometer is extremely useful since significant changes in these parameters are evident well before any significant changes in proximate and ultimate analyses, or calorific value are observed. On the contrary, swelling index is not considered a sensitive indicator of weathering. The sensitivity of the Gieseler plasticity test is also evident in studies on industrial coal stockpiles and laboratory experiments by Sen et al., (2009). Petrographic staining technique was found particularly suitable for the detection of weathering because it is independent of prior knowledge of the properties of the fresh coal and of the petrographic composition. The characteristics of the initial fresh coal, including rank and petrographic classification and estimates of volatile matter, calorific value and dilatation can be derived from weathered samples.

Marchioni (1983) investigated the effects of natural weathering on coal using petro-graphic, rheological, proximate and ultimate analytical techniques and found petrographic and rheological parameters the most sensitive indicators of weathering. The change in rheological properties can be determined by comparing the swelling index, dilation or Gieseler fluidity values of weathered coal with base values for fresh coal. Of all these variables, changes in the Gieseler maximum fluidity have been found the most reliable indicator of weathering. The quality of coke that can be produced from weathered coal or a blend containing weathered coal diminishes. Coke strength, coke stability, coke reactivity, coking rate and coke yield all decrease and the amount of coke breeze increases due to a reduction in coke strength (Crelin et al., 1979). Huffman et al., (1985) observed a reduction of nearly 10% in maximum fluidity of coal after just six hours of thermal oxidation at 105°C.

Huggins et al., (1985) in their studies of piles of two coal types stored at room temperature using Gieseler fluidity measurements found losses of 20-30% in 455 days. Larsen et al., (1986) proposed that the loss of donatable hydrogen is one of the key factors responsible for the low value of Free Swelling Index of coal due to aerial oxidation and can be restored by reduction with lithium aluminium hydride. Wu et al., (1988) investigated the effect of weathering on the plasticity, dilatation and floatation characteristics of a coking coal over a range of temperatures. The fluidity reduced to zero within 40 days when the coal was oxidized at 80°C but at 25°C the coal lost only about 50% of its fluidity after 280 days (Figure 7.6). The Swelling Index reduced by 70% at 40°C within 40 days but remained unchanged even after 280 days at 25°C Figure 7.7).

In a study by Gray and Lowenhaupt (1989), a high volatile bituminous coal was stored outdoor and weathering monitored monthly over a period of one year. Deterioration of both the Gieseler fluidity and Free Swelling Index were observed (Figure 7.8). Maximum fluidity decreased by 50% but heating value decreased by only 2% over the year. Coal grain size did not appear to have a significant effect. They also found that oxygen content of the coal increased from 9% to 11% (Figure 7.9). The results of Fryer and Szladow (1973) indicate that Canadian prime coking coals could lose 10-20% fluidity in the first week of storage (Figure 7.10).

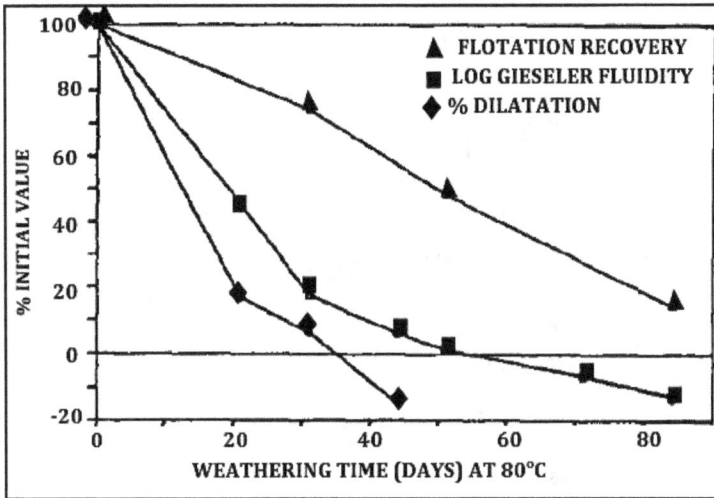

FIGURE 7.6 Thermoplastic properties of high volatile bituminous coking coal weathered at 80°C. *(Wu et al.,1988).*

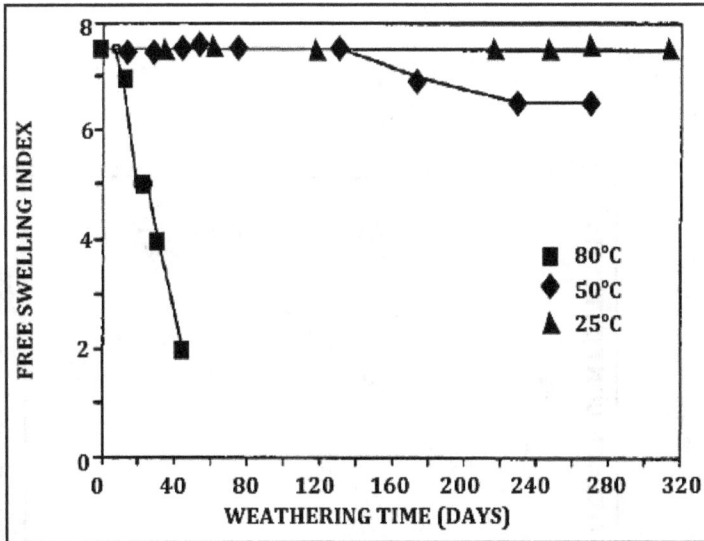

FIGURE 7.7 Free Swelling Index as a function of weathering time and temperature. *(Wu et al., 1988).*

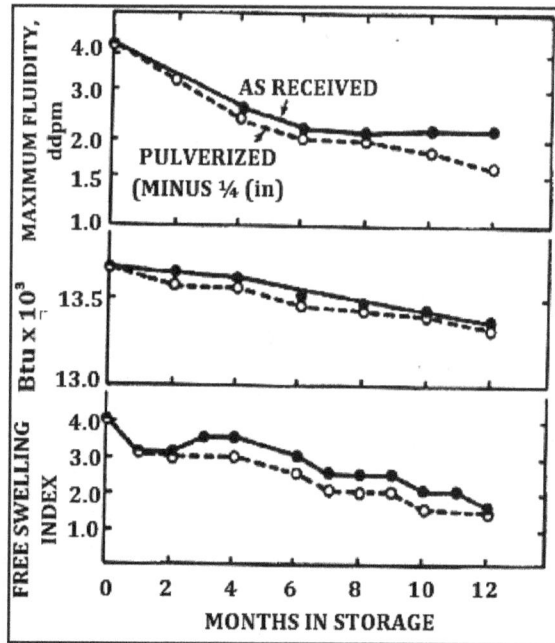

FIGURE 7.8 Fluidity, calorific value and FSI changes with weathering time. *(Gray et al., 1989).*

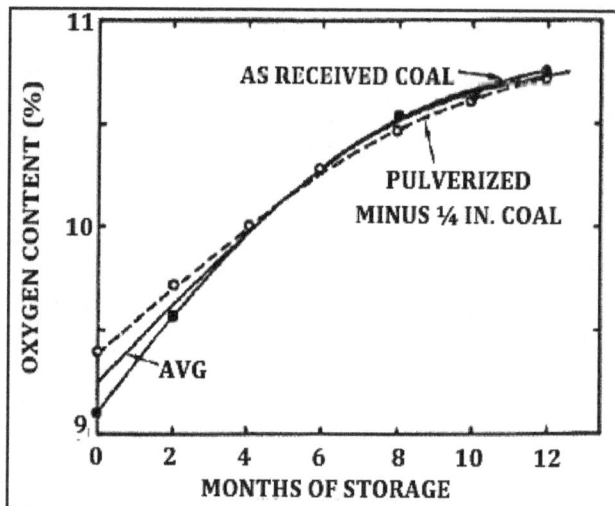

FIGURE 7.9 Oxygen content changes with weathering time. *(Gray et al., 1989).*

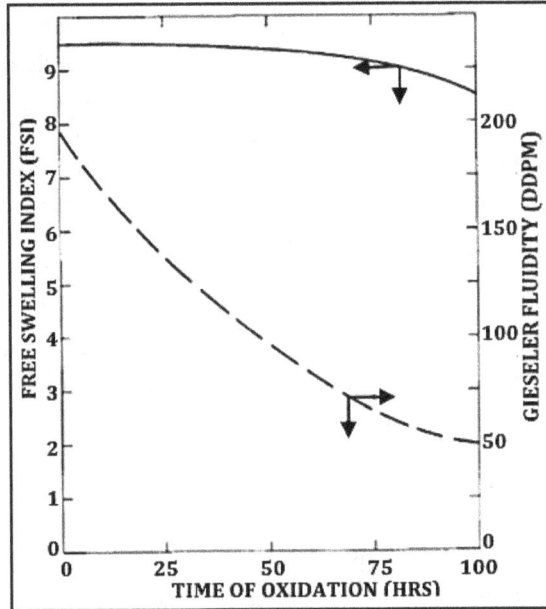

FIGURE 7.10 Change of Free Swelling Index and Gieseler maximum fluidity with time of oxidation of Canadian coals with air at 70°C. *(Fryer and Szladow., 1973).*

Because oxidation is a surface reaction, its rate under otherwise constant conditions will depend upon the rate at which oxygen can diffuse into a coal lump and this, in turn is largely determined by the surface area of the coal to which oxygen can have access. It is estimated that increases in oxygen contents of about 1% could totally destroy the coking properties of prime metallurgical coals. It has been found also that the rate of oxidative destruction of coking properties can vary widely in otherwise closely similar coals.

Seki et al., (1990) found that, Gieseler maximum fluidity and FSI decreased as a result of oxidation for all coals which they examined but the pattern and rate of decrease differed significantly among coal types. Iglesias et al., (1998) studied the weathering characteristics of a medium volatile bituminous coal and found that the coal lost 28% of its original calorific value in 14 days during oxidation at 200°C. They also observed a reduction of FSI from 8 to zero after six hours of oxidation which they correlated with corresponding significant reduction in alkyl groups.

Casal et al., (2003) attributed the loss of Gieseler fluidity of coking coal to concomitant decrease in methylene groups. Giroux et al., (2006) found that dilation of coking coal decreased with exposure time and the rate of loss increased with temperature. Coal stored at 40°C lost dilation totally within 20 weeks. By contrast, coal stored at -15°C showed no loss of dilation even after 60 weeks of storage. Coke reactivity increased and hot coke strength decreased due to weathering.

Another important work on coal weathering was carried out by Leeder et al., (2011). Coking coal was stored under three conditions: a small 10kg pile in open weather, a large 100-ton pile also in open weather and a 10kg sample stored under vacuum. Changes in Gieseler maximum fluidity were monitored over time. The maximum fluidity was degraded under all three conditions although, as might be expected, more slowly in the coal stored in-vacuo. After a year the vacuum stored sample had lost 50% of plasticity, the small exposed sample had lost 75%, while the bulk pile had lost 95% of maximum fluidity (Figure 7.11). Tests were also carried out on 20-metre high mine stockpiles and it was found that deterioration was highest for samples collected from or near the surface and tested at regular periods over several years. The core was significantly less affected by oxidation due apparently to the low oxygen penetration. Recent studies by Lee (2012) on the weathering of large, industrial coking coal stockpiles over a period of nine months, using plastometric tests and SEM analysis, have shown that conclusions on coal oxidation based on small scale laboratory tests may not be applicable to industrial situations. While it was found that fluidity, swelling characteristics, coke stability, coke hot strength and ash content were negatively affected by weathering, the rate was specific to each coal.

Also, a correlation between fluidity loss and rank as determined by vitrinite reflectance could not be established. This is in contrast with the widespread belief that high rank coals are more resistant to weathering than low rank coals because the bulk of the carbon content is tied up in aromatic bonding in the macerals and there are fewer active carbons in their structure. Lee also found that the narrowing of the melting temperature range of weathered components of a coking coal blend can decrease the quality of the coke, even though the quality of coke made from each weathered coal is not significantly affected by weathering. This is because of a decrease in the desired overlap between the melting range of the coking blend components. It should be noted however that mild oxidation can have a moderating effect on high-fluidity, high-swelling coking coal thereby making it more suitable as a constituent of a coking blend.

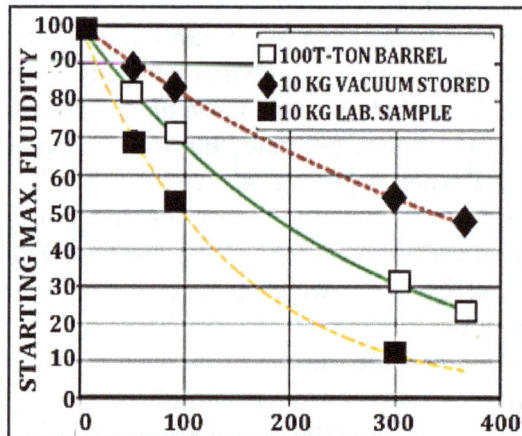

FIGURE 7.11   Impact of weathering on Gieseler maximum fluidity. *(Leeder et al.,2011).*

Studies carried out on weathered coking coal have shown that the loss of fluidity of coal could be reversed by chemical treatment. Ignasiak et al., (1972) attributed the loss of fluidity and caking properties due to aerial weathering of a high volatile bituminous coal to the formation of certain hydroxyl groups during oxidation and could be reversed by the addition of barium hydroxide. The work of Larsen et al., (1986) appears to support the fact that replacement of donatable hydrogen could restore thermoplasticity in coals, in this case, by reduction with lithium aluminium hydride. However the applicability of these methods on an industrial scale is very much in doubt.

It is clear from the foregoing discussion that there is no universally applicable theory on the effect of weathering and oxidation on the thermoplastic properties of coal. Furthermore, data obtained from weathering tests are only applicable to the specific coal tested. The common industrial practice is to mix weathered coking coal with several other coals of different rheological characteristics. Coking blends are designed carefully in order to achieve optimum maximum Gieseler fluidity, overall plastic range, and the appropriate overlap of fluidity temperature ranges of the blend components.

7.4.7 Effect of weathering on sulphur content

Weathering progressively reduces the total sulphur content of coal. Sulphate sulphur is leached from the coal and pyritic sulphur is oxidized to sulphuric acid which is also leached out of the coal pile. Pyritic and sulphate sulphur contents of coal exposed to weathering may be reduced by up to 50% within 18 months of exposure (Savage, 1951). Organic sulphur content is largely unaffected by weathering because it is part of the aromatic structure of the coal. Lee (2012) found no significant change in sulphur content of four coking coals stored in industrial piles for nine months. This conclusion was based on the total sulphur content of the coals studied which was less than 1%. When the sulphur content of coal is low, below about 1%, the sulphur is almost always predominantly organic and is not readily affected by weathering. Since Lee did not determine the nature or morphology of sulphur in the coal tested, it is safe to speculate that the sulphur was probably predominantly organic.

Organic sulphur is part of the chemical structure of the coal and is not removed by weathering or washing. It is also probable that the morphology and distribution of sulphur within a particular coal structure affects its behavior during weathering. Finely disseminated inorganic sulphur compounds within fine coal will oxidize more easily than coarse particles because of the relatively large surface area.

7.4.8 Effect of coal weathering on the environment.

Weathered coal piles pose considerable danger to the environment and human health. The sulphuric acid formed from oxidation of pyrites (see Equation 7.5) is leached out and degrades the environment (Figure 7.12). High concentrations of toxic and environmentally harmful gases including carbon monoxide, carbon dioxide, hydrocarbons, sulphuric and hydrogen gases which are released from spontaneous coal fires not only contribute to the degradation of the global ecosystem but also can cause human health problems particularly for mine workers and in populated areas around coal fields (IAEA, 2013). Many toxic chemical elements and compounds including mercury, lead, arsenic, benzene, toluene, xylene have been found in the environment around self-ignited coal fires (Heffern and Coates, 2004; Stracher, 2004;

FIGURE 7.12   An acid mine drainage *(AMD). (Pennsylvaniawatersheds.org).*

Stracher and Taylor, 2004; Nolta and Vice, 2004; Whitehouse and Mulyana, 2004, Sheail, 2005; Chatterjee, 2006; Pone et al., 2007). Also, when weathered coal is exposed to water through rain or the application of water spray, some organic and inorganic matter such as sulphuric acid, sulphur, calcium and magnesium leach out and cause environmental problems apart from possible loss in heating value of the coal. In extreme cases spontaneous ignition may occur in exposed coal stockpiles, with the resultant release of large volumes of carbon dioxide and carbon monoxide into the atmosphere (Pone et al., 2007; Carras et al., 2009).

7.4.9   Measurement of weathering of coal

Many methods of assessing the extent of weathering in coal have been developed over the last hundred years or so but most are not easy to apply in actual industrial situations. Methods include measurement of increase in acidity as a result of oxidation, identification of chemical compounds produced in the weathering process by measuring the ratio of aliphatic carbons to oxygenated complexes which decreases with weathering. Some methods involve determination of the ratio of the initial pyrite content of coal to the α-FeOOH which is the final product of pyrite weathering. Others involve determination of changes in important properties of coal including changes in the quantities of the gaseous products due to oxidation, in particular, CO and CO_2, and changes in the thermoplastic properties (fluidity, dilation or swelling properties) of coal. Other methods in common use include petrographic tests which evaluate changes in the reflectance and elasticity properties of the vitrinite content of weathered coal (de Korte, 2001), and measurement of changes in the wetting behavior of the coal surface (Fuerstenau and Dao, 1992; Fuerstenau et al., 1987; Sadowski et al., 1988; Sarikaya and Ozbayoglu, 1995).

A more direct method involves measurement of the oxygen functional groups by leaching the humic, carboxylic and phenolic acids formed on the coal surface and determination of the amount of base required to neutralize the acid (Fuerstenau et al., 1987; Yun and Meuzelaar, 1991). Another test developed by Lowenhaupt and Gray (1982) for coking coals involves boiling of a sample of coal in caustic solution. Any humic acids formed on the coal surface due to oxidation will dissolve in the caustic solution. The transmittance of the filtered slurry is compared with the value for unweathered coal. A drop in transmittance value of 20% or over is considered an indication of severe weathering which makes the coal unsuitable as a coking blend constituent. The method has been tested by Ingram and Rimstidt (1984). They found a very good correlation with results obtained by petrographic techniques. The method is relatively quick and considered a reliable indicator of weathering in coking coals. Tests in common use at the industrial level involve the determination of changes in the thermoplastic properties of weathered coal. Specific indicators are changes in the swelling characteristics as determined by the swelling Index method, reduction in Gieseler maximum fluidity and fluidity temperature range, and changes in the contraction and dilation characteristics, using the Ruhr of Audibert-Arnu dilatometer.

Most of the coal weathering tests in common use provide relative values and need to be compared with values for unweathered coal. Unfortunately even freshly mined coals may have weathered in-situ, hence there is no reliable reference point. Some recent investigations have shown that changes in the pH of rehydrated weathered coal can give a good measure of the extent of weathering (Yun et al., 1987; Jha, et al., 2014). Many of the tests in common use are empirical and do not take into account the process chemistry. Furthermore, most methods are based on analysis of very small laboratory samples in simulated environment which do not necessarily represent practical situations in which heterogeneous coal is stored in 20 to 30-ton piles. A few tests of note on industrial coal piles have been carried out recently by Leeder et al., 2011, and Lee, 2012).

7.5 SPONTANEOUS COMBUSTION OF COAL

Spontaneous combustion of coal has been known to be the cause of fires worldwide in coal mine stockpiles, industrial coal storage yards held for extended periods, and piles on ships during transportation for over a century (Figure 7.13). Literature on spontaneous combustion of coal is extensive (Wheeler, 1924; Van Doornum, 1954; Yohe, 1958; Guney and Hodges, 1969; Afonja, 1973; Manadevan and Ramlu, 1985; UNESCO, 2005; Stracher, 2004; Stracher et al., 2007, 2011, 2015; Nalbandian, 2010; Singh, 2013; Sloss 2015). Self ignition is particularly common in large, high coal stockpiles exposed to ventilation that is just enough to provide the oxygen necessary for oxidation reactions but not strong enough to evacuate the heat generated from the exothermic oxidation reactions. These conditions are common in coal stockpiles in mines, consumer stockyards and holds of ships during transportation.

7.5.1 Causes of coal self-ignition

Whenever a fresh coal surface is exposed to air the carbonaceous and inorganic constituents start to oxidize. The presence of moisture is a prerequisite for the oxidation reactions. It is estimated that the heat of wetting released due to the absorption of moisture by coal could be as high as 22 calories/g. (Fryer, 1973). Most of the oxidation reactions involved are exothermic and there is a significant amount of heat released (see also FFF, 2014; de Korte, 2014a-c, 2016).

(a)

(b)

(c)

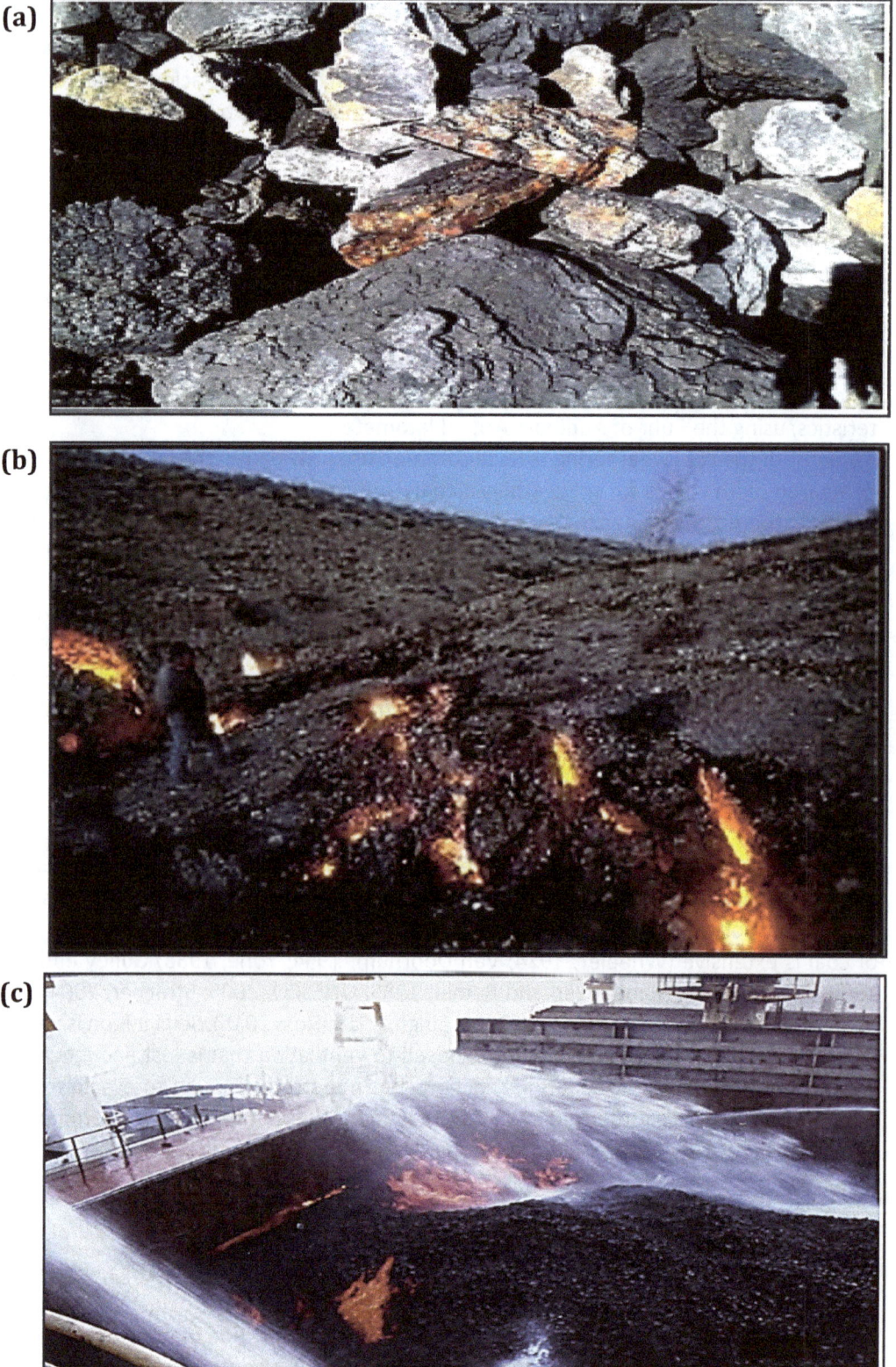

FIGURE 7.13 (a) Oxidized coal (b) Self-ignited coal pile in an opencast mine (c) Self-ignited coal in a ship's hold. *(slidesharecdn.com; de Korte,(2003).*

If the air flow in a coal pile is not sufficient to dissipate the heat generated by the exothermic reactions, heat builds up and temperature rises to a level in which spontaneous (endogenous) self-ignition can occur. This usually occurs in coal stockpiles exposed to an environment rich in moisture and oxygen. Self-ignition is most common in stockpiles at coal-fired power plants since they are the main users of low-grade coals which are stored in open stockyards over extended periods of time. Oxidation in coal stockpiles starts at the external layers of the stockpile and progresses inwards. If the stockpile consists of coarse, uniformly sized coal, the heat generated is carried away by the circulating air and coal piles can survive for decades without igniting. However, when the coal size variation in a pile is wide and a large quantity of fines is present, size segregation may occur and internal pores may be blocked by fines thereby preventing through flow of air.

Also, the large surface area of the fines enhances oxidation. The heat generated by oxidation is trapped, temperature build-up occurs and spontaneous combustion can occur because of the steep temperature gradient between the pile surface and the core. The larger the stockpile, the greater is the risk of spontaneous combustion. However, for large stockpiles, even sub-critical heating can cause substantial internal oxidation and significant changes in its physical and chemical characteristics, which may lead to significant deterioration in coal quality, particularly the heating value, coking quality and financial value of the stored coal.

According to Matoney et al. (1979) spontaneous combustion occurs in five stages:

i)   Oxidation of fresh coal pile starts slowly from the surface and
        gradually penetrates the pile, generating heat.
ii)   As the temperature rises beyond about 50°C oxidation rate
        increases rapidly.
iii)   At a temperature of about 140°C water vapour and carbon
        dioxide are liberated.
iv)   Liberation of gases proceeds rapidly until the temperature rises
        to around 230°C when ignition can occur.
v)   If the temperature rises to around 350°C spontaneous
        combustion will almost certainly occur.

Banerjee (1985), summarized the incubation-to-spontaneous ignition processes as shown in Figure 7.14. According to him, bituminous coals can ignite at temperatures as low as 200°C while anthracite coal may ignite at around 400°C. The incubation period between exposure and ultimate spontaneous ignition may range from weeks to years depending on the nature of the coal, the pile geometry, and the surrounding environmental conditions. The mechanisms which feature at various stages of the incubation period prior to ignition are still subjects of intensive research and many predisposing factors have been identified. These include intrinsic factors (properties of the coal and structure of the coal pile) such as inherent moisture content, particle size, particle size distribution, pile dimensions, chemical, petrographic and mineral matter constituents, volatile matter content, heat capacity, heat of reaction. Extrinsic factors (environmental and storage conditions) such as environmental air humidity, oxygen concentration, atmospheric temperature and pressure and the presence of certain bacteria also play a prominent role in weathering to the extent that spontaneous ignition can occur.

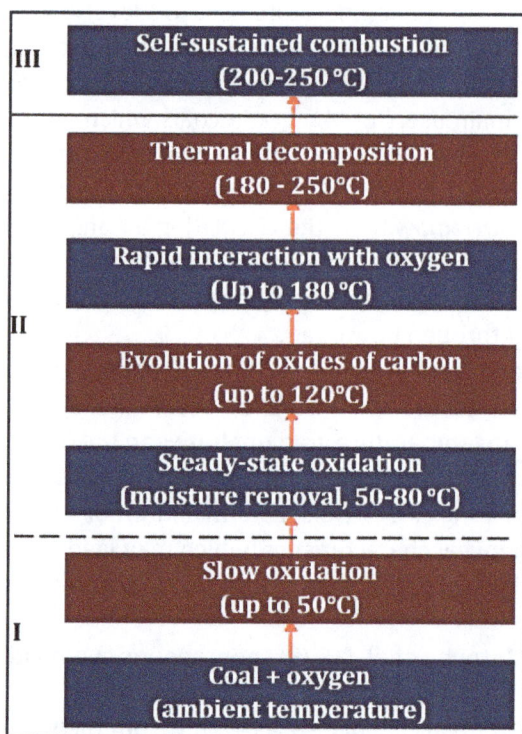

FIGURE 7.14 A model of sequential changes in the spontaneous combustion of coal. *(Banerjee, 1985).*

The self-heating of coal is due to a number of complex exothermic chemical reactions. Coal will continue to self-heat provided that there is a continuous air supply and the heat produced is not dissipated. The temperature of the coal increases due to self-heat until a plateau is reached, at which the temperature is temporarily stabilized. At this point, the heat generated by oxidation is used to vaporize the moisture in the coal. Once all the moisture has been vaporized, the temperature increases rapidly and the coal can self-ignite. On the other hand, dry material can readily ignite following the sorption of water.

In a recent study Avila (2012) identified two competing mechanisms in coal oxidation. The first stage of weathering is dominated by endothermic reactions in which oxygen molecules are adsorbed on the coal surface, forming coal-oxygen complexes, resulting in weight increases. However, at early stages (below 100°C), the temperature range in which this reaction occurs is masked by water desorption and available free reactive sites in the coal structure, that is, the initially available free active sites and those created by desorption of water molecules from the coal surface.

The second set of potential reactions occur at the higher temperatures and involve exothermic chemical reactions which generate the majority of the heat responsible for the overall temperature rise in the coal bed. The extent and temperature range over which this reaction occurs depend on the stability of the complex solids formed at the first stage. The main products of these reactions are carbon dioxide, carbon monoxide, hydrogen and hydrocarbons.

Avila concluded that the propensity of coal for spontaneous ignition depends on which of the two reactions governs the entire weathering process. In coals prone to self heating, the desorption of coal-oxygen complexes governs the entire weathering process. All coals have some tendency for spontaneous combustion and the oxidation mechanism is the same for all coals but the propensity varies with coal rank and type, and depends on the ease with which coal can absorb moisture and oxygen. High rank coals have low affinity for oxygen or moisture and therefore have low tendency to ignite spontaneously. For example, anthracite has very low predisposition to endogenous ignition.

On the other hand, low-rank and high volatile coals generally have high affinity for oxygen and moisture and therefore a high tendency for oxidation and spontaneous combustion. However, environmental factors and intrinsic variables can cause critical self-heating in all coals irrespective of rank. All ranks of coal including anthracite have been known to ignite spontaneously when stored under conditions which favour rapid oxidation. The oxygen: fixed carbon ratio appears to give a good indication of the propensity for self-ignition (Afonja, 1973). The higher the ratio, the greater the proneness to spontaneous combustion. The inorganic mineral matter content of coal such as pyrites can also contribute to oxidation during weathering by disintegrating and presenting new surface areas for oxidation, but the heat associated with the reactions involved is not considered to be significant except when the mineral content is high. When sulphur (in pyrite form) is present in significant quantities in finely divided form, it reacts with oxygen as shown in Equation 7.10.

$$2FeS_2 + 7O_2 + 16H_2O = 2H_2SO_4 + 2FeSO_4.7H_2O + 316 \text{ kcal} \qquad [7.10]$$

The volume of the products of the reaction is larger than the volume of the initial pyrite. This causes fractures around the pyrite which further expose new internal surfaces of the coal to oxidation. The contribution of pyrite oxidation to the total heat generated in weathering is estimated to be only about 10% (Wang, 2003; He and Wang 2006) and, in any case, the sulphur content of many coals is less than 1%. However coals containing 10-12% of finely divided sulphur have been known to ignite spontaneously due primarily to the high sulphur content. Sulphur content below 5% is not believed to contribute significantly to spontaneous ignition. Also, as discussed earlier, certain bacteria contribute to oxidation by digesting sulphur but, again the contribution to self-heating is considered negligible.

The absorption of moisture by coal is an exothermic reaction. However, the amount of heat liberated is a function of the inherent moisture of the coal and the ambient temperature to which it is exposed (Speight, 2015). The higher the inherent (equilibrium) moisture, the higher the heating tendency and an ambient temperature of about 20°C is considered the minimum for significant oxidation to occur. Dry coal will produce more heat when re-wetted than coal that already contains some moisture and this may contribute to increased oxidation and spontaneous combustion of the coal. Also, coal piles can survive winter storage but ignite spontaneously on a warm summer day.

7.5.2 Determination of self ignition propensity

The need for a reliable spontaneous ignition liability prediction method has been a subject of intensive research for over a hundred years, due to the numerous coal mine, coal storage and coal shipment fires in coal producing countries all over the world (see Afonja, 1973; Strichter, 2007). A number of techniques for assessing the susceptibility of coal to spontaneous combustion have been developed. These include the determination of crossing point temperature (CPT) (Ganguly et al., 1953; Banerjee et al. (1972); Didari et al., 2002; Ramlu, 2007), flamability temperature (Nimaje, 2010), wet oxidation potential (Tarafadar et al., 1989; Panigrahi and Ray, 2014), and thermogravimetry (Avila et al. 2014; Zhang et al., 2016). Comprehensive reviews of the most common methods have been carried out in recent years by Nelson and Chen (2007), Sen et al., (2009), and Novikov (2015).

Crossing Point Temperature (CPT) is the oldest, most widely used and simplest method, and involves the determination of the lowest temperature at which the internal temperature of the coal stockpile coincides with that of the environment (Figure 7.15). It is also called the relative ignition temperature and marks the beginning of accumulation of undissipated internal heat within the bed. The crossing point marks the beginning of rapid self heating of the coal sample due to internally generated heat of oxidation and ignition can occur at any time from the crossing point. Although this test is very widely used, it is not very reliable and can give very false results.

In many investigations, coals (in particular, high-moisture lignite) that are known to be highly prone to self ignition had high crossing point temperatures while low crossing points have been observed for coals known to be unreactive (Afonja, 1973; Avila, 2012). Clearly the CPT method does not take into account the effect of moisture evaporation which can occur over a wide temperature range and delay the onset of self heating, thereby giving an unrealistically high crossing point temperature for low-grade coals which usually have high moisture retention propensity. Another common and versatile method of evaluating spontaneous ignition propensity of coals involves the determination of the self-heating rate index. A dried coal sample is oxidized in an oxygen atmosphere under adiabatic conditions from 40 to 70°C (Figure 7.16). The average self heating rate (R_{70}) is calculated from the time taken for the temperature of the coal to rise between these two temperatures (Humphreys et al., 1981; Beamish et al., 2000, 2005, 2008). Heating curves for coals of different ranks are shown in Figures 7.17 and 7.18.

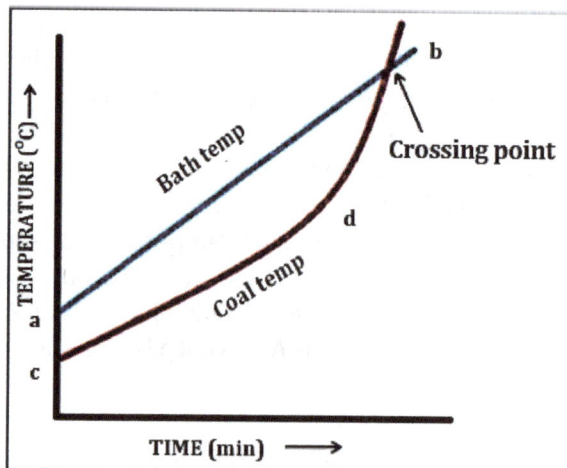

FIGURE 7.15 Crossing Point Temperature (CPT) for coal self-ignition, a-b = heating bath/furnace curve, c-d = coal sample temperature curve. *(Banerjee, 1985).*

The minimum self heating temperature (SHT) is the temperature at which there is a dramatic increase in the temperature rise rate of the coal sample. It marks the point at which moisture evaporation ceases and rapid self-heating begins, after which spontaneous self ignition can occur at any time. For most coals, the SHT is about 70°C and the self-heating rate index R_{70} is determined by calculating the rate of temperature rise between the starting temperature 40°C and the Self heating temperature 70°C. Using an extensive data base on coals of different ranks sourced from several countries, a significant relationship was established by Beamish and Arisoy (2008) between self heating rate and rank as determined from Suggate rank plot (Suggate 1998, 2000). (See Figure 7.19).

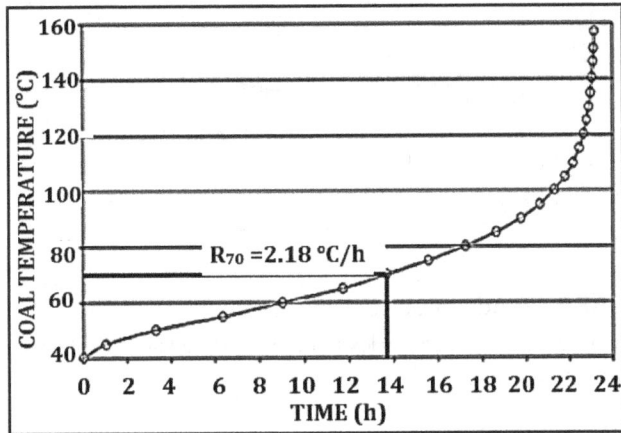

FIGURE 7.16 Determination of coal self heating rate index, R_{70} Index. *(Beamish and Arisoy, 2008).*

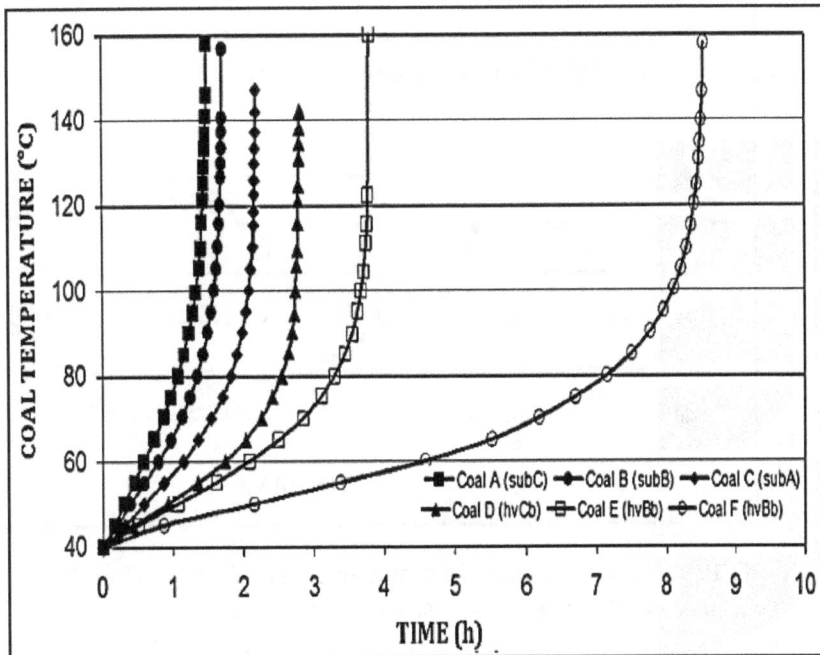

FIGURE 7.17 Adiabatic coal self-heating curves for low and medium rank coals. (Sub=sub-bituminous, b=bituminous; lv=low-volatile; mv=medium volatile; hv=high-volatile). *(Beamish and Arisoy, 2008).*

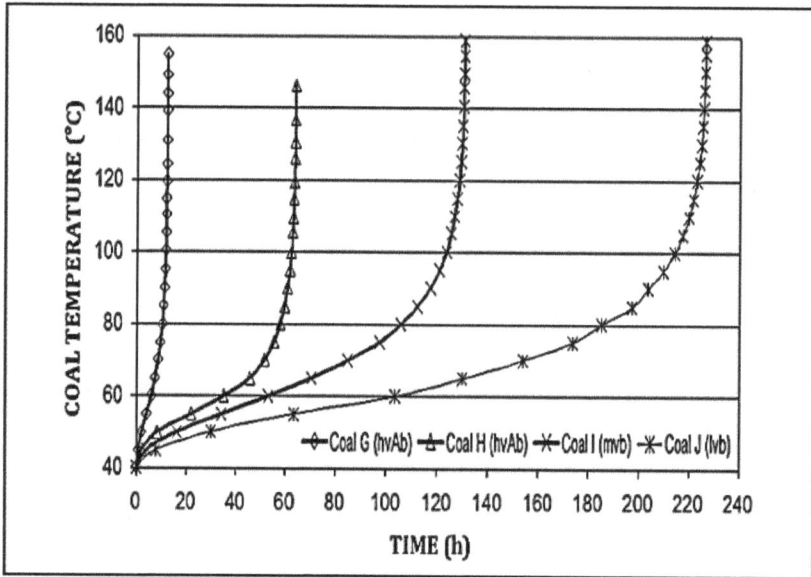

FIGURE 7.18 Adiabatic coal self-heating curves for high rank coals. (Sub=sub-bituminous, b=bituminous; lv=low-volatile; mv=medium volatile; hv=high-volatile). *(Beamish and Arisoy,2008).*

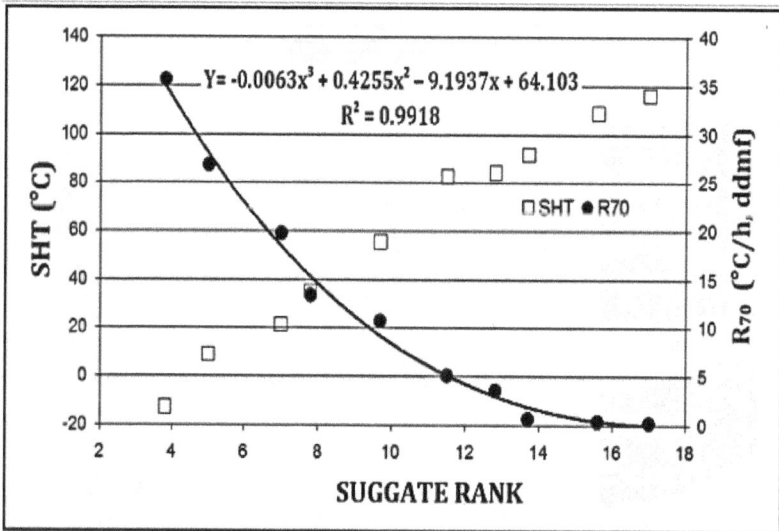

FIGURE 7.19 Relationship between coal rank and SHT and R_{70} values. *(Smith & Lazzara, 1987; Beamish and Arisoy, 2008).*

Coking coals were found to have lower self-heating rates than steam coals which is contrary to previous findings. This may be due to the higher vitrinite content of the coking coals compared with the steam coals which usually have lower vitrinite content but higher proportion of intertinite. In a more recent study, Beamish et al., (2015) found that coals of similar rank can have significantly different self-heating rates. Mineral matter in coal is a group of inorganic minerals present in coal which modify the coal behaviour in many coal utilization and conversion processes. They end up as ash in combustion processes. The effect of mineral matter on self-heating of coal depends on the type of mineral. Some minerals such as pyrites are involved in exothermic reactions which enhance self-heating, others such as calcium carbonate act as heat sink by using some of the heat generated by coal oxidation in endothermic reactions with oxygen, while others are involved in complex physico-chemical reactions which inhibit the access of oxygen to oxidation sites (Humphreys et al., 1981; Smith et al., 1988; Beamish and Arisoy, 2008).

Most of the methods in current use are empirical and involve very small samples tested in simulated environmental conditions. While results obtained from such tests may be useful in comparing the liability of different coals, they may not be applicable to real situations. However, large-scale experimental investigations of self heating of less than a tonne to over 3000 tonnes coal stockpiles are difficult and expensive and few results have been reported in the literature (Williams et al., 1983; Schmal, 1987; Kok et al., 1989; Romero and Miranda, 1991; Beamish and Hughes, 2009; Okten et al., 2009; Ozdeniz, 2010; Lee, 2012; Pak et al., 2015).

7.5.3 Prevention of spontaneous ignition

The most important variables that need to be controlled to prevent or minimize the risk of spontaneous combustion are humidity, aeration and particle size distribution. It is not possible to exclude air or moisture from a coal stockpile but storage conditions can be controlled such that exposure to moisture and strong wind is minimized and particle size variation and segregation in coal piles are reduced (de Korte, 2014). A mixture of coarse and fine coal particles will reduce aeration of coal piles while fine particles will increase the internal surface area and accelerate oxidation. Large, high-angle stockpiles will reduce the expulsion of heat generated within the coal pile and accelerate the build up of heat at the pile centre, thus increasing the propensity for self ignition. It is widely assumed that transport of oxygen and moisture into, and expulsion of internally generated heat from coal piles are controlled by diffusion and natural convection, both of which depend on differential temperature, pressure and humidity gradients between the center of the stockpile and the storage environment (Brooks and Glasser, 1986; Brooks et al., 1988, 1988; Bradshaw et al., 1991; Salinger et al., 1994). However, some studies have shown that wind-driven forced convection can greatly influence both processes and in fact may be the major factor which influences the initial heating of the coal pile (Krishnaswamy et al., 1996; Fierro et al., 1999).

Industrial-scale experimental investigations are difficult, expensive and few. Some studies on how to minimize weathering and the risk of spontaneous self-ignition in industrial coal stockpiles have been carried out over time. Methods evaluated include pile geometry, periodic compaction, storage in low-angle slope piles, storage in fenced and covered stockyards, and surface protection with ash-water slurries. Improper stockyard practices of filling and evacuation can also cause the production of fines and increase self-ignition propensity.

One of the most ambitious industrial-scale studies was carried by Fierro et al., (1999) on five large coal stockpiles of weight ranging from 2000 tonnes to over 3000 tonnes, each pile stored under different geometric, compaction and ventilation conditions over a period of up to 270 days (Figure 7.20). The pile covered with ash-water slurry (PCE) had the highest resistance to weathering with a loss in heating value of only 0.6% compared with coal piled in low angle slope (Pile C) which lost 14.2% and the unprotected pile which lost nearly 8%. Also, the average temperature of the wet slurry protected stockpile was below 40°C throughout the test period, compared with the unprotected coal which rose as high as 170°C.

A comprehensive review of the propensity of coal to self-heat and possibly self-ignite has been carried out by Nalbandian (2010). It included the methods of evaluating the propensity for coals to self-heat in stockpiles, during transit, storage in bunkers, as well as during utilization. The review also covered factors contributing to self heating and spontaneous ignition, methods for determining the conditions at which a coal pile could undergo spontaneous combustion, and prediction of safe storage time under a set of conditions. The review also discussed the significance of the greenhouse gas emissions (especially CO_2) resulting from the oxidation of coal during transportation, storage or utilization.

A simulation method developed recently by Kobe Steel, Japan involves prediction of the heat generation and dissipation inside an industrial coal pile from data generated in laboratory-scale tests (Pak et al., 2015). The simulation takes into account the chemical properties of the coal as well as extrinsic properties of the coal pile such as pile geometry and the environment. Using a 4m laboratory pile simulating an industrial stacking process, measurements were made of oxygen/carbon (O/C) ratio, particle size distribution, the evaporation/desorption behaviour of moisture in the coal, flow pattern of air in the pile, and pressure drop for each granule size distribution. The data was fed into the standard software package for unsteady-state fluid analysis (ANSYS Fluent). The result correlated very well with data obtained for a 15m pile.

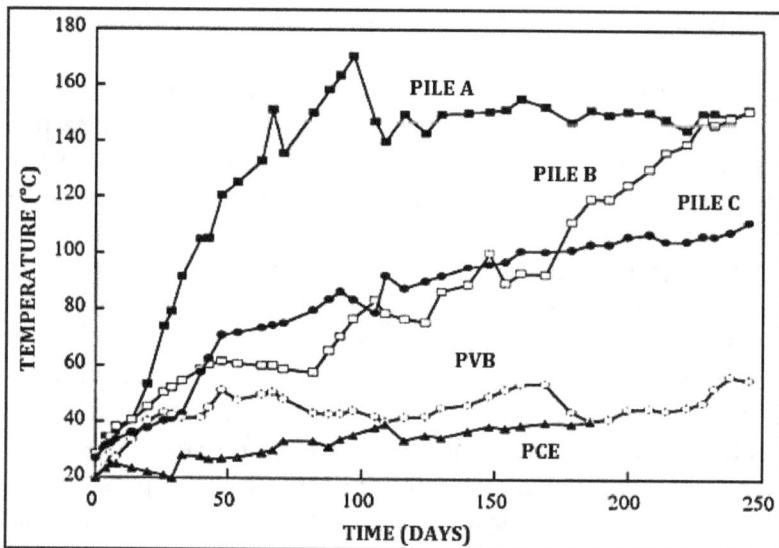

FIGURE 7.20 Variation of the average temperature with time for five industrial coal piles. PILE A = reference, no protection, PILE B = periodic compaction of slopes, PILE C = low angle WNW slope, PVB = wind screen, PCE = covered with fly ash-water slurry. *(Fierro et al., 1999).*

REFERENCES

Afonja, A. A. (1973). "An investigation of the spontaneous combustion propensity of Nigerian Coals." Research Report: IFE/CRL/NCC/7. Nigerian Coal Corporation.

Akgun, F and A. Arisoy (1994). "Effect of particle size on the spontaneous heating of a coal stockpile." Combustion and Flame, 99:137-146.

Arisoy, A. and Akgun. F. (2000). "Effect of Pile Height on Spontaneous Heating of Coal Stockpiles."Combustion Science and Technology, 153:1, pp. 157-168.

Arisoy, A., Beamish, B. B., and E. Cetegen (2006). "Modeling spontaneous combustion of coal."Turkish J. Eng. Env. Sci., Vol. 30, pp 193-201.

Alvarez, R.at al. (1998). "Weathering Study of an Industrial Coal Blend Used in Cokemaking." ISIJ. Vol. 38 No. 12, pp. 1332-1338.

Alvarez, R. et al. (2003). "Influence of coal weathering on coke quality." Ironmaking & Steelmakig Processes, Products and Applications, Volume 30, Issue 4, pp 307-312.

Avila, C. (2012). Predicting self-oxidation of coals and coal/biomass blends using thermal and optical methods. Ph.D. Thesis, University of Nottingham.

Avila, C., W. Tao and E. Lester (2014). "Estimating spontaneous combustion potential of coals using thermogravimetric analysis." Energy & Fuels (Impact Factor 2.79) 02/2014. DOI 10 1021/ef4021191.

Axelson, D. E., R. J. Mikula, and V. A. Munoz (1987). "Characterization of coal oxidation in-situ and on a stockpile." In J. A. Moulijn, K. A. Nater, and H. A. G. Chermin, Eds., International conference on coal science: Elsevier, pp. 419-422.

Banerjee, S. C., Nandy, D. K., Banerjee, D. K., and R. N. Chakravorty (1972). "Classification of coal with respect to their susceptibility to spontaneous combustion." Transactions of the Mining and Metallurgical Institute of Indi., Vol. 59, No. 2, pp. 15-31.

Banerjee, S. C. (1985). Spontaneous Combustion of Coal and Mine Fires. Oxford & IBH Publishing Co. Pvt. Ltd., 1st Ed.. pp. 1-38.

Battle, T., Srivastava, U., Kopfle, J., Hunter, R., NS j. Mclelland (2014). "The Direct Reduction of Iron." In Treatise on Process Metallurgy: Industrial Processes Part A. S. Seetharaman (Ed.), pp. 89-176. Elsevier.

Beafore, F. J., K. E. Cawiezel, and C. T. Montgomery (1984). "Oxidized coal": what it is and how it affects your preparation plant performance." Journal of Coal Quality, Vol. 3, pp. 17-23.

Beamish, B. B. and A. Arisoy (2008). "Effect of mineral matter on coal self-heating rate." Fuel, Vol. 87, pp. 125-130.

Beamish, B. B. and R. Hughes (2009). "Comparison of laboratory bulk coal spontaneous combustion testing and site experience – A case study from Spring Creek mine." Coal Operators' Conf., The AusMM Mawarra Branch.

Beamish, B. B. Modher, A., Barakat, J. D. & J. D. St George (2000). "Adiabatic testing procedures for determining the self-heating propensity of coal and sample ageing effects." Thermochimica Acta, 362, pp. 79-87.

Beamish, B. B. and Blazak, D. G. (2005a). "Relationship between ash content and R_{70} self-heating rate of Callide coal." Int. Jnl. Coal Geol., Vol. 64 (1-2), pp 126-132.

Beamish, B. B. Blazak, D. G., Hogarth, L. C. S., and I. Jabouri, (2005b). "R_{70} relationships and their interpretation at a mine site." Coal2005 Conference, Brisbane, QLD., 362, pp. 79-87.

Beamish, B. B. and W. Sainsbury, (2008). "Development of a site specific self-heating rate prediction equation for a high volatile bituminous coal." *Proc. 8th Coal Operators' Conf., Ed. N. Aziz*, pp. 161-165.

Beamish, B. B. and Theiler, J. (2015). "Contrast in self-heating rate behaviour for coals of similar rank." *2015 Coal Operators' Conference, University of Wollongong*, pp. 300-304.

Beier, E. (1994). " Long-term studies on the oxidation of coal and other substances in air." *Fuel*. Vol. 73, Issue 7, pp. 1179-1183.

Bend, S., Edwards, I. and Marsh, H (1989). "The effects of Oxidation and Weathering on Coal Combustion." *Am. Chem. Soc., Div. Fuel Chem.*, 34(3), 923.

Bend, S. L., I. A. S. Edwards, and H. Marsh (1991), "Effects of oxidation and weathering on char formation and coal combustion." In H. H. Schobert, K. D. Bartle, and L. J. Lynch, Eds., *Coal science II*: American Chemical Society Symposium Series 461, pp. 284-298.

Bhomwick, J. N., Mukherjee, P. N. and A. Lahiri (1959). "Studies on Oxidation of Coal in Low Temperatures." *Fuel*, Vol. 38, p. 469.

Bradshaw, S., Glasser, D., and K. Brooks. "Self-ignition and convection patterns in an infinite coal." *Chem. Eng. Commun.*, 105, p. 255.

Brooks, J. D. and T. P. Maher (1957). "Acidic oxygen-containing groups I coal." *Fuel*, Vol. 36, pp. 51-62.

Brooks, K. and Glasser, D. (1986). "A simplified model of spontaneous combustion in coal stockpiles." *Fuel*, 65, pp. 1035-1042.

Brooks, K., Balakotaiah, V., and D. Luss (1988a). "Effect of natural convection on spontaneous combustion of coal stockpiles. " *AIChE J.* 34 (3), p. 353.

Brooks, K., Bradshaw, S. and D. Glasser (1988b). "Spontaneous combustion of coal piles. A. unusual chemical reaction engineering problem." *Chem. Eng. Sci.*, Vol. 43,   p. 1239.

Butakova, V. I. (2013). " Initial stage in thee low-temperature oxidation of coal in air." *Coal and Chemistry*, Vol. 56, No.7, pp. 225-234.

Carras, J. N., Day, S. J., Saghafi, A., & D. J. Williams (2009). "Greenhouse gas emissions from low-temperature oxidation and spontaneous combustion at open-cut coal mines in Australia." *International Journal of Coal Geology*, Vol. 78, pp. 161-168.

Casal, M. D. Et al., (2003). "Thermal behaviour and kinetics of coal/biomass blends during co-combustion."*Fuel Process Technol.*, Vol. 84, p. 47

Chatterjee, R. S. (2006). " Coal fire mapping from satellite thermal IR data – a case example in Jhaaria Coalfield, Jharkhand, India." *ISPRS Journal of Photogrammetry and Remote Sensing*, Vol. 60, pp. 113-128.

Cimadevilla, J. L. G., Alvarez, R. and Pis, J. J., (2003). "Photoacoustic FT-IR study of weathered stockpiled coking coals." *Vibrational Spectroscopy*. Vol. 31, pp. 133-141.

Cimadevilla, J. L. G., Alvarez, R. and Pis, J. J., (2005a). "Influence of coal forced oxidation on technological properties of cokes produced at laboratory scale." *Fuel*. Vol. 87, pp. 1-10.

Cimadevilla, J. L. G., Alvarez, R. and Pis, J. J., (2005b). "Effect of coal weathering on technological properties of cokes produced at different scales." *Fuel*. Vol. 86, pp. 809-830.

Clemens, A. H., T. W. Matheson, and D. E. Rogers, (1991) "Low-temperature oxidation studies of dried New Zealand coals." *Fuel*, Vol. 70, pp. 215-221.

COALTECH (2016). *Prevention and Control of Spontaneous Combustion. Best Practice Guidelines for Surface Coal Mines in South Africa*. COALTECH Research Association. www.coaltech.co.za. (Accessed 8/8/2016).

Crelling, J. C., R. H. Schrader, and L. G. Benedict, (1979). "Effects of weathered coal on coking properties and coke quality" *Fuel*, Vol. 58, pp. 542-546.

De Korte, G. J. (2001). "Beneficiation of weathered coal." COALTECH 2020. Division of Mining Technology, CSIR, Pretoria, South Africa.

De Korte, J. (2014a). "Spontaneous combustion (self ignition)." *Fossil Fuel Foundation Symposium on Spontaneous Combustion.* Council for Scientific and Industrial Research, (CSIR), South Africa. www.fossilfuel.co.za.

De Korte, J. (2014b). "Managing spontaneous combustion of coal." *Fossil Fuel Foundation Symposium on Spontaneous Combustion.* Council for Scientific and Industrial Research, (CSIR), South Africa. www.fossilfuel.co.za.

De Korte, J. (2014c). "Tests to determine the spontaneous combustion propensity of coal*." Fossil Fuel Foundation Symposium on Spontaneous Combustion.* Council for Scientific and Industrial Research, (CSIR), South Africa. www.fossilfuel.co.za.

De Korte, J. (2016). "Methods for estimating the spontaneous combustion propensity of coal.*" Proc. XVIII International Coal Preparation Congress, St. Petersburg, Russia. Icpc-2016.com.*

Desna, N. A. and D. V. Miroshnichenko (2011). "Oxidized coal in coking: A review." *Coal, Coke and Chemistry,* Vol. 54, p. 139.

Didari, V., and E. Kaymakci (2002). "Relations between coal properties and spontaneous combustion parameters." *Turkish Jnl. Of Eng. & Env. Sc.* Vol. *26,* pp. 59-64.

Dryden, I. G. G. (1963). "Chemical constitution and reactions of coal." In *Chemistry of Coal Utilization.* Ed. H. H. Lowry, John Wiley and Sons, NY, pp. 232-295.

FFF (2014). "Spontaneous combustion – Part 1: Causes and Controls." *Fossil Fuel Foundation.* www.fossilfuel.co.za.

Fierro, V. (1999). " Prevention of spontaneous combustion in coal stockpiles." *Fuel Processing Technology.* Vol. 59 pp. 23-34.

Fredericks, P. M. Warrooke, P., and . A. Wilson (1983). "Chemical changes during natural oxidation of a high volatile bituminous coal." *Org. Geochem. ,* Vol. 5, No. 3, pp. 89-97.

Fuerstenau, D. W., Yang, G. C. C., and J. S. Laskowski (1987). "Oxidation phenomena in coal flotation, Part 1. Correlation between oxygen functional group concentration, immersion wettability and salt flotation response." *Coal Preparation,* Vol. 4, pp. 161-182.

Fuerstenau, D. W. and J. Dao (1992). "Characterization of coal oxidation and coal wetting behaviour by film flotation." *Coal Preparation,* Vol. 10, pp. 1-17.

Furimsky. E. MacPhee, J. A., Vancea, L. Ciavaglia, L. A. and B. N. Nandi (1983). *Fuel.* Vol. 62, p. 395.

Francis, W. and R. V. Wheeler (1926). "The spontaneous combustion of coal; The most readily oxidizable constituents of coal." *Safety I Mines Research Board.* Paper 28, 1926, 51 pp.

Francis, W. and H. M. Morris (1931). "Relationship between oxidability and composition of coal." *Bureau of Mines, US,* Bulletin p. 340.

Fryer, J. F., and A. J. Szladow (1973). "Storage of coal samples." *Info. Series ,* No. 66, Alberta Research Council.

Gala, H. B., Srivastava, R. D., Ree, K. H., and R. E. Hucko (1989). "Chemical Aspects of Coal Beneficiation." *Special Issue of Coal Preparation,* Vol. 7, pp. 1-28.

Ganguly, M. L., and N. G. Banerjee (1953). "Critical Oxidation and Ignition Temperature of Coal." *The Imma Review,* Vol. 2, No. 4.

Gerus-Piasecka, I., and Jasienko, S. (1989). *Koks Smola Gaz,* V. 34, p. 31.

Giroux, L. Kolijn, C. J., & F. S. Pichler (2006). "Storage of small samples of coking coal for thermal rheological tests." *Fuel Processing Technology,* Vol. 87, p. 547.

Given, P. (1984). " An essay on the organic geochemistry of coal." In Coal Science, Vol. 3. Ed. Gorbaty, M. L., Larsen, J. W. and Wender, I. Pp. 63-252. Academic Press.

Gray, R. J., Rhoades, A. H. and D. T. King, (1976) "Detection of oxidized coal and the effect of oxidation on the technological properties." *Transactions of SME,* Vol. 260, pp. 334-341.

Gray, R. J. and D. E. Lowenhaupt (1989). " Ageing and Weathering." *Sample Selection. Aging and Reactivity of Coal,* p. 255. R. Klein and Wellek, Eds. John Wiley and Sons, New York.

Green, U., Aizenshtat, Z., Ruthstein, S. and H. Cohen (2012). "Stable radicals formation in coals undergoing weathering: effect of coal rank." *Phys Chem Phys,* 14(37): 130, pp. 46-52.

Guney, M. and D. J. Hodges (1969). "Spontaneous heating of coal." *Colliery Guardian,* pp. 105-109, 173-177.

Guyot, F. and F. Pollard (1974). " Effects of oxidation on coal properties related to industrial use." *Australian Coal Industry Research Laboratories.,* V. 74, Issue 18 of P. R.

Hamza, H. A., Michaelian, K. H. and N. E. Andersen (1983). " A fundamental approach to beneficiation of fine oxidized coals." *Proc. International Conference on Coal Science. Center for Conference Management, Pittsburgh,* pp. 248-255.

He, Q. L. and D. M. Wang (2006). "Kinetics of oxidation and thermal degradation reaction of coal." *Jnl. of Sc. & Tech., Beijing,* Vol. 28, No. 1, pp. 1-4.

Heffern, E. L. and Coates, D. A. (2004). " Geologic history of natural coal-bed fires." *Int. Jnl. Of Coal Geol..,* Vol. 59, pp. 25-47.

Herring, J. R and F. J. Rich (1983). "Spontaneous coal combustion; mechanisms and prediction. *Proc. Int. Conf. on Coal Sc.,* Pittsburgh, Pa., International Energy Agency, pp. 753-756.

Hoffmann, H, and K. Hoehne (1954). Brennstoff-Chemie, Vol. 35, pp 202 and 302.

Hofrichter, M. and Fakoussa, R. N. (2001). "Microbial Degradation and Modification of Coal." *In Biopolymers,* Volume 1, Lignin, Humic Substances and Coal. Martin Hofrichter (Editor), pp. 393-407, Wiley-Blackwell.

Huffman, G. P. et al. (1985). "Comparative sensitivity of various analytical techniques to the low-temperature oxidation of coal." In Biopolymers, Volume 1, Lignin, *Fuel*), Vol. 64, pp. 849-856.

Huggins, F. E., Huffman , G. P. Kosmack, D. A., and D. E. Lowenhaupt (1980). "Mossbauer detection of goethite in coal and its potential as an indicator of coal oxidation." *International Journal of Coal Geology,* Vol. 1, pp. 75-81.

Huggins, F. E., Huffmann, C. P., and R. J. Lee (1982). *Coal and Coal Products: Analytical Characterization Techniques.* E. L. Fuller, Jr., Ed. ACS Symp. Series No. 205, Vol. 1, p. 239.

Huggins, F. E. and C. P. Huffmann (1983). *Int. Jnl. Coal Geol.,* Vol. 3, p. 157.

Huggins F. E., Dumnyre, G. R., Lin, M. C., and G. P. Huffmann (1985). "Storage of bituminous coal under argon." *Fuel,* Vol. 64, pp. 348-350.

Huggins, F.E., G. P. Huffman, G. R. Dunmyre, M.J . Nardozzi, and M. C. Lin (1987) "Low temperature oxidation of bituminous coals: its detection and effect on coal conversion."*Fuel Processing Technology,* V. 15, pp. 233-244.

Huggins, F. E., and G. P. Huffman, (1989). "Coal weathering and oxidation: the early stages, in C. R. Nelson, ed., Chemistry of coal weathering." *Coal Science and Technology,* v. 14, pp. 33-60. New York, Elsevier Science Publishers.

Humphreys, D., Rowlands, D. and J. F. Cudmore (1981). "Spontaneous combustion of some Queensland coals." *Proc. Ignitions, Explosions & Fires in Coal Mines Symposium,* pp. 5-1 – 5-19. AusIMM, Illawarra Branch, Melbourne, Australia.

IAEA (2013). "Gaseous emissions from coal stockpiles." International Atomic Energy Agency Clean Coal Centre, PF 13-1.

IEA (2007). *Tracking Industrial Energy Efficiency and CO_2 Emissions.* http:/www.iea.org/w/bookshop/add.aspx?id=298.

IEA (2009). *Technology transitions for industry – Strategies for the Next Industrial Revolution.* http://www.iea.org/w/bookshop.

Iglesias, M. J., Puente, G., Fuente, E., and J. J. Pis (1998). "Compositional and structural changes during aerial oxidation of coal and their relations with technological properties." *Vibrational Spectroscopy.* Vol. 17, pp. 41-52.

Ignasiak, B. S., Clugston, D. M. and D. S. Montgomery (1972). *Fuel,* Vol. *51,* p. 72.

Ingram, G. R., and J. D. Rimstidt (1984). "Natural weathering of coal." *Fuel,* V. 63, pp. 292-296.

IIP (2016). "Direct Reduced Iron. Industrial Efficiency Technology Database, Institute for Industrial Productivity." iipnetwork.org/content/direct-reduced-iron. Accessed 2016.

Itay, M., C. R. Hill and D. Glasser (1989). "A study of the low temperature oxidation of coal." Fuel Processing Tech., Vol. 21, pp. 81-97.

Jha, P. K., Das, T. K., and A. B. Soni (2014). "Effect of weathering on coal quality." *DOI: 10.7763/IPEDR,* V. 75, p. 31.

Jensen, E. J., Melnyk, N. and Berkowitz, N (1966). *Advances in Chemistry,* Series No. 55, American Chemical Society, Washington. DC, p. 621.

Jones, R. E. and D. T. A. Townsend (1949). "The oxidation of coal." *Journal of the Society of Chemical Industry,* Vol. 68, Issue 7, pp. 197-201. Published online: 21 April, 2010.

Joseph, J. T. and O. P. Mahajan (1991). "Effect of Air Oxidation on Aliphatic Structure of Coal.) *Coal Science II,* Chapter 23, pp. 299-312.

Kalemma. W. S. and G. R. Gavalas (1987). *Fuel,* Vol. 63, p. 640.

Keating J. C., Mahoney M. R., Le Bas A. D., Mcguine S. J., & Sakurovs, R. (2004). "Coal fluidity, Stockpiling and coke quality." *2nd International Meeting on ironmaking,* Victoria, BR. Vol. 2, Sep 12-15, 2004, pp. 761-770.

Kelekek, S. Salman, T. and G. W. Smith (19820). *Proc. 64th CIC Coal Symp.,* p. 145.

Kimpel, R. R. (1992). "Some results of various new chemical reagents for modifying flotation performance." *Coal Preparation,* Vol. 10, pp. 159-175.

Kohno, T. (1988). "Substitution and reduction of coking coal in cokemaking." *Coal and Coke Technical Exchange Session, International Iron and Steel Institute Committee on Technology, Brussels,* April 1988., pp. 85-101.

Kok, A., Bolt, N. and J. H. N. Jelgersma (1989). "Spontaneous heating and stockpiling losses of coal." Kema Sc. & Tech., Rep. 7 (2), p. 63.

Komaki, I, Iagaki, S. and T. Miura (2005*). Structure and Thermoplasticity of Coal.* New York, Nova Science Publishers Inc.

Krishnaswamy, S., R. D. Gunn, and P. K. Agarwal (1996). "Low-temperature oxidation of coal. 2. An experimental and modeling investigation using a fixed-bed isothermal flow reactor." *Fuel,* Vol. 75, pp. 344-352.

Labuschange, B. C. J. and J. G. Van der Watt (1983). *Intl. Conf. Coal Sci.,* P. 264.

La Grange, C. C. (1950). "Spontaneous heating of coal – summarized extracts from the literature." *Report No. 9 of 1950, Fuel Research Institute of South Africa.*

La Grange, C. C. (1951). "Storage of coal with special reference to quality deterioration and spontaneous combustion." *Report No. 28 of 1951, Fuel Research Institute of South Africa.*

Larsen, J. W., Lee, D., Schmidt, T., and A. Grint (1986). *Fuel,* Vol. 65, p. 595.

Lee, S. M. (2012). "Effect of Weathering on Coal Characteristics and Coke Qualities." unsworks.unsw.edu.au.

Leeder, W. R. et al., (1979). "Prediction of coke quality with reference to Canadian coals." *ISS-AIME Proc.,* Vol. 38, pp 385-397.

Leeder, R., Kolijn, C., Giroux, L., and T. Todoschuck (2011). "Deterioration of coking quality in samples and stockpiles." *Proceedings, METEC InSteelCon 2011, Dusseldorf, Germany.*

Loison, R., Foch, P. and A. Boyer (1989*). Coke quality and Production.* Butterworth, London, pp. 178-189.

Lowenhaupt D. E, and R. J. Gray (1980). "The alkali extraction test as a reliable method of detecting oxidized metallurgical coal*." International Journal of Coal Geology,* Vol. 1, pp. 63-73.

Lowenhaupt D. E, Griffiths P R, Fuller M P and I. P. Hamadeh (1982). "Spectroscopic study on the chemistry of coal oxidation." *Proc 41st Ironmaking Conf.* Vol. 41, p. 39.

Lowry, H. H. (Ed.) (1963). Chemistry of Coal Utilization, Supplementary Volume. John Wiley.

Mackowski, M. (1977). "Prediction methods I coal and coke microscopy." Journal of Microscopy, 109(1), pp. 119-137.

Mahajan, O. P., M. Komatsu and P. L. Walker (1980). "Low-temperature air oxidation of caking coals. 1. Effect of subsequent reactivity of chars produced." *Fuel,* Vol. 59, pp. 3-10.

Mahadevan, V. and M. A. Ramlu (1985). "Fire risk rating of coal mines due to spontaneous heating." *Jnl. Mines, Metals & Fuels,* pp. 357-362.

Marchioni, D. L. (1983). "The detection of weathering in coal by petrographic, rheologic and chemical methods." *Int. J. Coal Geol.,* Vol. 2, pp. 231-259.

Matoney, J. P., Bucklen, O. B., and P. G. Meikle (1979). "Coal storage and loading." *Coal Preparation, 4th Ed. American Institute of Mining, Metallurgical and Petroleum Engineers.*

Menendez, J. A., Alvarez, R. and J. J. Pis (1999). "Determination of metallurgical coke reactivity at INCAR: NSC and INCAR reactivity tests.*"* Ironmaking & Steelmaking, Vol. 26, pp. 117-121.

Midgalski, H. (1956). "Spontaneous ignition of anthracites." *Bergbautech.,* Vol. 6, pp 364-370; Chem. Abstr., 1956, Vol. 50, Col. 17381e.

Miroshnichenko, D. V. (2012). "Kinetic characteristics of coal oxidation." *Coke Chem.,* Vol. 55, no. 3, pp. 87-96.

Miyazu, T., Okuyama, Y., Sugimura, H. and K. Kumagai (1970*). J. Fuel Soc., Japan*, Vol. 49, p. 736.

Moroeng, O. M. (2015*). Spontaneous combustion of coal – South African Perspective.* M.Sc. Thesis, University of Pretoria, South Africa.

Murata, T. (1981). "Wettability of coal estimated from the contact angle." *Fuel*, Vol. 60, pp. 744-746.

Munzer, H. and W. Peters (1965). "Zur Kinetic der Kohloxidation im Temperaturbebeich 30-100°C." *Brenstoff-Chemie*. No. 12. Band 46.

Muzner, H. and Peters (1966). "Selbstentzundlichkeitsverhalten von Steinkohlen." *Erdol und Kohle-Erdgas-Petrochemie, XIX Jahrgang, No. 6.*

Nakamura, N., Togino, Y., and T. Tateoka (1977). "Behaviour of coke in large blast furnace." *Coal, Coke and Blast Furnace*. The Metals Society, London.

Nalbandian, H. (2010). "Propensity of coal to self-heat." *International Energy Agency Coal Research,* 2010-Clean coal technologies, 47p.

Neavel, R. C. (1982). " Coal Plasticity Mechanism: Inferences from liquefaction studies." In *Coal Science*. Vol. 1. Eds. M. L. Gorbaty, J. W. Larsen and L. Wender. Academic Press, New York.

Nelson, C. R. (1989). "Coal weathering: chemical processes and pathways." In C. R. Nelson, ed., *Chemistry of coal weathering*: New York, Elsevier Science Publishers, Coal Science and Technology, Vol. 14, pp. 1-32.

Nelson, M. I., and Chen, X. D., (2007). "Survey of experimental work on the self-heating and spontaneous combustion of coal." *In Stracher, G. B., ed.,Geology of Coal Fires: Studies from Around the World. Geological Society of America.*

Nolta, A. M. and D. H. Vice (2004). "Looking back at the Centralia coal fire: a synopsis of its present status." *Int. Jnl. Coal Geol.,* Vol. 59, pp. 99-106.

Novikov, E. A., Dobryakova, N. N., Shkuratnik, V. L., and S. A. Epshtein (2015). "Methods of coal oxidation estimation." *Gornyi Zhurnal (Mining Journal).,*. DOI 10. 17580/gzh.2015.05.06.

Ogunsola, O. I. and Mikula, R. J. (1992). "Thermal upgrading effect on the spontaneous combustion characteristics of Western Canadian low rank coals." *Fuel,* Vol. 71, pp. 3-8.

Okten, G. O. Kural and E. Algutkaplan (2009). "Storage of Coal: Problems and Precautions." In Energy Storage Systems – Vol. II pp. 172-186. Ed. Yalcin Gogus. UNESCO-EOLSS.

Ota, H., Takarada, T., Maeno, Y. and Nishimura, M. (2003). "Oxidation behaviour and caking property of coal under the atmosphere at low oxygen concentration." *ISIJ*, Vol. 43, No. 3, pp. 331-337.

Ozdeniz, H. (2009). "Statistical Modeling of Spontaneous Combustion in Industrial-Scale Coal Stockpiles." *Energy Sources, Part A: Recovery, Utilization and Environmental Effects*, Vol. 31, Issue 15, pp. 1368-1375.

Pak, H., Tada, T., Kikuchi, N., Shigehisa, T. and T. Higuchi (2015). "Evaluation of spontaneous combustion in stockpile of sub-bituminous coal." *Energy Sources*, Part A: Recovery, Utilization, and Environmental Effects, 32:7, 665-673, DOI: 10.1080/15567030802606129.

Panigrahi, D. C. and S. K. Ray (2014). "Assessment of self-heating susceptibility of indian coal seams – a neutral network approach." Arch. *Min. Sci.*, Vol. 59, No. 4, pp. 1061-1076.

Parr, S. and W. Wheeler (1908). "The deterioration of coal." *Journal of American Chemical Society*, Vol., 30, pp. 1027-1033.

Parr, S. and F. Kressman (1911). "The spontaneous combustion of coal." *Journal of Industrial Engineering Chemistry,* pp. 152-158.

Pisupati, S. V., A. W. Scaroni and P. G. Hatcher (1993a), " Devolatilization behaviour of naturally weathered and laboratory oxidised bituminous coals." *Fuel,* Vol. 72, pp. 165-173.

Pisupati, S. V., and A. W. Scaroni, (1993b), " Natural weathering and laboratory oxidation of bituminous coals: organic and inorganic structural changes." *Fuel,* Vol. 72, pp. 531-542.

Pisupati, S. V., A. W. Scaroni and R. D. Stoessner (1999). *Fuel Processing Tech.,*Vol. 28, p. 49.

Pone, J. D. N. Et al., (2007). "The spontaneous combustion of coal and its by-products in the Witbank and Sasolburg coalfields of South Africa." *J. Int. Jnl. Coal Geol.,* V. 72, pp. 124-140.

Prauuenyong, P. (2002). "Coal biodesulfurization processes." *Songklanakarin J. Sci. Technology,* 2002, 24(3): pp. 493-507.

Ramlu, M. A. (2007). *Mine disasters and mine rescue.* Oxford and IBH Publishing Co., New Delhi.)

Romero, C. and J. L. Miranda (1991). "Oxidacion y autocombustion de carbones en parque." 5° Informe, Proyecto 5th Report ECSC 7220-EA/752.

Sadowski, Z. at al. (1988). "Behaviour of oxidized coal during oil agglomeration." *Coal Preparation,* Vol. 6, pp. 17-34.

Salinger, A. G., Aris, R. & J. J. Derby (1994). "Modeling of the spontaneous ignition of coal stockpiles." *AIChE J.* 40 (6), p. 991.

Sarikaya, M. and G. Ozbayoglu (1995). "Flotation characteristics of oxidized coal." *Fuel ,* Vol. 74, No. 2, pp. 291-294.

Savage, W. H. D. (1951). "Weathering tests on coking coals." Report No. 2 of 1951, Fuel Research Institute of South Africa.

Seki, H., Ito, O. and M. Lino (1990). "Effect of mild oxidation of bituminous coals on caking properties." *Fuel,* Vol. 69, pp. 317-321.

Sen, R., Srivastava, S. K., and M. M. Singh (2009). "Aerial oxidation of coal – analytical methods, instrumental techniques and test methods: A survey." *India Jnl. of Chem. Tech.,* Vol. 16, pp. 103-135.

Schmal, D. (1987). *A model for the spontaneous heating of stored coal.* Ph.D. Thesis, University of Delft.

Schmidt, L. D. and J. L. Elder (1940). "Atmospheric oxidation of coal at moderate temperatures." *Ind. Eng. Chem.,* Vol. 32 (2), pp. 249-256.

Schmidt, L. D.(1945). In Chemistry of Coal Utilization, Vol. 1, p 627. H. H. Lowry, ed., John Wiley. *Ind. Eng. Chem.,* Vol. 32 (2), pp. 249-256.

Sheail, J. (2005). " Burning bings: a study of pollution management in mid-twentieth century Britain." *Journal of Historical Geography,* Vol. 31, pp. 134-148.

Singh, R. V., K. (2013). "Spontaneous heating and fire in coal mines." *Procedia Engineering,* Vol. 62, pp. 78-90. Elsevier.

Sloss, L. (2015). "Assessing and managing spontaneous combustion of coal." International Energy Agency Clean Coal Centre Rep. No. CCC/259.

Smith, A. C. and C. P. Lazzara (1987). "Inhibition of spontaneous combustion of coal." US Bureau of Mines Report RI 9196.

Smith, A. C. Miron, A. C. and P. Lazzara (1988). "Spontaneous combustion studies of US coals." US Bureau of Mines Report RI 9079.

Speight, James G. (2012). *The Chemistry and Technology of Coal.* CRC Press

Speight, J. G. (2013). *Coal-fired Power Generation.* John Wiley.

Speight, J. G. (2015). *Handbook of Coal Analysis.* John Wiley.

Stracher, G. B. (2004). "Coal fires burning around the world: a global catastrophe." *J. Chem. Int Jnl Geol.,* Vol. 59, pp. 1-6.

Stracher, G. B. and T. P. Taylor, (2004). " Coal fires burning out of control around the world: thermodynamic recipe for environmental catastrophe." *J. Chem. Int Jnl Geol.,* Vol. 59, pp. 7-17.

Stracher, G. B. (2007). "Geology of Coal Fires: Studies from Around the World." *Reviews in Engineering Geology,* V. XVIII, doi: 10.1130/2007.4118(06).

Stracher, G. B., Prakash, A., Sokol, E. V. (Eds.) (2011). *Coal and Peat Fires: A Global Perspective.* Vols. 1 & 2. Elsevier.

Stracher, G. B., A. Prakash and E. V. Sokol (Eds.), (2015). *Coal and Peat fires: A Global Perspective.* Vol. 3: Case Studies – Coal Fires. Elsevier.

Suggate, R. P. (1998). "Analytical variation in Australian coals related to coal rank and type." *Int. Jnl. Of Coal. Geol.,* 37:179-206.

Suggate, R. P. (2000). "The Rank (Sr)). scale: its basis and its application as a maturity index for all coals." *New Zealand Journal of Geology and Geophysics,* Vol. 43, pp. 521-553.

Swann, P. D. and D. G. Evans (1979) "Low temperature oxidation of brown coal: reaction with molecular oxygen at temperatures close to ambient." *Fuel,* Vol. 58, pp. 276-280.

Tarafdar, M. N. and D. Guha (1989). "Application of wet oxidation processes for the assessment of the spontaneous heating of coal." *Fuel,* Vol. 68, pp. 315-317.

Taweel, A., Farag, A. M., Kwak, J., Hamza, H. A. and M. Falk (1982). *Proc. 64th CIC Coal Symp.,* p. 125.

TSOP (2013*).* "Coal Weathering and Oxidation: Selected References*. The Society for Organic Petrology.* www.tsop.org.

Ulanovskii, M. L. (2012). " Kinetic research on coal oxidation: A review." *Coke Chem.,* vol. 55, no. 7, pp. 256–260.

UNESCO (2005). *Spontaneous Coal Seam Fires: Mitigating a Global Disaster.* Proc. Int Conf, Beijing, China.

van Doornum, G. A. W. (1954). "The spontaneous heating of coal."*J. Inst. Fuel,* Vol. 27, pp. 482-495.

van Krevelen, D. W. and Schuyer, J. (1957) *Coal science.* Elsevier publishing company.

van Krevelen, D. W. (1961; 1963). *Coal.* New York, Elsevier.

van Krevelen, D. W. (1981). Chemistry of coal weathering and spontaneous combustion, in Coal: typology - chemistry - physics - constitution: New York, Elsevier Scientific Publishing Company, pp. 238-262.

Wachowska, H. and W. Pawlak (1977). "Effect of coal oxidation on molecular weight of low-temperature tar." *Fuel,* Vol. 56, p. 342.

Wang, H., B. Z. Dlugorgoski and E. M. Kennedy (2003)." Coal oxidation at low temperature: oxygen, consumption, oxidation products, reaction mechanism and kinetic modeling." *Progress in Energy and Combustion Sciences,* Vol. 29, pp. 487-513.

Wang, Q. H. D. (2007). "Research on coal oxidation and pyrolysis reaction kinetics." *Jnl. Beijing Univ. of Sc. & Tech.,* Vol. 28, pp. 1-2, 6.

Wender, I, Heredy, L. A., Neuworth, M. B. and I. G. C. Dryden (1981). In *Chemistry of Coal Utilization*. (M. A. Elliot, ed.), 2nd Supplement, Ch. 8. Wiley, New York.

Wheeler, R. V, (1924). "The ignition of coal." *Fuel in Science and Practice,* Vol. III, pp. 366-370.

Whitehouse, A. E. and A. A. S. Mulyana (2004). "Coal fires in Indonesia." *International Journal of Coal Geology*, Vol. 59, pp. 91-97.

Wheeler, R. V. and T. G. Woodhouse (1932). " Effect of oxidation on coking properties of coal." *Fuel,* Vol. 11, pp. 44-55.

Williams, B. R. Williams, D. F, and D. Glasser (1989). "Maximum temperature rise in the waste dumps with porous surface covering." Contract Report to ISCOR.

Winschel, R. A. Robbins, G. A. and F. P. Burke (1988). "Coal liquefaction process solvent characterization and evaluation." Technical Progress Report, DOE/PC 70018-68.

Wu, M. M., G. A. Robbins, R. A. Winschel, and F. P. Burke (1988). "Low-temperature coal weathering: its chemical nature and effects on coal properties." *Energy and Fuels*, Vol. 2, pp. 150-157.

Yohe, G. R. (1958). "Oxidation of coal." Report of Investigations 201, Illinois State Geological Survey.

Yu, J. A. Tahmasebi, Y. Han, F. Yin and X. Li (2013). " A review of water in low rank coals: The existence, interaction with coal structure and effects on coal utilization." *Coal Processing Technology*, 106 (2013), pp. 9-20.

Yun, Y. et al. (1987). "Attempted development of a "Weathering Index for Argonne PCSP coals." *Am. Chem. Soc.Div. Fuel Chem.,* Vol. 32, No. 4, pp. 301-308.

Yun, Y. and H. L. C. Meuzelaar (1991). "Development of a reliable coal oxidation (weathering) index – Slurry pH and its applications." *Fuel Processing Technology,* Vol. 27, pp. 179-202.

Zhang, Y. (2016). "Evaluation of the susceptibility of coal to spontaneous combustion by a TG profile subtraction method." *Korean Jnl. Chem. Eng.,* Vol. 33, Issue 3, pp. 862-872.

Zuo, Q., G. Wang and X. Gao. (2012). "Experimental study on influence function of particle size at low-temperature oxidation of the coal." *Appd. Mech. & Matls.,* Vol. 214, pp. 515-519.

Part II

Coal Technology

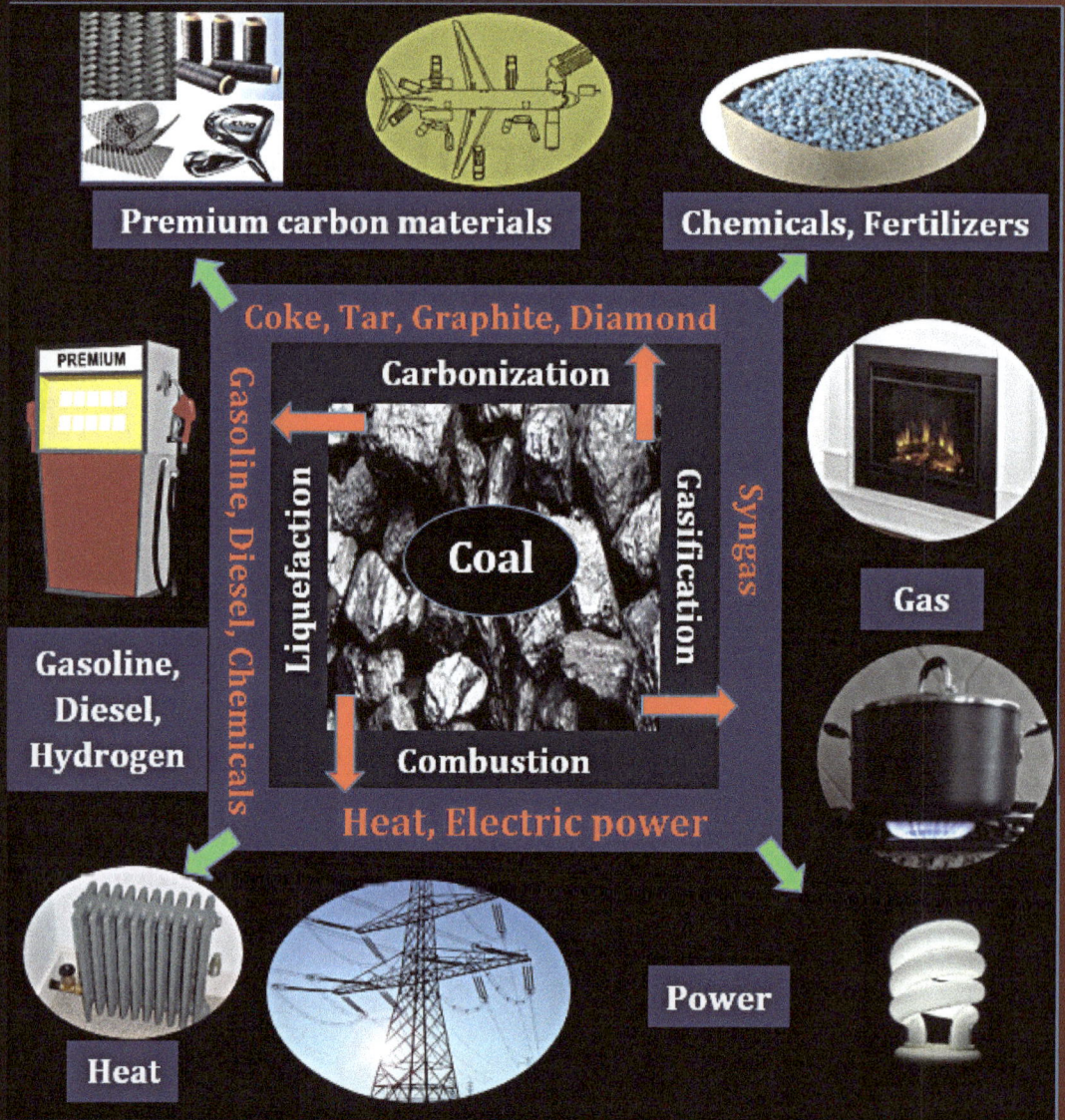

Premium carbon materials

Chemicals, Fertilizers

Coke, Tar, Graphite, Diamond

Carbonization

Gasoline, Diesel, Chemicals

Liquefaction

Coal

Gasification

Syngas

PREMIUM

Gasoline, Diesel, Hydrogen

Combustion

Gas

Heat, Electric power

Heat

Power

Coal conversion processes

8   Coal combustion and carbonization

8.1   INTRODUCTION

Coal is one of the most important natural resources of the world. Unlike oil and natural gas, it is abundant worldwide. It is also the fossil fuel with the widest global geographical distribution, deposits having been found in over a hundred countries, and around 40% of the global electrical energy supply derives from coal. Also, production of steel, one of the world's most important materials of construction depends critically on coal. Coal is also a major energy source for cement production. Furthermore coal is one of the primary energy sources of industrial and domestic heating and steam generation. In most of the utilization processes, coal is either burnt in excess air to produce heat energy or pyrolyzed in minimum air to produce coke used mainly for ironmaking. Apart from pyrolysis, coal can also be converted to chemicals by oxidation, hydrogenation, solvent extraction, hydrolysis, or halogenation. Coal is also processed to obtain many materials of prime industrial value including gasoline, diesel oil, chemicals and high technology carbon materials.

Virtually all the coal currently utilized in the world today is processed by heating. Processes include combustion, carbonization, gasification, and liquefaction. A good understanding of the behaviour of coal on heating is vital therefore in selecting suitable coals for a given application. In this chapter, the basic principles of coal conversion to heat and coke are discussed. Other coal conversion processes are treated in depth in later chapters.

8.2   COAL COMBUSTION

Coal combustion technology has advanced steadily since the nineteenth century when coal was burned on fixed stokers, to modern systems including moving grates, pulverized coal furnaces, cyclones, retort furnaces, rotary furnaces, and fluidized-beds. Most of the world's coal utilization is for fueling boilers of electric power plants, industrial processes, space heating, and also for heat supply to kilns used in producing cement, bricks, etc. Coal is also burned extensively for domestic cooking and heating. When coal is heated in air, it will ignite and burn, the combustion characteristics depending on the type of appliance in which it is burned and also on the type of coal. Burning appliances include up-draught, down-draught, stoker and fluidized bed combustors. The most important properties of coal for combustion are carbon content, volatile matter content, gross calorific value (heat value), swelling characteristics, and the fusion characteristics of the ash formed. Environmental pollution issues have also given prominence to new requirements, in particular, low sulphur content, which is also a requirement for metallurgical applications.

Virtually all coals including low-rank, non-caking, strongly caking, low volatile, medium volatile, high volatile coals are potentially suitable for combustion, with the exception of anthracite which has low reactivity and therefore does not burn easily. Coals with significant caking properties also cause operational problems, although they can be pre-oxidized prior to combustion to destroy the caking propensity. Reactive coals with low moisture, ash and sulphur contents and high ash fusion temperatures are the most preferred for combustion. By far the largest proportion (probably over 80%) of coals utilized in the world today is burned primarily to supply heat for electricity generation, steam-raising, or space heating. The balance is used to supply the required process energy for iron and steelmaking, metal founding, cement production burnt brick production, etc. When coal is heated in air, the constituent organic matter which is mainly carbon is pyrolyzed and the coal goes through three main stages as the temperature increases (Figure 8.1).

Stage 1
Coal pyrolyzes when heated in excess air and starts to release primary volatile matter (mainly carbon monoxide and methane) from around 350°C. Excess air of the order of 15-20% is always required for complete combustion because of the inhomogeneity of the coal pile.

Stage 2
As heating proceeds the inflammable gas products react with oxygen in the vicinity of the coal particles and form bright diffusion flames which set the coal particles burning. At around 550°C the initial gas release tails off but the gases ignite and further heat up the coal to form char. The start and end temperatures of primary volatile matter release as well as the volume released depend on the coal rank (Figure 8.2). In this temperature range some coals soften and become viscous while others do not.

Stage 3
From around 750°C the char starts to burn to provide further heat and is completely reduced to ash at about 1000°C. The chemical composition of the gases again depends on the coal rank. The main reactions taking place during coal combustion are shown in Table 8.1. Typical composition of combustion gases from two coals of different rank is shown in Table 8.2. Carbon oxides released by burning low rank coals is much higher compared with high rank coals while the tar produced is much lower.

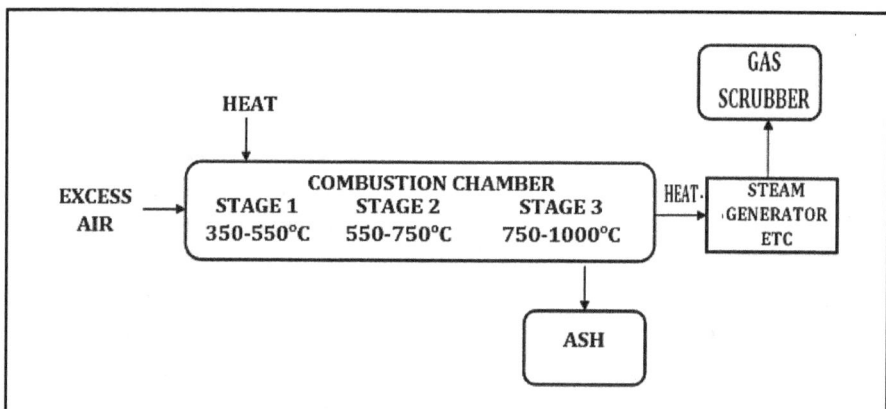

FIGURE 8.1 Stages in the coal combustion process.

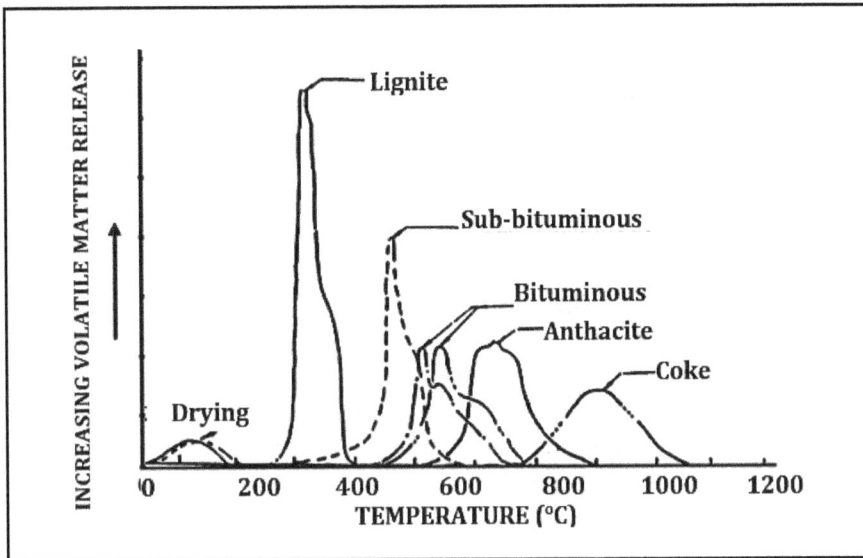

FIGURE 8.2 Schematic diagram showing dependence of volatile matter release during combustion of coal on rank and temperature. *(Amosov et al., 1957).*

TABLE 8.1 Main chemical reactions in coal.

$C + O_2$	$=$	$CO_2 - 393.5$ kJ/mole	(8.1)
$C + 0.5O_2$	$=$	$CO - 110.5$ kJ/mole	(8.2)
$C + CO_2$	$=$	$2CO + 172.5$ kJ/mole	(8.3)
$2CO + O_2$	$=$	$2CO_2 - 283.0$ kJ/mole	(8.4)
$2H_2 + O_2$	$=$	$2H_2O - 483.6$ kJ/mole	(8.5)

TABLE 8.2. Chemical composition of coal combustion gases for two coals.

Coal component	Lignite (%)	Bituminous Coal (%)
CO	10.1	2.9
CO_2	11.6	1.7
H_2O	12.4	9.6
CH_4	2.8	3.9
C_2H_4	0.8	0.7
C_2H_6	0.3	0.9
Tar	3.8	12.9

The lowest temperature at which coal can be ignited, known as ignition temperature varies with coal rank and volatile matter content. Low grade coals with high volatile matter content start to release volatile matter at lower temperatures compared with higher rank coals (Figure 8.2), hence they are easier to ignite. Typical ignition temperatures for different ranks of coal are shown in Table 8.3. The coal characteristics that are important in combustion are calorific value, grindability, combustion properties of the constituent macerals and ash properties. In recent times, potential anthropogenic emissions from combustion, in particular CO_2, SO_2 and Hg emissions, have also become important.

8.3   COAL CARBONIZATION

Coal carbonization is the process of heating coal in minimum air to yield almost pure carbon and gaseous by-products. The coal is normally ground to around minus 3mm particle size. The air supply should be just enough to pyrolyze the coal without igniting it. This process is used for the manufacture of smokeless coal, coke, synthetic gas and chemicals. The main stages of pyrolysis are shown in Figure 8.3.

TABLE 8.3 Ignition temperatures for different coal ranks.

Coal rank	Ignition temperature range (°C)
Lignite	250-400
Sub-bituminous	350-450
Bituminous	400-500
Anthracite	700-800

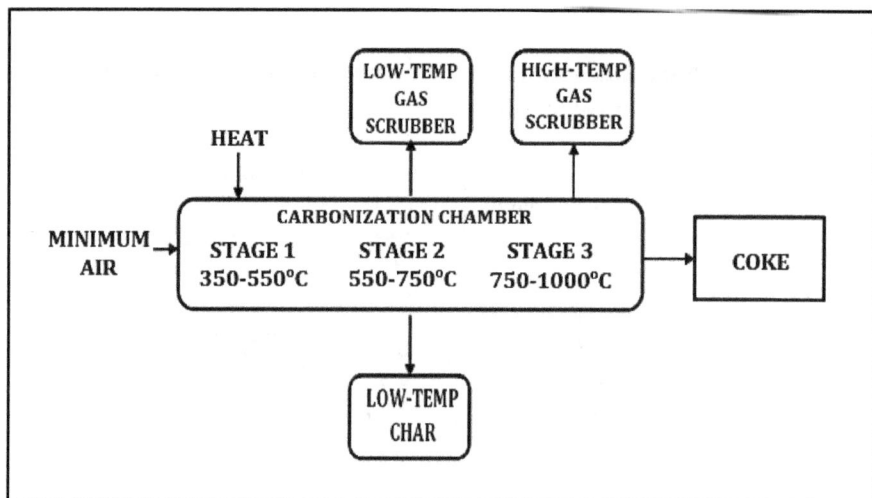

FIGURE 8.3 Coal carbonization process.

Stage 1.
At temperatures between about 330°C and 450°C coal softens, the particles become viscous and deformable. This plastic phase also known as mesophase is a transient phenomenon and further heating initiates a series of physico-chemical changes in the mass. Preliminary gases are released and the viscous mass starts to reconsolidate into a solid mass. Most of the tar volatilizes at this stage. The temperatures at which this stage starts and stops depend on the coal rank, type and to some extent the heating rate. Relatively few coals, all of them of the bituminous rank, go through this plastic (caking, fluid or meso) phase, most coals show little or no plasticity in this stage. The ability of coal to go through this plastic phase is determined by the nature and thermochemical properties of the petrographic vitrinite and exinite maceral contents. An optimal degree of plasticity is required for the coal to form coke of high strength and right quality for metallurgical applications. Non-agglomerating coals loosen up and also release tar and gases but they do not reconsolidate into a porous solid mass.

Stage 2
Between 550°C and 750°C hydrogen is released, the solid coal mass contracts and fractures. The mass produced at the end of this stage is solid and porous if the coal went through the plastic stage, if not, it retains the original granular form. In either case the coal retains around 20-25% of its original volatile matter content. The product at this stage is called char or semicoke and is used extensively as domestic and industrial fuel. Char produced from non-caking coals is either used as pulverized fuel for boilers or mixed with suitable binders, usually pitch and formed in rotary presses into smokeless coal briquettes which are used extensively worldwide as domestic and industrial fuels.

Stage 3
Temperatures higher than about 700°C promote further release of hydrogen and further contraction and fracture of the char mass. Above 800°C or so, carbon monoxide is released and at 950-1000°C most of the volatile matter has been expelled, leaving a porous mass called coke (or granular coke for non-coking coals) which is almost pure carbon, retaining only 2-3% of the volatile matter.

8.4 BASIC CHEMISTRY OF COAL COMBUSTION AND CARBONIZATION

The reactions between the organic and inorganic constituents of coal and oxygen are basic to both coal combustion and carbonization and the actual reactions that take place depend on the intimacy between the coal elements and oxygen. This in turn is influenced by the prevailing diffusion conditions and limitations. The chemical reactions are often summarized in one or two simple equations but the actual reactions can be very complex. Typical reactions are summarized in Table 8.4. The reactions taking place in the combustion or carbonization furnace which lead to the production of gases are so complex that detailed discussion is beyond the scope of this text. The thermal chemistry of coal has been well documented in many texts (see Lowry, 1963, Speight, 2015).

There is a significant difference in the chemistry of the gases produced at low temperature and high temperature carbonization, the production of hydrogen and carbon monoxide being much higher in high temperature carbonization (Table 8.5). In either process the gases are passed to the by-product separation plant for processing to retrieve soot and chemicals such as tar, ammonia, benzole, etc. The clean gas is recycled for pre-heating coal feed or stored in gas holders for use by other consumers. The gaseous by-products of coal combustion and carbonization are very rich sources of chemicals. Hundreds of important chemicals can be produced and indeed were being produced from coal combustion flue gases before they were replaced by petroleum-based feedstock. However, there has been a major revival in the coal-to-chemicals industry in the last decade or so. Chemical products from coal include polymer products, fertilizers, activated carbon, pharmaceuticals, textile dyes, food and wood preservatives, and several hundred other simple and complex chemicals. Coal combustion products such as fly ash, bottom ash, boiler slag are used as additives in concrete, as structural fill, road base material, land fill, mining fill, etc.

TABLE 8.4   Chemical reactions in coal combustion and carbonization.

Reactions involving carbon in coal	
$C + O_2 \rightarrow 2CO$	(8.6)
$2CO + O_2 \rightarrow 2CO_2$	(8.7)
$CO_2 + C \rightarrow 2CO$	(8.8)
$C + O_2 \rightarrow CO_2$	(8.9)
$C + H_2O \rightarrow CO + H_2$	(8.10)
Reactions involving other elements in coal	
$2H_2 + O_2 \rightarrow 2H_2O$	(8.11)
$S + O_2 \rightarrow SO_2$	(8.12)
$2SO_2 + O_2 \rightarrow 2SO_3$	(8.13)
$N + O_2 \rightarrow 2NO$	(8.14)
$2NO + O_2 \rightarrow 2NO_2$	(8.15)

TABLE 8.5 Typical coke and by-product yield as a function of coal carbonization temperature.

Gas Composition (%) and Yields	Low temperature carbonization 550-700°C	High temperature carbonization 1000-1200°C
CO	7.7	10.8
H_2	36	54
CO_2	4.9	2.8
CH_4	39	27
C_2H_6	4.9	0.4
Calorific value	29-27	22-20
Yield of gas (m³/ton)	165-175	300-357
Yield of coke (%)	75-80%	65-75%
Volatile Matter (%)	5-15	1-3
Mechanical strength	Poor	Good
Coke produced	Soft	Hard
Smoke produced	Smokeless	Smoky

Gasoline (petrol) and diesel can also be produced from coal or coal by-product gases by liquefaction. However, in the last six decades or so, petrochemicals have become the main sources of these products. This has greatly diminished the importance of coal as a source of chemicals. The ultimate depletion of oil and gas or strained supply of petroleum products could return coal to a dominant position once again, and this has started to happen already. Coal is much more abundant and more widely distributed than petroleum products and advances in coal carbonization, combustion, gasification, liquefaction and by-product processing technologies are greatly reducing the potential for environmental degradation which has been the main negative issue with coal utilization.

8.5 COAL FOR HEATING, STEAM AND ELECTRIC POWER GENERATION

Dependence on coal for heat and electric power generation varies between regions and countries. On the average, over 40% of the world's total heat and electric power generation is coal-based. However in many countries, dependence on coal is much higher, over 90% in some countries. Stringent environmental pollution control regulations in some countries have not had a major impact on global coal utilization. In fact, all available projections predict that coal will continue to play a major role in the world energy mix in the foreseeable future. Coal-fired energy generation comprises a coal combustion unit which produces heat for direct use in domestic heating, or for heating up water in a boiler to produce steam. The steam is either used for domestic or industrial heating, or to drive a steam turbine-generator system for the production of electric energy (Figure 8.4).

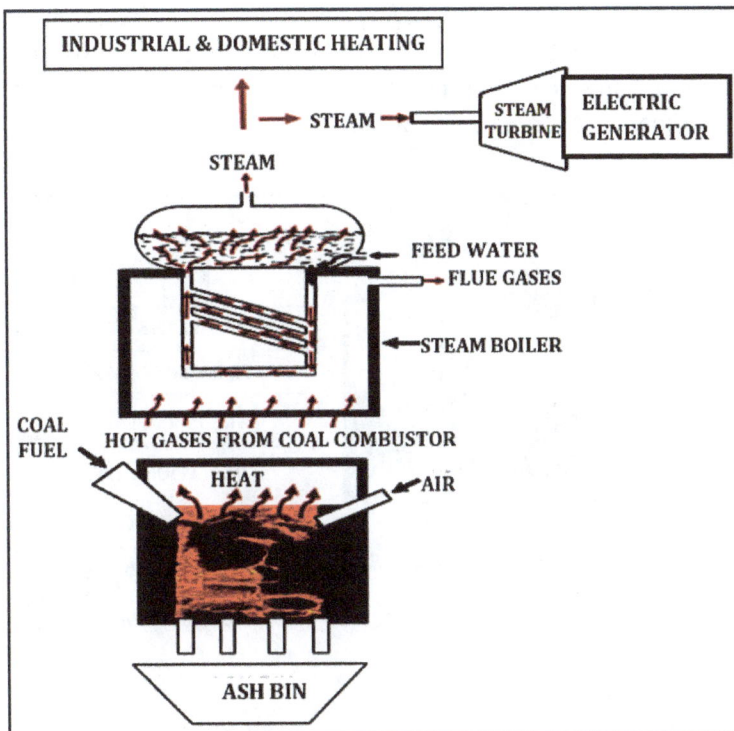

FIGURE 8.4 Schematics of a coal-fired heating and power generating system.

8.5.1   Fixed-bed coal furnace

The earliest coal combustor is a vessel with a horizontal grate on which lies a bed of coal. The coal bed is lit, and air is blown from above or below. This type of grate is still in use in small boilers and domestic stoves but it has been largely replaced in industrial boilers by moving grate, pulverized-coal or fluidized-bed combustors. A fixed-bed combustor is a simple refractory-lined furnace with a horizontal fixed-grate. Fresh lump coal is spread manually or by a mechanical spreader on the grate and ignited. Air is injected from above, from the side or below and the coal burns rapidly, heating up the refractory lining. The burning coal descends through the reducing and oxidation regions, and finally becomes ash which is removed from below (Figure 8.5). Once the burning becomes established, further burning is sustained by heat radiated from the refractory lining. It is best to inject the air from below because it recovers some heat from the hot ash on its way upwards. Carbon dioxide is produced in the oxidizing layer according to equation (8.16). If the fresh coal layer is thick, some carbon monoxide may be produced at the top layer according to equation (8.17). However, if enough oxygen is available the carbon monoxide will burn off.

$$C + O_2 = CO_2 \qquad\qquad [8.16]$$
$$CO_2 + C = 2CO \qquad\qquad [8.17]$$

At the lower temperatures, coal-oxygen complexes are formed in exothermic reactions which rapidly increase temperature in the furnace. As temperature rises the carbon-oxygen complexes decompose, forming CO_2 and H_2O molecules, and carboxyl, carbonyl and phenolic groups, with more heat generation. At high temperatures, these groups decompose to form hydrocarbons: ethane, methane, ethylene, propylene, etc. Finally, the aliphatic groups decompose and form CO, CO_2 and H_2O. Any sulphur in the coal will convert to sulphur dioxide. Excess air (at least 50% higher than the theoretical amount) must be supplied to achieve complete combustion since coal is non-homogeneous and enough oxygen must be available to promote intimate carbon-oxygen contact.

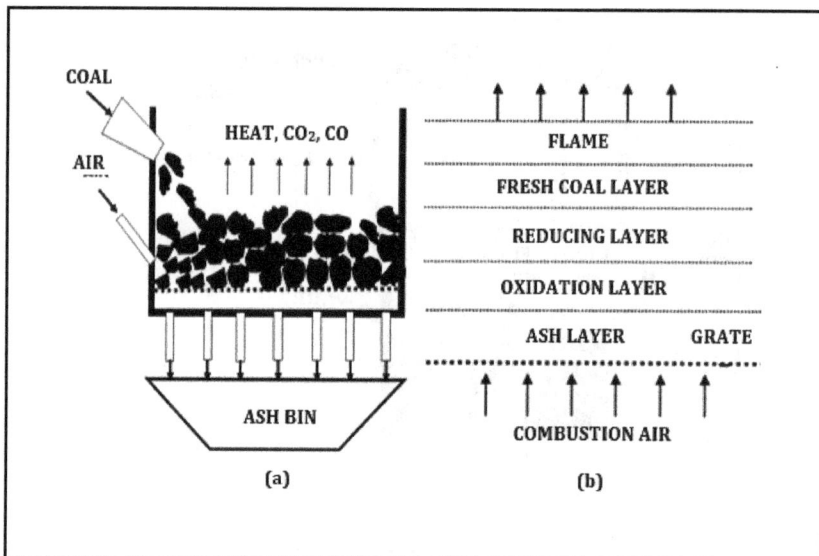

FIGURE 8.5 (a) Coal fixed-bed furnace (b) chemical reaction layers.

There are many variants of the fixed-bed coal furnace. For example, air may be introduced from above in a system known as down-draft system. The air travels in the same direction as the descent of coal and any volatile matter (including fine coal particles) that is evolved ahead of the flame front is swept back to the combustion zone and, if the conditions are right, it will burn, thus reducing pollutants in the flue gases. To achieve this, there must be sufficient oxygen available in the volatile matter, the temperature must be above 600°C, and the residence time in the combustion zone must be long enough (Speight, 2015).

8.5.2 Sloping/moving-grate coal-fired furnace

The sloping/moving-grate furnace works on the same principle as the fixed-grate system except that it comprises a sloping grate which feeds coal to the combustion zone by gravity, or a conveyor chain which moves continuously, feeding the combustion zone with fresh coal and discharging ash into a bin (Figure 8.6). The primary combustion air is supplied from underneath the fire bed to promote efficient contact with the coal bed. Additional air may be supplied above the coal bed to burn off combustible gases that have been released from the bed and generate additional heat.

8.5.3 Pulverized-coal fired furnace

Many modern coal-fired steam boilers use pulverized-coal furnaces with single or multiple burners and a wide variety of burner types Figure 8.7). The coal is ground to around minus 300 microns and a desirable size range which varies with coal rank. For bituminous coals, less than 2% should be +300 microns and 70-75% less than 75 microns. The pulverized coal in pre-heated air stream is supplied to a burner and ignited in the furnace. There are many different types of burners of different efficiencies, some using pure oxygen, some with inlets for secondary or tertiary air supply. Combustion temperature is between 1300-1700°C depending on the coal rank and burner design.

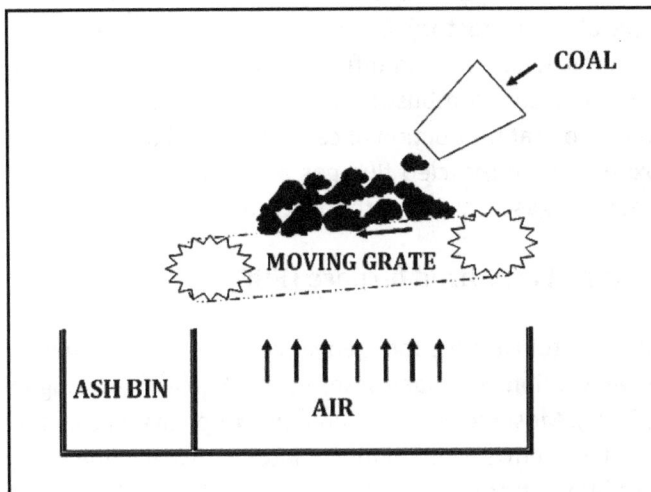

FIGURE 8.6 Solid coal combustion furnace.

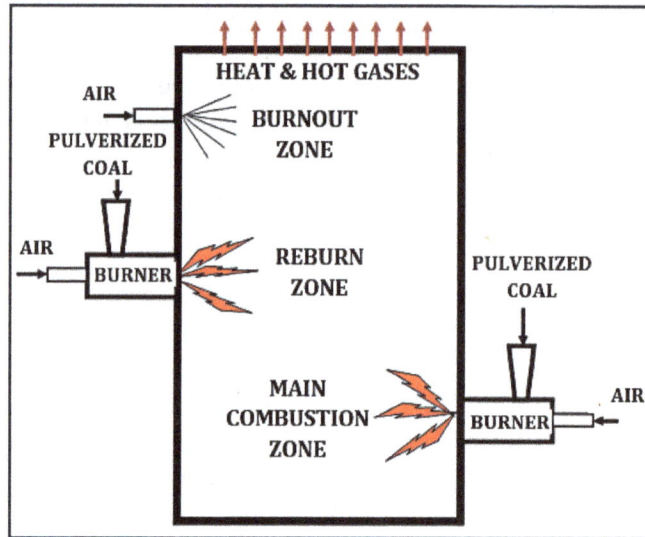

FIGURE 8.7 Pulverized-coal combustion furnace.

Pulverized coal furnaces have many advantages over fixed and moving grate types. For example, coal that has been ground to enhance cleaning can be transported as oil-coal slurry from the washery by pipeline and fed directly to boiler furnaces at power generating sites. Furthermore, process control is much more effective and efficient. Virtually all types of coals can be used in pulverized coal combustors and operating conditions are much more easily controlled to respond to changes in load. As many as four burners can be used tangentially to improve combustion efficiency and provide adequate heat for large furnaces. Furthermore, low NOx pulverized coal burners which can significantly reduce environmentally harmful compounds in combustion gases are now available (Ochi et al., 2009). Post combustion gases can be processed relatively easily to strip them of anthropogenic constituents.

In some pulverized-coal combustion systems known as oxy-fuel combustion systems, air is substituted with nearly pure oxygen (around 95% oxygen) but the furnaces require ultra-high-temperature refractory lining. The use of oxygen results in very high flame temperatures and high combustion efficiencies can be achieved. Most of the hydrocarbon gases produced in oxy-coal combustion are burnt to produce more heat, and oxidation of carbon is complete so that production of carbon monoxide is minimized or eliminated. Some systems mix oxy-fuel with recycled flue gas and, although the flame temperature may be reduced, the final flue gas is mainly carbon dioxide.

8.5.4   Fluidized-bed coal-fired furnace (FBC)

Fluidized-bed combustors are the most advanced furnaces for coal-fired boiler heat production and power generation and many systems are available although the basic principles are the same (Teir, 2002). Most modern coal-fired power plants feature fluidized-bed combustion. The basis of an FBC system is a bed of inert material, usually fine sand (medium) on a porous support, with air blown from below. At sufficiently high velocity the bed particles become suspended in the air stream, hence the name fluidized-bed. The bed is pre-heated to the ignition temperature range of coal (850-950°C).

Solid coal ground to 1-10mm size range or pulverized coal is injected into the bed and combustion occurs rapidly. The combustion chamber is shaped so that above a certain height the air velocity drops below that necessary to entrain the particles and most of the suspended particles drop back to the bed. Depending on the velocity of the combustion air, the bed exhibits different types of fluid-like behaviour as shown in Figure 8.8. Some systems have secondary air supply above the fluidizing bed to ignite and burn off hydrocarbons released in the combustion within the bed and produce additional heat.

There are many variants of the fluidized bed combustor which may be broadly classified into three main types: bubbling fluidized-bed (BFBC), circulating fluidized-bed (CFC), and pressurized fluidized-bed (PFBC). In the bubbling bed type, the fluidizing velocity is maintained at a value which is higher than the minimum but not high enough to carry the particles above the bed (typically 1.2 to 3.7m/sec). Circulating beds use higher velocities in the particle entrainment range to achieve intense mixing of coal, air and combustion products. Particles carried out of the bed are captured in a cyclone and returned to the bottom of the bed. It is also possible to remove sulphur in coal by mixing crushed coal with limestone to convert the sulphur into environmentally benign compounds which are removed with the ash.

Fluidized-bed combustion systems have many advantages over other systems. They are flexible and can burn all coal ranks and types. The combustor can be fed with a wide variety of coal particle sizes from less than 20 microns to 10 mm, and coal received from washeries can be used directly without further grinding. Heat and mass transfer between the fluidizing gas and the solid particles are very efficient and the temperature at which combustion takes place is considerably lower than the ash fusion temperature hence the usual problems of ash melting and clinkering are avoided. Process control of fluidized beds can be precise, flexible and responsive to load demand variations. Boiler water tubes can be inserted directly in the fluidized and above the bed to achieve more efficient heat transfer. Also, because fluidized-beds operate at much lower temperatures than pulverized-coal combustors (850-900°C compared with 1400-1500°C), production of anthropogenic nitrogen oxides is virtually eliminated.

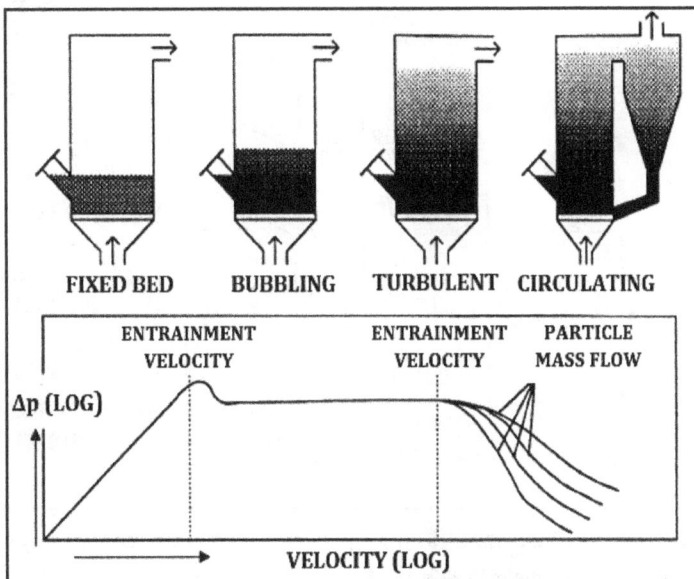

FIGURE 8.8 Fluidized-bed behaviour as a function of fluid velocity. *(Tier, 2002).*

8.5.5 Pressurized fluidized-bed coal combustors

The earliest fluidized-bed coal burning systems operate at atmospheric pressure and are becoming standard equipment in coal-fired heat and power generation systems. However, further development has produced even more efficient combustors. The first generation pressurized fluidized-bed combustor is an enhanced circulating fluidized-bed technology, using pressure vessel and pressurized air to promote higher entrainment, greater mixing and heat transfer between the fluid and solid particles. Contact between coal particles and hot medium is extended and combustion is much more efficient (Figure 8.9). Compared with the atmospheric fluidized-bed combustor, the pressurized type has a heat release rate per unit bed area several times higher and the bed height is 3-4 metres compared with 1 metre in non-pressurized fluidized beds systems.

Deep beds increase bed pressure and promote enhanced release of heat, leading to improved combustion efficiency. Steam can be generated in two water tube bundles, one inside the fluidized-bed and the other above it, and used to drive a steam turbine while the hot flue gases are stripped of particulate matter in hot cyclones and used to drive a gas turbine in a cogeneration power plant (see Chapter 13). Cogeneration systems featuring pressurized fluidized-bed combustors can achieve heat efficiencies of over 90%. The latest generation of pressurized fluidized-bed combustors incorporate two integrated pressurized combustion units. Coal-water slurry is injected into the first pressurized carbonizer which produces fuel gas known as synthetic or syngas, which is primarily a mixture of hydrogen and carbon monoxide. The syngas is stripped of particulate matter and other chemical compounds and fed to a high-temperature combustor that provides hot gases for driving gas turbine-generator systems.

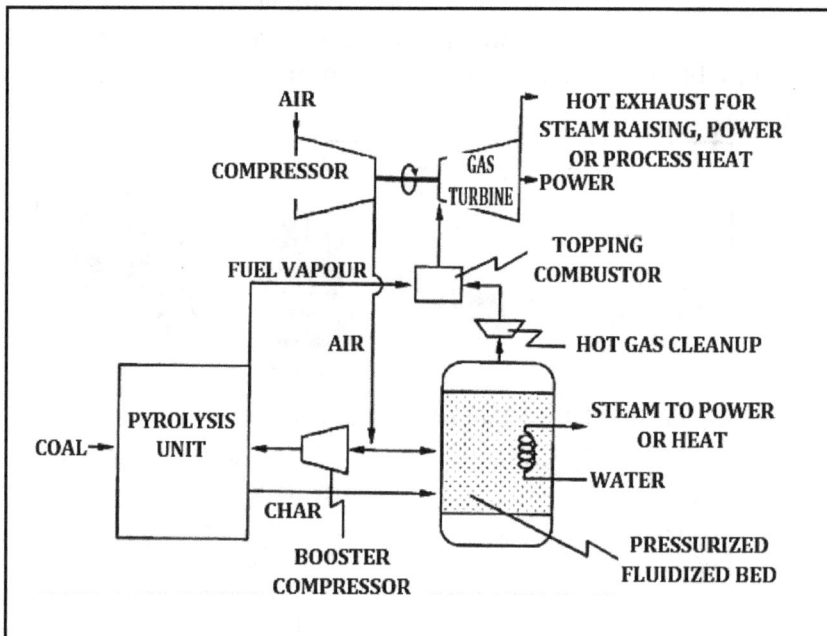

FIGURE 8.9 Pressurized fluidized-bed coal combustor in a cogeneration system. *(Robertson et al.1989).*

The residual char from the carbonizer is burned in a pressurized circulating fluidized-bed supplied with excess air to ensure complete combustion. Hot gases from the pressurized fluidized-bed also pass through a cleaning process into the gas turbine combustor. The high temperature of the gas turbine combustor (around 870°C) promotes the decomposition of nitrous oxide and NOx emissions are very low (Beer, 2000). The fact that two steam pipe bundles can be heated in the same bed increases efficiency significantly. The steam generated is used to drive a steam turbine in a cogeneration power system. Carbon monoxide produced in the first stage can also be converted to hydrogen and carbon dioxide if desired by reacting with high-pressure steam so that the final product is almost pure hydrogen.

8.5.6 Coal-fired combustor flue gas treatment

All modern coal combustors incorporate cleaning systems for flue gases. Starting with a heat exchanger which recovers and recycles as much of the waste heat trapped in the gases as possible. The gases then pass through a cyclone to separate particulate matter which is also recycled into the furnace. The gases from the cyclone pass through a series of chemical separation processes to remove amoniacal liquor, sulphur oxides and other chemical compounds, leaving mainly carbon dioxide which can be utilized, sequestrated or released into the atmosphere.

8.6 COAL USE IN IRON AND STEELMAKING

Around 15% of the world's coal production estimated at over 700 million tonnes a year is used by the steel industry and over 70% of the global steel output is produced in blast furnace-oxygen converter (BF-BOC) plants (WCA, 2014). Around 600kg of coke is required to produce 1 tonne of steel and about 800kg of coking coal is required to produce this amount of coke (see Figure 8.10). As a result of the extensive use of coal, the iron and steel industry accounts for 27% of carbon dioxide emissions from the global manufacturing sector (IEA, 2007). Annual world steel demand is expected to grow by nearly 60% by 2050 and the bulk of this growth will take place in China, India and other developing countries in Asia. This significant increase in steel production will inevitably drive a significant increase in the industry's absolute energy use and CO_2 emissions.

The primary problem of the blast furnace technology is that many blast furnaces in operation are old, small, costly and inefficient to operate compared with the newer, larger blast furnaces. An additional problem is the increasingly scarce and expensive sources of metallurgical grade coking coals. For these reasons, the more flexible, cost-effective direct reduction process (DRI) is becoming increasingly popular and many variants of the process have been developed. Direct reduction process installations are becoming increasingly widespread, particularly in developing counries because of its versatlity, flexibility on fuel, and the ability to start small and scale up. A detailed review and analysis of the proven and promising DRI technologies has been caried out recently and shows that the potential is not limited to developing economies. (Hasanbeigi et al., 2013; Battle et al., 2014; IIP, 2016; Midrex, 2016). About 30% of world steel production is by the scrap/DRI-electric arc furnace route and between 150kg and 800kg of coal is required to produce 1 tonne of steel by this route (the higher coal rate is for coal-based DRI production).

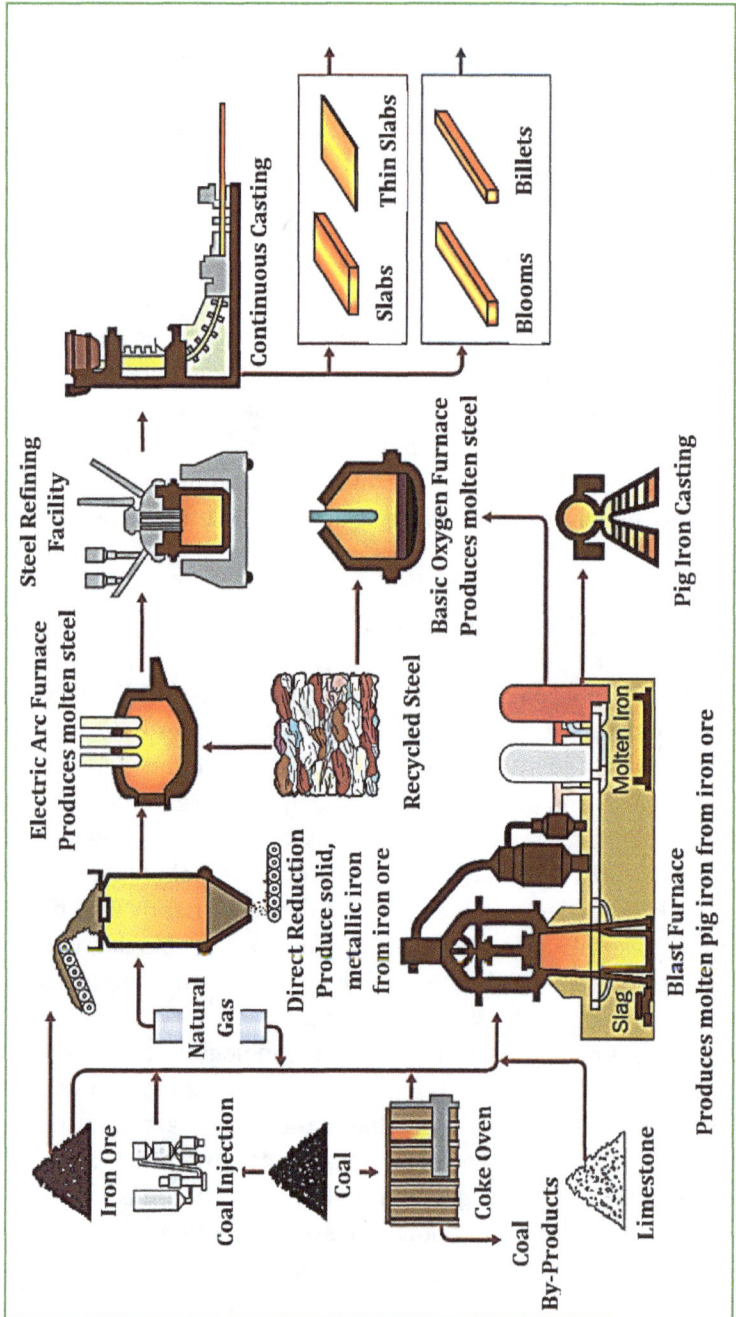

FIGURE 8.10 Steelmaking Flowlines *(AISI, 2010)*

Note: The Direct Reduction Unit can also operate on raw coal or coal-based syngas.

Electric arc furnace smelting is a very flexible method of steel production because it can operate on a wide variety of feed materials, from scrap iron and steel (SCRAP-EAF) to direct-reduced iron (DRI-EAF), molten iron, or mixtures of these raw materials. Coal is used mainly for electric power generation but also for the production of DRI. The total estimated coal requirements by the scrap/DRI-arc furnace steel prouction route is about 300 million tonnes a year. In effect, about one billion tonnes of coal is required annually by the steel industry.

8.6.1 Coal use in BF-BOC steelmaking process

The blast furnace-oxygen converter (BF-BOC) steelmaking process comprises a blast furnace for the production of pig iron which is refined into steel in an oxygen converter. Auxilliary units may include a sinter plant, a coke oven, an oxygen plant, a casting shop, primary rolling mills and a foundry. A typical integrated steel plant is shown in Figure 8.11. Iron ore (or sinter or both), coke, and limestone are charged at the top of the blast furnace in consecutive layers, starting with two or more layers of coke. Sinter is low-temperature agglomerated iron produced by firing fine ore, fine coke breeze from the coking plant, and limestone on a moving grate combustor. Sintering has the advantage of utilizing fine iron ore and coke which otherwise would have been discarded or compacted in an expensive briquetting process. However, not all integrated steel plants have sinter plants.

All materials charged into the blast furnace must be closely sized (8-10mm minimum size) in order to maintain adequate bed permeability and close interaction between the charge materials and the hot gases rising from the combustion zone below. It takes around 24 hours from first charge to first hot iron tap. Despite the fact that the blast furnace has been in existence for nearly three hundred years, the complex reactions that occur in different parts of the furnace and under what conditions are still subjects of intensive research. The reactions outlined in Table 8.6 (both direct and indirect) are regarded as possible or probable. There is no clear understanding which of the reactions take place concurrently, in what parts of the furnace, under what temperature, pressure conditions, and with what charge composition. In fact, blast furnace operation has become more of an art and successful operation depends on experience gained over many years on the same blast furnace.

There is no life span for a blast furnace. A typical furnace operates continuously twenty four hours a day, seven days a week for about five years after which it is shut down and the refractory lining is replaced. This cycle can be repeated indefinitely. In effect, uninterrupted supply of raw materials (in particular, coking coal) is critical to BF operation. Auxilliary equipment may be upgraded and new complementary technologies introduced from time to time but a blast furnace is hardly ever scrapped completely. Furthermore, most blast furnaces in operation worldwide already have well established and dependable sources of raw material supply and many years of experience gained on the same blast furnace.

8.6.2 Coal use in SCRAP-EAF steel production

Production of steel from virgin iron ore by the blast furnace route is capital intensive and very sensitive to the economy of scale. An output of one million tonnes of steel a year is considered the minimum for economic viability. Furthermore, the process cannot be shut down and re-started in response to fluctuations in demand for steel.

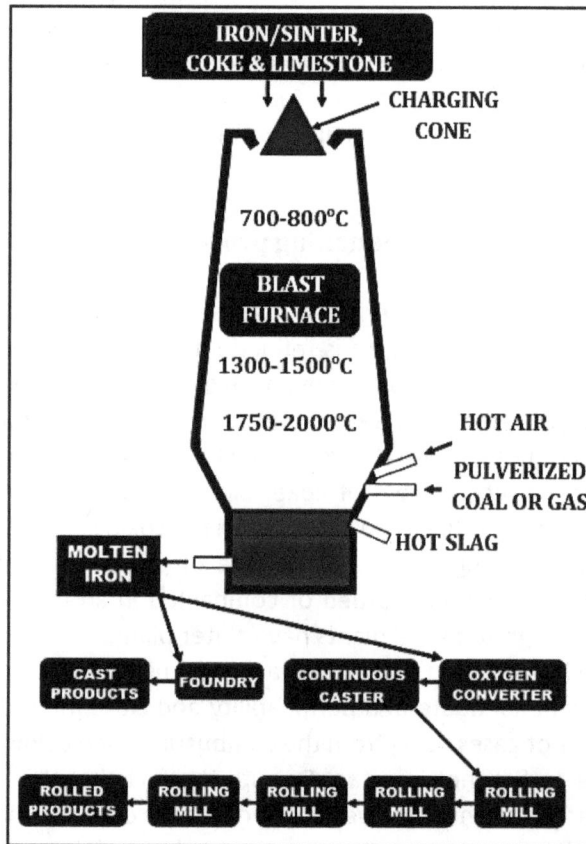

FIGURE 8.11 Schematics of an integrated BF-BOC steel plant.

TABLE 8.6 Iron direct and indirect reduction reactions.

(a)	(b)	(c)
$2C + O_2 = 2CO$ $3Fe_2O_3 + CO = 2Fe_3O_4 + CO_2$ $Fe_3O_4 + CO = 3FeO + CO_2$ $FeO + CO = Fe + CO_2$ $C + O_2 = 2CO$ $3Fe_2O_3 + H_2 = 2Fe_3O_4 + H_2O$ $Fe_3O_4 + H_2 = 3FeO + H_2O$ $Fe_3O_4 + H_2 = 3FeO + H_2O$ $FeO + H_2 = Fe + H_2O$	$6Fe_2O_3 + C = 4Fe_3O_4\ CO_2$ $2Fe_3O_4 + C = 6FeO + CO_2$ $2FeO + C = 2Fe + CO_2$ $3Fe_2O_3 + H_2 = 2Fe_3O_4 + H_2O$ $Fe_3O_4 + H_2 = 3FeO + H_2O$ $FeO + H_2 = Fe + H_2O$	$3Fe_2O_3 + CO = 2Fe_3O_4 + CO_2$ $3Fe_2O_3 + H_2 = 2Fe_3O_4 + H_2O$ $Fe_3O_4 + CO = 3FeO + CO_2$ $Fe_3O_4 + H_2 = 3FeO + H_2O$ $FeO + CO = Fe + CO_2$ $FeO + H_2 = Fe + H_2O$ $3Fe + CH_4 = Fe_3C + 2H_2$ $3Fe + 2CO = Fe_3C + CO_2$ $3Fe + CO + H_2 = Fe_3C + H_2O$
Indirect reduction (Blast furnace process with natural gas injection)	Direct reduction (solid iron-solid carbon reactions with hydrogen injection)	Direct reduction (gasified coal or natural gas)

On the contrary, a large number of scrap-based small electric arc furnaces of less than 50,000 tonnes capacity are operating economically in many parts of the world, particularly in developing countries. However, most of the plants operate on low-grade, poorly-sorted scrap and produce low grade rolled structural steel sections. The process is flexible on fuel and can respond to fluctuations in the steel market. Although electric arc furnaces are energy intensive, most of the electric power is generated from coal which is available in many countries.

8.6.3 Coal use in DIR-EAF steel production

The direct reduction process involves the reduction of lump or pelletized iron oxides with carbon in a shaft furnace. The required carbon reductant can be sourced from a wide variety of carbonaceous materials including any coal, natural gas, fuel oil, coal-based synthetic gas and biomass. If natural gas is the carbon source or coal is gasified, hydrogen is also available for the reduction reactions. The product is a spongy iron-rich solid mass (around 95% Fe) known as direct-reduced iron (DRI) or sponge iron which is melted and refined in an electric arc furnace, hence the name DIR-EAF process.

The basic difference in the chemistry of the blast furnace-oxygen converter steel production process (BF-BOC) also known as indirect reduction and the direct reduced iron-electric arc furnace process (DRI-EAF) is the nature of reactions that lead to the production of iron as summarized in Table 8.6. In the indirect process carbon supplied by coke is first oxidized to carbon monoxide which in turn reacts with the various types of iron ores to produce iron whereas carbon reacts directly with iron ore in the direct process. Modern blast furnaces inject natural gas or pulverized coal which supply hydrogen and carbon respectively to complement the iron reduction reactions.

The vast majority of electric arc furnaces worldwide operate entirely on scrap feed and only about 150kg of coal per tonne of steel produced is required to generate electric power. However, some arc furnaces now operate on direct-reduced iron or a mixture of DRI and scrap iron. DRI plants that use coal or coal-based syngas require around 800kg of coal to produce one tonne of steel. The DRI-electric arc steelmaking route (DRI-EAF) has many advantages over the blast furnace or EAF-Scrap routes. Blast furnaces require expensive and scarce coking coal and they have to be large, with a minimum annual output of no less than a million tonnes of pig iron to operate economically.

Although small EAF-Scrap plants are economically viable, they produce low grade steel because of the numerous impurities from scrap iron. In comparison, DRI-EAF plants producing less than a hundred thousand tonnes of high-grade steel are operating profitably in some countries. Furthermore, unlike the EAF-DRI process, the blast furnace process is inflexible and cannot respond quickly to the ups and downs of the world economy. The DRI-EAF process was first introduced in the 1950s and has developed very rapidly in the last two or three decades. It is based on a modular design concept and many small units can be integrated into a large unit, thereby making production highly flexible since individual units can be shut down as necessary. This is not possible in the blast furnace process.

The modular system of DRI-EAF plants also makes it possible to configure a wide variety of plant capacities, from very small to very large. Capacities can be upgraded simply by adding new modules. The process is also highly flexible on energy input. It can use natural gas or non-coking coal both of which are much more widely distributed geographically than coking coal. The coal used for the direct reduction process may be gasified to improve process efficiency and control. Furthermore, environmental pollution by a DRI-based steel plant is much lower compared with BF-plants. The only major downside of the process is the requirement of higher grade iron ore compared with the blast furnace. Most blast furnaces currently in operation are decades old and, in view of the unpredictable fluctuations in the global steel market, new investment is mainly in rehabilitating or upgrading existing facilities. This explains why, in spite of the numerous relative advantages of the direct reduction process, adoption has been restricted largely to developing countries. Countries such as India and South Africa with abundant reserves of coal prefer the coal-based DRI processes for new projects.

Many types of DRI plants have been developed but the natural gas-based Midrex and Hyl processes are the most popular, producing nearly 90% of the total world output of DRI. They are the preferred choice in emerging countries with abundant supply of natural gas, for example, the Middle East. All DRI-EAF plants currently in operation worldwide produce solid sponge iron which is melted and refined in the electric arc furnace but some processes are under development which are designed to produce molten iron thereby eliminating the melting stage in the arc furnace and conserving energy (Michishita, 2010). The basic direct reduction process involves an inefficient system of solid-solid reactions between carbon and iron ore (Table 8.6b). Furthermore, the process produces high pollutants. The most popular is the SL/RN process which involves reaction between solid iron ore and coal in a rotary furnace. It is an inefficient process but throughput can be increased by injection of hydrogen or inclusion of a coal gasifier unit. Atmospheric pollution is reduced by processing the kiln exit gas through a cleaning system before discharge to the atmosphere (Figure 8.12). In some installations the gas is utilized in power generation.

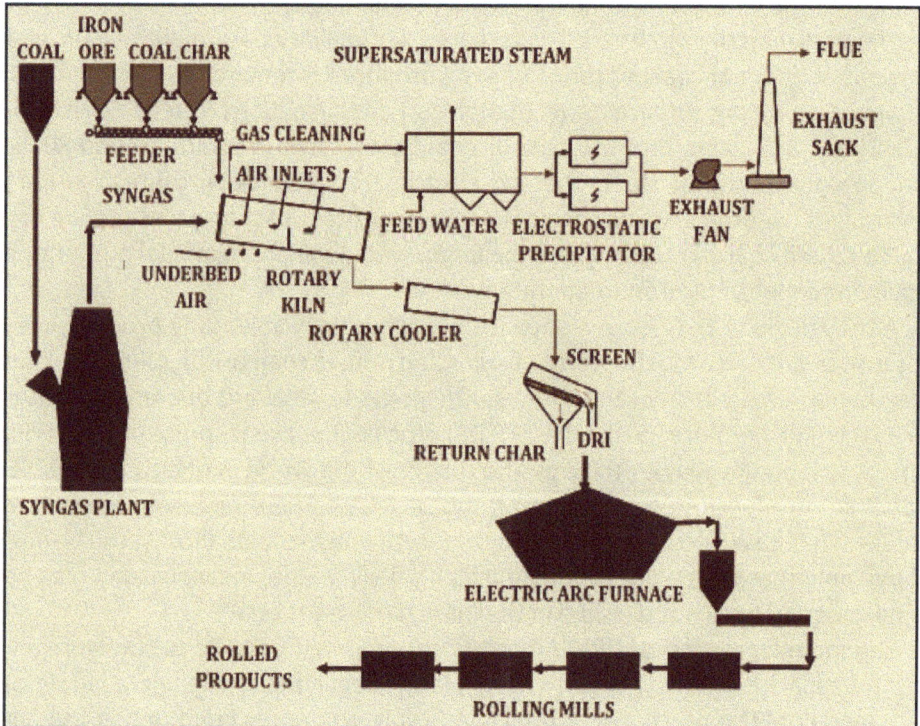

FIGURE 8.12 SL/RN DRI plant with a coal gasifier. *(Adapted from US DOE, 2003).*

The Midrex and Hyl processes use natural gas as reductant. The reactions involved are shown in Table 8.6c. The processes involve solid-gaseous reactions and are much more efficient compared with the SL/RN process. Furthermore, process control is much more precise and pollution is reduced significantly. The Midrex process which is the most popular is shown in Figure 8.13. The process is becoming increasingly popular because of its flexibility on fuel. Plants can be designed to use syngas from coal, coke oven gas and other coal-based sources. Furthermore, large capacity plants with up to 2.5-3 million tonnes/year output are already in operation.

Apart from the relative ease of operating a gas-based DRI plant and the higher efficiency compared with a coal-based plant , the later plant produces more pollutants and would have a problem meeting the increasingly stringent environmental pollution regulations in place in many developed countries. This has prompted the development of plants which integrate coal-to-syngas plants with DRI plants (Figure 8.14). The technology of producing synthetic gas from coal has been deployed already in electric power plants and reductions in pollution as high as 40% have been reported.

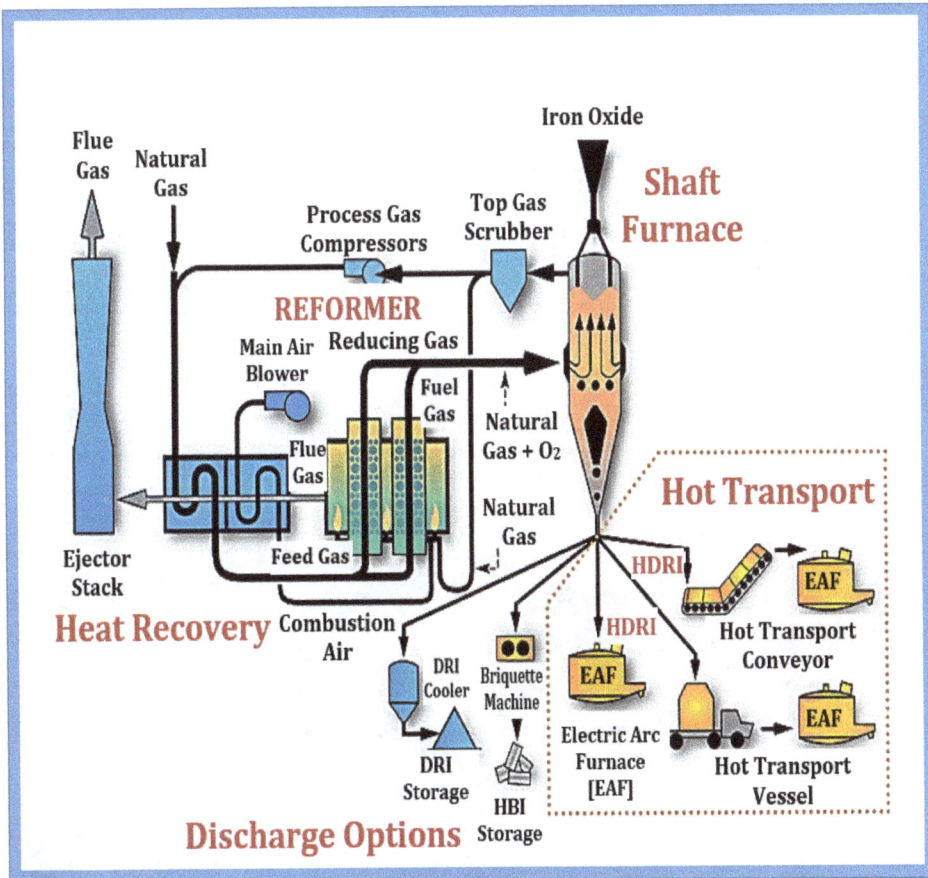

FIGURE 8.13 The Midrex process. *(midrex.com).*

A Midrex plant can produce several products including cold direct reduced iron (CDRI), hot direct reduced iron (HDRI) and hot briquetted iron (HDI). CDRI and HDI are fed to an electric arc furnace for further processing to steel. To conserve energy and reduce production costs significantly the plant can feed hot direct reduced iron (HDRI) at 650°C directly to an adjacent arc furnace. HBI can also be included in the charge of a blast furnace or an oxygen converter (Figure 8.14). Another major advantage of the Midrex process is that carbon dioxide emission from a Midrex/EAF process per tonne of steel produced is only about one-third of the amount emitted from a BF/BOF complex (IEA, 2009). The Midrex process represents over 70% of the installed DRI capacity worldwide.

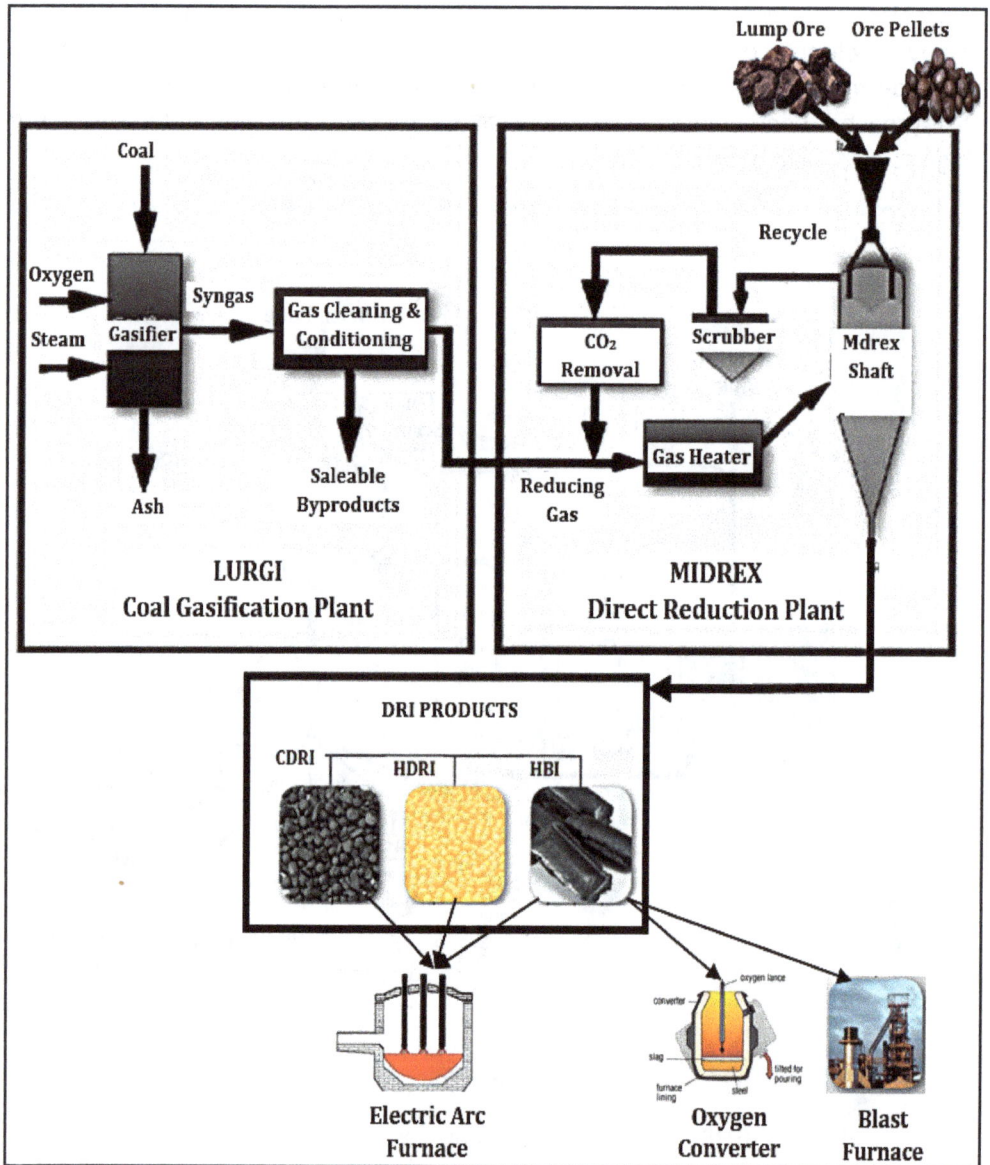

FIGURE 8.14   A Lurgi coal gasification plant supplying syngas to a Midrex direct reduction module. *(midrex.com)*.

Previous attempts to use CDRI in blast furnaces (BF), basic oxygen furnaces (BOF) and off-site electric arc furnaces (EAF) had not been successful because of the low strength and high reactivity of the product. However, a Midrex plant can now produce binderless hot briquetted iron (HBI) which is very strong and can be transported over long distances and stored. Many BFs, BOFs, and EAFs at distant locations are now using HBI (Figure 8.14). Most DRI plants produce solid, porous sponge iron which contains 93-94% iron, roughly the same as pig iron produced in blast furnaces. Some integrated DRI-EAF processes which will produce molten iron like the blast furnace are under development (Michishita, 2010). Many plants have briquetting or pettetizing units for agglomerating undersized sponge iron.

8.7 METALLURGICAL COKE PRODUCTION TECHNOLOGIES

The bulk of the coke produced worldwide is for use in the blast furnace but coke is also used in iron foundry cupolas and in some other thermal metal extraction processes. Nearly all metallurgical cokes are produced in beehive coke ovens, although some foundry coke is produced by formed-coke technology.

8.7.1 Beehive oven coke production

Coke fulfills three functions in the blast furnace: it provides a porous, permeable support for iron ore/sinter/limestone overburden; it supplies the heat for the chemical reactions; and produces the carbon monoxide required for the reduction of iron ore. Hence, coke must be strong and highly resistant to breakage during handling and passage down the blast furnace. Coke of blast furnace quality must have optimal internal porosity (around 50%); it must be closely sized to minimize blast resistance; and it should possess low to moderate reactivity to gases, in particular carbon dioxide in the blast furnace. These are often conflicting requirements the reconcilliation of which makes coke production technology very complex.

The earliest coke-fired blast furnaces (early 16th century) required about eight tonnes of coking coal to produce one tonne of iron but many technologies have been introduced to reduce the rate to less than one tonne of coking coal. These include preheating the blast tuyere air and injection of fine coal, oil or gas, or a mixture of fuels through the tuyeres to provide supplementary heat. Coke meant for use in foundry cupolas has higher strength and mean size specifications. The bulk of high-temperature coke produced worldwide is used for metallurgical smelting. The principal equipment is the coke oven which comprises an assembly of rectangular silica refractory brick-lined oven cells. A large number of cells, typically 48 cells, each of length of up to 17 metres, height 4-7 metres and width 0.4-0.5 metre, are arranged horizontally much in the same way as cells in an auto battery, hence the common name: coke oven battery (Figure 8.15(a). The wall separating adjacent ovens, and each end wall, are lined with a series of heating flues.

Each coke oven battery cell is charged from the top and holds up to 30 tonnes of coking coal. Removable doors at both ends facilitate the discharge of hot coke at one end by ram-pushing at the other. A refractory heating chamber is sandwiched between two cells and is heated by hot gases produced in combustion chambers underneath the cells (Figure 8.15(b)). Most coke ovens produce coke for the blast furnace but some also produce foundry coke (Figure 8.15(c).

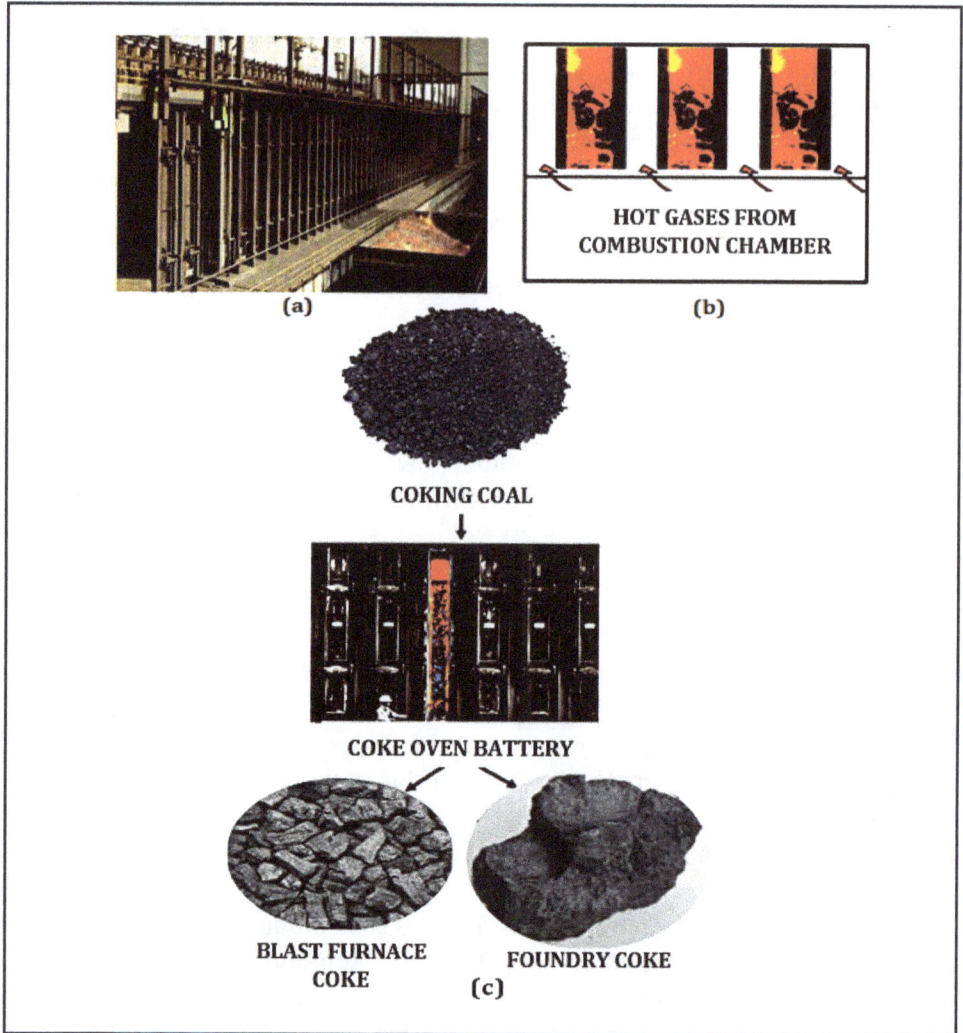

FIGURE 8.15   (a) Coke oven battery (b) Coke oven cells (c) Coke oven products.

Coking coal, ground and sized to minus 3mm is introduced by gravity through charging holes at the top of the oven, although some designs allow the side ram-charging of moist coal to increase bulk density which may also be increased by pre-drying the coal or mixing with oil. High bulk density improves the quality of the coke produced, and also increases oven throughput. Coking proceeds from the layer adjacent to the heating chamber at 900 to 1100°C, It takes 12 to 18 hours for the temperature at the centre of the charge to reach the skin temperature (25 to 30 hours for foundry coke), depending on the width of the cell, the coking blend, moisture content, and rate of heating. The heating rate in each coke battery cell is controlled, around 3°C/minute. At the end of the coking cycle, each cell door is opened at both ends and the hot, incandescent coke is ram-pushed from one end into a bin at the other end, and quenched with water or steam. Most coke oven designs have regenerative heat exchange chambers in which the combustion air or gas is pre-heated by hot flue gases exiting the coking cells.

Modern coke ovens are continuous operations which involve feeding of coking coal continuously at the top. The coal descends through the refractory lined oven which is heated through the sides up to 1000–1200°C at a controlled rate (3–5°C/minute) which ensures that the coal is fully carbonized and transformed into coke when it is discharged through the base into a sealed, steam purged cooling chamber. The steam produces water gas which dilutes the coal gas, producing mixed gases of varying calorific value between 17,000 and 19,000kJ/m³. The gases are recycled for preheating combustion air or coking coal charge.

8.7.2 Basic theory of metallurgical coking

The complex physical and chemical changes which coal undergoes on heating are still not clearly understood. However, from the results of numerous investigations over decades, five main stages of transformation are identifiable as shown schematically in Figure 8.16 (Hoffmann and Hoehne, 1954; van Krevelen, 1961, Lowry, 1963; Loison et al., 1989; Komaki et. al., 2005). The classification system based on coal thermoplasticity and the work of Hoffman and Hoehne was discussed briefly in Chapter 1.7.3. The following discussion is a summary of available theories behind this behaviour. Figure 1.12 is re-presented as 8.16 for ease of reference.

8.7.2.1 *Initial softening and contraction*

At a critical temperature, usually between 350 and 400°C depending on the coal rank, the outer layer of the coal softens and melts, creating a fusion zone. Gases, mainly saturated paraffin hydrocarbons with minor proportions of unsaturated hydrocarbons, hydrogen, carbon dioxide, carbon monoxide and hydrogen sulphide are produced, some of the gases escape and cause shrinkage in the coal. Relatively few coals go through this plastic phase, also known as mesophase.

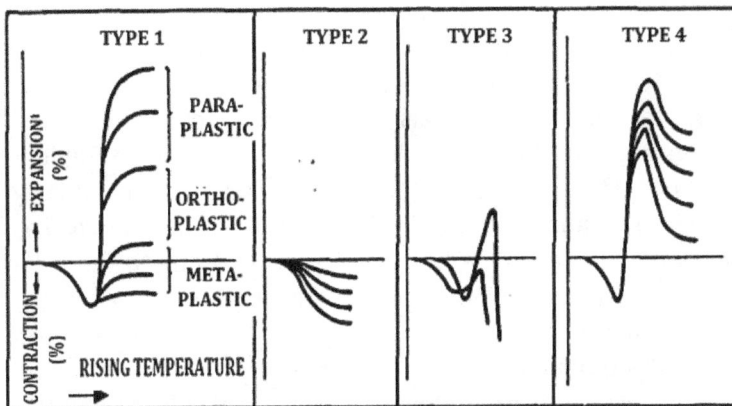

FIGURE 8.16 Thermoplastic behaviour of coals.
(Hofmann and Hoehne, 1954; van Krevelen, 1961, Lowry, 1963).

8.7.2.2 *Dilation*

As heating progresses, the fusion zone moves inwards, leaving an outer plastic shell which is impermeable to gas. The gases formed in the inner fusion zone become trapped and pressure builds up sufficiently to enable the gases force their way through the plastic layer, causing the coal to swell. The extent of dilation depends on the rate of degasification, the amount of gases released, and the rheological properties of the coal which determine the thickness of the plastic layer.

8.7.2.3 *Re-solidification*

At about 500-550°C the plastic layer starts to re-solidify due to the gradual disappearance of the fluid products partly by expulsion and also by decomposition. The residue is a black, very reactive, porous, caked or powdery char, often called semicoke.

8.7.2.4 *Post-solidification contraction*

Above 500-550°C the semi-coke starts to decompose, forming mainly hydrogen, methane and some carbon monoxide which are expelled. This secondary degasification causes the porous mass to contract.

8.7.2.5 *Fissuring*

As heating proceeds above 550°C , the fusion zone moves right to the center of the mass. A thermal stress gradient is set up accross the mass and this, combined with the differences in the contraction rates of the various components of the semi-coke, causes cracking of the solid semi-coke and further degasification progresses. The extent of fissuring is very much dependent on the plastic properties of the coal. At 900°C and above, the residue is a hard, fissured, less reactive coke or a powdery coke breeze, again depending on the plastic properties of the green coal. The residual volatile matter is usually of the order of 2%.

The schematic profiles shown in Figure 8.17 represent the thermoplastic behaviour of most coals on heating. Almost all coals contract with temperature rise to around 300°C but very few go through all the stages discussed above. Also the ability of a coal to form metallurgical coke of acceptable quality depends on the profile of its thermoplastic curve. Coals that show any dilation above zero after contraction are classified as caking but may not necessarily make good coke. The best coking coals belong to the ortho-plastic group and combine moderate plasticity, dilation and volatile matter.

Only coals with caking properties undergo some or all the stages described above and they are all of the bituminous rank. Apart from rank, other variables including maceral composition, heating rate, and pre-oxidation of the coal by pre-treatment or weathering also play an important role in determining the plastic behaviour of coal. Furthermore, the temperatures at which coal starts to soften and reconsolidate, the extent of contraction and expansion all vary between coals, even in the same rank. Since many coke ovens operate on blends of several coals, Figure 8.16 is very useful in identifying coals that are compatible and good coking blends can be designed when combined with plastometric tests (see Chapter 6).

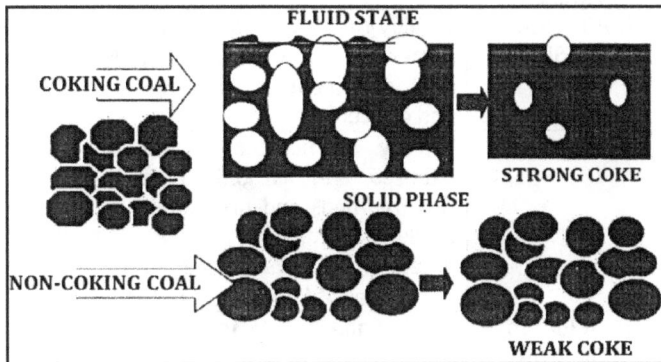

FIGURE 8.17 A schematic illustration of coke formation.
(Sakanishi et al., 2005).

8.7.3 Formed coke technology

The primary goal of high-temperature carbonization of coal is to produce metallurgical coke, the most expensive raw material input in blast furnace ironmaking. Coals that are suitable for coking are relatively rear and expensive. Intensive research is continuous on reducing the amount of coke consumption per tonne of iron produced through more precise process control and supplementation with cheaper and more readily available fuels. Considerable progress has been achieved through injection of supplementary fuels (fine coal, coke breeze, oil, gas) into the blast furnace, and incorporation of relatively inexpensive and widely available low-grade, pre-compacted weakly or non-caking coals in coking blends without compromising coke quality. Furthermore, it has been established that partial briquetting of coking coal blends could improve the bulk density by up to 15% thereby increasing furnace throughput and coke quality.

The development of processes that can produce formed-coke of metallurgical quality from non-coking coals has been a subject of intensive research since the 1950s and many processes have been developed and patented in many countries (USA, Japan, Russia, Poland, Germany) including FMC, BFL, CCT-FCP, BNR, Ancit, DKS, CFRI processes (Table 8.7). The product known as formed-coke or formcoke is made by compacting crushed coal or low-temperature char with or without a binder. The main differences between the various formed-coke processes are in the types of coals used, required pre-treatment of coal, method of compaction (hot briquetting, cold briquetting, palletizing), the carbonization process, binder additives, (coking coal, pitch, lime, bitumen), and the nature of pre- and post-compaction heat treatment. These include heat treatment in fluidized-beds, rotary kilns, multi-stage flash heaters, vertical retorts, vertical ovens, sand beds. For example, the FMC process converts sub-bituminous coal into pillow-shaped coke briquettes by charring and carbonizing the coal in multiple, staged fluidized-beds or other carbonizing reactors. The tars from the beds are captured and used as binders for briquetting the char. The briquettes are calcined in a shaft furnace.

TABLE 8.7   Formed-coke processes in operation or under development.

PROCESS	MAIN FEATURES
SAPOZNIKOV (RUSSIA)	Weakly-caking coals flash-heated in a multi-stage heater and briquetted at 400-500°C without the addition of a binder. Briquettes are carbonized at 850°C and gas-cooled.
FMC (USA)	Weakly or non-caking and sub-bituminous coals oxidized and carbonized at 850°C in multi-stage fluidized-beds and briquetted with 10-15% pitch. No binder is required if the volatile matter of the coal is greater than 35%. Briquettes are oxidized at 250°C, carbonized at 850°C and gas-cooled.
ICEM (ROMANIA)	Weakly and non-caking high volatile coal briquetted, using medium-caking gas coal and pitch as binder. Briquettes carbonized at 950 – 1000°C in shaft oven.
BFL (GERMANY)	high-volatile non-caking coal carbonized in fluidized-bed at 700°C and briquetted at 400 – 500°C using caking coal and recycled by-product as binder.
AUSCOKE (AUSTRALIA)	Weakly-caking low and medium-volatile coal mixed with pitch binder and cold-formed. Briquettes carbonized at 500°C in fluidized-sand bed and subsequently coked in directly-heated shaft oven at 1100°C.
JISF (JAPAN)	Continuous process. Low-volatile weakly caking coal mixed with coking coal, coke breeze and a binder is cold-formed in a briquetter. Briquettes are heat treated in an externally-heated twin-chamber shaft furnace at 1250 to 1300°C and water-quenched.
HBNPC (FRANCE)	Wide range of coals including non and strongly caking coals mixed with pitch binder, cold-formed and carbonized to 1000°C in a continuously or semi-continuously operated vertical oven with internal heating.
I.Ch.P.W. (POLAND)	Lump coal carbonized at 750 to 800°C, crushed and briquetted with 10% binder consisting of 70% pitch and 30% heavy oil. Briquettes oxidized at 200 to 300°C and water-quenched.

The BFL process which was developed in Germany in 1962 involves production of fine-grained char by low-temperature carbonization of low-grade coal in a fluidized-bed. The hot char is mixed with fine-grained binder caking coal at 750°C in a double screw mixer. The high char temperature brings the binder coal to the softening temperature range of around 450°C and the mixer feeds the coal to a double-roll press. The formed briquettes leaving the press at about 450°C must be cooled to room temperature under controlled conditions to obtain the desired formed coke quality that meets blast furnace specifications (Figure 8.18a). Japan, one of the foremost producers of steel in the world has been at the forefront of blast furnace trials of formed-coke because the country has only weakly-caking coals and is heavily dependent on imported coking coal. Japan imports approximately 90 types of coals from six countries and 10 to 15 different coals are blended for typical coke oven charges.

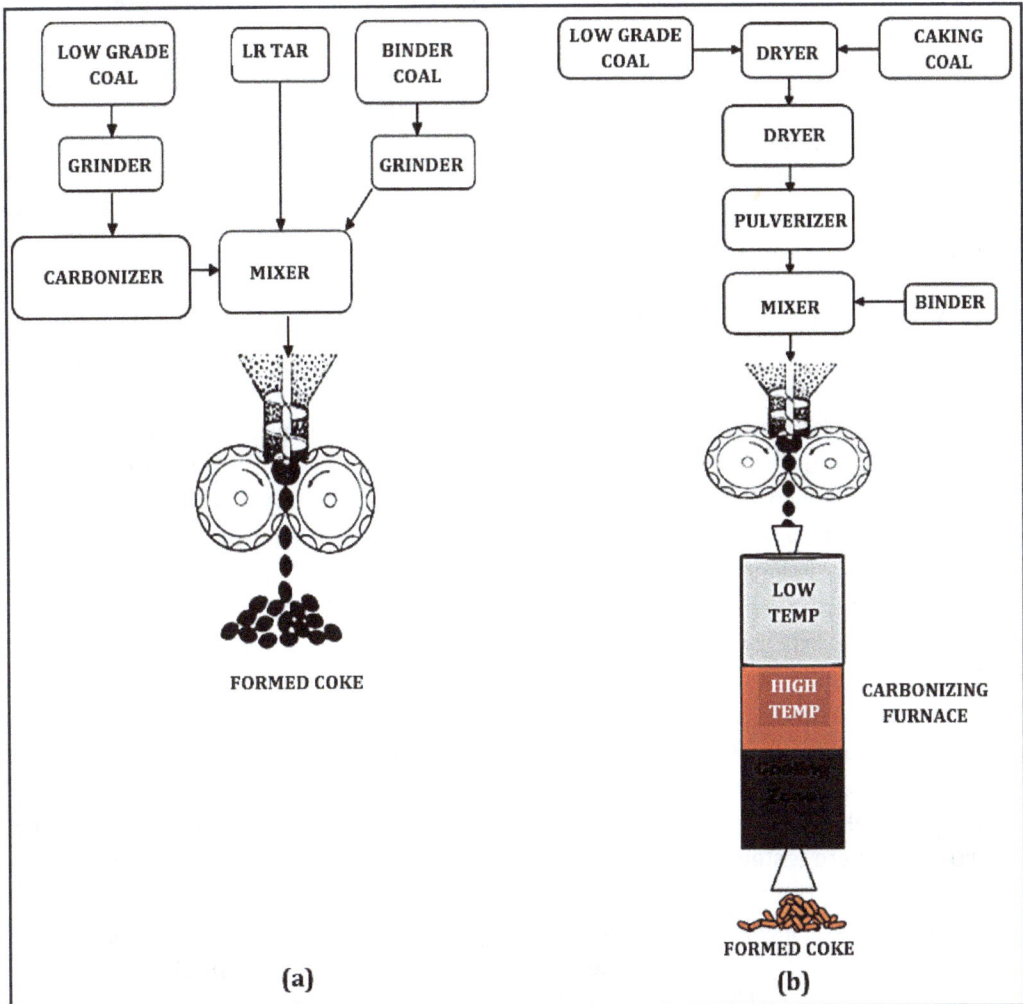

FIGURE 8.18 Schematic diagrams of formed coke processes
(a) BFL process (b) JISF continuous process.

In view of the increasing scarcity of prime coking coals and the continued dominance of the iron blast furnace technology, it is inevitable that formed-coke will in future become a viable substitute for metallurgical coke. The Japan Iron and Steel Federation continuous process (JISF) utilizes a mixture of high-grade and low-grade coals, with the high grade coal acting as binder. Both materials are crushed, mixed with a tar binder and formed into briquettes which are then heat treated in a three-stage carbonizer. The tar binder is produced by processing the flue gases from the carbonizer. The coal briquettes pass through a low-temperature conditioning zone in the carbonizer before being carbonized in the hot zone at around 1000°C. The third carbonizing stage involves controlled cooling in the third zone of the carbonizer (Figure 8.18b). One major advantage of the JISF process compared with the other batch processes is that it is continuous.

Although formed-coke is smokeless, it is of little use as domestic fuel because it has poor lightability. However it has great potential as a substitute for expensive metallurgical coke. Formed-coke technology has a number of potential advantages over conventional metallurgical coking technology. It is capable of utilizing a wide range of coals including up to 100% non-coking coals, thus widely extending the range of potential coals to include inexpensive low-grade, non-coking coals. These coals are much more abundant and widely available, thereby reducing operating costs. Furthermore, formed-coke is uniformly sized and creates more uniform, permeable support beds for iron ore and limestone. Also, it promotes more intimate interaction with the rising hot reducing gases. The size and shape of the briquettes can   also be varied to achieve optimum performance in the blast furnace. Formed-coke technology yields products of consistent quality and the potential for continuous, automated production is very high. Coke yield is higher than   in conventional coking, atmospheric pollution is considerably less, and capital investments and operating costs are less.

Another feature of great interest is the flexibility in the operation of formed coke processes, that is, the ability to start and stop production rapidly in response to changes in demand. Generally, formcoking allows better control of the final property of fuel than conventional coking. It is possible to produce formed-coke of comparable quality to oven coke but, in general, formed-coke is usually stronger, contains more volatile matter and is more reactive (Table 8.8). The higher reactivity is thought to be due to the extent of graphitization that has been attained during production rather than the effect of specific surface (Aderibigbe and Szekely, 1982). It is believed that increased reactivity due to higher specific surface is compensated for by loss in reactivity due to lower porosity or average pore diameter. This then suggests that formed-coke of comparable reactivity can be produced if more time is allowed for graphitization during production.

The higher reactivity of formed-coke is however not considered a disadvantage as regards its use as a blast furnace fuel. Formed-coke has been tested extensively in commercial blast furnaces though for relatively limited durations due to insufficient availability of the product (Ahland et al, 1971; Nekrasov, 1974; Dartnel, 1978; Kohna, 1988). Products of a 200t/day JISF pilot plant were tested continuously in a large 4,250m³ commercial blast furnace for 110 days (Kohna, 1988). The results of the JISF tests proved that formed-coke produced entirely from non and weakly caking coals can be used in commercial blast furnaces without any noticeable deterioration in performance indices. It was estimated also that formed-coke was about 15% cheaper in both facility and manufacturing costs than the conventional coking process. Furthermore, the process is continuous and therefore more amenable to automatic control. Other major advantages are the ability to utilize up to 100% non-coking coals, manpower savings, and lower environmental degradation.

TABLE 8.8 Comparison of the properties of oven coke and formed-coke. *(compiled from Afonja, 1979, Aderibigbe and Szekely, 1982).*

PROPERTY	OVEN COKE	FORMED COKE
Crushing strength (kN/m²)	6,895	12,000
Residual volatile matter (%, daf)	1.5-2	5-10
Density (g/m³)	1.1	1.42
Specific surface area (m³/g)	1.3-1.6	40
Porosity (%)	51-56	35-45
Average pore volume (μm)	9.6-10.8	2.07

Formed-coke is now being used in foundry cupolas in several countries including Poland and China, and some commercial blast furnaces routinely operate on coke replaced up to 30% by formed-coke. The main reason why formed-coke has not been adopted more widely is the general inertia in the steel industry and a reluctance to adopt new technologies that require steep learning curves, compared with the traditional coke technology which has been in operation for several hundred years. Furthermore, many of the coking coal mines worldwide are owned by steel plants or have long-term contracts with them, hence there is no incentive to adopt new coking technologies. Another problem with formed-coke is the fact that there is as yet no commercial plant capable of producing sufficiently large quantities for sustained evaluation in blast furnaces over a long period for proper evaluation of technical performance and process economics. Also, considering the volatility of the global steel market, making large Investments in new, unproven coking technologies is considered too risky.

8.7.4   Coke quality

As discussed above, coke fulfills three major functions in the blast furnace: it provides a strong, porous, permeable bed for the charge materials; it provides the heat for the reactions; it supplies carbon, a primary alloying element in iron and steel. To fulfill these functions, coke for ironmaking must meet stringent strength, abrasion resistance and porosity specifications. Furthermore, the coke must be carefully sized to obtain the desired bed permeability for the rising gases and optimum contact with the descending charge. Good metallurgical coke should contain no more than 1 or 2% residual volatile matter, the ash, moisture and sulphur contents should be low. Every percent reduction in the four variables significantly increases the value of coke. Furthermore, high ash and sulphur contents require complex slag composition and additional energy for fusion and removal. The chemical nature of the ash is also important since it largely determines the chemistry of the slag required for its removal as well as the removal of other gangue in the ores. The coke should have high carbon content and optimum reactivity to oxygen, carbon dioxide or moisture. The main quality parameters for fulfilling the above roles and methods of determination and control are outlined below (See also Osborne, 2013; Speight, 2015).

8.7.4.1 *Coke strength and abrasion resistance*

Coke strength in terms of its resistance to degradation by fracturing and abrasion is considered generally to be the most significant coke property in relation to blast furnace operation. The major methods of evaluation generally referred to as drum tests involve a tumbler or drum type test in which a weighed coke sample of a prescribed size distribution is tumbled in a steel drum for a specific number of revolutions. A size analysis is then carried out on the coke sample and the breakage resulting from the mechanical work is expressed in terms of the size reduction of the original sample. Coke strength is expressed in terms of the proportion of the original sample retained above a certain sieve size after testing while abrasion resistance is expressed in terms of the proportion below the sieve. The most common variants of this method are the American Society for Testing and Materials Tumbler Test for Coke (ASTM D5341-99), the Japanese Industrial Standards Test Method (JIS, 2150-72), the German Micum Drum Test (DIN 51717), and the International Standards Organization Micum Test Method (ISO R 556). Some of the important variables of the major tumbler tests are given in Table 8.9. The Drum and Tumbler tests are popular because they measure the resistance of coke to degradation which includes both the strength and abrasion resistance.

8.7.4.2 *Coke porosity*

One of the main challenges in metallurgical cokemaking is the reconciliation of coke strength with adequate porosity which should be at least 50% in order to maintain a permeable bed in the blast furnace and promote chemical reactions between the ascending hot gases and the descending blast furnace burden. Porosity is the fraction of the volume of coke that is occupied by pores. This parameter is measured by determination of the apparent densities of samples immersed in mercury and helium. The pore volume Vp is calculated from the formula:

$$V_p = 1/D_{Hg} - 1/D_{He} \hspace{4cm} [8.18]$$

Where D_{Hg} and D_{He} are the densities of mercury and helium respectively.

8.7.4.3 *Coke reactivity*

The descending coke in the middle and lower zones of the blast furnace burden contacts hot carbon dioxide gas rising from the combustion zone and chemical reactions occur which weaken and degrade the coke and, unless the coke is sufficiently strong, excess fines will be produced and can decrease the permeability of the blast furnace burden. Coke strength as determined by ambient-temperature drum tests indicates the ability of coke to withstand breakup at room temperature and reflects the ability of coke to withstand handling and the burden stresses in the upper part of the blast furnace. However, the parameter does not indicate the behavior of coke at high temperatures in a carbon dioxide environment. This has prompted the development of techniques which evaluate the resistance of coke under such conditions. The Nippon Steel Corporation (NSC) method (ASTM D 5341; ISO 18894) is the most widely used to assess the high-temperature reactivity of coke and involves laboratory-scale reaction of a coke sample in carbon dioxide environment at 1100°C for two hours.

TABLE 8.9. Major test methods for coke strength and abrasion characteristics.

TEST STANDARD	SAMPLE SIZE	SAMPLE WEIGHT	NO OF REVS	STRENGTH INDICES	TYPICAL RANGE FOR GOOD COKE
ASTM D5341-99	3x2 (in)	22 (lb)	600@ 20 rpm	Stability Factor	>60
ISO Micum R556	+ 60 (mm)	50 (kg)	100 @ 25 rpm	Hardness Factor M_{40} M_{10}	80-87 5-7
JIS K2151	+ 50 (mm)	10 (kg)	30/150 @ 15 rpm	D^{30} D^{150}	84-85 84-85

The weight loss percent (coke reactivity index, CRI) is determined and the residual coke is subjected to a drum test to determine a variable known as coke strength after reaction (CSR). Typical specifications for a blast furnace quality coke are shown in Table 8.10. These two variables are very useful in predicting the quality of coke obtainable from coking blends and will be discussed further in a later section.

8.7.5 Coal blending and blend design

All coking coals fall in the bituminous rank and caking/coking properties vary widely within the rank. A good coking coal should have moderate thermoplastic properties, medium volatile matter content and low ash and sulphur content. Only a very small proportion (less than 5%) of world coal reserves have coking quality and they constitute only about 20% of the bituminous coal rank. However, numerous projections indicate that there is enough to meet world requirements in the foreseeable future at the present and projected rates of depletion.

Despite the abundance of coking coals, it is becoming increasingly difficult for many steel plants to source suitable, reliable and inexpensive coking coals due to the inequitable geographical distribution of hard coal in general and coking coal in particular. For example, about 85% of the total world reserves of coal and 75% of coking coals are located in North America, China and the former Soviet Union. In view of the skewed geographical distribution of coking coal resources, good coking coals are not readily available and many coke ovens worldwide operate on imported coking coal (Japan imports most of its coking coal requirements). Even in countries where coking coal is available, only a small proportion falls in the class *of* prime coking coals (single coals that can form coke of acceptable blast furnace quality).

TABLE 8.10 Specifications of a typical prime coke.
(O'Donnell and Poveromo, 2000).

QUALITY PARAMETER	OPTIMAL VALUE
Mean Size	47-70
M40 (+60mm)	>78 - >88
M10 (+60mm)	<5 - <8
I40	53 - 55
I20	>77.5
DI 150/15	83 - 85
ASTM Stability	60 - 64
CSR	>60
CRI	20-30
Moisture (wt%)	1-6
Volatile matter (wt.% db)	< 1.0
Ash (wt.% db)	8-12
Sulphur (wt.% db)	0.5-0.9
Phosphorus (wt.% db)	0.02-0.06
Alkalies (wt.% db)	<0.3

Furthermore, the need to conserve high-cost, limited reserves of prime coking coals has strongly promoted the development of coal blending technologies for cokemaking which involves mixing several coals in a coke oven charge. Most coke ovens worldwide operate on blends of several different coals from the range of low-volatile, medium volatile, high volatile, weakly caking, medium caking or high caking coals. Blend components are carefully chosen and proportioned so that the technical behaviour of the blend on coking matches that of a prime coking coal as closely as possible (Table 8.10). Some plants also use low-temperature char of appropriate quality to replace medium volatile coal in the blend.

The need to blend several coals of different quality has prompted intensive research worldwide on coking blend design and the behavior of blend constituents on carbonization. A number of design principles are now in use to facilitate the reconciliation of several divergent coal quality variables in order to obtain a cost-effective blend which behaves like a prime coking coal as closely as possible. The carbonization behavior of a coking blend can now be predicted fairly accurately by a complex reconciliation of several microstructural and thermochemical properties of any coal or the components of a coking blend. The most important guiding principle of coal blend design is that the behaviour of the blend must be as close as possible to that of a prime coking coal, in particular, fluidity, dilation, post consolidation contraction, and swelling characteristics. Furthermore, there should be an overlap between the plastic temperature ranges of constituents of a coal blend. If one blend constituent has already resolidified before another becomes plastic, the resulting coke will be weak.

In order to achieve a good coking blend there must be a careful balancing of the various properties. The two most important variables are the degree of plasticity and the temperature range of each coal in a blend. It is possible therefore to blend several coals of extreme coking properties to obtain a good coking blend. For example, a coal that has high volatile matter content or high fluidity may be moderated with a coal of opposite properties. It should be noted however that coal quality variables are not simply additive and laboratory tests involving complex calculations are often required. Coking coals are also blended with low value carbon materials in order to lower the operation costs and conserve valuable coking coals. These include compacted non-coking coals, coal fines, coke breeze, coal tar, petroleum coke, asphaltene. Coke quality can be improved provided that these materials are added in the right amount, morphology, and to the right kind of blend (Vilia, 1990; Vilia and Cooper, 1994).

Coking coals also weather and oxidize in storage, the effect being most severe on the lowest rank coals. There could be major reductions in thermoplasticity as well as the plastic temperature range. Coke strength after reaction (CSR), coke stability, coke size, coke yield, coke oven wall pressure, are all affected negatively by overuse of weathered coal in a coking blend.

8.7.6 Coke quality prediction

Methods of predicting coke quality from the characteristics of blend components fall into two main categories: those that predict coke strength under high-temperature and carbon dioxide environmental conditions, and those designed to predict ambient temperature coke strength. The models that predict hot coke strength are considered more reliable since they are based on a simulation of the actual conditions in a hot blast furnace.

8.7.6.1 *Prediction of high-temperature coke quality*

The vital need for determination of hot coke strength became evident in a major investigation by Nippon Steel Corporation (NSC) in the 1970s which involved quenching three commercial blast furnaces while in operation, dissection and analysis of the burden at various sections of each furnace. The valuable data obtained from these experiments led to the development of a hot coke strength prediction method widely known as NSC model. The Nippon investigation involved the determination of two variables: the coke reaction index (CRI) and coke strength after reaction index (CSR). The two parameters were then used to develop curves from a database of a wide range of coals from which values for other coals and coking blends can be predicted.

The results of several investigations have identified CSR as a vital variable that must be optimized to obtain good blast furnace operation (Nakamura et al., 1977; Colleta et al., 1990; Hatano et al., 1990). Several authors have also reported that increased CSR results in a reduction in coke rate and an increase in blast furnace output (Hara et al., 1980; Rooney et al., 1987). It should be noted however that optimal values vary depending on the prevailing operating conditions of a particular blast furnace and different investigations have recommended CSR values between 45 and 60% to achieve good bed permeability. The CSR index is dependent on many factors including the original plant material that transformed into coal, depositional conditions and rank, and coking conditions (Figure 8.19). The results of several studies (Nakamura, 1977; Bernard et al., 1985; 1986) have demonstrated that coal characteristics largely determine CSR and one investigation determined that coking conditions contribute only 30% to the hot quality of coke (Nakamura et al., 1977). The NSC model which combines the measurement of coke reactivity (CRI) and post-reaction strength (CSR) has gained wide international recognition and is in routine use by the coking industry. It has been adopted as ASTM D-5341-93a) and ISO (18894) Standards. The system has been tested extensively for coke produced from a wide variety of coking coals and coking blends and found to be a very useful tool for evaluating coke quality and designing coking blends.

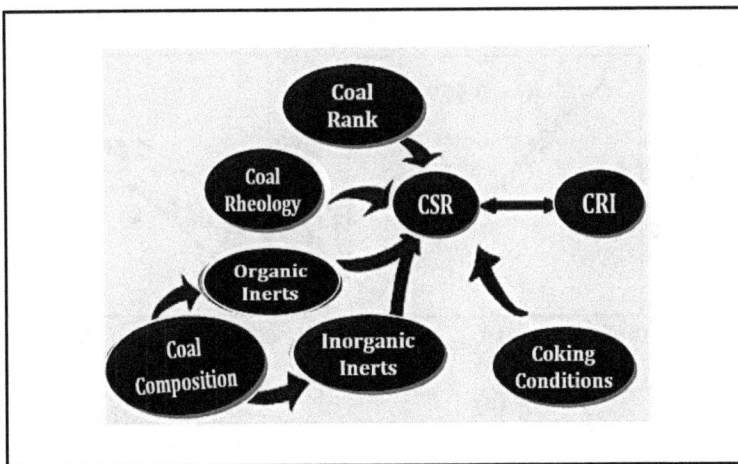

FIGURE 8.19 Main factors influencing coke strength after reaction (CSR).
(Diez et al., 2002).

A strong correlation has been found between hot coke strength (CSR) and coke reaction (CRI) as well as vitrinite reflectance (coal rank) (Figure 8.20). A good coke should have high hot strength and low reactivity. Excessive reactivity (high CRI) between coke and hot CO_2-rich oxidizing gases in the blast furnace will weaken and degrade the coke, resulting in low CSR and the production of fines which will impair the permeability of the burden (Diez, 2002). Many variants of the NSC method have been developed and are in use at different coke plants around the world. Some of the most widely used have been reviewed by Diez et al., (2002).

It should be noted that, in spite of the usefulness of CSR in predicting coke quality, its universal reliability has been questioned, justifiably since no two blast furnaces have identical operating conditions. Blast furnace temperature profiles, gas flow and composition vary and there are varying chemical reactions apart from reduction of iron ore, depending on the chemical composition of the burden. It should be expected therefore that CSR and CRI will vary between blast furnaces having different furnace characteristics and operational conditions, even when the coke rate is similar. However, many coke plants test their range of coals extensively and develop necessary modifications to the NSC system in designing coking blends which give consistently good coke quality.

Another hot coke strength prediction method is based on the mesophase theory which states that the reactions that are most critical to good coke production take place in the range between 350 and 550°C when many coals go through a plastic phase and reconsolidate into a porous mass called char, and carbon dioxide is produced. It is believed that, in the plastic range, an isotropic plastic mass is first formed, followed by formation of lamelar nematic liquid crystals. This polymeric phase is known as the mesophase and has properties that are intermediate between solids and liquids. If the fluidity in the phase is sufficiently high and the temperature range is adequate, there is coalescence of the crystals in the mesophase to form a large anisotropic carbon mass known as semicoke. The shape, size and texture of the anisotropic carbon formed from the mesophase largely determine the strength and thermal resistance of coke formed because it constitutes the optical texture of the pore-wall material (Grint and Marsh, 1981).

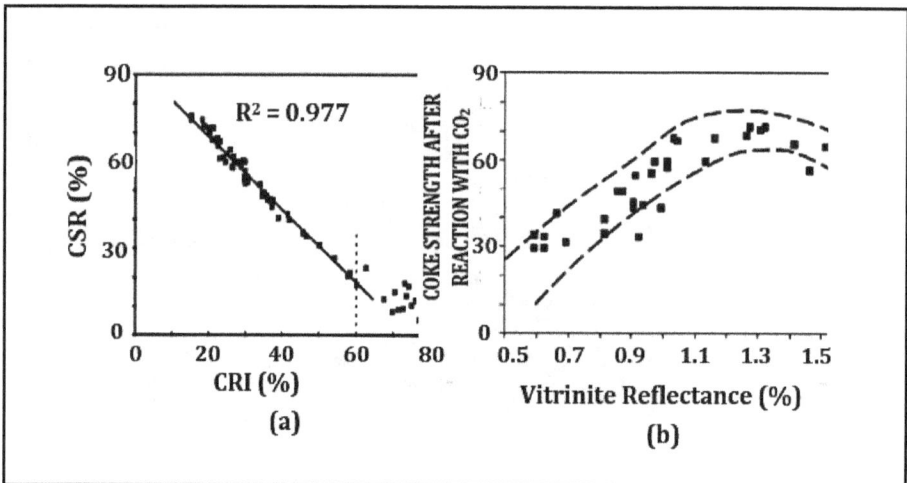

FIGURE 8.20 Relationship between hot coke strength (CSR) and (a) coke reaction index (CRI) and (b) coal rank based on vitrinite reflectance. *(Menendez et al., 1999; Nakamura et al.,1977).*

Coals of different ranks form cokes of different optical textures. The mesophase method of assessing coke strength and reactivity combines the Gieseler fluidity plastic range with a calculated catalytic index (CI) of carbon dioxide gasification which takes into account the combined effect of ash chemistry and sulphur (Valia, 1989). The variable CI is calculated as a product of ash content and weight percent ratio of the basic and acidic components of the ash as shown in Equation 8.19, and the CSR can be calculated from Equation 8.20.

$$CI = 9.64 \text{ alkali index} + 14.04 \text{ sulphur content} \qquad [8.19]$$

Where Alkali index = $\dfrac{\text{ash content} \times [(Fe_2O_3 + CaO + MgO + Na_2O + K_2O)}{(SiO_2 + Al_2O_3)]}$

$$CSR = 28.91 + 0.63PR - CI \qquad [8.20]$$

Where PR is the plastic temperature range.

A plot of coal plastic range versus calculated CSR is shown in Figure 8.21 from which CSR of single coals or coal blends of known plastic behavior can be derived. Since CSR is a linearly additive property, values for coke produced from blended coals can be predicted from the coal characteristics by applying the additivity law (for example, coals A and B or C and D). It is generally agreed that high levels of base oxides act as catalysts in the reaction of coke with carbon dioxide. Therefore, coal mineralogy, particularly the ash chemical constituents and refractoriness can affect coal fluidity significantly, and falsify results based on rank only. It is important therefore to consider including a basicity index (base/acid ratio) in calculations.

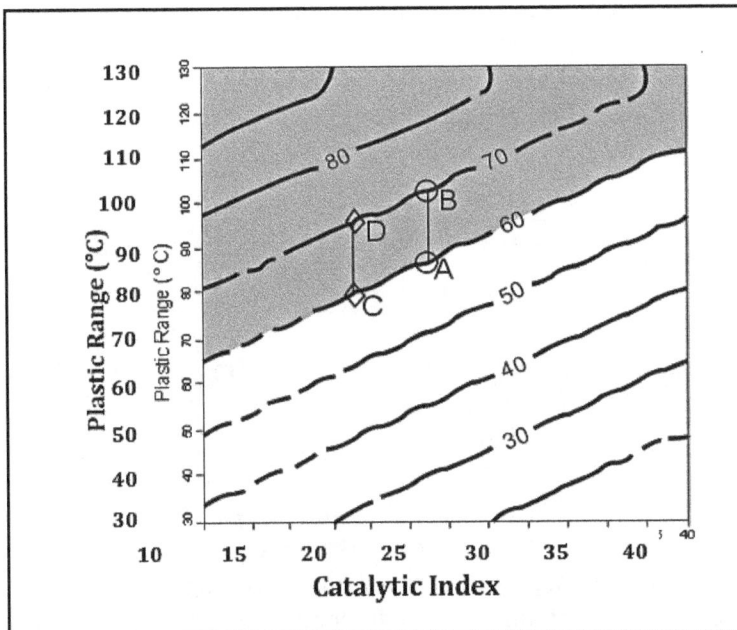

FIGURE 8.21 Prediction of CSR of coke from plastometric and ash analyses determination. *(Valia, 1989).*

8.7.6.2 Prediction of coke strength from maceral composition and characteristics

Coal contains petrographic macerals which are microscopically distinct and differ in chemical and thermal reaction characteristics. On the basis of reflectance measurements and morphological analysis, different macerals can be identified. They can be grouped into reactive, semi-inert and inert categories, based on their behavior during carbonization. The inert category includes inorganic components of coal. The reactive components exhibit fluidity during carbonization and act as binders, while the inerts act as fillers in the formation of the coke structure. Coking blend design based on petrographic analysis seeks to optimize the relative proportions of reactives to inerts in order to obtain the desired coke quality. Several techniques have been developed to predict hot coke strength from petrographic characterization of the maceral components of coal or a coking blend (Mackowsky, 1977). They are all based on the principle that the strength of coke produced from a coal or coal blend is dependent on the ratio of the reactive maceral components relative to the inert components. An optimal balance produces the strongest coke.

The main difference between the various methods that are based on this principle is the method of allocation of proportions of semifusinite which is partially reactive and partially inert. Also, while most of the methods ignore the influence of coal rank, others recognize the fact that optimal mixes may vary depending on the ranks of the coals in a coking blend. A study by Ammosov et al., (1957) was one of the earliest attempts to use maceral composition of coal and measurements of their reflectances to predict the coking capacity of a coal blend. Further studies by many authors have refined this pioneering work and resulted in several prediction techniques. A method known as the US Steel (USS) model developed by Schapiro et al., (1961), Schapiro and Gray (1964), Goldring, (1970) involves the calculation of a composition balance index (CBI) which represents the balance between the fusible (vitrinite, exinite and semi-fusinite) and non-fusible components of coal (inertinite, mineral matter and a portion of semi-inertinite) (Equation 8.21).

$$CBI = \frac{100 - P_t}{(P_1/M_1 + P_2/M_2 + ...P_N/M_N)} \qquad [8.21]$$

Where M_1, M_2......M_n are the experimentally determined ratios of reactives to inerts for optimum coke strength and are constants.

An ideal balance is allocated a value of 1, coals which are oversaturated with respect to inert constituents have CBI higher than 1 while undersaturated coals have values below 1 (Bustin et al., 1983). A strength index (SI) or rank index (RI) is also calculated by summing of the strength factors for each vitrinite class, related to the amount of inert components of the coal. If a coal blend contains P_t total reactives comprising P_1, P_2 .....P_n = volume percent of individual reactive types, then the Strength Index is given by Equation 8.22.

$$SI = \frac{(P_1 x K_1) + (P_2 x K_2) + ...(PnxKn)}{P_t} \qquad [8.22]$$

Where K_1, K_2........K_n are strength indices of reactive vitrinite types and are constants. The CBI and SI values can then be used to plot iso-stability curves shown in Figure 8.22. These curves are used as the basis of a relationship to calculate drum indices of coke and predict the coke strength after reaction (CSR) as well as the cold strength (ASTM stability factor). The correlation of stability factors calculated by Schapiro et al., and several other authors, using the USS model and those obtained by measurement from carbonized coal blends was very strong, around 0.98. It should be noted that a relatively small part of the total diagram represents prime coals hence the need for most coke plants to blend several coals. It is important to note also that the curves in the figure were developed from data obtained from blends of specific coals and may need to be adjusted for blends of coals from other sources. For example, the model was developed for coal blends with ash content less than 12% and has been found to be inaccurate for higher-ash blends or blends containing oxidized coals.

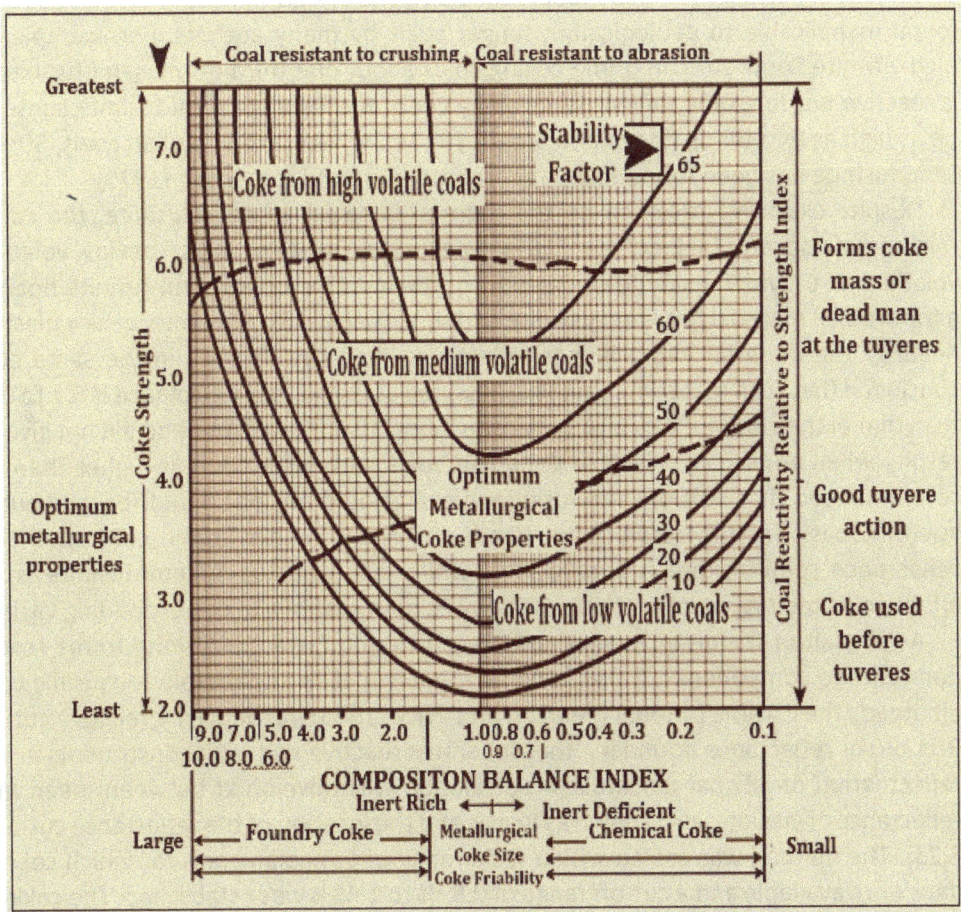

FIGURE 8.22 Relationship between coke strength and petrographic constituents of coal (USS Model). *(Schapiro et al., 1961; Zimmerman, 1979).*

The USS model has been tested extensively with carboniferous coals in the Northern Hemisphere and found to give fairly accurate prediction of coking capacity. However, it has been found to be less accurate for Southern Hemisphere coals, for example, Australian, Indian and South American coals, many of which have high inertinite content (Schapiro et al., 1961). The main limitation of the USS model is its method of determination of what components of coal are reactive or non-reactive. While the model allocates one-third of the semi-fusinite constituent as reactive and the balance together with the remaining inertinite and mineral matter as non-reactive, coals have been identified in which the total semi-fusinite is reactive (Steyn and Smith, 1977) or inert (Brown et al., 1964).

The structure and chemical nature of semi-fusinite are determined primarily by the nature and type of the original plant debris that metamorphosed into coal, but also on the depositional and environmental conditions prevailing during the processes of coalification. It should be expected therefore that the structure and thermal behavior of the maceral will vary from coal to coal and models which simply apportion one-third as reactive cannot be universally applicable. The USS model based on data for Carboniferous-Age US coals has been found inapplicable to geologically younger coals by many authors. For example, in some high-vitrinite South African prime coking coals about one third of the semi-fusinite content is reactive but for some low-vitrinte coking coals, the reactive semi-fusinite content could be as high as two thirds (Steyn and Smith, 1977). For Western Canadian coals, 50% reactive semi-fusinite has been found to be appropriate (Carr and Jorgensen (1975).

Despite extensive research on the thermal behavior of semi-fusinite, the controversy remains unresolved. Taylor et al., (1967) tested a wide range of coals from low volatile to high volatile and provided evidence in support of the inert nature of semi-fusinite but identified a transitional material between vitrinite and semi-fusinite that became weakly plastic during heating. The material had a higher reflectance than the vitrinite in the same coal. They concluded that such material should be classified as inert if the reflectance is 0.1 to 0.2 higher than that of the vitrinite. Carr and Jorgensen found that the USS method did not give accurate results when applied to Western Canadian coals with inert content higher than 20% and recommended that 50% of semi-fusinite in coal should be considered reactive but gave no scientific basis for the assumption. Nandi and Montgomery (1975) confirmed that high-reflectance semi-fusinite is totally inert while low-reflectance semi-fusinite is reactive, although its melting point is about 80°C lower than that of the corresponding vitrinite.

As a result of the disagreement on the contribution of semi-fusinite to the reactive and non-reactive components of coal, several models that are applicable to specific coals have emerged. The CANMET model (Pearson and Price, 1985) introduced a reactive cut-off value Rcut-off) or reflectance boundary that separates reactive and inert constituents in a random reflectogram of all coal macerals. A correlation was developed between mean maximum reflectance of vitrinite also known as Romax and the location of the reflectance cut-off (Figure 8.23). The method was tested with a wide range of Canadian coals for which coke stability data were available and a cut-off range of 0.89% to 1.63% was established. The coking quality of new coals or coal blends can then be determined from the correlation curves established for the reference coals.

The work of Pearson and Pearce (1985) has demonstrated the complexity of the status of semi-fusinite in coal. They found that the cut-off reflectance value is a function of both rank and the vitrinite content of the coals. Among coals of the same rank, the position of the reflectance boundary varies inversely with vitrinite content, migrating between low values

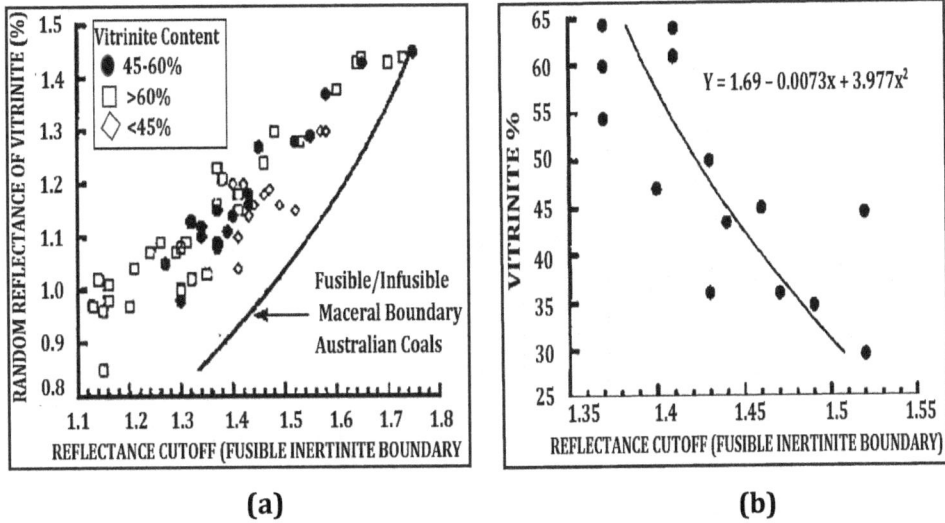

FIGURE 8.23 Correlations between (a) Romax and (b) volume% of vitrinite and fusible inertinite boundary reflectance cut-off for Western Canadian coals. *(Pearson and Pearce, 1998).*

for vitrinite-rich coals to higher values for inertinite-rich coals. It is therefore not enough to apportion semi-fusinite between reactive and inert and reach conclusions on the carbonization behavior simply on the basis of reflectance as suggested by other authors. The amount of associated vitrinite should be taken into account.

8.7.6.3 *Prediction of ambient-temperature coke quality*

Coke strength prediction methods in this group are based on coal Gieseler fluidity or Ruhr/Audibert-Arnu dilation, both of which are measurements of the physico-chemical changes in the mesophase zone in which coking coal softens, releases some gases, then reconsolidates into a solid mass. It has been established in numerous investigations that it is in this phase where almost all the physical and chemical processes which control coke quality take place. Since coking coals of different ranks exhibit plastic behavior over different temperature ranges (Figure 8.24), the production of good coke from coking blends depends critically on the extent of overlap of the plastic range of the different coals in the blend. The longer the overlap the greater the interaction between particles and the number of chemical bonds formed in the plastic zone.

A Japanese method uses the mean maximum reflectance of coal as a measure of rank. A curve widely known as the Miyazu curve or coking window model, relating maximum fluidity to mean maximum vitrinite reflectance of a coal or coal blend is shown in Figure 8.25. The location of each coal on the curve gives valuable information on its potential role in a coking blend. An ideal prime coking coal/coal blend window is identified and coal blends whose overall properties fit into the window usually produce good coke. Based on the 'coking window', a good coking blend should have a mean maximum reflectance (Romax) in the range of 1.2-1.3% and a Gieseler maximum fluidity between 200 and 1000 ddpm.

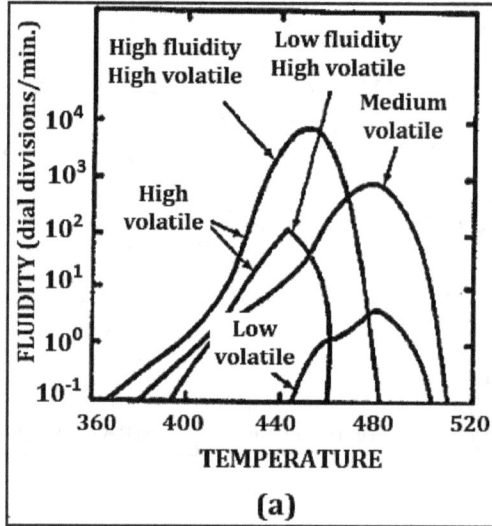

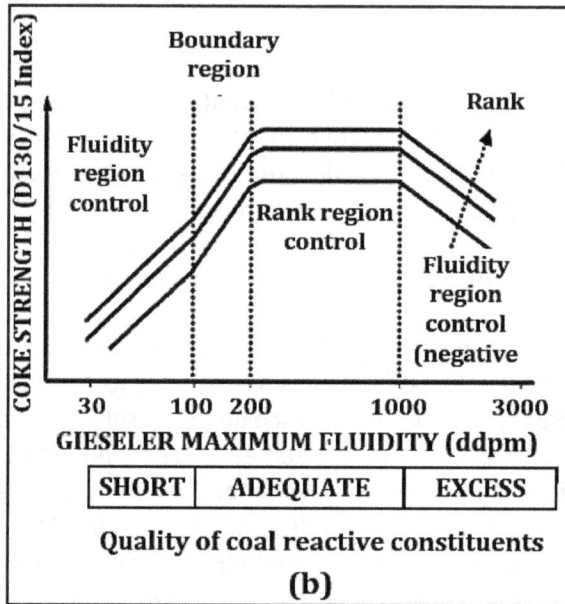

FIGURE 8.24 (a) Thermo-fluidity curves for different ranks of coal (b) Diagram showing the dominant regions of fluidity and the rank of coal that control coke strength. [*(a) Gray et al.,1989; (b) Miura, 1978; Morishita et al., 1986*].

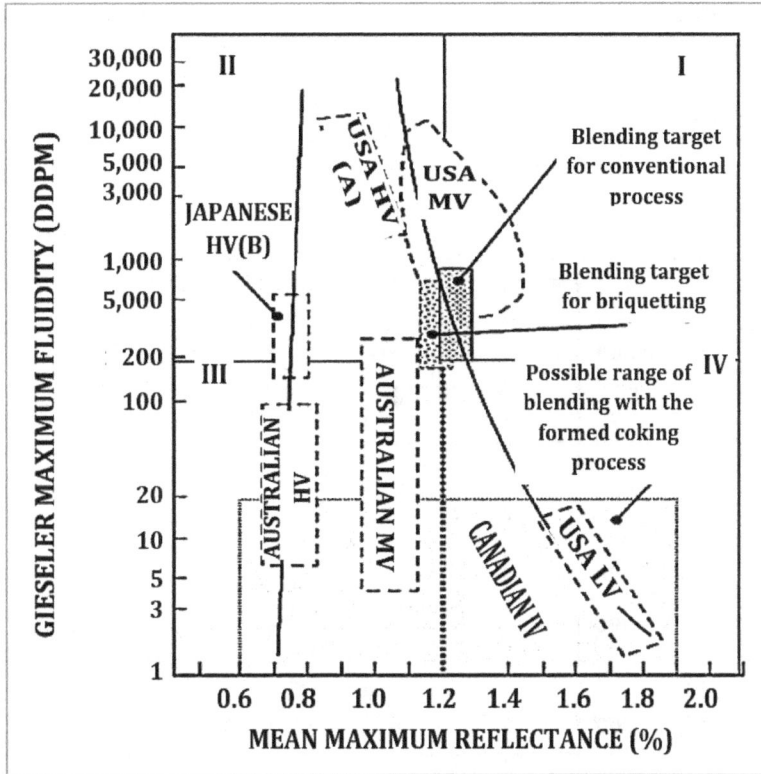

FIGURE 8.25 Relationship between Gieseler maximum fluidity and maximum reflectance of vitrinite for coal blending target (MOF diagram). *(Miyazu et al., 1974).*

The coking window model has been tested extensively using Japanese, American, Australian and Canadian coals (Miyazu et al., 1970) and British, US, Canadian, Australian, Russian and Nigerian coals (Afonja, 1979, 1996) (Figure 8.26). It can be used to predict the individual behavior of a coal in a coal blend and its contribution to the overall quality of coke produced from the blend. For example, coals in quadrants I and II (Figure 8.25) provide fluidity to the blend and largely determine the CSR, coals in group IV largely determine the coke strength while those in group III act as fillers. Careful balancing of the proportions of these different groups in a coking blend using the curve can produce good quality blast furnace coke from a wide range of coking blends. Also, it is possible to incorporate up to 10% inexpensive, finely ground non-coking coals in a coking blend, using this coking window (Afonja, 1983). Fluidity and reflectance values have also been used to predict stability for coals and coking blends as shown in Figure 8.27 (Leeder et al., 1979).

Another Japanese method for ambient temperature prediction of coke quality involves the calculation of a caking capacity index known as the Simonis G factor from the dilation curve of coal or coking blend, using a simple equation (Simonis et al., 1966; Gibson and Gregory, 1977, 1978).

The Simonis method is based on the dilation and contraction that occur in the plastic zone (which is an indication of the viscosity and resistance of the plastic layer to penetration of escaping volatile matter). The G factor is calculated from the variables shown in Figure 8.28 using Equation 8.23.

$$G = \frac{(E+N)}{2} \cdot \frac{(c+d)}{N.c + E.d}$$

[8.23]

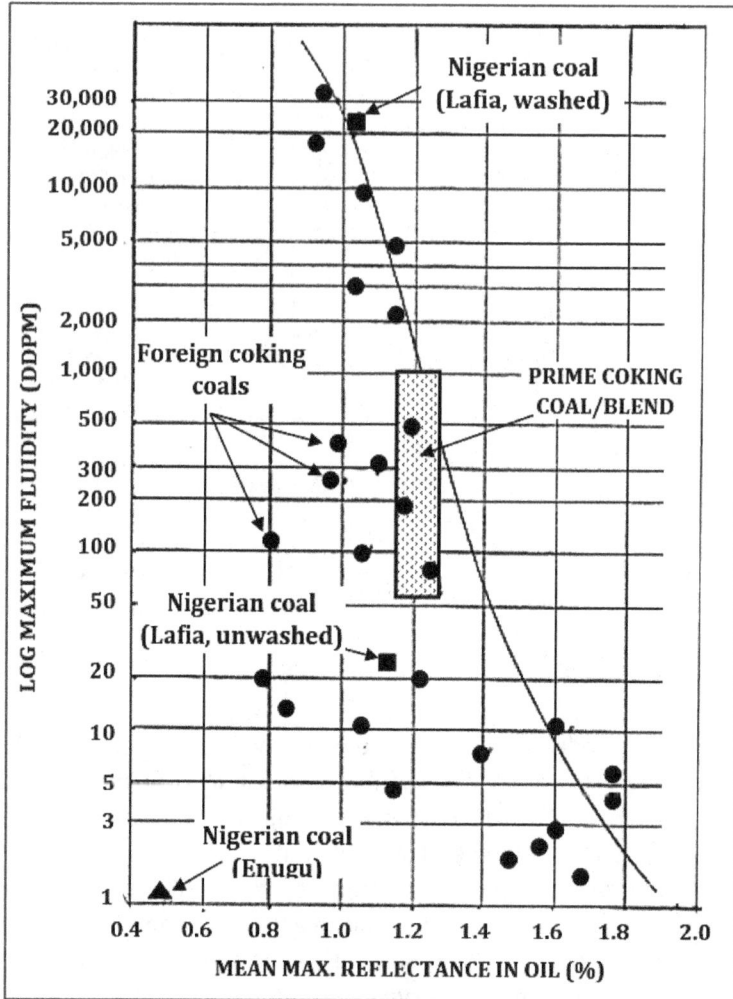

FIGURE 8.26 Prediction of coke strength from plasticity and reflectance measurements. *(Afonja, 1979; 1996).*

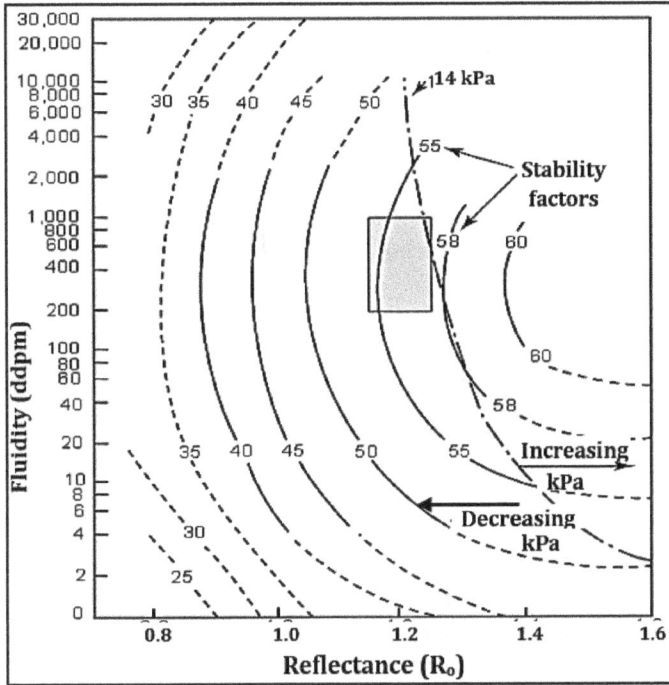

FIGURE 8.27 Prediction of coke stability factors from plasticity and reflectance measurements. *(Leeder et al., 1979).*

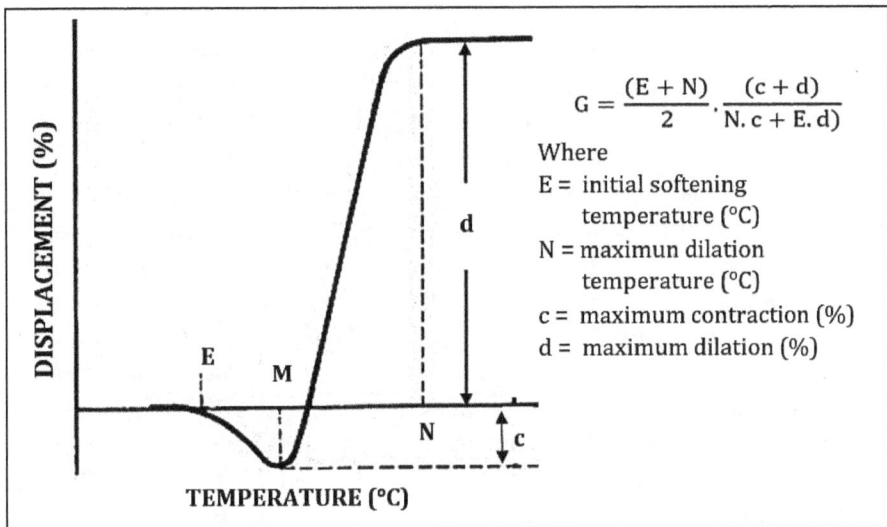

$$G = \frac{(E + N)}{2} \cdot \frac{(c + d)}{N.c + E.d}$$

Where

E = initial softening
   temperature (°C)

N = maximun dilation
   temperature (°C)

c = maximum contraction (%)

d = maximum dilation (%)

FIGURE 8.28 Prediction of coke strength from dilation measurements.
(Simonis et al., 1966).

The Micum drum strength M_{40} and M_{10} can then be calculated from Equations 8.24 (a)-(c).

$$M_{40} = A + BK + CS \qquad\qquad [8.24a]$$

Where A, B and C are strength coefficients, K is the coking conditions factor and S the size distribution factor, A is a rank factor, B and C are coefficients based o volatile matter and the G factor.

$$M_{10} = M_0 + M_1K + M_2\,KS + M_3\,KS^2 \qquad\qquad [8.24b]$$

Where the coefficients M_0, M_1, … are dependent on the volatile matter content, V and the dilatometer G value according to the equation:

$$M_i = C_{i,0}V + C_{i,1}V^2G + C_{i,2}V^3G^2 \qquad\qquad [8.24c]$$

The Simonis model was applied to coal blends containing 20–29% volatile matter and found to give reliable prediction of M_{40} coke strength with a standard deviation between predicted and measured values of 2.4 units. However, when it was applied to British coals of higher volatile matter content, the results were of a similar level of accuracy for coals containing 20 – 30% volatile matter but disappointing for higher-volatile coals, the standard deviation being as high as 13.3. The accuracy of M_{10} prediction was even less, due probably to the relatively high inertinite content of British coals. The Simonis M_{40} equation was modified, based on results of coking tests as shown in Equation 8.25.

$$M_{40} = 103.9 + 24.8G + \frac{1.196V^5}{10^6} + \frac{2.57V^2}{T} - \frac{88V}{T} \qquad\qquad [8.25]$$

Where V = volatile matter of blend (dry, ash free %); T = time (hr) of carbonization to a center temperature of 900°C in an oven of average width of 46 cm.

The G-Factor for most medium and strongly caking coals usually falls between 0.9 and 1.15 and coking blends should be designed to have G values in this range. G Factor is equal to 1 if contraction equals dilation, higher if contraction is less than dilation, and less than 1 if the contraction is greater than dilation. For a given volatile matter there is an optimum G Factor (Figure 8.29). It has been shown that G Factor is linearly additive for coking blends and can be used to predict coke strength (Ramachandran, 1984, Adeleke et al., 2009). The Simonis model is only applicable to a certain range of coals characterized by volatile matter contents between 18 and 35 wt.% dry, ash-free and inert contents below 20%.

Rheological behavior of coal is strongly dependent of rank as well as the chemical and petrographic composition. Coals of different geological age have also been found to have different rheological properties. Western Canadian coals generally have lower rheological properties compared with Australian or US Carboniferous coals of similar rank. This should be noted when predictions of coking quality of coals or coal blends are based solely on fluidity data. Western Canadian coals with fluidity in the range of 3-10 dial divisions minute (ddpm) have been found to make high quality coke compared with 200-500 ddpm recommended for Japanese coals (Figure 8.25). Similar discrepancies have been found with dilation characteristics. Canadian coals have been found to exhibit three to four times lower Ruhr dilation values when compared with US coals of the same rank. The Miyazu coking window and Simonis G Factor should therefore be taken as applicable to coals that fall within the established framework of the investigation that established the optimum values. Pilot plant tests are needed for blends of coals from other sources (Leeder et al., 1979).

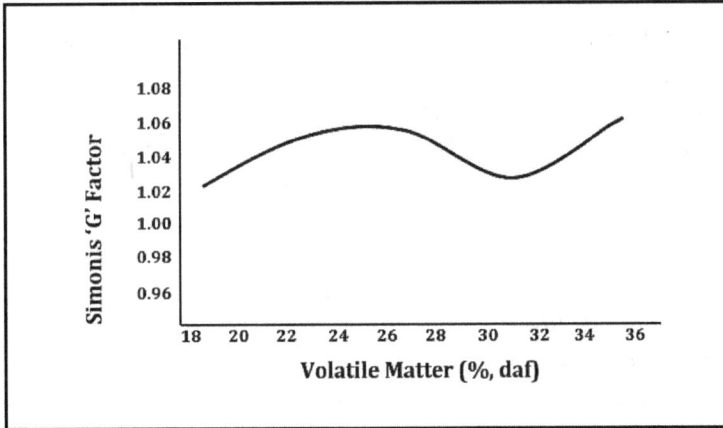

FIGURE 8.29 Relationship between Simonis G factor and volatile matter of coal. *(Simonis et. A\al., 1966).*

It should be noted that all the coke strength prediction methods discussed above were developed using specific coals and have given divergent results for other coals. Hence, each model should be considered as applicable within the spectrum of coals and process variables in which it was developed. None of these models is universally applicable but many coke oven plants have developed a variant of one method or the other (in particular, the hot coke strength prediction models) using coals from their databases. Hence there are now many variants of the established models adapted for local use by coke plants all over the world.

REFERENCES

Adeleke, A. O., Olulana, A. B., Adahams, R. S. and S. A. Ibitoye (2009). " A prediction of Micum strength of metallurgical coke using Ruhr dilatometer parameters of parent coals," Petroleum and Coal, Vol. 51 (2), pp. 75-79.

Aderibigbe, D. A. and Szekely, J. (1982). "Studies in coke reactivity: Part 2, Mathematical model of reaction with allowance fore pore diffusion and experimental verification." Ironmaking and Steelmaking, Vol. 9, No. 1, pp. 32-40.

Afonja, A. (1979). "Further studies on the Lafia coal deposit (Phase II." Research Report No.IFE/CHE//CO/11. Nigerian Steel Development Authority. 172p.

Afonja, A. A. (1983). "Optimization of the use of a sub-bituminous coal in conventional and form-coking." Fuel Processing Technology., 7(3), pp. 293 - 314.

Afonja, A. (1996). "Production of metallurgical coke from non-coking coals." In Nigerian Coal: A Resource for Energy and Investments. Okolo, H. And M. Mkpadi, Eds., pp. 89-109. Raw Materials Research Council of Nigeria.

Ahland, E., Bock, B. Jagnow, H. J. and Lehman, J. (1971). "Hot briquetting processes for the production of fuel for the metallurgical industry." Proceedings of the 12th Annual Conference of the Institute of Briquetting and Agglomeration, Vancouver, Canada.

AISI (2010). "Technology Roadmap Research Program for the Steel Industry." American Iron and Steel Institute, www.steel.org.

Amosov, I. I., Eremin, I. V., Sukhenko, S. F. and L. S. Oshurkova (1957). "Calculation of coking charges on basis of petrographic characteristics of coals." Coke Chem. U.S.S.R.,

Vol. 2, pp. 9-12.

ASTM D5341-99 Standard Test Method for Measuring Coke Reactivity.

Battle, T., Srivastava, U., Kopfle, J., Hunter, R. an J. McClelland (2014). "The Direct Reduction of Iron." In Treatise on Process Metallurgy, Industrial Processes Part A. Seshadri Seetharaman (Ed.). Elsevier.

Beer, J. M. (2000). "Combustion technology developments in power generation in response to environmental challenges." *Progress in Energy Combustion Science*, Vol. 26 (2000), pp. 301-327.

Bernard, A. Duchene, J. M. and D. Isler (1985). Etude de la gazefication du coke." *Rev. Met. CIT,* Vol. 82, pp. 849-860.

Bernard, A. Duchene, J. M., Vogt, D., Jeulin, D., Steiler, J. M., Bourrat, X., and J. N. Rouzaud (1986). "Effect of rank and of interactions of coals on experimental coke properties." Proc. Ironmaking & Steelmaking Conf., ISS-AIME, pp. 211-219.

Brown, H. R., Taylor, G. H. and A. C. Cook (1964). "Prediction of coke strength from the rank and petrographic composition of Australian coals." *Fuel*, Vol. 43, pp. 43-54.

Bustin, R. M., A. R Cameron (1983). *Coal Petrology, its principles, methods and applications.* 2nd edition: Geological Association of Canada. Short Course Notes 3, 230p.

Carr, A. J. and Jorgensen, J. G. (1975). "An estimate of the effective reactive/inert ratio of semifusinite in Western Canadian coals." CANMET Rep. 75-141, 12 pp.

Colleta, A., Barnaba, P., Federico, G. (1990)."Influence of coal properties on hig coke quality for blast furnace." Proc. 49th Ironmaking Conf., ISS-AIME, pp. 243-252.

Dartnel, J. (1978). "Coke in the blast furnace." Coal, Coke and Blast furnace. Metal Society Conference, Middlesborough.

Diez, M. A, Alvarez, R., and C. Barriocanal, (2002). "Coal for metallurgical production: predictions of coke quality and future requirements for cokemaking." International Journal of Coal Geology 50 (2002) pp. 389-412. Elsevier.

DIN 51717 (2016). *Testing of solid fuels – Determination of mechanical strength of coke.*

Gibson, J. and D. H. Gregory (1978). "Selection of coals and blend preparation for optimum coke quality." *The Cove Oven Managers' Association (COMA) Year-Book, Mexborough,* pp. 159-182.

Goldring, D. C.(1970). "Petrography of coal and its applications to coking." *Coke in ironmaking*. The Iron and Steel Institute, pp. 11-29.

Gray, R. J. and D. E. Lowenhaupt (1989). "Aging and Weathering." *Sample, Selection, Aging and Reactivity of Coal*, pp. 225-234.

Grint, A. And H. Marsh (1981). "Carbonization of coal blends: mesophase formation and coke properties." *Fuel*, Vol. 60, Issue 12, pp. 1115-1120.

Hara, Y., Mikuni, O., Yamamoto, H. and Y. Yanamaka, (1980). "The assessment of coke quality with particular emphasis on sampling technique." In Lu, W. K. (Ed.) *McMaster Symposium on Iron and Steelmaking*, McMaster University, Vol. 8, pp. 4.1 – 4.38.

Hasanbeigi, A. and L. Price (2013). "Emerging Energy-Efficiency and CO_2 Emissions-Reduction Technologies for Industry: A Review of Technologies for Alternative Ironmaking and Pulp and Paper Industry." ACEEE Summer Study on Energy Efficiency in Industry. Aceee.orgfiles/proceedings/2013/data/papers/1_027.pdf

Hatano, M., Miyazaki, T. and Y. Iwanaga (1990). "Analysis of blast furnace operation based on the mathematical model of evaluation of coke properties." *Coke Reactivity and its effect on Blast Furnace Operation.* The Iron and Steel Society, pp. 21-34.

Hoffman, H. and K. Hoehne (1954). *Brennstoff-Chemie*, Vol. 35, p. 302.

IEA (2007). *Tracking Industrial Energy Efficiency and CO₂ Emissions.* International Energy Agency. www.iea.org/w/bookshop/add.aspx?id=298.

IEA (2009). *Energy Technology Transitions for Industry – Stages for the Next Industrial Revolution.* International Energy Agency, www.iea.org.

IIP (2016). *Direct Reduced Iron. Industrial Efficiency Technology Database.* Institute for Industrial Productivity. Iipnetwork.org/content/direct-reduced-iron. Accessed 2016.

ISO 556 (1980}. *Coke (greater than 20mm in size} – Determination of mechanical strength.*

ISO 18894 (2006). *Coke, Determination of coke reactivity index (CRI).*

JIS 2150-72 (1972). *Testing Method for Coke Strength.*

Kohna, T. (1988). "Substitution and reduction of coking coal in cokemaking." *Coal and Coke Technical Exchange Session, International Iron and Steel Institute Committee on Technology,* Brussels, April 1988., pp. 85-101.

Komaki, I., Itagaki, S. and T. Miura (2005).*Structure and Thermoplasticity of Coal.* New York: Nova Science Publishers Inc.

Leeder, W. R. et al., "Prediction of Coke Quality with reference to Canadian Coals." *ISS-AIME Proc.,* Vol. 38, pp. 385-397.

Loison, R., Foch, P. and A. Boyer (1989). *Coke, Quality and Production.* Butterworth, London.

Lowry, H. H. (Ed.) (1963). *Chemistry of Coal Utilization, Supplementary Volume.* John Wiley.

Mackowski, M. (1977). "Prediction methods in coal and coke microscopy." *Journal of Microscopy,* 109(1), pp. 119-137.

Menendez, J. A., Alvarez, R., and J. J. Pis (1999). "Determination of metallurgical coke reactivity at INCR: NSC an ECE-INCAR reactivity tests." *Ironmaking and Steelmaking,* Vol. 26, pp. 117-121.

Michishita, H. And H. Tanaka (2010). "Prospects for Coal-based Direct Reduction Process." *Kobelco Technology Review* No. 29, Dec. 2010.

Midrex (2016). *2015 World Direct Reduction Statistics.* www.midrex.com.

Miura, Y. (1978). "The science of cokemaking technology and its development in Japan." *The Coke Oven Managers' Association (COMA) Year-Book,* Mexborough, pp. 292-311.

Miyazu, T. Okuyama, Y., Sugimura, H. and M. Kumagai (1970). J. *Fuel Soc.,* Japan.

Miyazu, T. (1974). "The evaluation and design of blends using many kinds of coals for cokemaking." *Int. Iron and Steel Congress,* Dusseldorf, Pap 1...2.2.1.

Morishita, N., Tsukada, K., Suzuki, N., and K. Nemoto (1986). "Development of automatic coal/coke microscopic analyzer and its application to cokemaking." Proc., 45th Ironmaking Conference, ISS_AIME, pp. 203-209.

Nakamura, N., Togina, Y. and T. Tateoka (1977). Behaviour of coke in large blast furnace." *Coal, Coke and Blast Furnace.* The Metals Society, London, pp. 1-18.

Nandy, B. N. and D. S. Montgomery (1975). "Nature and Thermal behaviour of semi-fusinite in Cretaceous coals from Western Canada*." Fuel,* Vol. 54, pp. 193-196.

Nekrasov, Z. I., Ketov, K. I., Gladkov, N. A., Zhembus, M. D., and Goncharov, V. F. (1974). "Some results from test melts with formed coke." Steel in the USSR, Vol. 29, pp 556-568.

Ochi, Kenichi et al., (2009). "Latest Low-Nox Combustion Technology for Pulverized-coal-fired Boilers." Hitachi Review Vol. 58 No. 5, pp. 187-193.

O'Donnell, E. M. and J. J. Poveromo (2000). "Coke Quality Requirements from a North American Perspective." *Proc., Paris2000*, pp. 12-19. European Ironmaking and Cokemaking Congress, Paris.

Osborne, D. (2013). *The Coal Handbook: Towards Cleaner Production.* Elsevier

Pearson, D. E. and J. T. Price (1985). "Reactivity of inertinite (coal-typing) of Western Canadian coking coal." *Proc. Int. Conf. on Coal Science.* Pergamon press, Sydney, p 907.

Pearson, D. (1998). "Fusible inertinites in coking coals*." ICST Itonmaking Conference Proceedings*, pp. 753-755.

Ramachandran, P. (1984). *Dilation Characteristics of Western Canadian Bituminous Coals."* M. S. Thesis, University of Waterloo, Ontario, p. 236.

Robertson, A., R. et al., (1989). *Second Generation Pressurized Fluidized Bed Combustion Plant.* Foster Wheeler Dev. Corp. Report to the US DOE DE-AC-21-86MC21023.

Rooney, K. A., Gregory, O. R., Nightingale, R. J., and K. Price (1987). "Developments in coke quality and the improved performance of a 300 cubic meter blast furnace." 1st Cokemaking Cong., Essen, Vol. 1, C3, 17pp.

Sakanishi, K. and M. Nishimura (2005). "Formation mechanisms of Coke Texture during Resolidification." *Structure and Thermoplasticity of Coal,* pp. 119-131.

Schapiro, N., Gray, R. J. and G. R. Euser, (1961). "Recent developments in coal petrography." *Proc, Blast Furn., Coke Oven and Raw Matls. Committee, American Institute of Mining Engineers (AIME),* Vol. 20, pp. 89-112.

Schapiro, N. and Gray, R. J. (1964). "The use of coal petrography in coke making." *Jnl., Institute of Fuel,* Vol. 9(30), pp. 234-242.

Speight, J. G. (2015). *Handbook of Coal Analysis,* John Wiley. (Second Edition, 2015).

Simonis, V., Gnuschke, G. and K., G. Beck. (1966). "On the preparation of optimum blends of coking coal to achieve maximum M-40 hardness values of coke." *Gluckauf-Forschungshefte,* Vol. 27, p. 105*.*

Steyn, J. G. D. and Smith, C. H. (1977). " Coal petrography in the evaluation of South African coals*." Coal, Gold and Base Minerals of South Africa,* pp. 107-117.

Taylor, G. H., M. Th. Mackowsky and B. Alpern (1967). "The behaviour of inertinite during carbonization." *Fuel,* Vol. 46, pp. 431-440.

Teir, S. (2002). "Modern Boiler Types and Applications." Helsinki University of Technology Energy Engineering and Environmental Protection Publications Steam Boiler eBook.

US DOE (2003). "SL/RN Process of Direct Reduction of iron." U.S. Department of Energy, Washington, DC. (See http://ietd.iipnetwork.org/content/ -process.

van Krevelen, D. W. (1961;1963). *Coal. New York, Elsevier.*

Valia, H. S. et al. (1989). "Production and Use of High CSR Coke at Inland Steel Company. " *ISS-AIME Prc.*, Vol. 48, pp. 136-146.

Vilia, H. S. (1990). "Effects of Coal Oxidation on Cokemaking." *ISS-AIME Proc.,* Vol. 49, pp. 199-209.

Vilia, H. S. and W. Hooper (1994). "Use of Reverts and Non-Coking Coals in Metallurgical Coke-making." *ISS-AIME Proc.*, Vol. 53, pp. 89-104.

World Coal Association (2014) "Coal Facts, 2014." www.worldcoal.org

Zimmerman, R. E. (1979). "Evaluating and Testing the Coking Properties of Coal." *Miller Freeman*, San Francisco, 1979, p. 144.

9   Coal gasification

9.1 INTRODUCTION

Archeological evidence supports the common belief that extraction of flammable gas from coal (coal pyrolysis) was practiced from ancient times. In 1609, Jan Baptista Van Helmont, a Belgian chemist discovered that gas could be produced from heating wood or coal and, for the next two centuries, the process was developed and refined by several others. However the practice only became widespread towards the end of the eighteenth century (Higman and Tam, 2013; WCA, 2014).

The first commercial coal gasification plant began operation in 1813 and supplied coal gas, also known as town gas for street lighting in London, United Kingdom and, within a few years, the technology spread to other parts of Europe. The process has remained a vital technology for the supply of industrial and domestic energy, and production of prime industrial chemicals despite the emergence of alternative technologies. Currently, around a quarter and a third of the world's requirements of synthetic ammonia and methanol respectively is produced from coal gasification and, with nearly a thousand new plants planned or under construction worldwide, coal gasification will remain a vital technology in the foreseeable future. Increasing concerns about the anthropogenic effect of burning coal directly for power generation have prompted a rapid proliferation of coal-to-syngas systems now widely deployed as part of power generating plants or chemicals production complexes. This development has been propelled by the relatively lower carbon footprint of syngas compared with direct coal combustion and the ease with which by-products can be captured and disposed of safely.

9.2   COAL GASIFICATION TECHNOLOGIES

The reactions that occur in a coal gasifier and the types of products depend on the coal rank and process operating conditions, in particular, temperature and C:O ratio. Volatile matter released can account for as much as 50% of the heating value of coal and for up to 70% of the weight loss. The chemical composition of the gas, tar and liquor produced also varies depending on both coal rank and type, and the maximum temperature of carbonization. Gas yield is highest in low rank coals and lowest in high-rank coals. On the other hand tar yield is lowest in low-rank coals and highest in high-volatile bituminous coals. Also, coals which have high liptinite maceral content have high yield of gas and tar. Coal is a large polymeric matrix comprising mostly aromatic ring systems cross-linked by aliphatic and ether bridges (Figure 9.1). The degree of aromaticity increases with rank.

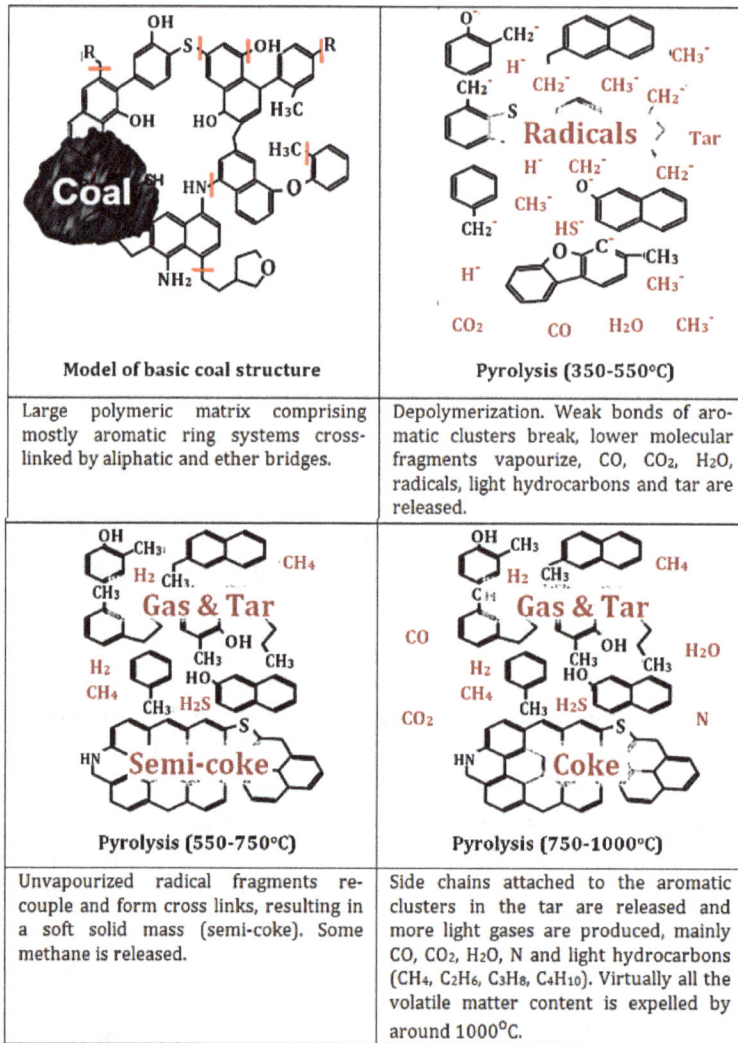

Model of basic coal structure	Pyrolysis (350-550°C)
Large polymeric matrix comprising mostly aromatic ring systems cross-linked by aliphatic and ether bridges.	Depolymerization. Weak bonds of aromatic clusters break, lower molecular fragments vapourize, CO, CO_2, H_2O, radicals, light hydrocarbons and tar are released.
Pyrolysis (550-750°C)	Pyrolysis (750-1000°C)
Unvapourized radical fragments re-couple and form cross links, resulting in a soft solid mass (semi-coke). Some methane is released.	Side chains attached to the aromatic clusters in the tar are released and more light gases are produced, mainly CO, CO_2, H_2O, N and light hydrocarbons (CH_4, C_2H_6, C_3H_8, C_4H_{10}). Virtually all the volatile matter content is expelled by around 1000°C.

FIGURE 9.1 Changes in coal molecular structure during pyrolysis.
(Adapted from Zhenyu et. Al., 2012).

Coal pyrolysis and the accompanying devolatilization are very complex processes because of many parallel and cross-linking reactions at different rates and in different temperature regimes. However, they can be classified broadly into two steps: primary and secondary pyrolysis. In the temperature range 350°C to around 700°C, depolymerization and cross-linking reactions occur. Bonds linking aromatic clusters break, lower molecular fragments vaporize and condense from the flue gases as liquid tar at room temperature. Some CO, CO_2, H_2O and light hydrocarbons are produced as a result of the detachment of some functional groups linked to the aromatic clusters. At around 500-550°C, the unvapourized aromatic fragments re-consolidate and form cross links in the solid residue (char or semicoke) structure. These reactions mark the beginning and end of what is commonly referred to as the plastic or rheological phase in coal carbonization technology. The beginning and end temperatures of the phase, the range, the extent of fluidity and swelling in the temperature range are all critical parameters in the evaluation of coal for metallurgical applications, and are dependent on the coal rank and type. In the temperature range of around 750°C to 850°C more cross links are broken and high-temperature tar is produced. Further increase in temperature causes side chains attached to the aromatic clusters in the tar to be released and more light gases are produced, mainly CO, CO_2, H_2O, N and light hydrocarbons (CH_4, C_2H_6, C_3H_8, C_4H_{10}). Virtually all the volatile matter content is expelled by around 1000°C. The residue from this secondary pyrolysis is coke and contains only 1-2% volatile matter. The quantity and chemical composition of the gases and tar released during coal pyrolysis depend on the rank and type of coal, pyrolysis temperature, as well as the heating rate. High volatile coals yield more gas and tar while rapid heating promotes faster degasification (Figure 9.2).

The modern coal gasification converts coal into combustible gas, widely known as synthetic gas or syngas which has much higher calorific value than the traditional coal gas because of the lower nitrogen content. The same process can be applied to any other material containing carbon such as petroleum coke, biomass, heavy oil. There are two basic types of gasification: coal pyrolysis (low-temperature gasification), and gasification (high-temperature gasification)

9.2.1 Low-temperature gasification of coal

When coal is heated in minimum air up to 650 to 750°C it becomes incandescent but does not burn. Gases start to be released from around 250-300°C depending on the rank of coal. Tar is first released but as temperature rises, the coal bed becomes incandescent and reactions between carbon, oxygen or hydrogen occur, leading to the production of mainly carbon monoxide but also some hydrocarbons, mainly methane. The coal-oxygen reaction is exothermic and self-sustaining. The gas produced , known as coal or town gas has low calorific value of 3.5-6 MJ/m^2 due to the high content of nitrogen, typically around 55% which comes from the injected air.

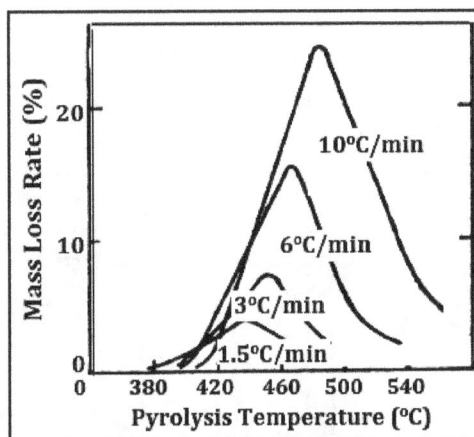

FIGURE 9.2 Effect of heating rate on devolatilization of coal during pyrolysis. *(Zhenyu Liu et al., 2012).*

However the introduction of steam into the input air results in a reaction between steam and carbon with the production of hydrogen and carbon monoxide (Figure 9.3). The C-H (water gas) reaction is endothermic and air and steam are introduced alternately to the incandescent bed to sustain the gasification process. The gas produced is known as water gas or blue water gas and, compared with coal gas, has a much higher calorific value of 11-12 MJ/m³. Some plants use coke but the heating value of the gas produced is lower because raw coal also produces combustible hydrocarbons which boost the heating value of the gas produced. Production of coal gas has become obsolete but water gas is still in use in many parts of the world especially in countries that have abundant coal resources. The two most popular types of low temperature coal gasification technologies are the Koppers and Lurgi processes.

The fundamental chemical reactions occurring in the gasification process are summarized in Table 9.1. The most important reaction is the gasification (water gas) reaction between carbon and steam that produces hydrogen and carbon monoxide (Equation 9.3). However it is an endothermic reaction and needs the energy released from the exothermic combustion reactions between carbon and oxygen (Equations 9.1 and 9.2) to proceed. Some carbon monoxide produced may also react with steam to produce more hydrogen (Equation 9.5). The hydrogenating gasification reaction between carbon and hydrogen (Equation 9.6) may occur resulting in the production of methane while carbon monoxide may also react with hydrogen to produce methane as well (Equation 9.7).

Blue water (town) gas can be produced from any coal rank, from anthracite to lignite (brown) coal and the fuel dominated the world's energy scene for several hundred years. There were many types of blue water gas plants and most operated at atmospheric pressure but some advanced Lurgi plants which are still in use operate at high pressure and produce town gas with calorific value as high as 20 MJ/m³. The post-gasification residue known as semi-coke or char is also a valuable smokeless fuel. Developments in the petroleum industry since the 1950s have provided cheaper alternative fuels (notably natural and liquefied petroleum gas) and virtually eliminated the production of town gas, although it is still being produced in some developing countries. Also, the process has evolved into the modern coal-to-syngas technology which is now the starting point for many modern coal-powered power plants all over the world.

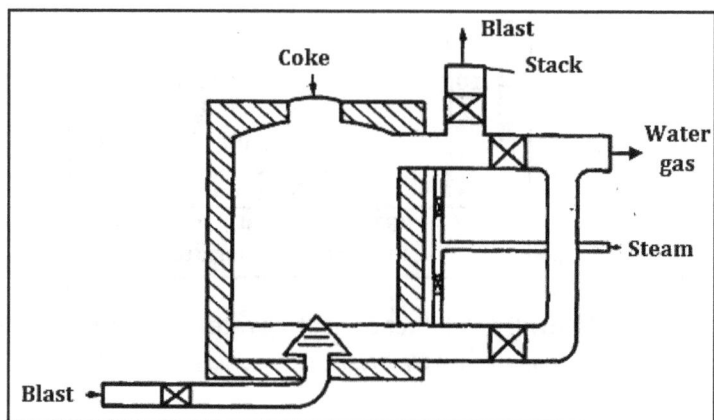

FIGURE 9.3 Blue water gas reactor. *(Francis, 1965).*

TABLE 9.1 (a) Producer and water gas reactions
(b) Typical chemical composition of producer gas and water

(a)

$C + O_2 \rightarrow CO_2$	$\Delta H_r = -393.4$ MJ/kmol	[9.1]
$C + \frac{1}{2} O_2 \rightarrow CO$	$\Delta H_r = -111.4$ MJ/kmol	[9.2]
$C + H_2O \rightarrow H_2 + CO$	$\Delta H_r = 130.5$ MJ/kmol	[9.3]
$C + CO_2 \leftrightarrow 2CO$	$\Delta H_r = 170.7$ MJ/kmol	[9.4]
$CO + H_2O \leftrightarrow H_2 + CO_2$	$\Delta H_r = -40.2$ MJ/kmol	[9.5]
$C + 2H_2 \rightarrow CH_4$	$\Delta H_r = -74.7$ MJ/kmol	[9.6]
$CO + 3H_2 \rightarrow CH_4 + H_2O$	$\Delta H = -206$ MJ/kmol (kJ/mol)	[9.7]

(b)	**Producer Gas**	**Water Gas**
Carbon monoxide (%)	29	40
Hydrogen (%)	10.5	50
Carbon dioxide	5.5	5
Nitrogen	55	5

Gasifiers operate at higher temperatures and the gas produced is of higher calorific value. It is also a precursor to the production of hydrogen and many valuable industrial chemicals.

9.2.2 Smokeless coal fuel production

Coal briquetting has been practiced in many parts of the world since the nineteenth century. In fact, the first patent for a briquetting machine taken out in 1848 in the United States was for "the formation of small particles of any variety of coal into solid lumps by pressure." The technology quickly became widespread, applied to a wide variety of materials from minerals to pharmaceutical products. In the earliest coal briquetting processes coal was crushed, sized by screening, mixed with a binder before compaction in a high-pressure briquetting press. The products were used as industrial and domestic fuel. The binder used initially was about 10% of starch or 6% molten asphalt, both of which produced considerable smoke during burning. Further development led to the introduction of low-temperature carbonization processes with auxiliary facilities for the removal of the tar in coal and processing to obtain smokeless pitch binder.

Although coal briquetting specifically for production of fuel has been virtually phased out in many developed countries because of the wide availability of petroleum-based alternative fuels, the technology is still widely practiced in many countries of Asia and Europe. Also, valuable coal mine fines which would otherwise have been discarded, coal transported to consumers through pipelines, coal that had to be finely ground to remove sulphur, and char residues of low-temperature coal gasification are still being agglomerated by briquetting. . Furthermore, some iron blast furnaces and coke ovens in Europe and Asia use carbonized coal briquettes (formed-coke) as charge blend components and some foundry cupolas operate entirely on formed-coke.

Char (semicoke) produced from low-temperature carbonization of caking coal is solid-spongy, dull gray, porous and friable, with 10 – 25% volatile matter retention depending on the rank of the feed coal. When heated it releases copious hydrogen, it is easily lightable and burns freely with smokeless, almost non-luminous flame. These features make it an ideal domestic fuel. Liquid and gaseous fuels have replaced char in some parts of the world, in particular, the United States. However, in Europe and China, much of domestic heating is still from coal char briquettes. Some industrial boilers are also fired by char. Chars are used as reducing agents in copper and phosphorus production as a cheaper replacement for traditional high-temperature coke, and for the production of electrodes.

Chars also have good potential as feedstock for producer gas plants, and for the production of hydrogen and methane. Char is too reactive and structurally weak for direct use in blast furnaces but some quantities are used in coal blends with considerable degree of success. Char residue from low-temperature pyrolysis of non-caking low-rank and high-rank coals is similar chemically to the residue from caking coals but decrepitates into a granular form and is of little value as a fuel unless it is mixed with a binder and agglomerated. High-rank coals (anthracite and semi-anthracite) are also pyrolyzed, not for gas production since the volatile matter is very low (6-8% compared with up to 50% in some low-grade coals), but for the production of high-grade char.

The appropriate temperature of carbonization for char production depends on the type of coal and carbonizing technology adopted. Fluidized-bed carbonizers are the most efficient since they can operate at the lower temperatures of 500 to 600°C, carbonization time is shorter, and heat required is internally generated. The product of this process is also granular. Chars from low-grade, non-caking coal and anthracite pyrolysis both of which are in granular form are mixed with a binder (caking coal, pitch, petroleum bitumen, starch, molasses, etc) and compacted into briquettes. Some hydrated lime may be added to remove sulphur and some clay may also be added to improve compaction. It is possible to briquette some coals without the use of a binder if the pressure is high enough or if the coal has some rheological properties and is pre-heated to around 450°C, followed by briquetting at very high pressure while there is some fluidity.

Formed semi-coke briquettes may be calcined in an oven at around 450°C to remove any residual tarry volatile matter and improve briquette strength. The product of smokeless char compaction is widely known as smokeless fuel being marketed under a very wide variety of proprietary names. It is used widely in both developed and developing countries for domestic and Industrial heating and barbecues (phurnacite, ancit, taybrite, etc). A schematic representation of the smokeless fuel production technology is presented in Figure 9.4.

9.3 BY-PRODUCT COAL GASES

Combustible gas is produced during the carbonization of coal in coke ovens, and also in the iron-making blast furnace. Comparative values of typical compositions of gases from coal are given in Table 9.2. Clearly, coke oven gas has the highest heating value owing to the relatively high content of hydrogen and methane. However, it also contains coal dust which needs to be separated by filtration. As mentioned earlier, carbonization of coal specifically for the production of producer and blue water gas has become obsolete although it is still being practiced in some developing countries. Also, the gases are by-products of smokeless fuel production and are still used for steam raising, home heating, etc. on the other hand coke oven gas is a very valuable fuel, used in many industrial applications.

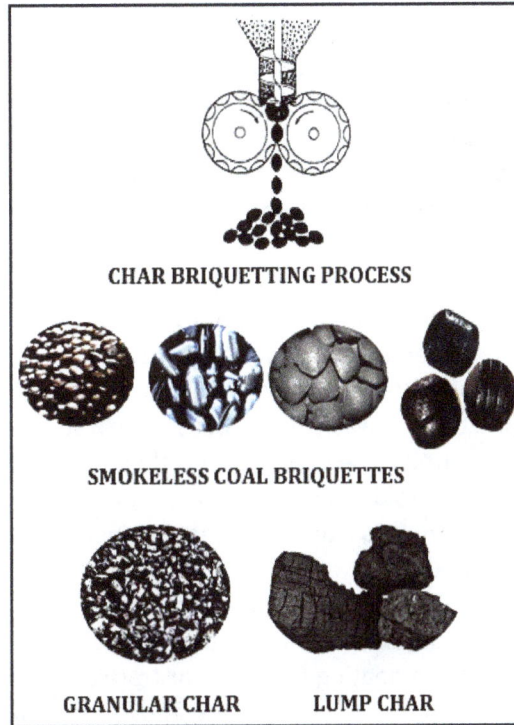

FIGURE 9.4 Smokeless coal and char products.

TABLE 9.2 Typical composition of gases from coal carbonization.

COMPOSITION	PRODUCER GAS	BLUE WATER GAS	COKE OVEN GAS	BLAST FURNACE GAS
CO_2	5	4.7	2	11
CO	25	41.0	8	27
H_2	15	49.0	55	2
CH_4	2	0.8	31	-
N_2	53	4.5	4	60
Calorific Value (kJ/m^3)	5,885	10,990	19,370	3,725

9.3.1   Coke oven gas

Although the operation of the coke oven is cyclic, a typical battery contains a sufficiently large number of ovens to produce an essentially continuous flow of raw gas which is processed to recover valuable by-products. Coke oven gas accounts for 20 to 35% by weight of the initial coal charge and is composed of moisture, tar, light oils (primarily benzene, toluene and xylene), heavy hydrocarbons and many other chemical compounds. The gas exits the oven at 750 to 870°C and, aided by exhausters, moves through a stripping system in which it is rapidly cooled to about 100°C by spraying with recycled flushing amoniacal liquor. The tar in the gas precipitates in a series of cooled tanks and serves as a medium for the condensation of other vapours, collectively known as liquor. The crude tar is separated from the liquor in a decanter and further processed to yield purified tar and tar derivatives. The gas then passes through another set of tar tanks or through electrostatic precipitators to remove any residual tar. Only a small amount of the liquor is condensed in the tar extractor and the gas is further processed by passing it through cooled water to recover the ammonia as an aqueous solution, or as ammonium sulphate by passing it through a saturator containing a solution of 5 to 10% sulphuric acid. The ammonium sulphate is crystallized and dried.

The gas exiting from the saturator at about 60°C is cooled to about 25°C in a series of condensers to remove water, benzene and naphthalene. The gas then passes into a light oil scrubber in which it is sprayed with heavy petroleum fraction (wash or straw oil). Most of the light oil is removed and the rich wash oil is processed in a stripping unit with steam to recover the light oil and water vapour. The stripped oil is then recycled to the light oil scrubber. The recovered light oil is either sold as crude or further processed to recover benzene, toluene, xylene and solvent naphtha. Gas that has been stripped of tar, ammonia and light oil is further processed to remove SOx gases, and the final product is clean gas of high calorific value of around 17 to 20 MJ/m³. Coke oven gas has high calorific value due to the high hydrogen and methane contents. The gas is used extensively in steel plants for firing billet re-heating furnaces, gas pre-heating furnaces, or to raise steam for power generation. Not all coke ovens process the flue gas as described above. Many plants sell off recovered crude tar and ammoniacal liquor to independent refiners, others do not process by-product gas at all and may recycle part of the gas to preheat coal charge while the balance is burned to produce steam and electric power.

9.3.2   Blast furnace gas

Blast furnace gas contains more carbon monoxide and nitrogen, and much lower hydrogen and methane compared with coke oven gas, hence its significantly lower heating value (Table 9.2). The gas contains considerable dust which is usually filtered out before being used mainly in heating regenerative heat exchange units for pre-heating blast gas, or it may be enriched with coke oven gas to raise its calorific value for use in re-heating furnaces or steam raising for power generation.

9.4   HIGH-TEMPERATURE GASIFICATION OF COAL (SYNGAS PRODUCTION)

Production of town gas from coal has become obsolete due to the availability of natural gas and increasing pollution control regulations which have made production unprofitable because of the need to install expensive gas cleaning processes. However, there has been a renewed interest in the development of environmentally friendly technologies for the conversion of coal to low-pollution, combustible gas known as synthetic natural gas or syngas. Coal gasification has become one of the most versatile and clean ways to convert coal into a valuable gas which fuels electric power generation and serves as feedstock for a wide range of valuable products including hydrogen and industrial chemicals (Figure 9.5). Coal syngas is currently produced in gasifiers but gasification in-situ of coal beds is also becoming common.

Syngas has many advantages over coal as fuel. It is very similar to natural gas and the carbon footprint is around 40% less than that of coal. Furthermore, production of other anthropogenic gases, in particular SOx and NOx gases is extremely low and both can be captured and processed into valuable chemicals. If oxygen is used instead of air in a coal gasifier, carbon dioxide produced is in a concentrated form at high pressure, hence can be captured and sequestrated relatively easily. Another major advantage of syngas is the relative ease of handling and transportation through pipelines. This greatly enhances the value of syngas produced in-situ from coal seams underground. A syngas plant is also easily deployed as part of a highly efficient electric power and heat production system, for example, an Integrated Gasification Combined Cycle (IGCC) power plant, with adequate provision for capturing and processing anthropogenic by-products. Increasingly stringent pollution limits and natural gas price increases, coupled with the world's extensive and geographically widespread coal reserves have generated renewed interest in the centuries old coal gasification technology in the last two decades or so. Also, the gas is a rich source of chemicals, including hydrogen, methanol and ammonia.

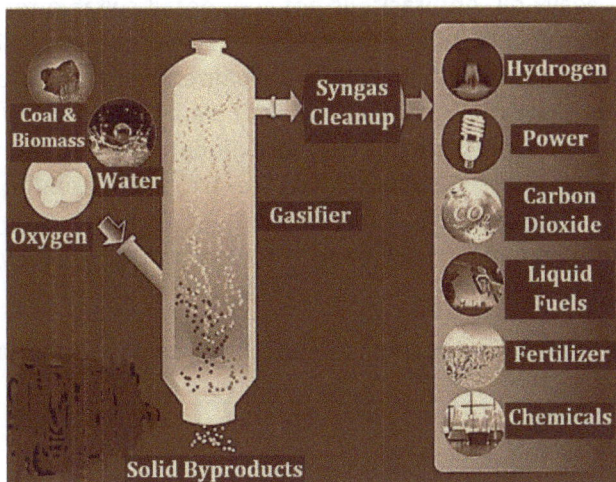

FIGURE 9.5 Syngas production and utilization. *(netl.doe.gov).*

9.4.1   Above-Ground coal gasification

The core of a syngas plant is the gasifier which is a high-pressure, high-temperature vessel. A carefully controlled amount of oxygen and steam are introduced into a coal bed in the gasifier to facilitate partial combustion of coal and produce the energy required for the endothermic reactions between carbon and steam. There are many types of gasifiers with different configurations and process details. Coal feed may be in dry or coal-water slurry form and operating conditions may be pressurized, non-pressurized, autothermal, or alothermal. In the autothermal process, heat required for gasification reactions is supplied internally by partial oxidation of coal with air or oxygen/steam mixture whereas in the alothermal process, heat for gasification reactions is supplied from an external source. The chemical composition of the syngas produced depends on the coal type, operating conditions (temperature, pressure, heating rate, particle size, residence time) and gasifier design (Table (9.3).

Unlike combustion and carbonization of coal which take place under oxidizing conditions, coal gasification for the production of syngas occurs under reducing conditions. When coal is reacted with steam at high temperature and pressure in a pressure vessel (gasifier), the product is a mixture of carbon monoxide and hydrogen. The gas has good combustion properties similar to natural gas although lower in heating value. Any type of coal (both caking and non-caking) can be gasified and the main focus currently is on coals that are not suitable for other applications and therefore have low market value in order to mitigate the added cost of gasification.

Coal produced cheaply by opencast mining is particularly suitable and high ash coals (20-50% wt), low-rank coals (brown coals and lignite) are of special interest for gasification. It should be noted however that capital and operating costs per unit energy output are higher for low rank coals (Table 9.4). Other gasification feedstocks include municipal waste, biomass, agro-waste and heavy oil. Anthracite is not considered suitable for gasification because of the low volatile matter content and the more valuable potential applications. Different types of gasifiers in operation include moving bed, fluidized bed and entrained-flow systems. The most common are shown in Figure 9.6. The older plants (Koppers-Totzek and Wrinkler types) operate at atmospheric pressure but the more modern plants operate at elevated pressure. These include Lurgi, British Gas/Lurgi (BGL), Texaco, Shell, Prenflo, and High Temperature Wrinkler (HTW) processes. The current status of coal gasification technology and new processes under development have been reviewed in depth recently by Cortes et al., (2009).

TABLE 9.3 Composition of gases from coal gasification. *(Perry, 1998).*

Raw gas technology	H_2O	H_2	CO	CO_2	CH_4	COS	H_2S	NH_3 + HCN	N_2 + AR
	(% volume)								
Entrained flow	2	26.7	63.1	1.5	0.03	0.1	1.3	0.02	5.2
Moving bed	5.1	52.2	29.5	5.6	4.4	0.04	0.9	0.5	1.5
Fluidized bed	4.4	27.7	54.6	4.7	5.8	0.1	1.3	0.08	1.7

TABLE 9.4 Performance and costs of syngas production from coals of different qualities. *(iea etsap, 2010).*

Performance	Bituminous	Sub-bituminous
Efficiency (%)	74.7	72.8
Specific capital cost $/GJ	13.5	17.2
Variable operating and maintenance costs ($/GJ)	1.4	1.6
Coal cost, input ($/GJ)	1.3	0.9
Production cost ($/GJ)	15.6	19.3

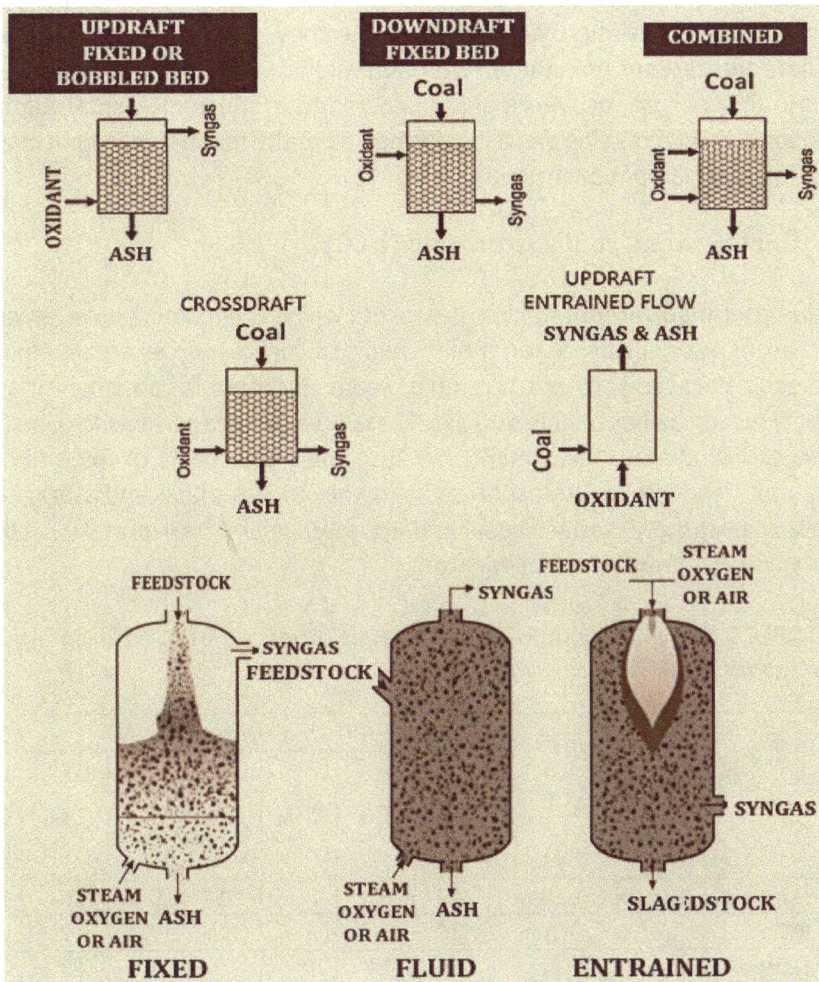

FIGURE 9.6 Types of gasifiers in common use. *(Partly from Kerester, 2012).*

The Lurgi and BGL types which use pressurized fluidized beds produce syngas that contains methane which can be converted to carbon monoxide and hydrogen for downstream ammonia or methanol synthesis (Hiller et al., (2012). The processes that use entrained bed in which coal dust and gasification agent (oxygen or air) are introduced into the gasifier co-currently, produce methane-free syngas. Steam or coal-water slurry feeding may be added to increase the hydrogen content of the syngas. These include the Koppers-Totzek and Texaco processes. Because they operate at higher temperatures, the syngas produced has higher carbon monoxide content compared with fluidized bed processes. Typical chemical compositions of syngas produced by different processes are shown in Table 9.5.

Syngas is primarily a mixture of carbon monoxide and hydrogen but other gases may be present and the overall chemical composition may vary depending on the operating conditions of the gasifier and the type of coal feedstock. The amount of carbon dioxide produced also depends on the process and may vary from below 5% to over 30%. The prevailing reducing atmosphere in the gasifier prevents the formation of polluting nitrogen oxides but promotes reactions between nitrogen and hydrogen to produce ammonia, a very valuable chemical that is recovered for use as feedstock for fertilizer production and many other chemical products. Modern gasifiers have extensive auxiliary units in which the gas is stripped of particulate and gaseous pollutants. Hydrogen sulphide and carbonyl sulphide produced from sulphur in the coal are recovered and processed to produce elemental sulphur or sulphuric acid. Solid by-products - slag and ash - formed from the mineral matter in coal are processed into useful materials of construction.

9.4.2   Underground coal gasification (UCG)

Coal can be gasified without mining, a process which is becoming increasingly attractive for many reasons. A significant amount of the world's coal resources are unmineable by current technologies because they are too deep, seam thickness is too small or sporadic, or the geological setting makes mining difficult. These resources can be gasified in-situ. Other major advantages include low Investment because the major costs of mine operation, (mining labour, coal handling and transportation to the mine surface and storage), mine surface facilities, coal washery, surface coal handling, storage and transportation, combustion, dust recovery and ash disposal are eliminated.

TABLE 9.5 Typical composition of syngas produced from different processes.
(Extracted from Hiller et al., 2012, Table 1).

Process	Content (vol%)						
	H_2	CO	CH_4	C_{2+}	N_2+ Ar	CO_2	H_2S + COS
Lurgi pressure gasification	43	12	11.5	1	0.3	32	0.2
BGL	28	57	7.1	0.4	4.2	3	0.3
Koppers-Totzek	30	55	0.1		1.4	13.2	0.3
Texaco	37	41.3	0.1		0.6	20.7	0.3

Furthermore, the carbon footprint of UCG is reduced significantly because surface gasification and associated anthropogenic emissions are eliminated and the syngas generated comes to the surface at a high pressure hence it can often be used directly in many applications without the need for further pressurization. Also, because of the high pressure, carbon dioxide capture is enhanced. Another advantage of underground gasification is the fact that the syngas produced can be transported in pipes to consumers and potential for environmental pollution is significantly lower than that of coal. Also, an underground coal gasification facility can include an above-ground electricity generation plant that feeds power to a distribution grid.

The projected worldwide increase in coal usage is 55% by 2030 and the bulk of the increase to be accounted for by emerging countries. Underground coal gasification (UGC) is expected to play a prime role because 85% of known coal reserves are inaccessible using surface mining techniques (Lauder, 2014). An underground coal gasification setup has two wells drilled on either side of a coal seam, one for injecting oxygen, air and steam to start off the chemical reactions, while the other is for evacuating the syngas produced (Figure 9.7). The underground processing environment acts as a pressure vessel without the need for application of additional pressure and the chemical reactions are controlled in the same manner as for above-ground gasification.

When a UCG seam is depleted the cavity created is filled with saline water sourced from underground and capped, and new wells are drilled in a different section of the coal seam. The potential of these cavities for the storage of sequestrated carbon dioxide is also being actively evaluated. Depth of operation of UCG is currently limited to around 250 metres but seams which would be considered too deep for conventional underground mining can also be exploited. The high conversion efficiency is also a major asset in favour of UGC. Around 80% of the original heating value of the coal is recoverable through underground gasification.

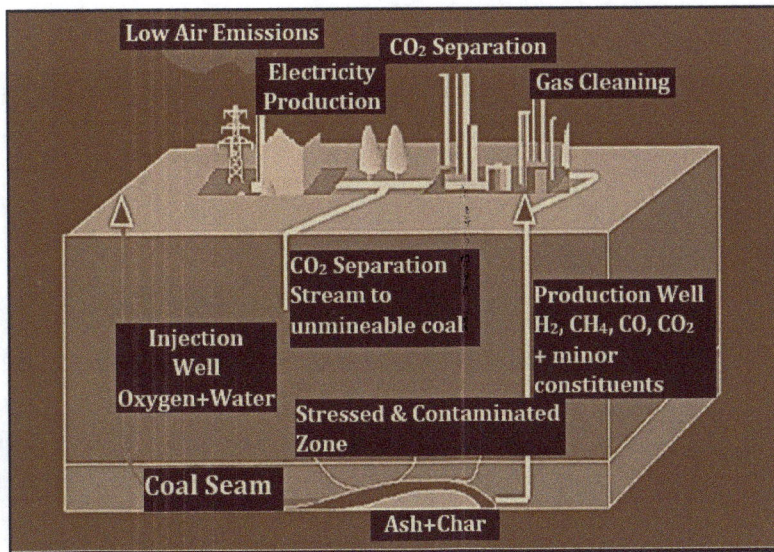

FIGURE 9.7 Underground coal gasification-electric power generation system. *(UCG Engineering Ltd., 2006).*

Underground coal gasification was developed over a century ago in the United Kingdom and was first deployed commercially in the former Soviet Union in the 1950s but proliferation was stunted by the discovery of petroleum and natural gas. There has been renewed interest in recent years and multiple UCG commercial projects are in various stages of development in around 50 countries, led by Australia, China and South Africa. The current status of UCG technology development and deployment had been reviewed by several authors recently (Burton et al., 2008; Brown, 2012; Bhutto et al., 2013).

There are three basic types of underground coal gasification technologies and many variants. The oldest method is the vertical shaft system, also known as the Soviet method. Although this system is relatively easy to deploy, it is only suitable for relatively shallow seams and the gas produced is of low quality because of the low pressure. The enhanced vertical system features wells linked by horizontal boreholes. The vertical shafts are deeper and linked horizontally either by hydraulic fracturing of the seam to create a series of physical cracks in the coal, linking the boreholes, or drilling connecting boreholes.

The latest and most advanced system known as CRIP (Controlled Retractable Injection Point) can inject air and oxygen at any point in a coal seam, thereby making it possible to exploit deep seams which could not be economically exploited by other UCG methods. The CRIP configuration was developed by Ergo Exergy to establish a channel between the injection well and production well. The channel may exist naturally if the coal seam is permeable or there are cracks linking the two areas of the seam. However, in many cases such a link needs to be established. The CRIP process links a vertical production well and the injection well. The two are connected horizontally by directional drilling, thus creating a gasification cavity at the end of the injection well (Figure 9.8). The CRIP process enables the gasification front to be shifted by retracting the combined steam and oxygen injection front once the coal near the cavity is used up. The syngas produced is rich in hydrogen and other combustible gases (carbon monoxide, methane and higher hydrocarbons) are converted to more hydrogen in the above-ground cleaning plant.

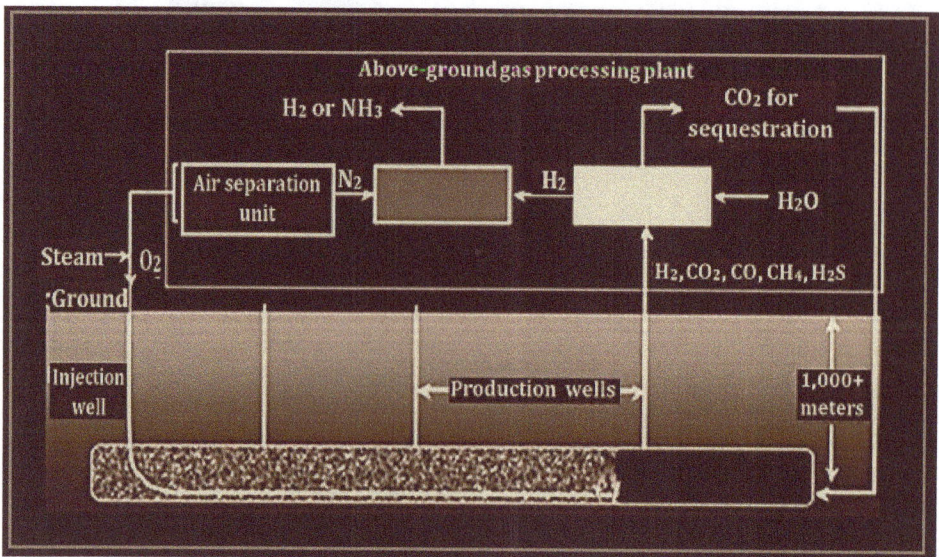

FIGURE 9.8 Schematic of the CRIP Process. *(Burton et al., 2008).*

Another process (EUCG) which has been developed, also by Ergo Exergy is reported to be more efficient than the CRIP process because it uses in part the natural pathways that already exist in the coal seam. Very little is known about the details of the new process, especially how it differs from the earlier CRIP process, because it is a proprietary process. However, it has been tested extensively in Australia and is being considered for several new projects in other countries.

Any coal from bituminous to lignite, including high-ash coals can be gasified in-situ. However there are geological and hydrogeological considerations that determine the suitability of a coal deposit for UCG exploitation. Shallow seams are not considered suitable because of high gas losses, potential ground subsidence, and contamination of ground waters. A depth of 100 meters is often considered the minimum. Seams below 3 metres thick are uneconomical to gasify in situ unless there are several seams in the same environment. Deep seams yield syngas with higher heating value because of the much higher pressures in the reaction zone which promote reaction between carbon monoxide and hydrogen, resulting in higher methane content. Syngas with calorific value of around 15 Mj/m^3 has been produced from deep seams using the CRIP method.

Underground coal gasification has significantly less environmental footprint than surface gasification because of the reduced percentage of the pollutants such as particulates, ash, and the electronegative metals (mercury, arsenic, lead) in the syngas that exits the production well. Furthermore, Conversion of syngas carbon monoxide to hydrogen combined with high-pressure CO_2 capture and storage (also known as pre-combustion capture) is much more cost effective than post-combustion CO_2 capture. However, there are concerns about possible negative impact on agriculture and groundwater (Lloyd-Smith, 2015).

The economics of a UCG project would depend largely on the relative cost of producing alternative energy sources, in particular, natural gas. For example, just a few years ago, UCG was being actively promoted as the gasification technology of the future in the US. However, the price of natural gas dropped from around $210 per thousand cubic meters (TCM) in 2006 to about $70/TCM in 2012 as a result of the deployment of new hydraulic fracturing (fracking) technologies which currently account for about 50% of US oil and gas production (USEIA, 2014). This development has greatly altered the economics and potential of UCG in the US. Countries that have abundant coal resources but no petroleum or natural gas have the best chances of deploying UCGs economically. This explains why Australia, New Zealand, South Africa, China and India are at the forefront of UCG research and development. However, there are several challenges, potential limitations and concerns for the proliferation of UCG, the most important being the potential aquifer contamination and ground subsidence, both of which are current subjects of intensive research.

Although many pilot plants have been operating for many years in different countries, there is not enough commercial experience worldwide to form a basis for realistic assessment of the economics of the UCG technologies. However, the available experience from China and Australia in the continuous operation of many small plants over many years and plans to build commercial plants support the conclusion of many studies that UCG economics are potentially favourable. The potential for deployment is also high in many developing countries which have abundant coal resources but no petroleum or natural gas resources (Burton et al., 2008) Ideally, a UCG project should be located in the proximity of syngas consumers in order to eliminate the cost of conversion to a suitable form for transportation over long distances. Deployment of UCG as part of an on-site integrated power and heat co-production systems, for example the Integrated Gasification Combined Cycle (IGCC) system is one of the best options since the final electric power product can be connected to a power grid. Syngas can also be converted to methanol on site and transported to consumers like petroleum products. Although all ranks of coal are potentially suitable for underground coal gasification, lignite and sub-bituminous coals are particularly suitable because of the relatively high seam porosity which facilitates linkage of injection and production wells and increases the rate of gasification by making reactant transport easier. Bituminous coals tend to swell and occur in thinner seams and both cause significant operational constraints (Stephens et al., 1985 a&b).

Attempts to gasify anthracite coal seams in-situ have so far been unsuccessful and, in any case, anthracite has more valuable applications. It has also been suggested that the impurities in lower rank coals improve the kinetics of gasification by acting as catalysts for the combustion process. On the other hand, high coal bed porosity and permeability increase the influx of water, and increase product gas losses (Grens, 1985; Creedy and Garner, 2004; Burton et al., 2008). Underground gasification of coal seams less than 2 metres thick is considered uneconomical because more wells are required. Also, thermal efficiency is low with such seams because of high heat losses, and the quality of gas produced is lower. The nature of the coal seam overburden is also important. Strong, dry rock overburden is less permeable and therefore reduces gas and heat losses.

Deep seams increase drilling and well linkage costs but produce higher quality gas at higher pressures and are less prone to subsidence and aquifer contamination. In general, exploitation of coal seams deeper than around 200 metres (similar depths of long-wall underground mining) will have minimal surface expression. If the syngas is to be used to fuel a gas turbine in an integrated power generation system, the high pressure in deep seams is an advantage because further compression may not be required. It is important therefore that these variables are considered in evaluating the process technology and economics of a potential UCG site.

9.4.3   Basic chemistry of syngas production

The chemistry of coal gasification is apparently straightforward as shown in Table 9.6. The exothermic reactions occur rapidly and supply the heat for the reducing reactions, hence the process is largely self-sustaining. The oxygen: carbon ratio in the gasifier is strictly controlled to ensure that only partial oxidation of carbon occurs (Equation 9.8). The carbon-oxygen reactions are highly exothermic and provide the heat required for the endothermic syngas production reactions (Equations 9.10-9.12). The actual reactions which occur in a gasifier depend on the type of gasifier, the O/C ratio, temperature, pressure, heating ratio, thermodynamic equilibrium and variable reaction kinetics in a particular zone of the reactor.

Depending on the gasifier technology employed and the operating conditions, significant amounts of H_2O, CO_2 and CH_4 can be present in the synthesis gas and the CO/H_2 ratio may vary. The current status of coal gasification technology and the state of knowledge on the complex reactions involved have been reviewed by several authors (Higman and van der Burgt, 2003, 2008; Cortes et al., 2009; Higman and Tam, 2013). The gasification process is basically simple but the strict process control required to achieve the right distribution of reactants in the gasification zone and the required set of reactions needed to produce syngas of desired quality can introduce considerable complexity.

TABLE 9.6 Basic chemical reactions in syngas production from coal. *(Adapted from Cortes et. al., 2009).*

Type of reaction	Chemical reactions	Heat of reaction (298°K)	
Reactions in the solid phase			
Partial oxidation	$C + \frac{1}{2} O_2 \rightarrow CO$	$\Delta h = -123$ kJ/mol	[9.8]
Combustion	$C + O_2 \rightarrow CO_2$	$\Delta h = -406$ kJ/mol	[9.9]
Gasification with steam	$C + H_2O \rightarrow CO + H_2$	$\Delta h = +118.9$ kJ/mol	[9.10]
Gasification with steam	$C + 2H_2O \rightarrow CO_2 + 2H_2$	$\Delta h = +78$ kJ/mol	[9.11]
Boudouard	$C + CO_2 \rightarrow 2CO$	$\Delta h = +159.7$ kJ/mol	[9.12]
Hydrogenation	$C + 2H_2 \rightarrow CH_4$	$\Delta h = -88.4$ kJ/mol	[9.13]
Reactions in the gaseous phase			
Partial oxidation	$CO + \frac{1}{2} O_2 \rightarrow CO_2$	$\Delta h = -283$ kJ/mol	[9.14]
Water shift	$CO + H_2O \rightarrow CO_2 + H_2$	$\Delta h = -40.9$ kJ/mol	[9.15]
Methanation	$CO + 3H_2 \rightarrow CH_4 + H_2O$	$\Delta h = -206.3$ kJ/mol	[9.16]
Hydrogen oxidation	$H_2 + \frac{1}{2} O_2 \rightarrow H_2O$	$\Delta h = -285.8$ kJ/mol	[9.17]
Methanation	$CO + 2H_2 \rightarrow CH_4 + CO_2$	$\Delta h = -247$ kJ/mol	[9.18]

Control of the UCG process is also more difficult compared with surface gasification in view of the fact that the reactions involved are inherently unsteady-state and both the flow rate and the heating value of the product gas fluctuate over time. Critical process parameters can only be estimated from temperature measurements and analysis of the quality of syngas produced.

9.4.4   Coal syngas utilization

Raw syngas produced by direct gasification of coal contains mainly carbon monoxide and hydrogen (Table 9.7), but chemical composition can vary widely depending on type of gasifier, process variables, and whether the gas is shift-reacted. Blue water gas and coke oven by-product gas are also syngases but have different chemical composition. However, because they are at low pressure they are often not considered as syngas. They also contain impurities which need to be removed before the gases are used as fuel, and the cleaning processes have been discussed in Section 9.3.

Raw syngas contains particulate and gaseous impurities many of which are anthropogenic or capable of causing equipment damage in processes that use syngas, hence cleaning is usually required. However the type of impurity and amount vary widely depending on the type of coal and gasifier. Furthermore underground gasifiers produce purer syngas because a significant amount of impurities remain trapped underground. Strict environmental regulations in many countries set acceptable limits for greenhouse gases that can be discharged into the atmosphere by electric power and other industrial plants, hence existing plants are being retrofitted with more efficient burners and auxiliary gas cleaning plants to capture particulate and anthropogenic emissions from flue gases.

TABLE 9.7 Typical chemical composition of syngas and natural gas.

	Natural gas	Syngas	Coke oven gas
Methane CH_4 (Vol.%)	70-90	0.03-11.5	28-32
Ethane C_2H_6 (Vol.%)	0-20		1-2**
Propane C_3H_8 (Vol.%)	0-8		
Carbon monoxide CO (Vol.%)		12-63	5-8
Carbon dioxide CO_2 (Vol.%)	0-8*	1.5-32	2-5
Hydrogen H_2 (Vol.%)		26-53	42-55
Nitrogen N_2 (Vol.%)	0-5	0.3-1.5	3-13
Ammonia NH_3 (Vol.%)			1-1.5
Hydrogen sulphide H_2S (Vol.%)	0-5	1-1.5	0-1
Oxygen O_2 (Vol.%)	0-0.2		0-0.5
Calorific value (MJ/m^3)	37-43	4-20	19-22
* Includes propane, butane, and higher hydocarbons **Includes other higher hydrocarbons			

Technologies are now available for the removal of most of the pollutants in syngas and will be discussed in some depth in a later chapter. Coal gas is no longer produced except in a few developing countries or as by-product in smokeless fuel manufacturing. Coke oven and blast furnace gases are recycled within the plant for supplementary heating. However, syngas has a wide variety of applications, from heat and power generation to hydrogen, chemicals and liquid fuels production. Syngas can also be upgraded to the level of natural gas in terms of heating value. Clean syngas obtained by gasifying coal has a significantly lower calorific value than natural gas but can also be considerably cheaper depending on local availability of either or both primary energy sources. The fact that any grade of coal can be processed into syngas and the relative abundance of coal worldwide make coal-to-syngas technology very competitive in many local situations.

Syngas is particularly valuable for the production of electric power, steam, liquid fuels, hydrogen, and a wide variety of chemicals (Figure 9.9). The ability to produce a number of high-value products at the same time with reduced undesirable emission in a polygeneration configuration also helps to reduce capital and operating cost requirements. Coal gasification is also of particular interest in many countries that have large coal resources (China, South Africa, India, Australia, United States). Although the United States also has large natural gas resources, increasing prices due to the deployment of expensive fracturing technologies are making syngas competitive and interest in coal gasification is rising.

China alone has over one hundred coal gasification plants that convert coal (and other carbonaceous materials) into electric power, industrial heat, a variety of chemicals, fertilizers, and gaseous or liquid fuels, and projections (considering planned plants and plants under construction) show that the plant population in China may double within a decade (Higman and Tam, 2013). Although syngas can be used as a stand-alone fuel, its heating value is low, approximately half that of natural gas.

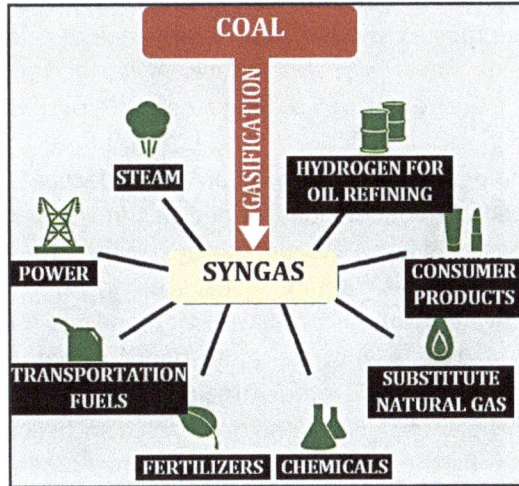

FIGURE 9.9 Coal gasification products. *(Adapted from Kerester, 2012).*

As a result, syngas has greater value as feedstock for the production of liquid fuels, industrial chemicals, and clean hydrogen, particularly if natural gas is also locally available (Dodds and Mcdowall, 2012). The coal gasification process produces a synthesis gas that can be converted to liquid fuel by further processing over appropriate catalysts. The unconverted gas can be recycled to the synthesis reactor to recover more of the coal energy, or used to generate electricity in a gas turbine combined cycle. The gas must be de-sulphurized prior to liquefaction to avoid catalyst poisoning. It is also preferable to remove the CO_2 at the same time to improve productivity (Larson and Tingjin, 2003).

9.5 THE FUTURE OF COAL GASIFICATION

Coal gasification has been practiced widely for well over a century and, with the current and projected dominant position of coal in the world primary energy scene, coal gasification/liquefaction is widely regarded as the only realistic way of minimizing the carbon footprint of coal. There are over 2,000 coal/biomass gasifiers operating worldwide and around 700 at the planning or construction stage (Higman, 2014). Most of the plants are located in Asia/Australia, most are for the production of chemicals and around 90% are coal-based (Figures 9.10- 9.12). There is little doubt that coal gasification will continue to proliferate in spite of competing petroleum options because of the abundance and relatively wide geographical spread of coal resources. Another development which may make coal gasification more attractive in future is the increasing concern about hydraulic fracturing (fracking), a technique adopted for the recovery of gas and oil from shale rock. The process involves drilling down into shale rocks vertically or, more commonly, horizontally and injecting high-pressure mixture of water, sand and chemicals to release the gas inside (Figure 9.13).

The major advantage of fracking is the ability to recover oil and gas deposits from rocks and shales which are too deep for regular recovery techniques. Hydraulic fracturing has been in use for over five decades and involves drilling a vertical well to reach oil and gas deposits at great depths. The well is encased in steel and cement and fracking fluid is pumped down into the well at extremely high pressures to create fractures and fissures in the surrounding rock through which oil and gas can flow. The fracking fluid which is a mixture of water, chemicals, sand, ceramics and other proprietary additives not only fractures the rocks but also forces the fluids upwards to the surface. A new technology introduced in the 1990s involves drilling of a horizontal well as long a nearly 2km through the rock layer to improve production.

Fracking technology has been used in Europe for decades but the adoption in the United States is more recent. The technology helped to transform the country from a net energy importer to an exporter in just a decade, having fully met the internal requirements. In spite of the enormous benefits of fracking, there are concerns. Recent research has provided evidence of groundwater contamination with methane and fracking chemicals. Furthermore, there is evidence of release into the atmosphere of potentially carcinogenic chemicals, including benzene, toluene and other organic compounds. Observed increases in earth tremors in areas where fracking is practiced are also believed to be due to geophysical disturbances caused by the technology. Furthermore, proliferation of fracking technology tends to drive down the cost of natural gas and production can become uneconomical in times of global depression in oil and gas prices. This is in fact happening right now and many oil and gas rigs in the USA are being shut down.

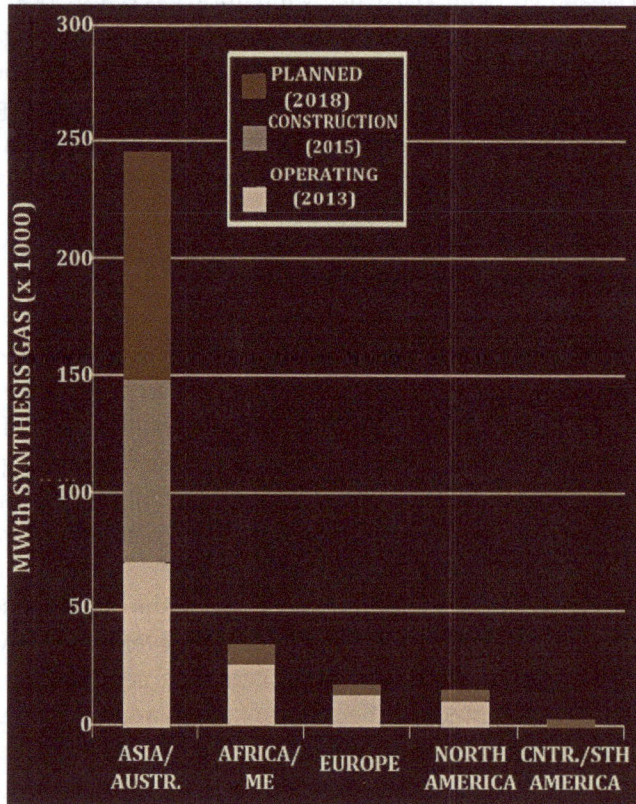

FIGURE 9.10 Gasification by region. *(Higman, 2014).*

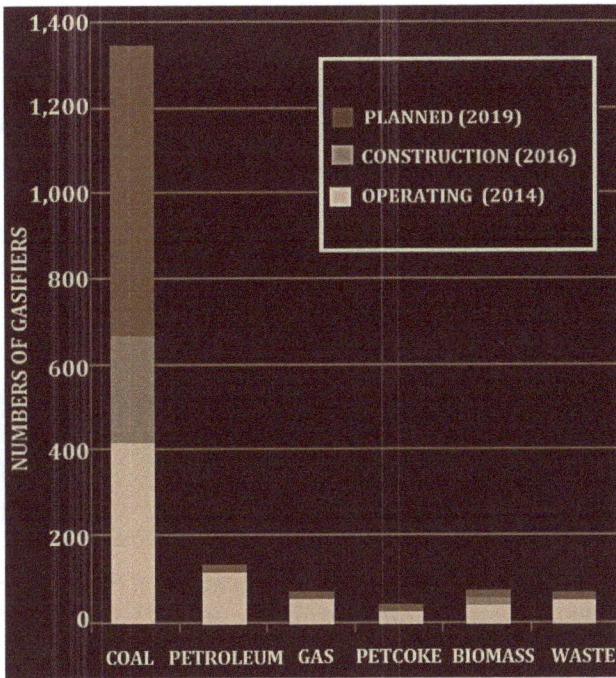

FIGURE 9.11 Gasification by primary feedstocks. *(Higman, 2014).*

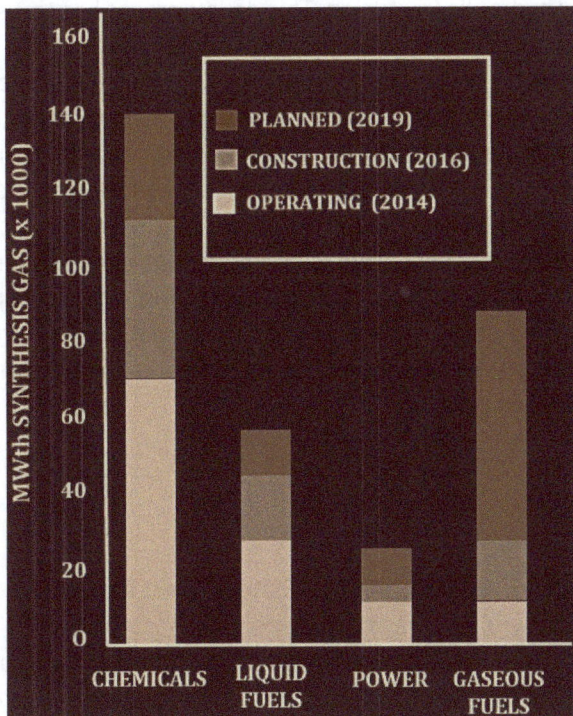

FIGURE 9.12 Gasification by end use. *(Higman, 2014).*

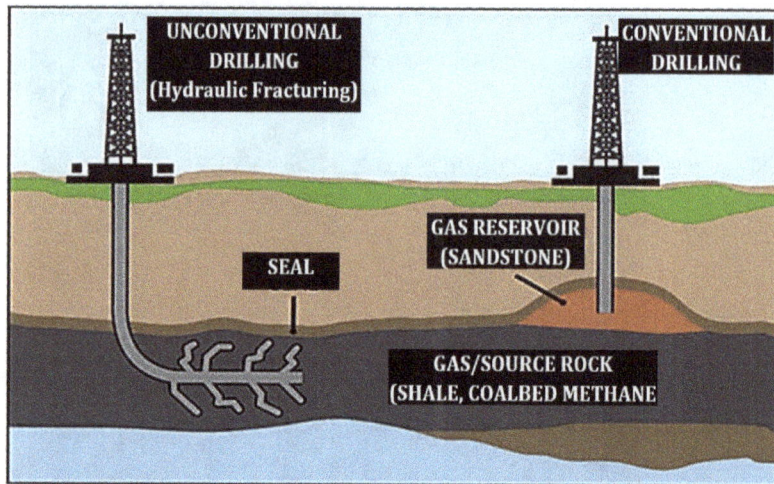

FIGURE 9.13 Conventional and unconventional (fracking) gas production technologies. *(www.nytimes.com).*

Although hydraulic fracturing has been practiced in the US since the late 1940s, proliferation has been energized only in the last decade or so. In 2013, around 70% of gas and 45% of oil were produced by hydraulic fracturing. This development has transformed the country from a net energy importer to the world's leading producer of combined oil and gas (USEIA, 2013). However, it has greatly depressed local prices. One positive aspect is the greatly increasing prospect for exports of refined petroleum products and natural gas, considering the relatively high gas prices in Europe and other countries. For example the current price of gas in the US is only about 30% of the price in Europe. Canada and Mexico are linked to the US by gas pipelines and liquefied natural gas (LNG) terminals export gas to many other countries, including Japan which is actively replacing idle nuclear power plants with natural gas-fired plants. Also, North Korea, China and other Asian countries are importing LNG from the US. The increased production of natural gas at low prices has also greatly benefited local end users which depend on natural gas as fuel or feedstock, in particular, electric power plants and the petrochemicals industry.

REFERENCES

Burton, E., Friedman, J., and R. Upadhye (2008). " Best Practices in Underground Coal Gasification." Lawrence Livermore National Laboratory.

Brown, K. M. 2012. "In-Situ Coal Gasification : An Emerging Technology." Journal American Society of Mining and Reclamation, 2012 Volume 1, Issue 1, pp. 103-122.

Bhutto, A. W., Bazmi, A. A., and G. Zahedi (2013) "Underground coal gasification: From fundamentals to applications." Progress in Energy and Combustion Science 39(2013), pp. 189-214.

Cortes, G., E. Tzias, et al., (2009). "Technologies for coal based hydrogen and electricity co-production power plants with CO_2 capture." *Luxemburgh, Institute for Energy, European Commission.*

Creedy, D. P. and K. Garner (2004) . "Clean Energy from Underground Coal Gasification in China." DTI Cleaner Coal Technology Transfer Programme, Report No. COAL R250 DTI/Pub URN 03/1611.

Dodds, P. E. and W. Mcdowall (2012). " A review of hydrogen delivery technologies for energy system models." www. bartlett. ucl. ac. uk/ energy/ research/ themes/ energy-systems/ hydrogen/ WP7_ Dodds_ Delivery. Pdf.

Francis, W. (1965). *Fuels and Fuel Technology,* Volume Two. Pergamon Press, London.

Grens, E. A. (1985). "Effect of non-uniform bed properties on cavity wall recession." Lawrence Livermore National Laboratory, Livermore, CA. UCRL-92487.

Higman, C. and M. van der Burgt (2003).*Gasification*. Burlington, Gulf, USA Professional Publication.

Higman, C. and M. van der Burgt (2008).*Gasification*. Burlington, Ma. Elsevier Science, 2nd Edition.

Higman, C., and S. Tam (2013). "Advances in Coal Gasification , Hydrogenation, and Gas Treating for the Production of Chemical Fuels." Chemical reviews, Vol. 114, pp. 1673-1708. - ACS Publications.

Higman, C. (2014). "State of the Gasification Industry: Worldwide Gasification Database, 2014 Update." Gasification Technologies Conference, Washington DC.

Hiller, H., Heimert, R., Stonner, H. (2012). "Gas Production, 1. Introduction." In Ullmann's Encyclopedia of Industrial Chemistry, Volume 16, pp. 403-421. Wiley-VCH.

IEA ETSAP (2010). *Syngas Production from Coal.* International Energy Agency Engineering Technology Systems Analysis Programme (IEA ETSAP) Technology Brief S01-May, 2010. www.etsap.org.

Kerester, A (2012). "Gasification, A Problem Solving Technology." Gasification Technologies Council. akerester@gasification.org.

Larson, E. D. and R. Tingjin (2003) "Synthetic fuel production by indirect coal liquefaction." *Energy for Sustainable Development, Volume VII No. 4, December 2003.*

Lauder, J. (2014). "Development of the UCG Industry and Future Commercial Growth." Future Fuel Foundation (FFF) Workshop, Johannesburg, August, 2014.

Lloyd-Smith, M. (2015). "Underground Coal Gasification (UGC). National Toxics Network (NTN). www.ntn.org.au

Perry, R. H. and D. W. Green (1998). *Perry's Chemical Engineers' Handbook,*1st Edition, McGraw-Hill Professional.

Stephens, D. R., Hill, R. W., and I. Y. Borg (1985a) "Underground Coal Gasification Review." Lawrence Livermore National laboratory, Livermore, CA-URCL-92068.

Stephens, D. R., Hill, R. W., and I. Y. Borg (1985b) "Underground Coal Gasification Annual Report." Lawrence Livermore National laboratory, Livermore, CA-URCL-50032-84.

Stephens, D. R., Hill, R. W., and I. Y. Borg (1985c) "Status of Underground Coal Gasification Review." American Association of Petroleum Geologists Bulletin, Vol. 69, No. 2, pp. 301-319.

UGC Engineering (2006). Underground Coal Gasification: Basic Concepts. www.coal-ucg.com/concept.html.

US EIA, (2013). "U.S. to become world's leading producer of oil and gas hydrocarbons in 2013." 4 October 2013.www.eia.gov.

USEIA (2014). *Natural gas gross withdrawals and production.* United States Energy Information Administration. Retrieved from www.eia.gov/dnav/ng/ng_prod_sum_dcu_smi_m.htm

WCA (2014) "Chemical Engineers mark coal gasification anniversary." World Coal Association. worldcoal.com.

Zhenyu Liu et. Al., 2012). "Analysis of Coal Pyrolysis Technologies from a Radical Reaction Point of View." *2012 APEC Symposium on Energy Efficiency of Low Rank Coal*. July 5-6, 2012, Beijing, China.

10 Coal liquefaction

10.1 INTRODUCTION

The production of coal-derived liquids dates back to the 1840s in Germany and the United Kingdom when gaseous by-products of iron and steel plant coke ovens were processed to obtain a variety of liquids used as solvents, wood preservatives, coal tar dyes and fuels. The liquids were also prime aromatic feedstocks for the petrochemical industry. However, attempts to produce liquid fuels directly from coal yielded little result. The first major breakthrough in coal liquefaction was the discovery in 1902 by scientists working simultaneously in France and Germany. They had developed a process for producing methane by reacting carbon monoxide and hydrogen under pressure in the presence of a nickel catalyst.

In 1914, a German scientist Friedrich Bergius succeeded in producing petrol and diesel by reacting coal dissolved in heavy oil with hydrogen under a pressure of 700 atmospheres and a temperature of 400°C in a 40-litre vessel, using iron oxide as catalyst. The major constraint to further development was building a pressure vessel that could withstand the harsh operating conditions. Two German Scientists (Franz Fischer and Hans Tropsch) developed a process that could operate at lower temperature and pressure by using a coal-derived gas (H_2 + CO) instead of coal and then performing synthesis using an iron catalyst and they were granted a patent in 1925. Further development of coal liquefaction which had been initiated in several countries was stalled by the discovery of large oil fields in several parts of the world, in particular, the USA. In the late 1920s, Germany developed two processes known as direct and indirect liquefaction for conversion of coal to obtain a wide range of hydrocarbon transportation fuels. Products included gasoline, diesel oil and jet fuel, oxygenated compounds such as methanol (alcohol fuel), or oxygenated fuel additives such as olefins and paraffinic wax. The first commercial plant began operation in Leuna, Germany in 1927 and about thirty plants built within a few years produced over 90% of the country's transportation fuel requirement for the German World War II effort.

Subsequent development of the indirect technology of coal liquefaction has been pioneered by South Africa, resulting in many innovations but further development of the direct liquefaction process which was pioneered by the UK and the US became stalled due to increasing access to petroleum products. The current and projected dominant role of coal in the world's primary energy equation has prompted renewed interest in clean coal processing, in particular gasification and liquefaction. The two liquefaction processes that were invented in the 1920s have developed side by side ever since. The direct coal liquefaction (DCL) process involves the reaction of pulverized coal dissolved in oil or water with hydrogen under high temperature and pressure. The indirect coal liquefaction (ICL) process involves gasification of coal under high temperature and pressure to produce syngas followed by conversion to liquid products. The product of either process is reacted with hydrogen in a high temperature, high pressure reactor in the presence of a catalyst. The process variables and catalysts differ significantly, so are the products. While the DCL yields high-quality liquid fuels, the product of the ICL route has to be refined further to obtain liquid fuels. The ExxonMobil *MTG* route is a variant of the ICL process and involves the processing of syngas in a reactor to obtain methane which is then refined to obtain liquid fuels. The three processes are shown schematically in Figure 10.1. The current status and future prospects of coal liquefaction have been discussed in several recent publications (Kaneko et al., 2012; Akash, 2013).The indirect coal liquefaction process has evolved over many decades and, although the initial deployment of commercial plants was mainly in South Africa, plants are now operating in China and many other countries.

The economics of production of liquid fuels from coal is uncertain. Both the DCL and ICL processes require heavy capital investments and profitability is closely tied to prevailing oil prices or relative local availability of petroleum products and abundant coal resources. In view of the frequent instability of the global oil market, investment in coal liquefaction is considered highly risky.

10.2   DIRECT COAL LIQUEFACTION [DCL]

The direct coal liquefaction process (also known as hydro-liquefaction) involves direct hydrogenation of coal at high temperatures and pressures in the presence of a catalyst, creating lighter, more stable oil molecules and simultaneously removing sulphur, nitrogen and ash to produce clean fuel. The process involves the breakdown of the organic structure of coal with solvents or catalysts at high temperatures and pressures. Reactions are carried out in a two-stage reactor and involve dissolution of coal in a solvent mixed with a catalyst in the first reactor at about 500°C and between 100 and 1,000 bar pressure while hydrogen is added (Figure 10.2). Hydrocracking is completed also in the presence of a catalyst in the second reactor operating at a higher temperature and lower pressure.

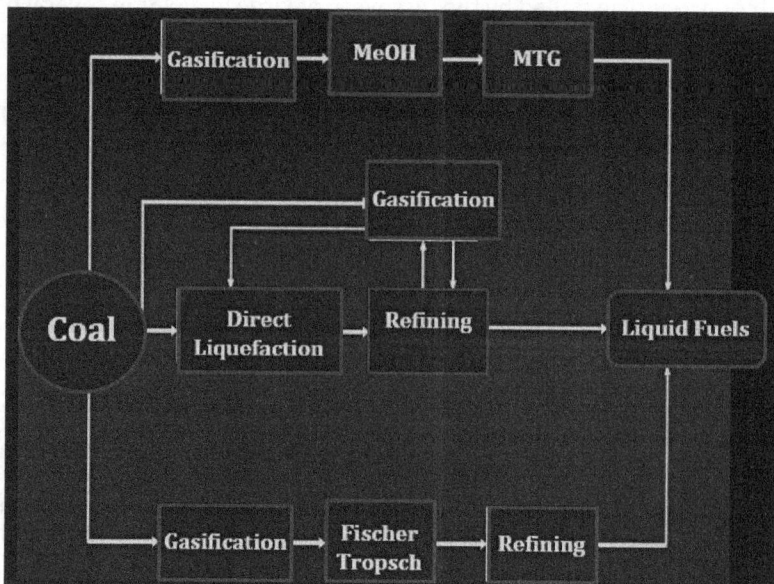

FIGURE 10.1 Alternative routes for coal to liquids and chemicals. *(ExxonMobil, 2008).*

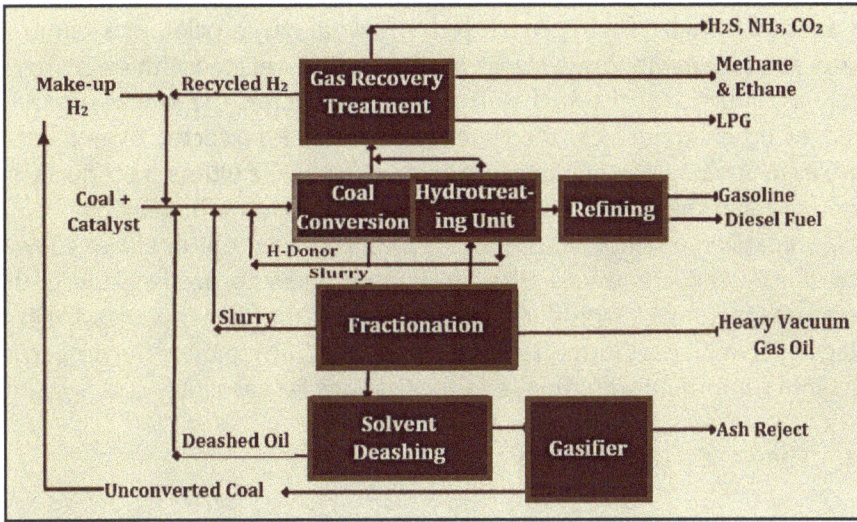

FIGURE 10.2 The DOE-HTL direct coal liquefaction process. *(netl.doe.gov)*.

Earlier processes used only one reactor, usually including an integrated online hydrotreating reactor to upgrade the distillates but this system has become obsolete. The solution must be desulphurized prior to the second stage to avoid poisoning of the catalyst. The original process used recycled coal-derived heavy process liquid but many variants have been developed based on the same chemical reaction principle, the main differences being in the reactor design and configuration, the process temperatures and pressures, the solvents, and the catalysts (mostly iron-based). While the DCL process is simple compared with the indirect method, a source of clean hydrogen is required and may be provided by gasifying additional coal feed or biomass. The heavy residue produced by the DCL reactor may also be mixed with the feed. The DCL process results in a relatively wide hydrocarbon product range consisting of a variety of molecular weights and forms, with aromatics dominating. Accordingly, the product requires substantial upgrading to yield acceptable transportation fuels.

Many two-stage DCL processes were developed in different countries but most have been abandoned. A notable exception is the DOE-HTL process which involves the co-processing of coal and non-coal derived low grade liquid hydrocarbon such as bitumen, heavy crude oil or tar by-product from conventional crude oil processing (Figure 10.2). The DOE-HTL reactor may be single or multi stage and there is no solvent recycling. The underlying objective is to refine the solvent and liquefy coal simultaneously, thereby improving the process economics. Conversion of the coal and heavy oil ranges from 80% to 92%. The DOE-HTL process was licensed to Shenhua Corporation of China and a commercial plant producing 1.08 million tonnes per year of liquid products including diesel oil, liquefied petroleum gas (LPG) and naphtha (petroleum ether), has been in continuous, stable and profitable operation in Inner Mongolia since 2010. The Mongolian plant is being expanded already to 3 million tonnes and will ultimately have a capacity of 5 million tonnes a year in the third phase. Another plant with a capacity of 3 million tonnes a year is under construction in Northwestern Xinjiang. Several other countries are considering adopting the DCL process combined with an overground or underground coal gasification hydrogen-producing system.

10.3   INDIRECT COAL LIQUEFACTION [ICL]

In the indirect process, coal is converted by partial oxygenation or steam reforming into hydrogen and carbon monoxide (syngas) which is shift-reacted with steam to raise the H:CO ratio to 2 as required for the Fischer-Tropsch (F-T) reactor. The syngas is desulphurized and stripped of other impurities, then reacted in the F-T reactor over a catalyst, usually cobalt-based, at moderate temperature and pressure. The ultimate products depend on the catalyst used and the process conditions, and may be paraffins, olefinic hydrocarbons or alcohol, primarily methanol. Further processing is carried out in either a low-temperature reactor which produces waxes that are hydrocracked to produce diesel oil, or a high-temperature reactor which produces olefins ($C_3 - C_{11}$) for further processing in an isomerization stage or cold-process isomerizing hydrocracking (HDI) to bring the properties of the final product into conformity with required specifications for gasoline.

10.3.1   The SASOL CTL Process

The indirect coal liquefaction (ICL) process principle is also applicable to the conversion of natural gas and coal-derived synthetic gas to hydrocarbon fuel. South Africa (SASOL) adopted the German Fischer-Tropsch (F-T) indirect process in the early 1940s for the production of transportation fuel from the country's abundant coal resources, in response to the worldwide isolation faced as a result of apartheid and the consequent lack of access to a reliable source of crude oil. From the start, the primary focus of SASOL was coal-to-liquids (CTL) and much of the progress achieved so far worldwide in coal gasification, reactor design, coal-to-chemicals, and catalysis has been due to the extensive research and development effort of SASOL. The first SASOL commercial plant was built in 1955 and two more plants were commissioned in the early 1980s. The three plants produced about 60% of the country's transport fuel requirements and operated at about 37- 50% efficiency. The latest SASOL CTL plant in Secunda is the world's largest and processes 40 million tonnes of coal a year into around 160,000 barrels/day of liquid fuels, and a wide range of chemicals. The process is shown schematically in Figure 10.3.

The core of the SASOL indirect CTL process is the fixed-bed, dry bottom (FBDB) gasifier which produces crude syngas from coal, using iron-based catalysts. The syngas is processed to recover chemicals (tar and oil, phenols, ammonia, sulphur), and the clean syngas can be processed to produce clean hydrogen or chemicals. Production can be switched between the two products relatively easily. A major advantage of the SASOL CTL technology is the ability to process low grade, high-ash coals which are widely available in around a hundred countries. Typical CTL plants are fuel-flexible and can process virtually any carbonaceous materials including natural gas, biomass, heavy shale oil, tar. This explains why the SASOL CTL technology now dominates coal-to-liquid and coal-to-gas projects worldwide.

10.3.2   ExxonMobil Methanol to Gasoline process (MTG)

The ExxonMobil methanol-to-gasoline technology is a variant of the indirect coal liquefaction process and is similar to the Fischer-Tropsch technology in that coal is first gasified to produce syngas. The main differences are in the follow-up processing and the types of products (Figure 10.4).

Like the F-T synthesis technology, the MTG system can also process natural gas. While the F-T synthesis produces an F-T crude which is further processed to produce mainly diesel, jet fuel, gasoline and naphtha, the MTG synthesis yields methanol for further processing to obtain mainly gasoline. Either process (and also the direct liquefaction process) can produce high-purity hydrogen. In the MTG synthesis unit, pre-heated methanol is fed to a fixed-bed reactor system. Reactions convert methanol into a mixture of C_1 to C_{10} hydrocarbons, approximately 90% of which are in the gasoline boiling range with 87% octane value. The only other products are fuel gas and free water. The raw gasoline produced is separated into liquefied petroleum gas (LPG), light gasoline and heavy gasoline. The heavy gasoline is hydro-treated to reduce the durene content and recombined with light gasoline to obtain gasoline in the octane range 87 to 92.

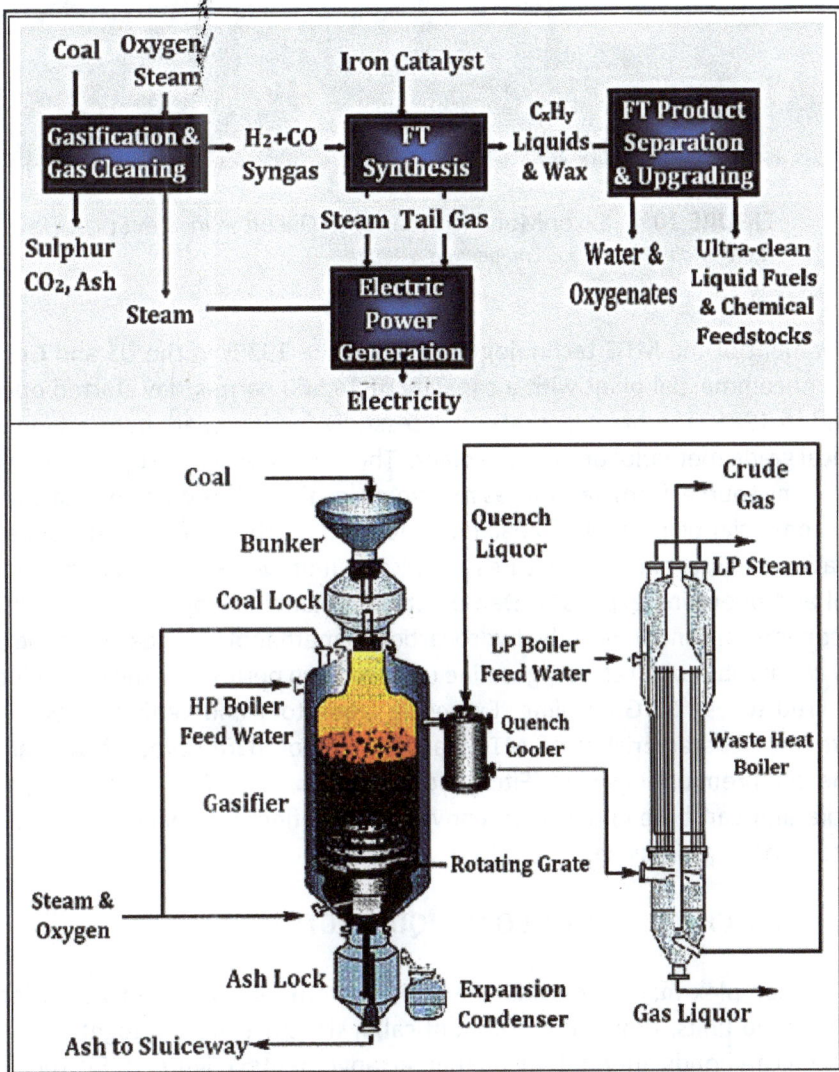

FIGURE 10.3 SASOL coal-to-liquid indirect process. *(Mangena, 2012).*

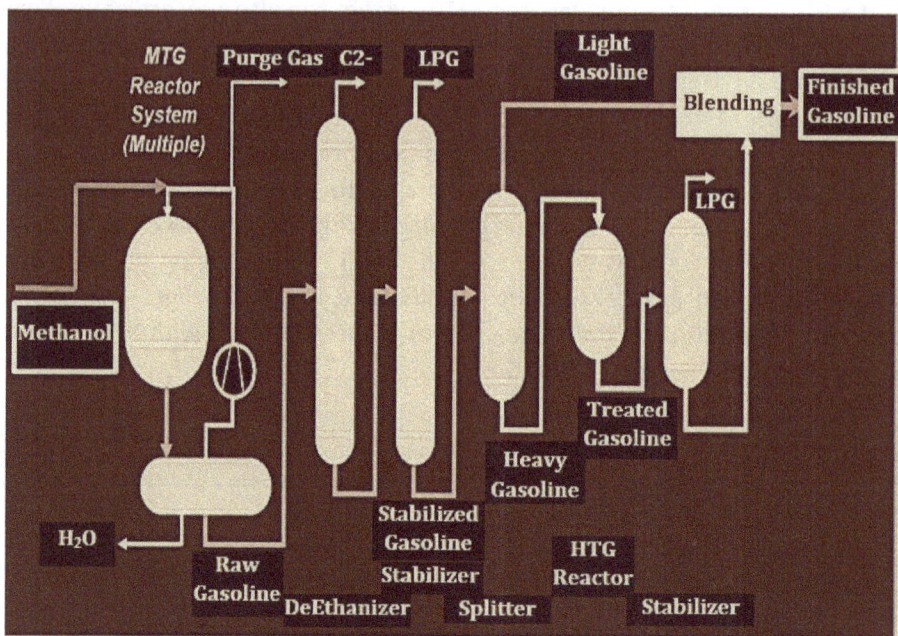

FIGURE 10.4 ExxonMobil Methanol to Gasoline indirect process.
(ExxonMobil Research & Engineering, 2009a).

Development of the MTG technology started in the 1970s in the US and Germany and the first semi-commercial plant with a capacity of 14,500 barrels/day started operation in New Zealand in 1985. The plant operated successfully for ten years before being converted to chemical grade methanol production plant. The MTG process can operate on methanol from virtually any source, from natural gas reforming, coal gasification, to biomass conversion. The first commercial plant in New Zealand processed methanol from natural gas. The second generation MTG technology features significant improvements including a reduction in both capital and operating expenses. Demonstration plants are operating in China and the US. MTG converts around 90% of the hydrocarbon in methanol to a clean gasoline product which is fully compatible with refinery gasoline derived from petroleum and no engine modification is required to use MTG gasoline. Extensive laboratory and vehicle tests have shown the performance characteristics of MTG gasoline to compare favourably in all aspects with commercial premium gasoline (Fitch and Lee, 1981; ExxonMobil, 2009b). MTG gasoline is versatile and can be blended with conventional refinery gasoline or sold separately with minimal further processing.

10.4   BASIC CHEMISTRY OF COAL LIQUEFACTION

Coal is a complex macromolecular network system which contains a variety of complex molecular sub-units, similar but not identical in structure, linked together by chemical and non-chemical bonds in random spatial arrangements (Figure 10.5). The structural units have condensed aromatic rings as their cores, the number generally increasing with coal rank.

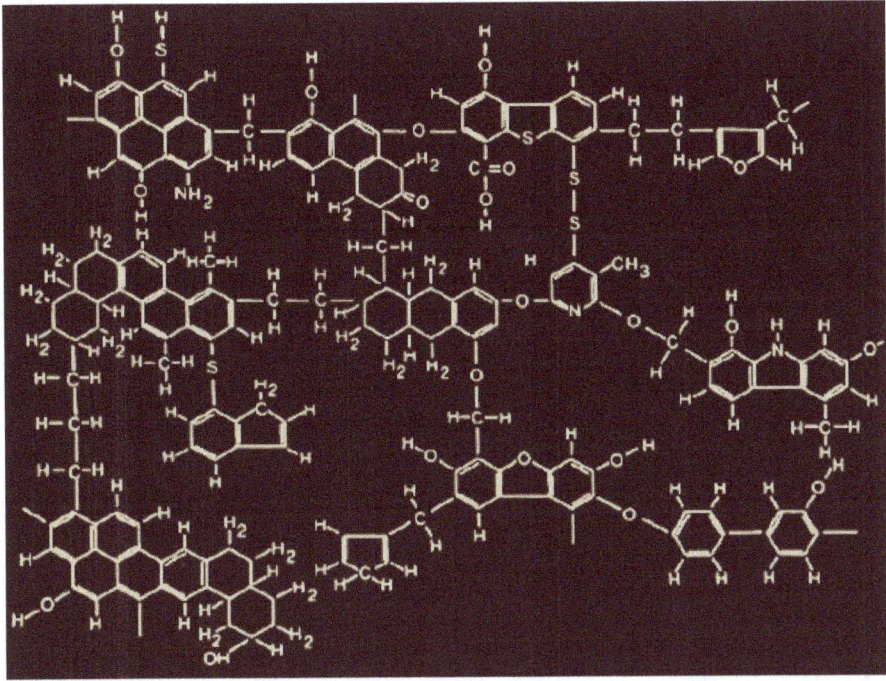

FIGURE 10.5 Molecular structure of a bituminous coal. *(www.wou.edu).*

Each structural unit in coal has its peripheral units consisting of alkyls, mainly methyl and ethylene, and functional groups, mainly phenolicyclic hydroxides and carbonyls. Heteroatoms such as oxygen, nitrogen and sulphur exist in chemically bonded state. Also, some amount of low-molecule compounds can be found scattered in the high-molecule structures. Coal structure contains two major phases: organic phase made up of macerals, and an inorganic phase made up mainly of mineral matter which is largely inert. The building blocks are primarily aromatic and hydro-aromatic, with hetero-atoms (oxygen, nitrogen, sulphur, hydrogen) at the edges.

The building blocks are liberated during heat treatment of coal at high temperatures. The ratio of aromatic and hydro-aromatic macromolecules will vary with coal rank. The processes of coal gasification or liquefaction start with thermal or chemical breakdown of as many molecular bonds as possible to free them for hydrogenation. There is a substantial overlap in the chemistry of coal gasification and coal liquefaction. The direct liquefaction process requires that the coal is gasified to produce syngas and shift reacted to produce clean hydrogen for hydrocracking dissolved coal, while the gasification process is also part of the indirect liquefaction process. Hence the coal gasification chemistry discussed in Chapter 9 also applies to both processes. The main difference in the chemistry of the direct and indirect liquefaction processes is that coal is dissolved in a solvent and reacted with (usually coal-derived) hydrogen in the direct process. The indirect process involves promoting a reaction between carbon monoxide and hydrogen in gasified coal under high temperature and pressure. The production of liquid fuels from coal is essentially a process of adding hydrogen to coal to improve the atomic hydrogen/carbon ratio to the level of methane which is a primary feedstock for a wide range of liquid fuel and chemicals production processes.

The H/C atomic ratio is in the range of 0.3 to 0.9 for different coal ranks compared with 4 for methane. At the same time, the molecular weight of the product is reduced significantly relative to that of the feed. In general, coal liquefaction involves two processes: thermal decomposition in which the complex molecular structure of coal is broken down either by gasification or dissolution in a solvent; and hydrogenation, a process of increasing the H/C ratio by reacting syngas or dissolved coal with hydrogen under high temperature and pressure. Although the process of hydrogenation is often represented by a simple equation, the process is quite complex and involves a series of sub-reactions (Lee Tingchen, 2009; Akash, 2013).

10.4.1   Basic chemistry of direct coal liquefaction (DCL)

Coal is dissolved in a solvent mixed with a catalyst and the solution is heated in a pressure reactor to around 400°C under a high pressure, with a constant supply of hydrogen. As temperature rises the weak van der Waals and hydrogen bonds break up. At the higher temperatures, bond rupture extends to the more stable bonds and the amount of radical fragments in the solution increases with a reduction in the mean molecular size of the constituents of the solution. The hydrogen reaction involving carbon and hydrogen molecules occurs in accordance with Equation 10.1.

$$C_{coal} + 2H_2 = CH_4 \hspace{6cm} [10.1]$$

Although, in theory, most coals from lignite to bituminous can be processed by direct liquefaction, yield varies widely. Higher rank coals are less reactive and anthracites are essentially non-reactive. The yield of liquids depends on the type of coal feed, in particular, the carbon content (Figure (10.6), the petrographic constituents and liquefaction conditions. Most of the early studies on direct coal liquefaction identified bituminous coals as the best feedstock because of the high yield of liquid hydrocarbon products. However, more recent studies have concluded that the older lignites and younger bituminous coals with H/C ratio of between 0.8 and 0.9 are the best precursors.

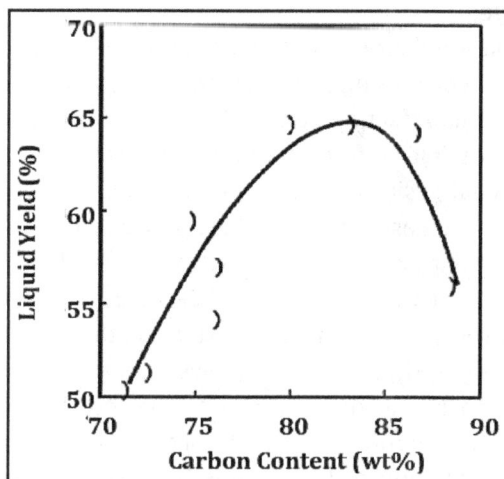

FIGURE 10.6 Relationship between carbon content and liquid yield in direct liquefaction of coal. *(Li Tingchen. 2009).*

Although younger lignites have higher H:C ratios, oxygen, aliphatic carbon and moisture content are usually very high and the liquid produced will have high moisture content, hence low yield. On the other hand, older coals tend to be less reactive. Coals with carbon content in the range 78-85% appear to be the best (Figure 10.6). Liquefaction of one tonne of coal yields around half a tonne of liquids, with molecular structures similar to those of aromatics.

The petrographic constituents of coal also have a significant effect on response to direct liquefaction. Vitrinite and clarinite have good solubility in solvents, fusinite is insoluble. The relative proportions of these macerals vary with coal rank and type. It can be concluded from the above discussion that the suitability of a coal for direct liquefaction depends on an optimum mix of variables, in particular, H:C atomic ratio and the reactivity of the maceral constituents. Coals with H:C ratios below 0.6 are considered unsuitable because of the potentially high oxygen and moisture contents which affect yield negatively. Coals with unreactive vitrinite or high content of inertinite are also unsuitable. Although hydrogenation of coal dissolved in a solvent is more efficient than hydrogenation of gasified coal, the quality of liquid fuel is lower and further processing (hydrocracking or addition of hydrogen over a catalyst) is required to produce fuel of required specifications. Furthermore, processing slurries at high temperature and pressure presents formidable process control and solid/liquid separation problems.

10.4.2   F-T Indirect coal liquefaction (ICL)

The indirect liquefaction process starts with a syngas that has been produced by gasifying coal (or other carbonaceous materials). Syngas is primarily a mixture of carbon monoxide and hydrogen, with an H:C ratio of around 2 to 2.5. The raw syngas is stripped of impurities (mainly sulphur and carbon dioxide) in a series of absorbers and adsorbers and the H:C ratio is increased to around 3.0 (the minimum required for the second stage of the liquefaction process) by shift reacting the syngas. The clean, hydrogen-rich syngas is reacted with more hydrogen at high temperature of around 300°C and pressure of about 20 bars in an F-T reactor in the presence of a catalyst to produce hydrocarbons in accordance with Equation 10.2.

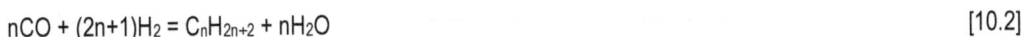

$$nCO + (2n+1)H_2 = C_nH_{2n+2} + nH_2O \qquad [10.2]$$

The product of the F-T reactor passes through an isomerization stage or cold-process isomerizing hydrocracking. The products are a wide range of liquid fuels including light, high quality gasoline, high cetane number distillates, diesel oil, jet fuel, lubrication products, waxes, and paraffinic naphtha which is a valuable feedstock for the production of a wide range of chemicals. The first stage of the F-T coal-to-liquids (CTL) process involves gasification of coal hence the process is also suitable for conversion of any hydrocarbon gas to liquids (GTL).

10.4.3   MTG Indirect coal liquefaction (ICL)

The MTG process converts syngas to methanol in accordance with Equation 10.3. Methanol is first dehydrated to dimethyl ether (DME) in an exothermic reaction. The equilibrium mixture of methanol, DME and water is then mixed with recycled gas and passed over a proprietary zeolite-based catalyst to produce hydrocarbons, mainly light olefins $C_2 - C_4$) and water (Zhao et al., 2006).

$$CO + 2H_2 \rightarrow CH_3OH \qquad [10.3]$$

A final reaction step produces a mixture of higher olefins, n/iso-paraffins, aromatics and naphthalenes. The reaction path is shown in Table 10.1. Most of the hydrocarbon products are in the gasoline range. The raw gasoline is distilled to remove all non-hydrocarbons and light hydrocarbons. Methane and propane are also removed in a de-ethanizer. Further treatment in a stabilizer removes propane and part of the butane which are discharged as part of the fuel gas. Finally the gasoline passes through a splitter where it is separated into light and heavy gasoline fractions. The reactions could be controlled to produce either gasoline or light olefins (Mcgihon and Tabak, 2008). The products of the F-T process depend on the process details and the catalyst, and F-T liquids contain a range of hydrocarbons, alcohols and other oxygenates that require further refining processes to convert to conventional fuels. On the contrary, the MTG technology produces a conventional gasoline and some liquefied petroleum gas directly from methanol (Table 10.2).

TABLE 10.1 Chemistry of MTG coal to liquid technology. *(Helton and Hindman, 2014).*

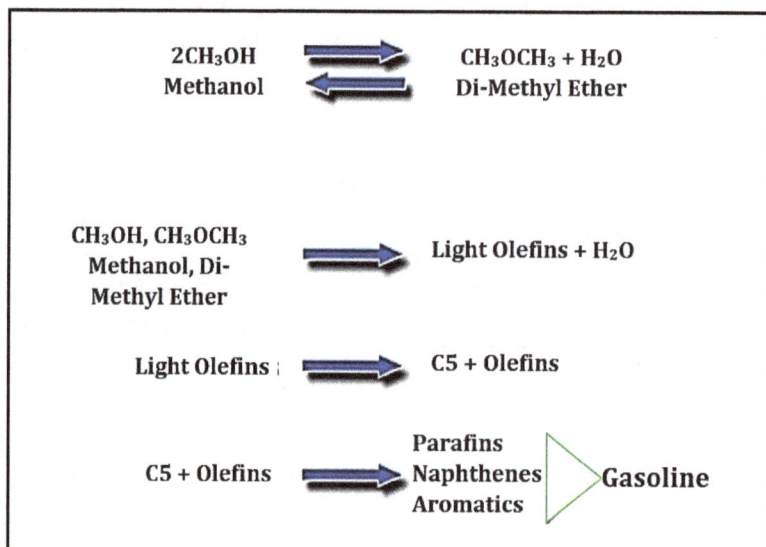

TABLE 10.2 Products of F-T and MTG indirect processes. *(ExxonMobil Research & Engineering, 2009a).*

Components	Fischer-Tropsch Co Catalyst at 220°C	Fischer-Tropsch Fe Catalyst at 340°C	MTG
Fuel Gas	6	15	1.1
LPG	6	23	10.0
Naphtha	19	36	
Gasoline			88.8
Distillate/Diesel	22	16	
Fuel Oil/Wax	46	5	
Oxygenates	1	5	

Although both The F-T and MTG processes start with the conversion of coal to syngas, their respective products are very different. The F-T process produces a broad spectrum of straight chain paraffinic hydrocarbons that require upgrading to produce diesel fuel, high cetane number distillates, lubricant feedstock and paraffinic naphtha for petrochemical applications. In contrast, low sulphur, low benzene, high quality gasoline is the main product of the MTG process.

Available data on F-T and MTG technologies indicate that the required capital investments for the two routes are similar and coal preparation and gasification account for 65 – 75% of the overall capital expenditure (DOE-NETL, 2007). In effect, since the overall capital investment for either process is dominated by syngas production, technology selection between the two processes will depend more on factors such as the desired products. Another significant difference between the two processes which may be considered in the choice of either technology is the fact that the MTG system can operate as two separate units which can operate out of phase. In effect, the methanol production and MTG synthesis units can be configured as separate plants. In an F-T process, the coal gasification unit is linked by syngas to the F-T unit and an F-T plant would be difficult to operate as separate independent gasification and hydrogenation units.

10.5   THE FUTURE OF COAL LIQUEFACTION

Despite the fact that the first patent on coal liquefaction in 1913 was for the direct process (the patent for the indirect process was granted in 1925) and despite its relative advantages, the indirect process is the more developed. To date only one commercial direct liquefaction plant operating is operating in China, commissioned in 2010, compared with indirect liquefaction plants which have been in operation for around six decades. Direct liquefaction of coal by dissolution in a solvent is inherently more efficient than gasification and the thrust of research in Europe and North America was in this direction up to the 1980s when interest declined due to wide availability of petroleum products at competitive prices. On the other hand, the development of the indirect process was driven by a need to produce liquid fuel from abundant local coal resources in the absence of other alternatives.

Up to 2009 all commercial indirect coal liquefaction plants in the world were located in South Africa. However, four demonstration plants have been established in China in the last few years and several projects are under development in Australia, India. Mongolia, Botswana, Russia and USA. The first MTG Coal-to-Liquid 2,500 barrels/day gasoline production demonstration plant began operating in 2009 in Shaanxi province, China as part of a demonstration complex which includes coal gasification, gas clean-up and methanol synthesis. More plants are planned or under construction, five in the US and several more in China and other countries (El-Malki and Hindman, 2014). Ten new MTG plants have come on stream in the last five years. A new plant with an output capacity of 4 million tonnes of liquids (diesel oil, naphtha, and liquefied gas) a year from 20 million tonnes of coal was commissioned in Ningxia Hui in 2016. China completed a coal-to-olefins plants with a combined capacity of 11.2 million tonnes a year in 2016. The country's Five-Year Plan targets 13 million tonnes of coal-to-liquids and 17 billion cubic metres of coal-to-gas per year by 2020.

The future of CTL as a widely accepted fuel production process is uncertain due primarily to its dependence on the fluctuating fortunes of alternative petroleum energy sources. While many estimates have shown that CTL becomes economically competitive when the cost of crude oil rises above around $45-60/barrel (Miller, 2007) (it has fallen below this value in recent months), other factors may enhance or deter proliferation of the process. Factors which may retard the spread of coal liquefaction include relative availability and pricing of natural gas, crude oil and liquefied natural gas, all of which are direct competitors. For example, the recent agreement signed between China and Russia for the supply of Russian oil and gas, coupled with worldwide pressure on China to mitigate environmental pollution will inevitably slow down the pace of CTL development in China. Intensive research is ongoing in many countries on the direct coal liquefaction process which is considered to be more efficient than the indirect process. Also, the liquid products are of much higher quality and can be used directly for most stationary fuel applications. They have similar characteristics to crude oil and can be processed directly or mixed with crude oil as feedstock for a conventional refinery.

The required coal feed for direct coal liquefaction is about 30% less than for the indirect process for the same liquid output. Both processes produce ultra low sulphur fuel but also produce 7 to 10 times more carbon dioxide compared with crude oil. On the other hand, the indirect process produces higher cetane distillates suitable as feedstock for the petrochemical industry. Furthermore carbon dioxide can be captured relatively easily at the coal gasification stage. Co-plants comprising a direct and an indirect liquefaction line are under development in China with a view to producing distillates that can be blended to obtain the desired cetane level. The coal liquefaction process is currently only economically viable in countries which have abundant, low-quality and cheap coal resources and where crude oil is either not available or has to be imported at high prices. However, many countries are engaged in active research and development in anticipation of future crude oil supply problems.

The industrialized and emerging countries currently source the bulk of their crude oil requirements from politically unstable regions of the world and in any case, the world reserves are expected to satisfy the projected demand in the next thirty years or so after which demand may outstrip supply. In contrast, many countries have abundant resources of coal or have easy access to deposits in stable parts of the world. The need for local primary energy independence has motivated intensive research on clean coal technologies, including coal liquefaction and it is possible that investments in coal liquefaction plants may rise in the foreseeable future.

REFERENCES

Akash, B. A. (2013). "Thermochemical Liquefaction of coal." *Int. Jnl. of Thermal & Environmental Engineering,* Vol. 5 No. 1, pp. 51-60.

Doe/NETL 2007). *Baseline Technical and Economic Assessment of Commercial Scale Fischer-Tropsch Liquids Facility.* DOE/NETL-2007/1260, Final Report, April, 2007.

El Malki, E. and M. Hindman (2014). " Methanol To Gasoline (MTG) Technology." ExxonMobil Research and Engineering (EMRE). GTL North America, March 12-13, 2014.

ExxonMobil (2008). *An Alternative Route for Coal To Liquid Fuel.* Coal Gasification Technologies Conference, October, 2008.

ExxonMobil (2009a). *An alternative for Liquid Fuel Production.* ExxonMobil Research and Engineering (EMRE). Gasification Technology Conference, October, 2009.

ExxonMobil (2009b). *"Methanol to Gasoline (MTG) Production of Clean Gasoline from Coal."*. ExxonMobil Research and Engineering (EMRE). www.zeogas.com.

Fitch, F. B. and W. Lee (1981). "Methanol-to-gasoline, An alternative route to high quality gasoline." *SEA Technical Paper Series* 811403.

Helton, T. and Hindman M. (2014) "Methanol to Gasoline Technology: An Alternative for Liquid Fuel Production." *Technology Forum 2014*. GTLTechForum.com.

Higman, C. (2014). State of the Gasification Industry: Worldwide Gasification Database, 2014 Update." Gasification Technologies Conference, Washington DC.

Johnson Matthey Methanation Process (2014). www.jpmprotech.com

Kaneko, T., Derbyshire, F., Makino, E., Tamura, M., and K. Li (2012). *Coal Liquefaction*. Ullmann's Encyclopedia of Industry Chemistry.

Li Tingchen (2009). "Coal Liquefaction." *Coal, Oil Shale, Natural Bitumen, Heavy oil and Peat* Vol. 1. pp. 403-424. Encyclopedia of Life Support Systems (EOLSS). www. eolss.net

Mangena, S. (2012). "Coal Gasification and Liquefaction – SA Experiences and Opportunities." *4th EU-South Africa Clean Coal Working Group Meeting*, November, 2012.

McGihon, R. D., and S. A. Tabak (2008). "Coal to Clean Gasoline." HydrocarbonEngineering. www.hydrocarbonengineering.com

Miller, C. L. (2007). "Coal Conversion – Pathway to Alternative Fuels." 2007 EIA Energy Outlook Modeling and Data Conference, Washington D.C., March 28th , 2007.

Zhao, Xinjin, McGihon, R. D., and Tabak, S. A. (2006). "Coal To Clean Gasoline." *Hydrocarbon Engineering*, September, 2006.

COAL CHEMICALS TREE

Every branch/leaf is a valuable chemical or precursor
to premium materials

(Source: U.S. Geological Survey)

11 Coal conversion to premium chemicals and materials

11.1 INTRODUCTION

Coke produced by carbonizing coal has been used for producing iron and steel for three centuries. A highly viscous oil known as coal tar was removed from the by-product gas by condensation from the high-temperature cokemaking oven, not because anyone knew what to do with it but so that it would not clog the chimneys. By the end of the 19th century low-temperature coal carbonization processes had been developed and were producing town gas, also with tar as a by-product.

Scientists in several countries studied the coal tar in depth and identified three different types of oil: light, middle and heavy fractions. They also found that tars produced from low-temperature and high-temperature carbonization of coal had different chemical composition. The thrust of research was to characterize these tar products with a view to producing pharmaceutical drugs from the different fractions. Textile dye was probably the first chemical ever produced from coal and the discovery was purely accidental. In the 1850s, British chemist William Perking was experimenting on the production of the anti-malaria drug, quinine from coal tar. He accidentally dropped a piece of cloth in a jar containing coal tar overnight and, next morning, he found it had turned purple. Previously dyes had been produced from rare plants and trees.

Working with his German teacher, Perkin identified a component of coal tar, alizarin as the active chemical. By the 1870s scientists working independently in England, Germany and France had found a way of producing alizarin from anthracine obtained from coal tar and were producing synthetic dyes in different colours. Bayer Germany set up the first commercial plant to produce dyes in the mid 1870s. Germany sustained its leadership in coke oven by-product utilization through research on utilization of other chemicals in the by-product gas, in particular, carbon and hydrogen for the production of industrial value products. Fischer and Tropsch were granted a patent in 1925 to produce fuel oil from coal and by the 1930s, several commercial plants had been established for the production of automotive and aviation fuel from coal. By the 1950s hundreds of chemicals were being produced from coke oven by-product gas.

Interest in processing coal to chemicals receded in the late 1950s due to the availability of alternative petroleum and natural gas. However, there has been renewed interest in the last decade or so, driven by a concern that oil reserves (much of it located in politically unstable regions of the world) would soon prove insufficient or unavailable. Furthermore, traditional petroleum sources of chemical feedstocks are in steep and likely irreversible decline.

Oil resources are located in relatively few countries compared with coal which is abundant in many countries of the world. China is leading the world in the production of chemicals from coal and, in just over a decade, has established over a hundred coal-to-chemical plants, all based on coal gasification or processing of coke oven by-product gas. China has the world's third largest proven coal reserves, holding around 13% of the global total. Coal has several advantages as a feedstock for aromatic specialty chemicals and carbon materials. In most coals the major fraction of carbon is in aromatic structures which are dominated by polycyclic as well as monocyclic aromatic ring systems. Hence coal may be better for aromatic chemicals production than alternative feedstocks in which much of the carbon is aliphatic (IEA, 2005). For example, naphthalene derivatives are becoming increasingly important monomers for new generation polymeric materials.

There are three basic routes from coal to chemicals: recovery of coal tar from coke oven by-product gas; conversion of coal to acetylene; and coal conversion to syngas (Figure 11.1). These three primary products are major building blocks for the synthesis of hundreds of industrial and pharmaceutical chemicals and premium industrial materials. Apart from fuels and chemicals, many high-value industrial materials are produced from coal. These include calcium carbide, activated carbon, carbon fibre, graphene, synthetic industrial graphite, and diamond. Graphite and diamond are both allotropes of carbon and occur in nature but are believed to be of different origins. Graphite is the highest rank of coal. Like coal, it is formed from buried plant debris but there are significant differences in the geophysical conditions (see Chapter 1).

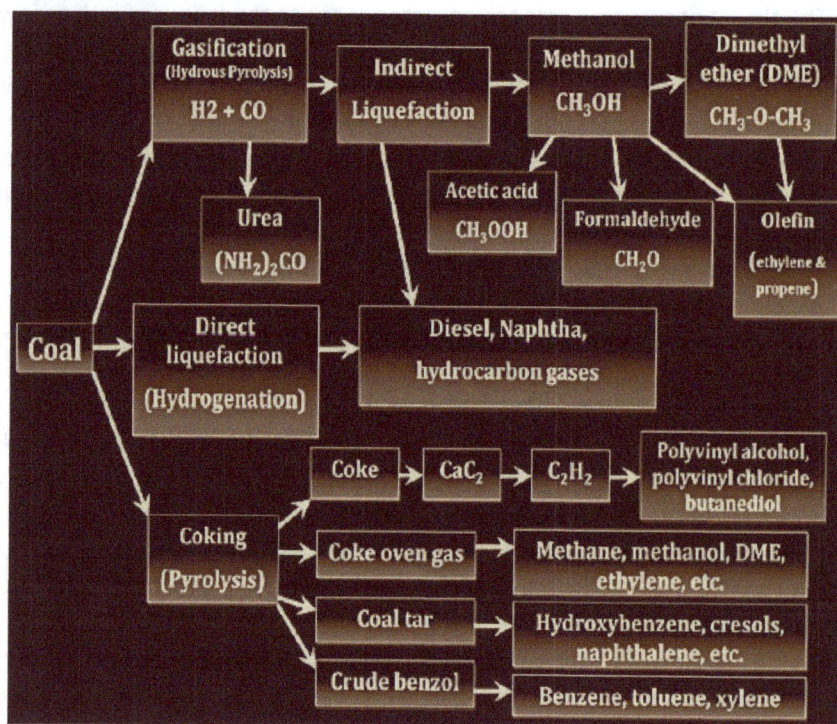

FIGURE 11.1 Options for coal conversion to premium chemicals and materials.
(Adapted from Yang and Jackson, 2012).

11.2   CHEMICALS FROM COKE AND BY-PRODUCT OVEN GAS

Processing by-product coke oven gas is an old technology being revitalized in the light of current favourable economics in some countries, in particular, China. Another reason for revival is the increasingly stringent environmental regulations in some countries which restrict industrial by-products that can be released into the atmosphere. It should be noted also that coke oven by-product gas is not considered a major source of chemicals because the main sources (tar, ammonia liquor and crude benzole) constitute less than 5 % of the coke oven output. Consequently coal is not carbonized specifically for the production of chemicals but most coke ovens have tar, ammonia liquor the crude benzol recovery units. Some also have tar distillation plants but most plants sell off these crude chemicals to independent refineries.

11.2.1   Acetylene from coke

Calcium carbide is produced by reacting coke and limestone in an electric arc furnace at around 2200°C. The calcium carbide is then hydrolyzed to produce acetylene (Equations 11.1 & 11.2a). Calcium carbide is also reacted with nitrogen to form calcium cyanamide, a fertilizer also known as nitrolim (Equation 11.2b).

$$CaO + 3\,C \rightarrow CaC_2 + CO \qquad\qquad [11.1]$$
$$CaC_2 + 2H_2O \rightarrow Ca(OH)_2 + C_2H_2 \qquad\qquad [11.2a]$$
$$CaC_2 + N_2 \rightarrow CaCN_2 + C \qquad\qquad [11.2b]$$

Apart from extensive use in welding and cutting, acetylene is a feedstock for many chemicals including vinyl chloride monomer (VCM), butanediol (BDO), vinyl acetate monomer (VAM) and acrylic acid (AA). These chemicals are in turn used to produce polyvinyl alcohol (PVA), polyvinyl chloride (PVC) and other valuable polymers. The availability of natural gas from the 1950s provided a cheaper alternative route to these chemicals, and, with the exception of BDO, production of chemicals from acetylene declined. However, China has revived and upgraded this technology in the last decade or so and nearly 90% of China's total VCM capacity is based on coal-derived acetylene (Plotkin, 2012). In the late 1990s, acetylene was largely replaced by olefin feedstocks such as ethylene and propylene, because of its high cost of production and safety issues involving handling acetylene at high pressures. However, coal-to-acetylene technology has been revived and upgraded in recent years, and is used extensively particularly in countries with vast cheap coal resources.

11.2.2   Chemicals from light oil and coal tar

Crude tar condensed from coke oven flue gas is distilled to obtain light, medium and heavy oil. The light oil comprises aromatics, mainly benzene, toluene and xylene (BTX), benzene being the predominant chemical. Tar, naphthalene and ammonia are also recovered from the flue gas. Benzene is a precursor for tens of thousands of chemical products including cosmetics, drugs, pesticides, lubricants, dyes, explosives, detergents, nylon, polymers, plastics (Plotkin, 2012; Wittcoff et al., 2013). China is one of the world's leading producers and around 27% of production is coal-based.

Low-temperature tars differ in yield and chemical composition from tars obtained from high-temperature carbonization. Properties of low-temperature tars vary widely within the temperature range 450–700°C, due to aromatization which begins around 600°C, hence use of the term 'low-temperature' usually refers to tar produced at 500°C or below. Wide variations in quantity and properties of tar produced from coal carbonization also occur depending on the nature of the feed coal, the carbonizing equipment, and the process conditions, in particular the heating programme, the residence time and the highest temperature attained. The main constituents of low-temperature tars are low-boiling phenols, paraffins, olefins and naphtenes. High-temperature tars are mainly olefins, paraffins, pyridines, hydrocarbons and aromatics and are used for the manufacture of hydroxybenzene, cresols, naphthalene, derivatives of naphthalene, etc. High-temperature tar is a by-product of coke ovens, captured by condensation from the flue gases. It is usually sold as crude tar to tar refineries where it is distilled to obtain valuable chemicals.

About 300 different chemicals can be distilled from high-temperature coal tar. Some of the major constituents are the aromatics and heterocyclic compounds including naphthalene, phenanthrene, anthracene, benzene, toluene, xylene, carbazole, pyridine, phenol, cresol, etc. However, only about 25 coal tar chemicals are considered economically viable. These include napthalene , phenanthrene and anthracene. Other chemicals from high-temperature tar include carbazole, cresols and xylenols. The residue of the tar distillation is pitch ($\approx$ 50%) which is further processed into high quality pitch of desired chemical and physical properties for use as briquetting binders, electrode binders, etc. The chemical composition of a typical high-temperature coal tar is shown in Table 11.1.

11.3   SYNGAS FROM COAL

The production of syngas from coal has been discussed in depth in Chapter 9. There are many different technologies for gasification of coal. All of them carbonize coal in a reactor by supplying air or oxygen. The main differences are in the type of reactor, process variables (temperature, pressure, number of stages, etc). The syngas produced contains mainly carbon monoxide and hydrogen and the H:C ratio of most products is between 0.6 to 0.8 unless the gas has been shift-reacted. Chemical compositions of synthesis gas produced from some commercial gasifiers are shown in Table 11.2.

TABLE 11.1 Typical chemical composition of high-temperature coal tar. *(Adapted from Mukhulyonov et al., 1974).*

CONSTITUENTS	CONTENTS (%)
Naphthalene	5-10
Phenanthrene	4-6
Carbazole	1-2
Anthracene	0.5-1.5
Phenol	0.2-0.5
Crezol	0.6-1.2
Pyridine Compounds	0.5-1.5
Pitch	$\approx$50

TABLE 11.2 Synthesis gas compositions produced from some commercial coal gasifiers. *(Kang and Lee, 2013).*

GASIFIER	TRIG	BGL	SHELL	TEXACO	E-GAS
CO (mol %)	39.7	54.3	57.2	34.4	42.0
H_2 (mol %)	28.5	29.7	29.0	33.5	33.2
CO_2 (mol %)	14.3	4.3	2.1	15.1	9.8
CH_4 (mol %)	4.3	5.9	0	0.1	0.4
H_2O (mol %)	12.6	0.2	3.6	14.3	12.2
Other Gases* (mol %)	0.6	5.6	8.1	2.6	2.4
H_2/CO ratio	0.72	0.55	0.51	0.97	0.79
CO_2/CO ratio	0.36	0.08	0.04	0.44	0.23
*Other gases contain N_2, Ar, NH_3 and sulphur					

There are around twenty different technologies for processing coal syngas into liquid fuels and chemicals, all of them featuring proprietary process design and catalysts. The oldest of the technologies, Fischer-Troppch (F-T) synthesis process produces different mixes of chemicals depending on the process variables, the catalyst used, and the intended end products. SASOL of South Africa has been operating F-T for decades producing different chemicals and fuels (Figure 11.2).

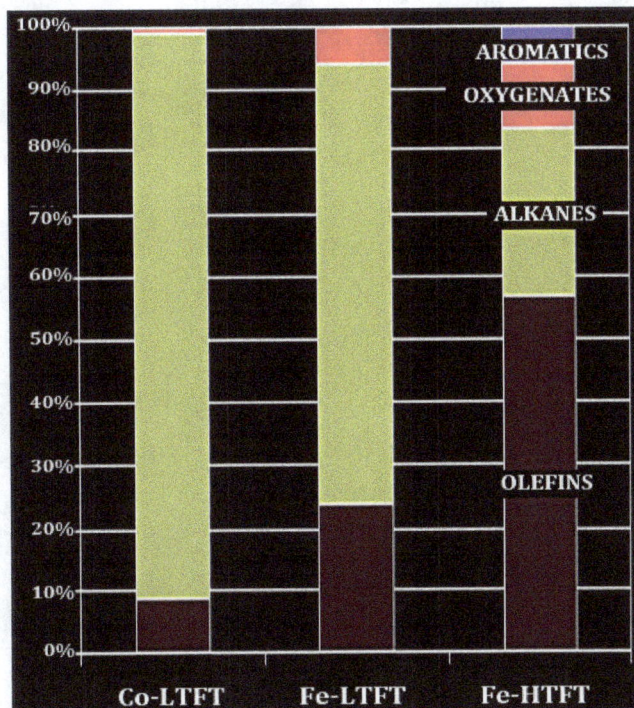

FIGURE 11.2 SASOL F-T chemical synthesizers. (Co-LT = cobalt-low temp; Fe-LT = Ferrous-low temp; Fe-HT = Ferrous high temp). *(Adapted from Gibson, 2007).*

There are three variants of the F-T technology which produce fundamentally different types of hydrocarbons: Cobalt low-temperature Fischer-Tropsch (Co-LTFT), Iron low-temperature Fischer-Tropsch (Fe-LTFT), and Iron high-temperature Fischer-Ttopsch (Fe-HTFT). Apart from its direct utility as fuel, syngas is a feedstock for the manufacture of many chemicals including hydrogen, methanol, liquid fuels, and a wide range of chemicals (Figure 11.3) (Sheldon, 1983; NETL.DOE, 2017). Some of the important chemicals produced from syngas are discussed briefly in the following sections.

11.4   COAL CONVERSION TO HYDROGEN

Hydrogen is a valuable chemical and fuel, and is widely regarded as a fuel of the future because it produces no environmentally harmful residue on combustion. Furthermore it can be produced from many renewable sources. It is used in a combustion turbine or fuel cell to produce power, or utilized as feedstock for the production of a wide range of valuable chemicals. The most common method of producing hydrogen is by steam-reforming natural gas or syngas produced from coal or biomass. However, the process also results in greenhouse gas emissions. An alternative process of producing hydrogen is by electrolysis of water. An electric current is passed between two electrodes submerged in water. Hydrogen migrates to the cathode and oxygen accumulates at the anode. This process produces no greenhouse gases but it has a carbon footprint resulting from the source of electricity - coal, gas, hydro, nuclear, solar, wind power all have total-life cycle carbon footprints (see Chapter 12).

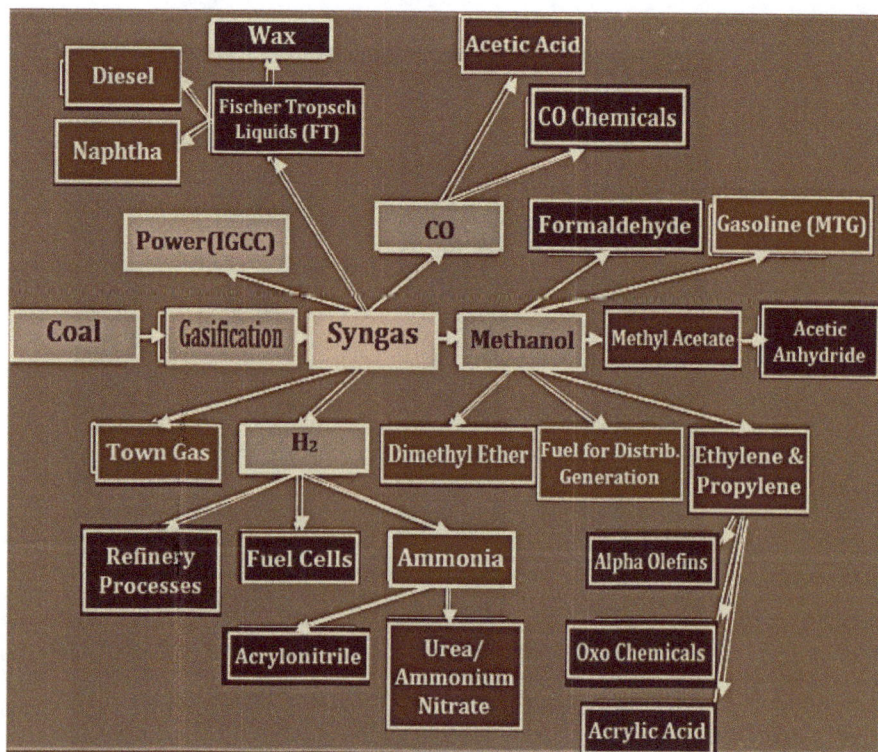

FIGURE 11.3 Chemicals from coal syngas. *(Adapted from netl.doe.gov).*

Syngas produced from coal or biomass by oxidation contains primarily carbon monoxide and hydrogen. The synthesis gas is cleaned and stripped of impurities, and further reacted with steam to produce additional hydrogen in a water gas shift (WGS) reaction which involves a reaction between carbon monoxide and steam in the presence of a catalyst to produce hydrogen as shown in Equation 11.3.

$$CO + H_2O = H_2 + CO_2 \qquad\qquad [11.3]$$

Depending on the operating conditions some concurrent reactions may occur between carbon, carbon monoxide and steam to form methane. The water gas shift reaction is usually carried out in two stages, the first at 300°C to 500°C depending on the catalyst used. Iron oxide based catalyst in a two-stage reactor is the most common configuration in order to conserve expensive catalyst, and CO concentration can be reduced to around 3% mol. Lower temperatures can favour higher CO conversion hence the second stage is carried out at 200°C to 270°C, using a copper-zinc-aluminium catalyst. The final CO concentration is around 0.3% mol. Many catalysts, for example, copper-zinc-aluminium catalysts, are susceptible to sulphur damage hence pre-desulphurization may be required before the water gas shift process (Clean Gas Shift). Some catalysts, in particular, cobalt-molybdenum based catalysts can tolerate sulphur, hence are used when the gas is not desulphurized before WGS reaction (Raw Gas Shift).

The final syngas contains mainly hydrogen, carbon dioxide and water vapour. Hydrogen is separated by passing the gas through adsorbers to remove carbon dioxide and the captured CO_2 is utilized or sequestered. Pressure swing adsorption (PSA) carbon dioxide capture system is the most common method but improved membrane and cryogenic systems are also under development. A typical syngas-hydrogen production system is shown schematically in Figure 11.4. Gasification of coal can produce high-purity hydrogen (90-99%) for a wide variety of applications. Gas turbines use hydrogen as fuel for electric power generation and petroleum refineries use the gas for removing sulphur, nitrogen and other impurities from intermediates and finished products. Heavy distillates and oils produced by refineries are hydrocracked and hydrotreated to obtain light products; high-purity hydrogen is required for solid oxide fuel cells (SOFC); and hydrogen-fueled automobiles are already in commercial operation. Also, hydrogen is used extensively in the production of many chemicals including ammonia, acrylonitrile, and fertilizers (ammonium nitrate and urea).

11.5   COAL CONVERSION TO METHANOL [CTM] AND OLEFINS [MTO]

Methane has been produced from coal-based syngas since the 1980s by ExxonMobil, the primary objective being the production of automotive fuels. Coal is gasified to produce syngas which is mainly a mixture of carbon monoxide and hydrogen. The gas is shift-reacted with steam to achieve the desired H:C ratio, cleaned, and synthesized to obtain methane, one of the major industrial chemicals. The methane is processed further in an MTG reactor to produce liquid fuels. Methane is also a high-value industrial chemical and primary feedstock for the production of a wide range of chemicals including methanol, dimethyl ether (DME), acetic acid, methyl acetate, and formaldehyde. About a third of the world's total production of methanol is used for the manufacture of formaldehyde (Figure 11.5). Ethylene and propylene which are feedstocks for the production of alpha olefins, oxo chemicals, acrylic acid, etc, are also produced from methanol. China currently produces around 25% of the world's total annual requirements of methanol. In the process of developing the technology for conversion of syngas to methane, it was found that some quantities of ethylene and propylene were also produced under certain operating conditions. Ethylene and propylene are primary building blocks for a wide range of polymers. Since then intensive research in several countries has resulted in the development of viable methane-to-olefin (MTO) technologies to produce either or both chemicals (Figure 11.6). Olefins, also commonly known as alkenes are raw materials for the production many chemicals, including ketones, carboxylic acids, ethylene and alcohols. The first commercial methanol-to-olefins plant was built in China by Shenhua and began operation in 2010 and another ten plants or so had come on stream by 2016.

11.6   ACETIC ACID FROM COAL

Acetic acid is an important industrial chemical, used for the production of many premium chemicals and materials. Perhaps the most important use is for the production of cellulose acetate for photographic film, polyvinyl acetate for wood glue, many polymers and synthetic fibres, and many domestic products and food additives, including descaling agents and vinegar. Acetic acid can be produced by bacterial fermentation of a suitable alcoholic agro-product such as grain, rice, potato, fruit to produce vinegar which is weak acetic acid. However, the bulk of industrial acetic acid is produced by carbonylation of methanol produced from various sources, notably from coal-sourced syngas.

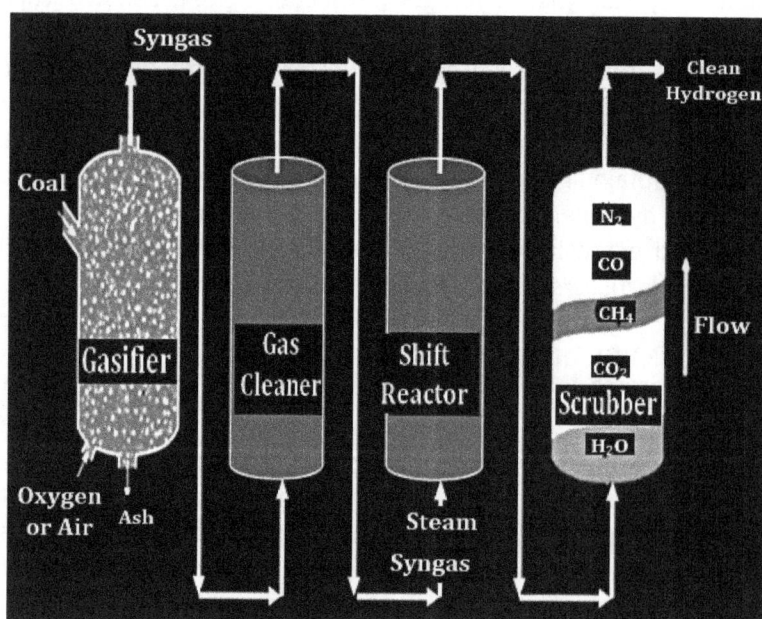

FIGURE 11.4   A syngas-to-hydrogen production system with PSA CO_2 capture.

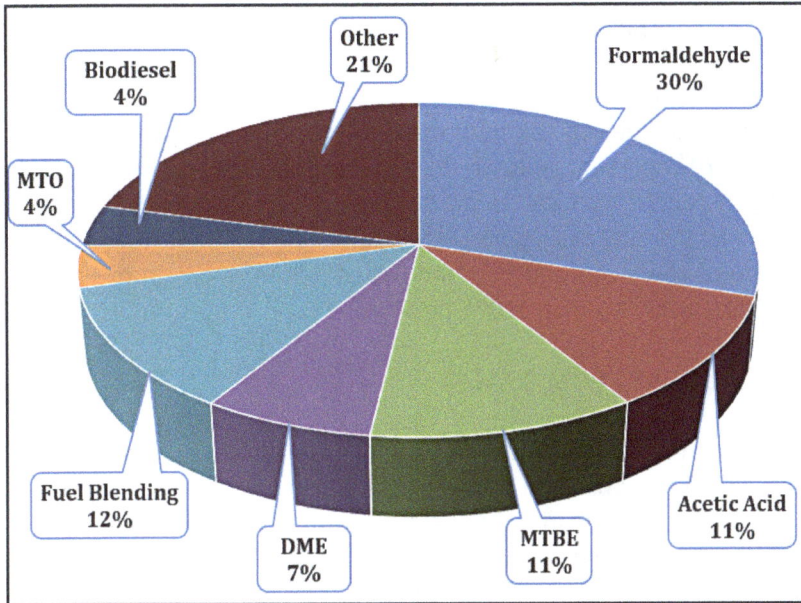

FIGURE 11.5 Global demand and end-use for methanol. *(Tweddle, 2014).*

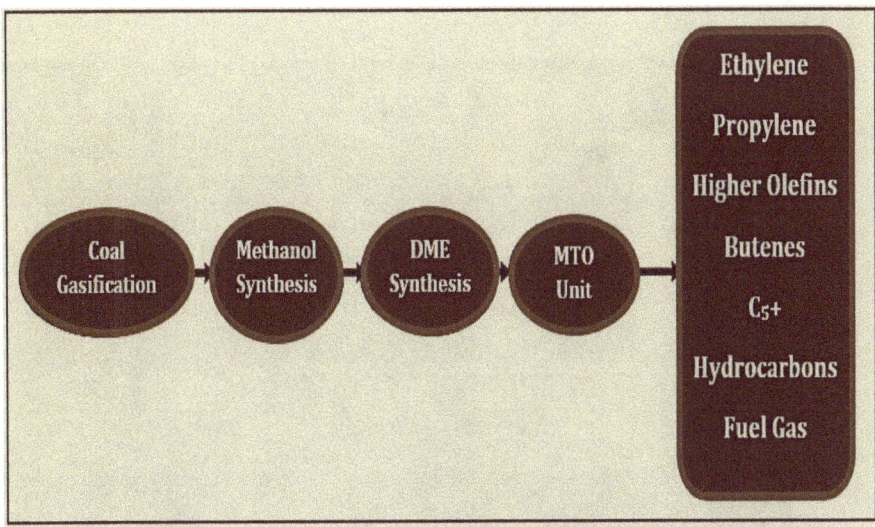

FIGURE 11.6 Coal conversion to methanol and olefins.

The methanol carbonylation process involves reacting methanol with carbon monoxide in the presence of a catalyst (Equation 11.4).

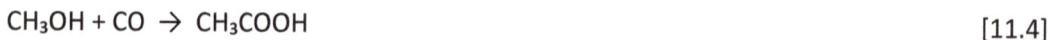

$$CH_3OH + CO \rightarrow CH_3COOH$$ [11.4]

The acetic acid may be reacted with methyl alcohol to produce methyl acetate and acetic anhydride, two important industrial chemicals. Another important industrial chemical, formaldehyde can be produced by partial dehydrogenation and oxidation of methyl alcohol. The Eastman Kodak coal gasification plant at Kingsport, Tennessee, USA has been producing a wide range of chemicals and premium industrial materials from coal-derived syngas and methanol since 1983 (Nexant, 2006). See Figure 11.7.

11.7 FERTILIZERS FROM COAL

Plants are made up of four basic elements: carbon, hydrogen, oxygen and nitrogen and the same four elements constitute the bulk of coal which is formed from buried plant debris. Carbon is obtained from carbon dioxide exhausted by human and animal respiration and oxygen is readily available in the atmosphere. However, although nitrogen is also readily available, it is not in a form that is useable by plants. Plant-bacteria interaction is required to fix the element into useable form by converting it into ammonia which plants can readily assimilate. Nitrogen is the building block of chlorophyll, the primary chemical required for photosynthesis and it is estimated that around 70% of the global farmland is nitrogen-deficient. Natural gas is an excellent feedstock for nitrogen-based fertilizers but countries which have no reserves are turning to syngas from coal which is much more abundant.

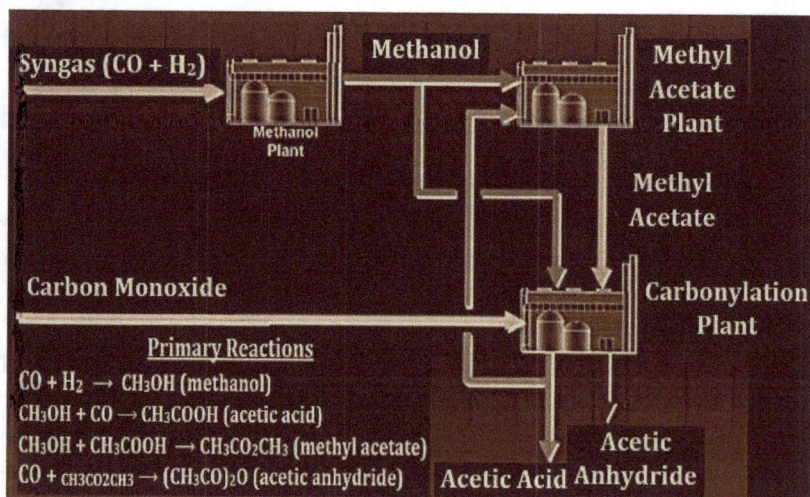

FIGURE 11.7 Eastman Coal to Acetic Acid and Derivative Chemistry. *(netl.doe.gov).*

11.7.1 *Ammonia from coal*

Natural soil contains micro-organisms which are capable of converting atmospheric nitrogen to ammonia. These include free-living soil bacteria and bacteria that form associative relationships with plants, and, most importantly, bacteria that form symbioses with legumes and other plants. Nitrogen in form of ammonia is the most important plant nutrient and a critical limiting element since it is a major component of chlorophyl, the most important pigment needed for photosynthesis. It is also vital for the production of amino acids, the key building blocks of proteins, DNA, and other important biomolecules (Wagner, 2011). Phosphorus and potassium are the two other main plant macronutrients. Hydrogen in shifted coal syngas can be separated and reacted with nitrogen sourced from the atmosphere to obtain ammonia in accordance with equation 11.5.

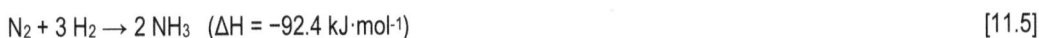

$$N_2 + 3\,H_2 \rightarrow 2\,NH_3 \quad (\Delta H = -92.4\ kJ{\cdot}mol^{-1}) \hspace{3cm} [11.5]$$

Ammonia is sometimes injected directly into the soil as fertilizer but the compound is used mainly as feedstock for all other nitrogen fertilizers such as anhydrous ammonium nitrate (NH_4NO_3) and urea ($CO(NH_2)_2$).

11.7.2 *Gypsum from coal*

Gypsum or hydrated calcium sulphate ($CaSO_4{\cdot}2H_2O$) is a naturally occurring mineral used in many industrial and building construction applications, but it is also a major fertilizer as a source of sulphur needed for plant growth. Synthetic gypsum is produced from sulphur recovered from flue gases of coal combustion and carbonizing furnaces by passing the gases through a slurry of hydrated lime. The calcium sulphite ($CaSO_3{\cdot}0.5H_2O$) formed is oxidized to form gypsum which is treated to remove contaminants and partially dehydrated. Gypsum is a major soil conditioner because of its relative solubility compared with other calcium-rich conditioners such as limestone. The compound readily dissociates into calcium (Ca^{2+}) and sulphate (SO_4^{2-}) ions in the soil. The presence of these ions in the soil helps to reduce aluminium toxicity in acid soils and to supply calcium and sulfur for plant nutrition (Chen and Dick, 2011). Some natural and synthetic sources of gypsum also contain other chemical compounds such as calcium carbonate (agricultural lime), calcium oxide (burned lime), or calcium hydroxide (hydrated lime), all of which have a liming effect and can be used to adjust the soil pH. (OSU, 2005). Synthetic gypsum contains more soil macronutrients than natural gypsum (Table 11.3).

11.8 SYNTHETIC NATURAL GAS, MONOETHYLENE GLYCOL [MEG] AND ETHANOL FROM COAL

Syngas derived from coal is very versatile as a source of energy for a wide range of applications including power generation and steelmaking. It is also a major feedstock for the production of many important chemicals and premium materials and many coal-to-chemicals plants have been commissioned in the last two decades or so, mostly in China but also in the United States and South Africa.

TABLE 11.3 Concentrations of soil macro- and micronutrients in synthetic and natural graphite. *(Extracted from Table 3, OSU, 2005).*

Soil nutrient	Synthetic gypsum	Natural gypsum
Calcium (%)	23.0	19.1
Magnesium (%)	0.03	1.35
Sulphur (%)	18.7	15.1
Boron (ppm)	26.7	9.4
Iron (ppm)	264	1045
Manganese (ppm)	5.5	14.6
Phosphorus (ppm)	16.7	30.6

11.8.1   Syngas-to-SNG (Synthetic natural gas)

One of the major deficiencies of syngas is the low calorific value which is only about half that of natural gas. Depending on the technology adopted, calorific value of syngas varies from $4\text{-}20 MJ/m^3$ compared with $37\text{-}43 MJ/m^3$ for natural gas. In effect, nearly twice as much of syngas compared with natural gas is required to produce the same amount of heat. The main reason for the low calorific value of syngas relative to natural gas is the fact that natural gas contains 70-90 vol% methane compared with syngas which is made up of 10-65% carbon monoxide and 25-55% hydrogen. It should be noted however that there is no unique specification for natural gas since composition depends on its geographic and geologic origin. Although methane is always the predominant hydrocarbon in natural gas, chemical composition and physical properties can vary widely.

Technologies have been developed to upgrade syngas by processing it through a methanation reactor to convert the hydrogen, carbon monoxide and carbon dioxide contents to methane. The gas is passed over a catalyst (usually a nickel-based catalyst) in a high temperature, high pressure reactor (Weiss et al., 2008; Sudiro et al., 2010; Liu, 2012; Jensen at al., 2011; Li, 2014; Ronsch et al., 2016). Impurities, in particular sulphur and ammonia must be removed from syngas prior to methanation. Water in the product gas is removed by condensation and the gas goes through a purification process for the production of high-purity methane.

The methanation process reactions are complex but the main ones are summarized in Equations 11.6 and 11.7. Both reactions are strongly exothermic and the water gas shift reaction (Equation 11.8) is mildly exothermic. The feasibility of CO_2 methanation reaction (Equation 11.7) depends on the thermodynamic equilibrium of the other two reactions and does not occur as long as the CO concentration is greater than a few hundred ppm.

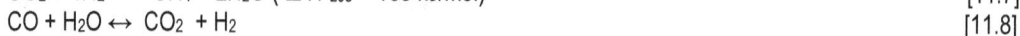

$$CO + 3H_2 \leftrightarrow CH_4 + H_2O \ (-\Delta H^o_{298} = 206 \ kJ/mol) \tag{11.6}$$
$$CO_2 + 4H_2 \leftrightarrow CH_4 + 2H_2O \ (-\Delta H^o_{298} = 165 \ kJ/mol) \tag{11.7}$$
$$CO + H_2O \leftrightarrow CO_2 + H_2 \tag{11.8}$$

The high heat of reaction in the methanation process results in a large potential adiabatic temperature and one of the main challenges is the control of the resultant high temperatures to prevent damage of the catalyst and improve methane yield. The optimum temperature and pressure for enhanced methanation are about 300°C and 15 atmospheres respectively. A typical flow diagram for the methanation process is shown in Figure 11.8.

The Johnson Matthey technology uses two reactors to achieve temperature control (Johnson Matthey, 2015). In the first reactor, purified syngas from the gasifier is heated through interchange with hot gases leaving the methanator reaction stages, then split into two streams. The first stream is mixed with recycle gas before being fed to the first methanator reactor in which both the CO shift and methanation reactions take place simultaneously over a catalyst. Both reactions in the first methanator are strongly exothermic and the heat generated is partially recovered by cooling the exit gas and using the recovered heat to generate superheated steam for the shift reaction. The temperature in the first reactor is moderated by varying the flow rate of the cooled exit gas recycled back to the inlet. The cooled gas from the first methanation reactor is mixed with the second stream of purified syngas and fed to the second reactor, and heat in the exit gases is also recovered by cooling. The crude SNG then passes through several stages of purification to upgrade the methane content and reduce the carbon monoxide and hydrogen contents.

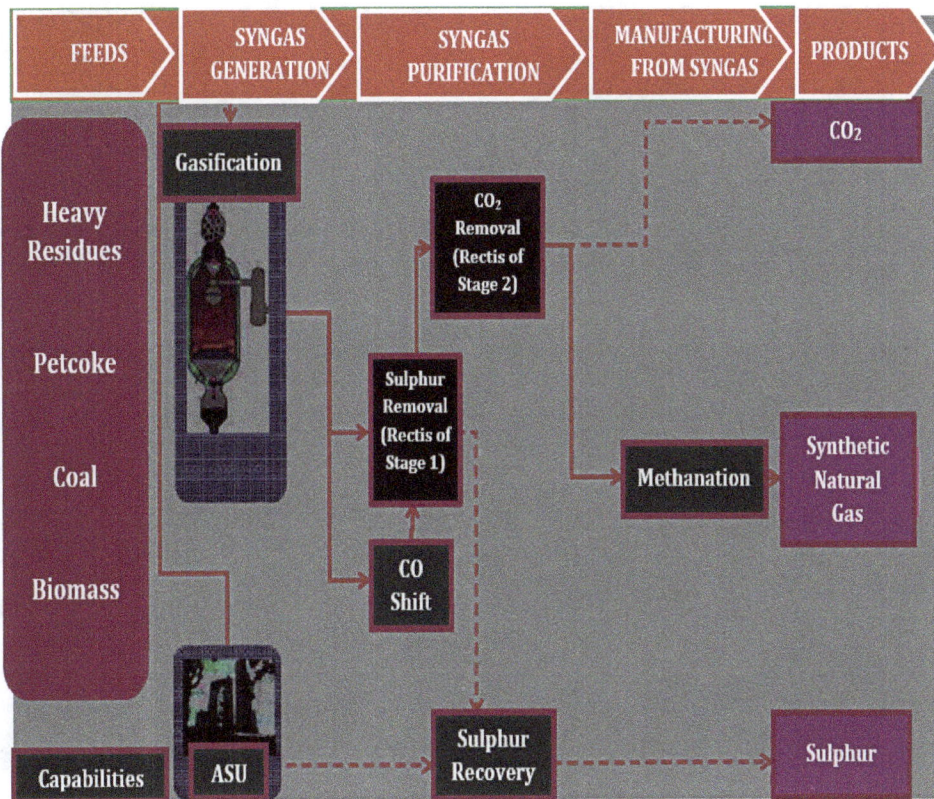

FIGURE 11.8 The Lurgi methanation technology for production of SNG from coal. *(Weiss et al., 2008).*

The TREMP (Topsøe high temperature methanation process) technology adopts three reactors to achieve control of the process thermodynamics (Figure 11.9). A proprietary high-temperature catalyst in the first reactor allows for high temperatures, with exit temperature as high as 700°C. A low-temperature catalyst is installed in the third reactor and the inlet temperature could be as low as 200°C. The catalyst in the intermediate reactor is chosen on a case by case basis (Wix et al., 2007). The reverse reaction of Equation 11.6 is the steam reforming process for producing syngas from methane. Although it is considered preferable to remove carbon dioxide from syngas before methanation, some CO_2 is still produced during the process.

The production of carbon dioxide and transformation to methane are believed to be initiated by a reverse shift conversion reaction with hydrogen to yield carbon monoxide and steam (Equation 11.8). The carbon monoxide produced then reacts with hydrogen to yield methane. Addition of hydrogen to the syngas results in higher methane yield (Kang and Lee, 2013). The synthetic natural gas (SNG) produced compares favourably with natural gas in terms of composition and heat value. The TREMP high temperature methanation process can produce SNG containing 94-96% methane (Udengaard et al., 2006; Wix et al., 2007). However, specific process conditions, and configuration of the methanation stages depend on the composition of the feed syngas and the required product specification. About ten plants are either in operation or under construction in China (Chunqi Li, 2014).

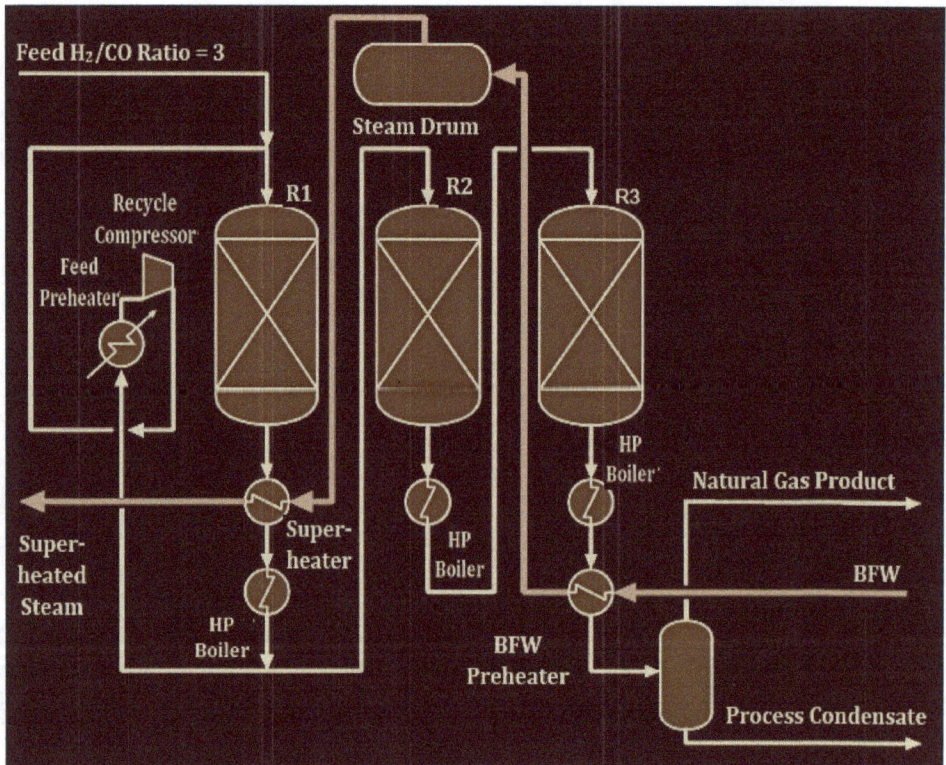

FIGURE 11.9 The TREMP high temperature methanation process flow diagram.
(Wix et al., 2007).

The ratio of hydrogen to carbon monoxide in syngas largely determines the heating value, the higher, the ratio, the higher the heating value. The H_2:CO ratio of syngas for conversion to SNG should be at least 3 and the gas must be shift reacted to achieve this ratio. Most commercial coal gasification processes produce syngas with H_2:CO ratios less than 1, with the exception of the Lurgi pressure gasification technology syngas which contains a significant amount of methane and has a ratio of around 3. It should be noted however that the desirable ratio depends on the ultimate use of the gas (Table 11.4).

11.8.2 Syngas-to-MEG (monoethylene glycol) and ethanol)

Mono ethylene glycol (MEG) is a key industrial chemical and is also a precursor in the production of polyesters for many applications including fibers, film, solid-state resins, antifreeze, etc. MEG is produced from ethylene by direct oxidation of ethylene to obtain ethylene oxide (EO) which is also a versatile intermediate for other chemicals including di- and triethylene glycols, ethanol amines, ethoxylates and glycol ethers. Ethylene oxide is then thermally hydrolyzed to obtain MEG. Traditionally, MEG is produced from petroleum-sourced ethylene or bio-ethanol but new technologies for production from syngas have emerged. A commercial plant is running in China and twelve more are either under construction or planned.

Carbon monoxide from syngas is processed through a series of complex oxalate intermediates in the presence of catalysts, at high pressures of the order of 1300-7000 atmospheres at a temperature of around 200°C. There are many variants but in the Chinese process carbon monoxide is converted to dimethyl oxalate in a carbonylation process, followed by hydrogenation, both in the presence of catalysts. Both the CO an H_2 are sourced from syngas. The crude MEG produced goes through a purification process. Apart from this process of direct synthesis, MEG can also be produced from coal-derived methanol. Ethanol currently produced primarily from food-related raw materials (corn and sugar cane) can now be produced by catalytic conversion of methanol. Selection of the appropriate catalysis is critical to this process.

TABLE 11.4   Hydrogen:carbon ratios of different ranks of coal and hydrogen:carbon monoxide ratios required for the production of some chemicals. *(Littlewood, 1977).*

H/C ATOMIC RATIO OF COAL & NATURAL GAS		H₂/CO RATIOS FOR APPLICATIONS	
Peat	1.15	Synthetic natural gas	3.0
Lignite	0.86	Methanol	2.0
Bituminous coal	0.80	Ethanol	2.0
Anthracite	0.38	Acetic acid	1.0
Natural gas ((LNG)	3-3.5	Glycol	1.5
Petroleum gas (LPG)	2.5-2.7	Acetic ether	1.5
Methane	4.0	Acetaldehyde	1.5

11.9   CONVERSION OF COAL TO HIGH-VALUE INDUSTRIAL MATERIALS

Apart from fuels and chemicals, many high-value industrial materials are produced from coal. These include calcium carbide, activated carbon, carbon fibre, industrial graphite, industrial diamond. High value nanotechnological materials such as nano carbon fibre, nanotubes, fullerenes and graphenes are also produced from synthetic, coal-derived graphite. Some of the coal-derived premium industrial materials are shown in Figure 11.10.

11.9.1   Synthetic graphite and diamond from coal

Carbon is an exceptionally versatile element, capable of forming a wide range of allotropes, from the most common form (soft graphite) to the hardest known material (diamond), depending on the crystallographic arrangement of the atoms. Carbon is the base for DNA and all life on Earth. Graphite and diamond, both allotropes of carbon, occur in nature but are believed to be of different origins. Graphite is the highest rank of coal. Like coal, it is formed from buried plant debris but there are significant differences in the geophysical conditions as discussed in Chapter 1.

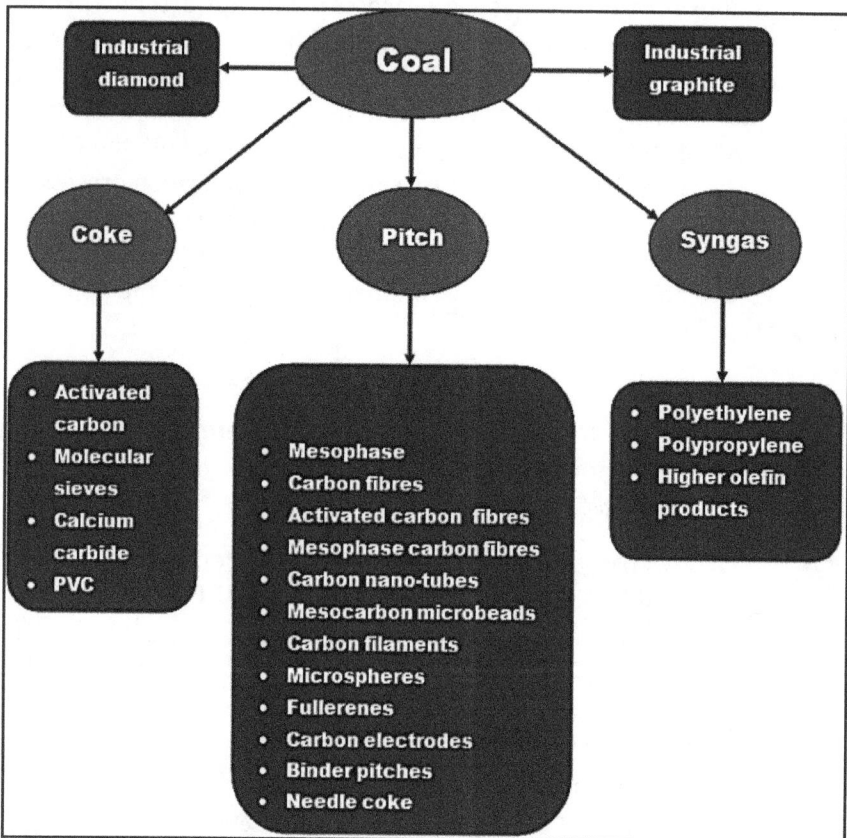

FIGURE 11.10 Coal-derived premium industrial materials.

Graphite is found in deep deposits in rocky areas and the unique structure compared with coals of other ranks is due to the exceptionally high pressure and temperature resulting from the high, rocky overburden. Natural graphite is relatively rare, hence it is usually not considered as an energy source.

Diamond is believed to have been formed from carbon-containing organic minerals at exceptionally high temperatures and pressures, up to 200 kilometres beneath the Earth's surface over periods from 1 billion to 3 billion years ago. They are brought nearer to the Earth's surface by volcanic eruptions. The rarity and exceptional luster of diamond on polishing make it perhaps the most valuable gem in the jewelry industry. Diamond and graphite are the purest, naturally occurring forms of carbon. However, they have very different techno-logical properties. Diamond is the hardest known material, electrically inert and an excellent insulator, while graphite is one of the softest yet cohesive industrial materials. It is also conductive in the planar direction of the atomic structure. The two materials are chemically identical (both are composed of carbon atoms), but with different physical structure (Figure 11.11).

Both graphite and diamond can be produced from any amorphous natural carbonaceous material but the carbon content and purity must be sufficiently high. Also, both processes involve high temperature an pressure. Precursors to graphite are many but the most common are pitches derived from petroleum, coal, natural and synthetic organic materials. The material can also be manufactured by the direct precipitation of graphitic carbon from pyrolysis of a carbonaceous gas such as acetylene. Synthetic diamond is produced by processing high-purity graphite or any other high-purity carbon source, including gas mixtures under high temperature and pressure.

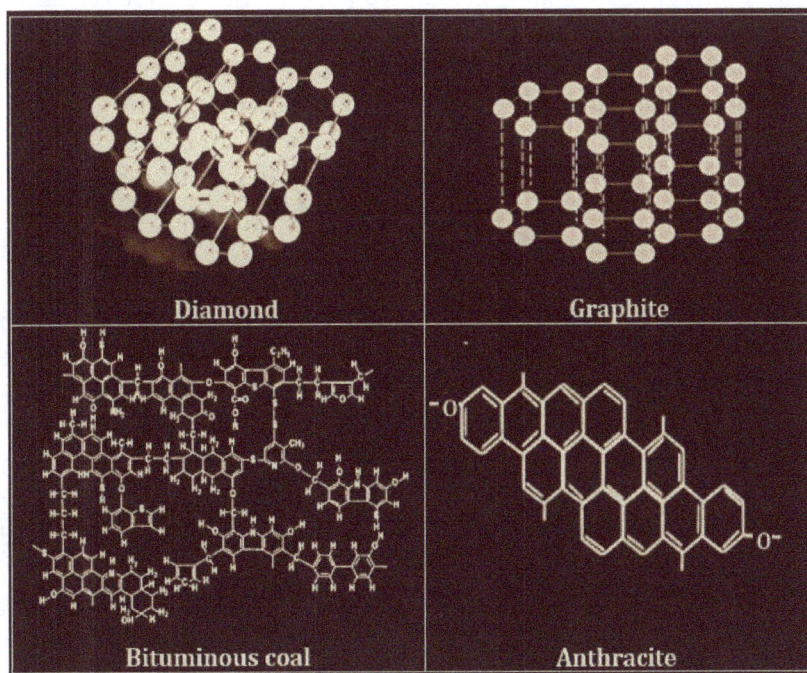

FIGURE 11.11 Structures of diamond, graphite and coal.

11.9.1.1 *Synthetic Graphite from coal-derived tar*

The strong planar covalent bonds combined with the weak inter-planar bonds make graphite an excellent high-temperature lubricant, since strong and heat-resistant atomic planes can adhere to two mating faces, yet slide over each other. While all the carbon atoms in diamond are held together by strong covalent bonds, in graphite, the covalent structure is restricted to planar single atom thick horizontal sheets known as graphenes stacked vertically and held together by weak van der Waals forces. The fact that the planar sheets of graphite can slide over one another relatively easily makes the material one of the softest materials (1-2 on the hardness scale compared with diamond's maximum of 10), while the strong planar bonds make it highly resistant to temperature.

Only three of the four atoms on the horizontal planes of graphite are involved in covalent bonds, the fourth being free in an electron cloud which makes graphite electrically conductive (diamond is non-conductive). There are very few materials which combine these incongruous material properties and this makes graphite a very special and versatile material. The basal plane thermal conductivity of graphite is comparable to diamond which has the highest thermal conductivity of any bulk material. Typical natural and synthetic graphite are imperfect due to crystalline imperfections and inclusions of impurities. Graphite can be synthesized by simulating the very high temperature and pressure conditions under which the material was formed naturally. Synthetic graphite can be manufactured by high-temperature treatment of amorphous carbon and precursors are many: any coal, coal tar pitch, petroleum coke, natural and synthetic organic materials (Asbury Carbons, 2006).

Most of the world's current commercial production processes use coal tar pitch or calcined petroleum coke as precursors. Both materials are composed of highly graphitizable forms of carbon. Synthetic graphite can be manufactured in varying degrees of technological specifications depending on application, and in any number of forms including solid articles of varied shape and size, granular and powder forms. China dominates the global natural graphite production, accounting for 70-80% of total supply, and largely regulates prices through export controls. The price of natural graphite per tonne has more than doubled over the past few years. Recent developments in information and transportation technologies have propelled graphite to the forefront by featuring the material in energy storage devices including fuel cells, supercapacitors, lithium-ion batteries for laptops, cell phones, electric cars, air planes, etc.

Although synthetic graphite is much more expensive than natural graphite despite the severely restricted geographical distribution of natural graphite, synthetic graphite is preferred because of its superior properties, in particular, consistency, purity and reliability. Synthetic graphite typically contains greater than 99% carbon compared with natural graphite of 90-95%. Furthermore synthetic graphite contains very low impurities. However, several processes have been developed to purify natural graphite to levels comparable with synthetic graphite. These include hydrofluoric acid, hydrochloric acid or caustic leaching, and high-temperature thermal treatment.

Graphite is a very versatile industrial material. It is used for high-temperature industrial furnace lining, foundry melting crucibles, electrodes for electric arc furnaces, fuel cell bi-polar plates, melting crucibles, manufacture of high-temperature lubricants, high temperature coatings, corrosion prevention coatings, electrolytic processes, conductive filters, carbon fibres. It is a precursor to the production of graphene, synthetic diamond, activated carbon, molecular sieves, rubber and plastic compounds, drilling materials, and many more high-technology materials. Graphite is an important structural component in nuclear reactors, used as moderator for controlling nuclear reaction rates by slowing down fast neutrons to an appropriate energy level that can be absorbed for activating uranium isotope.

11.9.1.2 *Synthetic diamond from coal-derived graphite*

Synthetic diamond, also known as cultured diamond, is produced by several different methods but the most common are the high-pressure/high-temperature (HPHT) and Chemical vapour deposition (CVD) processes. The HPHT process is the more common because it is much cheaper although it is more complex. In the process, a capsule is filled with a mixture of a pure carbon source, (usually graphite derived from coal tar pitch) and nickel. The capsule is heated to about 1500°C under very high pressure between the anvils of a press. The nickel melts and dissolves the graphite, also acting as a catalyst for the conversion of the graphite into diamond which is deposited on small diamond seeds placed at the bottom of the press. The seeds grow gradually as the process progresses, forming a large synthetic diamond (IDL, 2009).

In the CVD process, a carbon-rich gas, usually methane (CH_4) is ionized in a chamber, using a high energy source, including microwave, electric arc, laser, hot filament, or electron beam. The gas ionizes into carbon and hydrogen. The carbon vapour deposits as diamond on a substrate, usually silicon held at around 800°C. The CVD process has a relative advantage over the HPHT technology because it does not require high pressure and larger diamond crystals can be grown on a wide variety of substrates. Also, impurity content can be strictly controlled but the quality of diamond produced is inferior.

Although natural diamond has the highest thermal conductivity of any known material because of the very strong covalent bonding, it is electrically inert because it does not have any free electrons. However, synthetic diamond can be made electrically conductive by doping with some elements, in particular, boron can render the material electrically conductive or superconductive. Industrial use of natural diamond is severely limited by cost, despite its excellent hardness properties, and use is restricted mainly to the production of expensive jewelry. However, synthetic (industrial) diamond is used extensively for cheap costume jewelry, industrial cutting blades and saws, machine cutting and drilling tool bits, abrasive and polishing grits and powders. Boron-doped synthetic diamond is also used in some electronic applications.

11.9.2 Calcium carbide from coal-derived coke

As discussed briefly earlier, calcium carbide is produced by reacting coke with limestone in an electric arc at high temperature, around 2000°C. Hydrolysis of calcium carbide produces acetylene in accordance with following Equations 11.8 a&b.

$$CaO + 3C \rightarrow CaC_2 + CO \quad\quad [11.8a]$$
$$CaC_2 + 2H_2O \rightarrow Ca(OH)_2 + C_2H_2 \quad\quad [11.8b]$$
$$CaC_2 + N_2 \rightarrow CaCN_2 + C \quad\quad [11.8c]$$

Calcium carbide is a feedstock for the production of acetylene used extensively in industry for welding, for the production of polyvinyl chloride (PVC), the third most widely used polymer after polyethylene and polypropylene. Calcium carbide also reacts with nitrogen at high temperatures to form calcium cyanamide (Equation 11.8c), a high-quality fertilizer. It is also hydrolyzed to form hydrogen cyanamide used extensively in agriculture and pharmaceuticals. Calcium carbide is used in miner's head lamps, for desulphurization of iron and steel, and many other applications including ripening of fruits.

11.9.3   Activated carbon and molecular sieves from coal-derived graphite

Activated carbon is a carbonaceous material that has been processed to obtain a highly developed internal structure and large internal surface area. Most carbonaceous materials have a certain degree of porosity and an internal surface area in the range of 10 to 15m²/g. Controlled treatment with steam at high temperature promotes internal oxidation of carbon atoms and extended internal surface area of between 600 and 1,200m²/g depending on the carbonaceous material and the activation process conditions. Virtually all materials containing a high fixed carbon can potentially be converted into activated carbon, for example, wood and bone charcoal, all types of coal, coconut shell, rice husk, corn cobs, walnut shell, peat, waste tyres, petroleum pitch. However, raw material supply is a major limitation since activated carbon yield is only about 10%. In effect, one thousand tonnes of raw material would be required to produce only one hundred tonnes of activated carbon.

Most of the world's requirements of activated carbon is presently produced from anthracite, bituminous coal or wood although specialized types are also produced from rice husk, coconut shell, etc. Coal is ground and mixed with a suitable binder and briquetted or extruded into pellets. In some processes a blend of coking and non-coking coals is constituted, carefully proportioned to obtain optimal rheological properties during hot briquetting, followed by low-temperature carbonization at 500 to 600°C. The product is activated in a furnace, usually vertical retort, kiln, rotary furnace, hearth furnace or fluidized-bed at a temperature of 800 to 1000°C. Full carbonization takes place in the upper and middle parts of the furnace and air is blown in to oxidize the carbon monoxide and hydrogen produced into carbon dioxide and water vapour.

Some fine materials such as grain husk and sawdust may be activated with a dehydrating chemical such as phosphoric acid, sulphuric acid or zinc chloride. The raw material and chemical are mixed intimately and carbonized at 500 to 600°C. Higher temperatures of 800 to 1000°C are required when phosphoric acid is used as the activating agent. The activated product is then washed, dried, ground and sized as desired. The reactions between carbon and oxygen are exothermic and provide heat that makes the process self-sustaining. Steam is supplied to the lower part of the furnace where activation takes place. The control of the process, in particular the residence time in the activation zone largely determines the quality of activated carbon produced. Post activation processing includes crushing and screening to obtain desired mesh size ranges.

Activated carbon has a very wide range of industrial applications, from refrigerator deodorizers, gas absorption, water purification, decaffeination, pharmaceuticals, gold purification, metal extraction, sewage treatment, decolorizers, respiratory devices, to nuclear power containment systems. A large variety of activated carbons is available for different applications. A primary requirement is that the internal structure must be accessible to the passage of a fluid or vapour. Easy access to internal pores of activated carbon requires a highly developed internal network of pores of varying diameters classified as micropores (<40 Angstroms), mesopores (40-5000 Angstroms) and macropores (5000-20000 Angstroms). Activated carbon is produced in pelletized, extruded or granular form. The activation technologies required to obtain a particular internal structure have been reviewed exhaustively by CameronCarbon (2006). In general, the molecular size of the fluid to be absorbed and the internal pore size of activated coals should be comparable for effective adsorption. For example, activated carbon produced from coal or wood has relatively large pore structure and would be unsuitable for adsorbing fine molecular gases such as Krypton of Xenon but are excellent decolorizers and deodorizers. On the other hand, activated products of grain husk, coconut shell have fine pores and are suitable for adsorbing fine molecular gases.

11.9.4 Carbon fibre from coal-derived graphite

The roots of carbon fibre can be traced to the late nineteenth century when Thomas Edison made a filament for his electric bulb invention by baking cotton thread (a natural linear cellulosic polymer made of repeating units of glucose) into an all-carbon fibre. The fibre connected two carbon electrodes and an electrical discharge between them caused the carbon fibre filament to glow. The modern carbon fibre was developed by Union Carbide in the late 1950s. Graphite can be processed into extremely thin, largely defect-free and exceptionally strong, high modulus fibre, 5 – 10µm thick (less than a tenth the thickness of a human hair). The material is then woven into mats or embedded in polymer matrices to form composites.

The processing of graphite at high temperature causes the carbon layers to lie linearly parallel with the fibre axis. The high strength-to-weight ratio of these materials has opened up a wide spectrum of applications from the early 1960s, from high-temperature gaskets to fishing and sports equipment, packaging materials, automobile components, etc. However, the (Young's) modulus was considered too low for high-integrity applications. Further developments led to the invention for processing rayon by heat-treating at temperatures of the order of 3000°C while stretching the fibre during heating. The process resulted in a tenfold increase in Young's modulus and opened up a wide range of new, mission-critical applications by the mid-1960s, particularly in aerospace, aeronautic, military, transportation, recreation and sporting industries.

Carbon fibre is manufactured by controlled carbonization of organic fibres known as precursors. The heat treatment removes the oxygen, nitrogen and hydrogen to form carbon fibres. Suitable precursors are fibres that already have highly oriented crystal structures which are maintained and improved in response to controlled heat treatment, mechanical stretching and stabilization treatments (LichaoFeng et al., 2014; Park, 2015). Heat treatment temperatures range from 1000 to 3000°C. The lowest temperature produces low modulus, low strength materials and both modulus and strength increase with temperature. The carbon content of carbon fibres ranges from around 90% to above 99%. Fibres with carbon content in excess of 99% are described as graphite fibres.

There are many types of precursors to carbon fibre but the most common are polyacrylonitrile (PAN), coal or petroleum tar pitch and phenolic fibres. Production of carbon fibre from either precursor involves similar processes: spinning into fibres, stabilizing, carbonizing, and graphitizing (Figure 11.12). However, there are significant differences in proprietary process details. For example the PAN-CF process includes simultaneous tensile stretching and low-temperature heat treatment, followed by carbonization at around 1000°C in an inert atmosphere, and graphitization treatment at temperatures between 1500 and 3000°C depending on the desired quality of carbon fibre (Nippon, 2011; Inagaki et al., 2012, 2014). The processes of producing carbon fibre from coal or petroleum tar pitch start with conditioning of the pitch by adjusting the molecular weight, viscosity and crystal orientation through heat treatment, followed by spinning or drawing to achieve a highly directional crystal orientation. The fibre is stabilized by heat treatment at temperatures between 250 and 400°C to obtain the desired shape, and then carbonized at temperatures between 1000 and 1500°C (Figure 11.12).

There are two main types of pitches in use for the manufacture of carbon fibres: low molecular weight isotropic pitch and high molecular weight anisotropic (mesophase) pitch (Figure 11.13). Isotropic pitch is melt-spun at high strain rates to align the molecules parallel to the fibre axis and the fibre is rapidly cooled and oxidized at temperatures below 100°C to render it infusible, promote cross-linking and stabilize the fibre. However, on carbonization, relaxation of molecules takes place and the end product fibre contains fibres with no significant orientation. Apart from the fact that the oxidation process is lengthy and expensive, response to the graphitization treatment is low, hence this step is omitted and fibres produced are of low quality. They are known as general purpose carbon fibres (GPCF) and used mainly as filters, and in fibre-plastic composites as thermal insulation materials.

Anisotropic pitch is preferred for the production of high-performance carbon fibre (HPCF) because thermal treatment at around 350°C produces a mesophase pitch which contains both isotropic and anisotropic phases. Spinning orients the anisotropic molecules parallel to the fibre axis but the orientation of the isotropic component is largely unchanged. Cooling to below the softening point causes the isotropic component to set and carbonization at around 1000°C followed by graphitization at around 3000°C produces high quality carbon fibre. One major advantage of this process compared with the PAN process is that no tension is required during the stabilization or the graphitization processes.

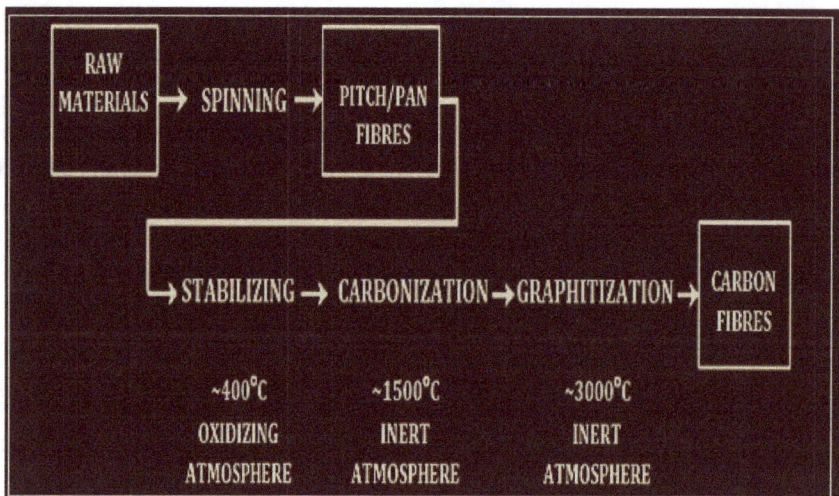

FIGURE 11.12 Carbon fibre production processes. *(Mitsubishi INC).*

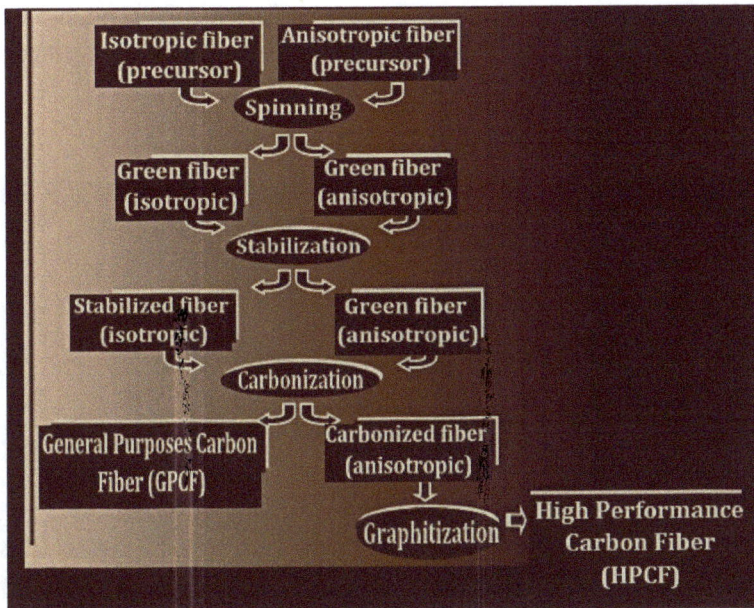

FIGURE 11.13 Carbon fibres from coal or petroleum pitch. *(Ania, 2007).*

The Pitch-CF process is relatively easy because the pitch is made up of condensed polycyclic aromatic hydrocarbons while PAN-CF is a straight-chain aliphatic polymer (Figure 11.14). Pitch precursor has significant advantages over resin: it has higher graphitizability, carbon yield is higher, it develops open porosity on carbonization, and a wide variety of fibres can be made to meet structural, conductivity, chemical resistance requirements at competitive costs. The PAN-CF process is more complicated and more expensive but produces carbon fibres of very high tensile strength and integrity. It is difficult and expensive to produce materials with a combination of high strength and modulus. On the other hand, pitch-based carbon fibres with a combination of high modulus and medium strength can be produced relatively easily and more cheaply. PAN-CF products are used for the more critical applications, in particular, aerospace, aircraft, nuclear engineering and transportation industries because of their combination of high tensile strength and high tensile modulus (Figure 11.15).

Pitch-based fibres are more suitable for a much wider range of applications requiring different strength-modulus ratios, including sporting goods, machine tools, construction, packaging. Pitch-based carbon fibre manufacture has several advantages over resin precursors. Carbon yield is higher, pitch graphitizes relatively easily and develops open porosity on carbonization. Furthermore, the pitch-based process is cheaper and a wide variety of structures, strength and heat conduction properties can be developed by controlling the graphite crystal (Arai, 2001; Ania, 2007). Pitch-based carbon fibres can be produced with a rare combination of weight-stiffness ratio, high thermal conductivity, low thermal expansion and low frictional properties. Fibres with combined high-modulus and high heat conduction which cannot be produced from the PAN process have been produced successfully from pitch (Figure 11.16).

There has been a rapid advancement in Pitch-CF technologies in recent years and high-integrity products for mission critical applications are now available. Pitch-based carbon fibres are unique for their ability to combine high modulus and thermal conductivity. While PAN-based carbon fibres have near zero thermal conductivity, pitch-based fibres having a wide range of thermal conductivities can be produced and are finding increasing applications in electronic equipment devices (Figure 11.17) (Nippon, 2001, 2011).

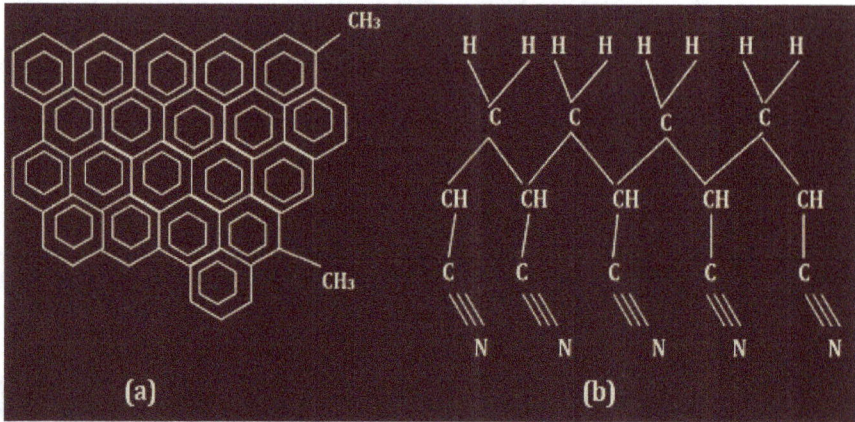

FIGURE 11.14 Molecular structure of carbon fibre precursors (a) Coal tar pitch. Condensed polycyclic aromatic hydrocarbons (b) Polyacrylonitrile (PAN) straight-chain aliphatic polymer.

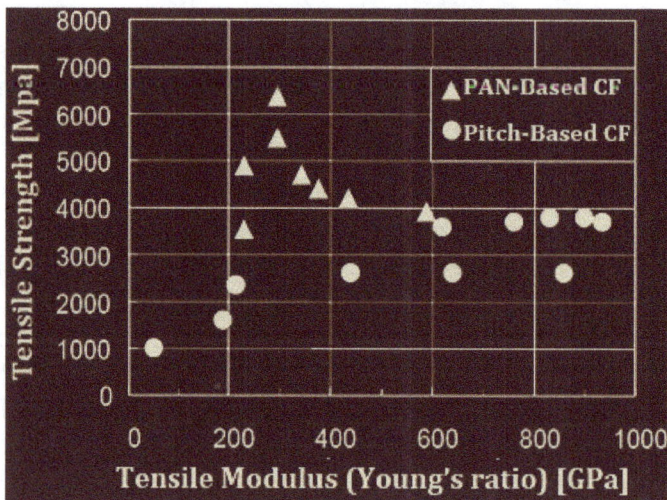

FIGURE 11.15 Comparison of mechanical properties between PAN- and pitch-based carbon fibres. *(Source: Nippon Steel News No. 383, 2011).*

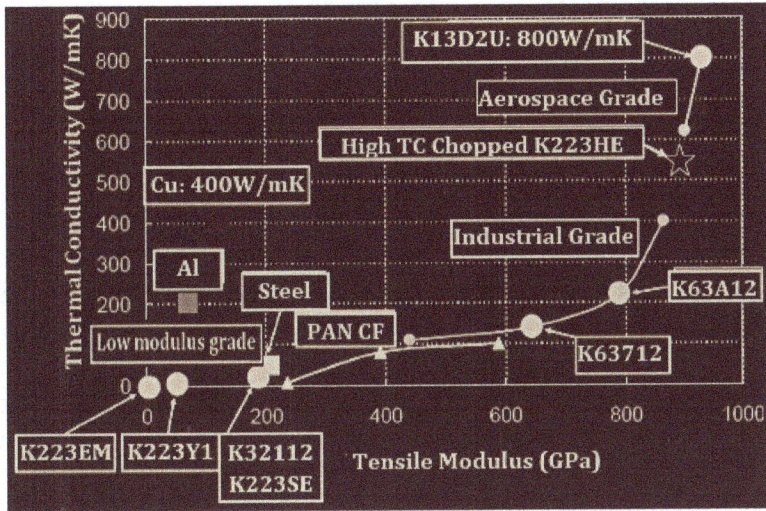

FIGURE 11.16 Combined strength and thermal properties of carbon fibres and other structural materials. *(Mitsubishi Plastics INC).*

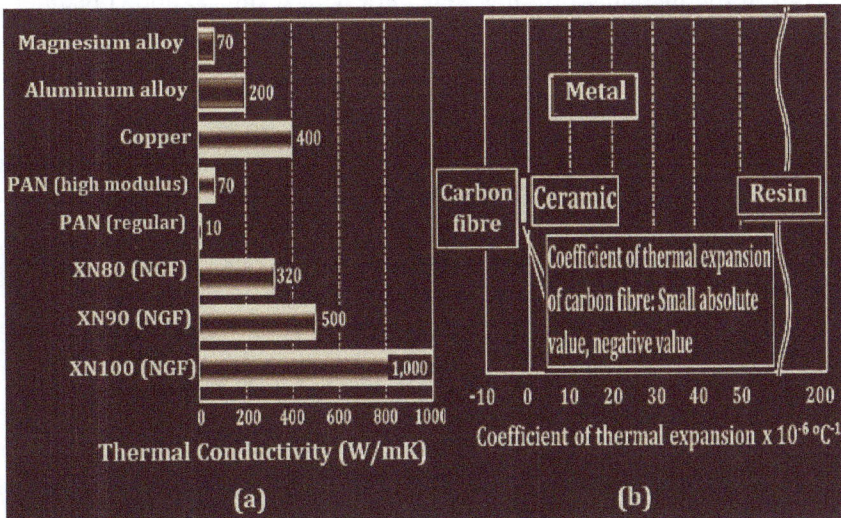

FIGURE 11.17 (a) Thermal properties of coal pitch-based carbon fibres (XN80-100) and other materials (b) Coefficient of thermal expansion of various materials. *(Nippon Steel).*

Activated carbon fibres have also been found to have superior adsorption characteristics compared with coke-based granular activated carbon (GAC) because of their relatively high surface area and controllable pore size and distribution. The pore surface chemistry and functional groups of activated carbon fibre can be designed to impact high adsorptivity kinetics and selectivity characteristics, and relative ease of regeneration. Adsorbents are used widely in water and air purification, chemical separation processes, biomedical and pharmaceutical technologies, energy technologies.

Two of the main parameters for assessing the utility of a material in structural engineering design are the strength to weight ratio and the stiffness to weight ratio, particularly in applications where weight matters. Carbon fibres are usually classified in ascending order of strength modulus, as standard, intermediate, high and ultra-high. Compared with typical metals and polymers in which properties are the same along all axes (isotropism), carbon fibre properties are highly directional (anisotropism) and maximum along the main axis. If loaded along the main axis of the crystal alignment, the tensile strength/unit weight of the strongest carbon fibre is up to ten times higher than for steel and eight times for aluminium. Additionally, their fatigue properties are superior to all known metallic structures, and they are one of the most corrosion-resistant materials available, when coupled with the proper resins.

However, carbon fibres are very brittle if loaded across the main fibre axis or subjected to bending stresses. The sheet-like aggregations allow easy crack propagation. On bending, the fiber fails at very low strain and cannot withstand oxidative environments. In order to exploit the excellent tensile characteristics and avoid the weaknesses, carbon fibres are used directly only when they will be subjected to zero or purely tensile loading, otherwise they are woven into uni-directional yarns, fabrics and mats or embedded in epoxy resins, or graphite and shaped to form strong and chemically resistant composite materials which are used extensively in the manufacture of a wide variety of technical and consumer goods. (Figure 11.18).

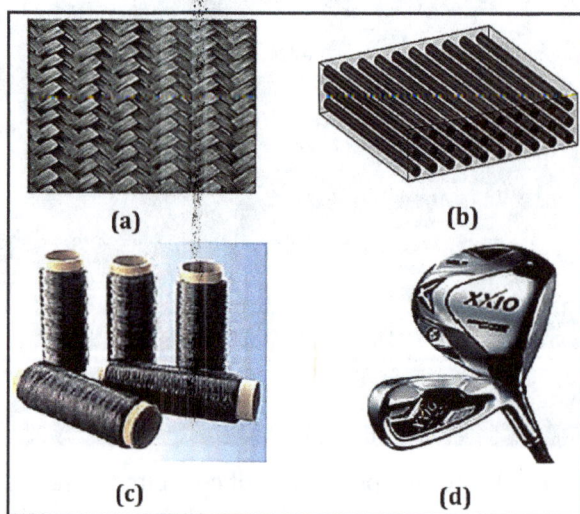

FIGURE 11.18 (a) Woven carbon fibre; (b) carbon fibres embedded in epoxy resin (c) carbon fibre yarn (d) carbon fibre golf club.

Carbon fibers are used in composites with a lightweight matrix, usually a thermoset resin such as epoxy (CFRP). However, some other thermosetting and thermoplastic polymers such as polyester, vinyl ester or nylon are also used. Carbon fibres are often mixed with other fibres such as glass, polyethylene, aluminium fibres in some composites. Carbon fibres are often embedded in graphite matrix to form composites known as carbon-carbon (C-C) composites or graphite fibre-reinforced polymer (GFRP). These composites have exceptional technical properties including high temperature resistance, high thermal stability, high strength to weight ratio, high chemical inertness and corrosion resistance, high fatigue resistance, high resistance to thermal shock. However, these C-C composite materials are very expensive and were developed for critical mission applications such as manufacture of nose and wing components of intercontinental ballistic missiles and space shuttle aircraft, fire-retardant materials, etc.

Carbon fiber composites are ideally suited to applications where high strength to weight, high strength to stiffness, chemical inertness (except to oxygen), high damping capability and outstanding fatigue characteristics are desirable, important or critical requirements. However, the physico-chemical properties of the matrix determine the ultimate technical properties of the composite to a significant extent. Composites with exceptional wear resistance have also been developed for mission-critical brake system pads. Composites made by embedding carbon fibre in silicon carbide are now being used also for brake disks and pads of racing cars and aircraft because of their high wear resistance and thermal stability. Both the Boeing 787 and Airbus A380 aircraft have over 50% carbon fibre reinforced polymer (CFRP) components, including the fuselage and critical components such as wing sections. Other uses of carbon fibres include aerospace equipment, satellite components, military and commercial aircraft structures and components, including high-speed aircraft skin structures, automotive components, computer casing, lithium-ion batteries, sporting goods, canoes, speed boats, ship components, storage tanks and structural reinforcement in construction materials.

The biocompatibility of C-C composite due to its chemical inertness has also opened up a wide range of applications in chemical, nuclear and biomedical technologies including process plant components, protective wear, x-ray equipment, human bio-inserts such as internal drug release capsules, medical prosthesis and implants, heart pacemakers, etc. Carbon-carbon composites have some major disadvantages: they have low impact resistance and are sensitive to high temperature oxidation. The break-up of the space shuttle Columbia during re-entry into the Earth's atmosphere in 2003 was believed to have been due to a crack in a reinforced carbon-carbon (RCC) protective wing shield of the shuttle orbiter which was caused by the high-speed impact of a falling piece of foam insulation that disengaged from the space shuttle external tank during launching. A hole possibly up to 0.3 metre wide caused by the impact seriously breached the thermal protection system of the spacecraft's wing leading edge during re-entry into the Earth's atmosphere by allowing hot air *to* penetrate the substructure of the wing. It was estimated that the aluminium alloy substructure (melting point around 600°C) may have been subjected to temperatures around 4000°C, hence the disintegration of the spacecraft (CAIB, 2003).

11.9.5 Graphene from coal tar and petroleum pitch

Graphite is made up of planar layers of covalent-bonded carbon atoms, each layer known as graphene (Figure 11.19). In effect, graphene is a one-atom thick planar sheet of graphite and materials made up of several layers of graphene are known as nanoplatelets. Graphene is the strongest material to date and studies have shown that its breaking strength to weight ratio is around 100-300 times that of steel (Lee et al., 2008). The source of the strength of nanoplatelets is the exceptionally strong planar aromatic bonds. The material conducts heat and electricity better than any other known material, due to the incredibly fast rate at which electrons can move through the structure. Graphene is also known to be the most reactive form of carbon, around a hundred times more reactive than thicker sheets (Gabrielsen, 2013). The unique technological properties make graphene a potential building block for a wide variety of new materials, from structural carbon fibres, fulerenes, to electronics and biomedical materials. It is also one of the most studied new materials currently.

The fact that graphene can be embedded in other matrices renders it exceptionally versatile because materials with combined strength and flexibility, exceptionally high thermal and electrical conductivity can be produced. However, like most other carbon materials, graphene has relatively low fracture toughness, only 10-20% of the range for metallic materials, depending on the impurity content. Although the possible existence of graphene has been known and researched since the 1940s, the major breakthrough came in 2004 when Andre Geim of the University of Manchester, UK succeeded in peeling away layers of graphite until a single layer was produced (University of Manchester, 2014). The work evolved from many previous studies on graphite structure by various authors for decades. This monumental achievement won Geim and Konstantin Novoselov the 2010 Nobel Prize for physics (Nobel Foundation, 2010; The Royal Swedish Academy of Science, 2010).

Many methods of producing graphene have been developed or are under development and most start with graphite produced from coal tar or petroleum pitch. The most common method known as mechanical cleavage involves systematic removal of atomic layers of graphite. Another increasingly popular method which is claimed to produce higher quality, mono-layer graphene involves the deposition of one layer of graphite onto another substrate. While there are currently very few commercial applications of graphene, there is little doubt that the material will dominate the family on new generation materials of the 21st century, based on the extent of intensive research and development worldwide (Patel, 2009; Bae et al., 2010; EU, 2013).

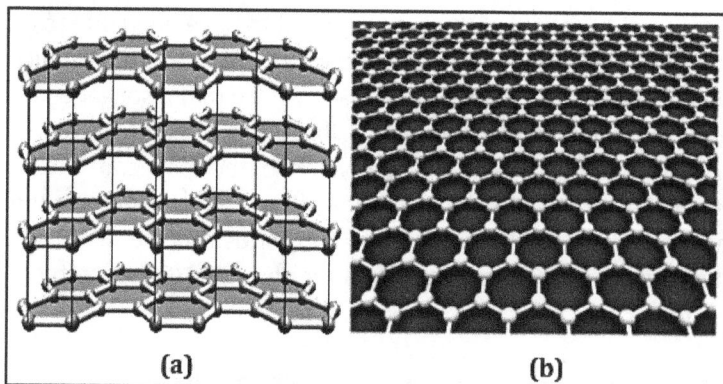

FIGURE 11.19   Crystal structure of (a) Graphite (b) Graphene.

Current efforts focus on methods of producing the material in commercial quantities, a key propellant for future applications (Nuwer, 2014). Potential areas of application of graphene include lightweight, strong structural composite materials, highly conductive composites, energy conversion and storage, photovoltaics, transparent screens, filtration/desalination, paints and coatings, high-temperature lubricants, semiconductors, transistors, high-conductivity supercapacitors, photodetectors, biomedical engineering and mobile and flexible/stretchable electronic devices (Palucka, 2013; Njoroge, 2014; Bennington-Castro, 2014; EU 2013); Graphene-Info, 2016).

Graphene holds great promise for biomedical engineering due to its exceptional combination of properties, in particular, strength, mechanical and chemical stability, biocompatibility and conductivity. Areas of intensive studies include neurological research, orthopedics, investigative medicine and drug therapy. It has been used as a reinforcing agent to improve the mechanical properties of biodegradable polymeric nanocomposites for engineering bone tissue application. Graphene solutions are being used as contrast agents in many medical procedures including medical diagnosis and disease detection, MRI-CT scan, tomography, DNA sequencing and cancer treatment. Other areas of application include tissue engineering, as reinforcement for biodegradable polymeric nanocomposites (Lalwani et al., 2013, 2014; Johnson, 2014).

Graphene is also being used in stem cell engineering. The European Commission "Graphene and Human Brain" project funds over two hundred research institutes working on various aspects of application of graphene-based electronics and information technologies in neuroscience and neuroinformatics, disease and drug simulation, drug delivery, medical diagnosis, etc. (EU, 2013; Majumder at al., 2014). Pure graphene is impermeable to all gases because its atoms are surrounded by dense electron clouds which are responsible for the exceptionally high thermal and electrical conductivity. However, doping with other elements or controlled oxidation by reacting with an acid produces graphene oxide.

Advances are being made in the production of graphene and graphene oxide impregnated ultra-thin filtration membranes capable of separating carbon dioxide from hydrogen, thus opening up a new potential use for carbon capture in coal combustion and hydrogen production from coal syngas. Other researchers are investigating the filtration properties of graphene doped with other elements. Another major development in the field of graphene technology is the successful production of graphene-metal composites with dramatically improved properties compared with untreated metals. Nanolayered graphene-copper and graphene-nickel composites have been produced by vacuum deposition of alternating layers of metal and monolayer graphene.

While the combination of graphene and metal makes no significant difference to the weight of the metal, the mechanical strength of the copper-graphene composite increases by around 500%. The strength increase for the graphene-nickel composite is around 200% (Kim et al., 2013). It is also possible to moderate the exceptional heat and thermal conductivity of graphene by doping with elements. Commercial availability of these technologies could have profound impact on structural material applications where weight matters such as aircraft and spacecraft components, and electronics. One of the major problems restricting wider deployment of the supercapacitor is the limited amount of energy storage per unit weight, around 150 Farads per gram. Extensive research is ongoing on the use of graphene as conductors in supercapacitors. The exceptionally high relative surface area compared with conventional activated carbon conductors could increase energy storage capacity per unit weight four-fold. This would greatly expand current fields of application beyond expensive industrial/transportation energy supply and backups to consumer electronic devices which can be charged at a much higher rate than current lithium-iron batteries, with major extension in the length of time between charges.

11.9.6 Carbon nanotubes from coal-derived graphenes

Carbon nanotubes are single-walled (SWNT) or multi-walled (MWNT) graphenes in form of tubes of nanometre size diameter (Figure 11.20a). Weak van der Waal forces hold the tubes together while the very strong covalent bonds along the length of the tube confer exceptional tensile strength. Typical carbon nanotubes have extreme aspect ratios, with diameters ranging from 1nm up to 50nm and lengths from a few microns to several centimetres. Although carbon nanotubes are stacks of 2-dimensional monoatomic graphene sheets, the tube curvature and quantum confinement in the circumferential direction lead to a wide variety of technical properties that are different from those of graphene sheet.

The fact that nanotubes can be rolled from a graphene in many ways means that tubes of different patterns and orientations of the atom hexagons can be achieved (Figure 11.20 (b). Carbon nanotubes combine outstanding mechanical, thermal and electrical properties which commend them for a wide range of applications such as components of material composites, electronics, optics. Electrons only move along the tube's axis in carbon nanotubes because of the nanoscale cross-section, hence electrical conductivity is mono-directional.

There are many methods for the production of carbon nanotubes on a commercial scale. These include chemical vapour deposition (CVD), catalyst chemical vapour deposition (CCVD), template-directed thermal plasma deposition, laser ablasion or arc discharge methods (Golnabi, 2012, Zhu et al., 2012).

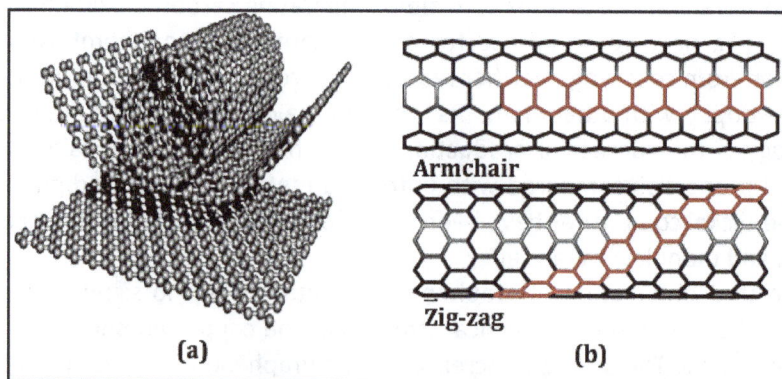

FIGURE 11.20 (a) Schematic diagram of multi-walled carbon nanotube being formed from monoatomic sheet of graphene. (b) Two different lattice orientations of carbon nanotube. *(Sources: (a) what-when-how.com; (b) phycomp.technion.ac).*

Much of the commercial production currently is restricted to bulk material made up of a disorganized mix of fragments of nanotubes used in CN-polymer composites to improve mechanical, thermal and electrical properties of the composite (Zhang et al., 2007; Garcia et al., 2008). These composites are used extensively in sports equipment, shells of speedboats, wind turbine blades, etc. Extensive research is ongoing to produce long, individual single and multi-walled nanotubes of exceptionally high strength and toughness. CN-metal and CN-ceramic composites are also being developed. These developments will greatly extend the range of potential applications in a very wide range of technologies including structural, biomedical, electrical, electronic, information technology, transistor and semiconductors, energy storage, solar cells, supercapacitors, water treatment, etc.

The combined high strength-weight ratio and excellent energy absorption capacity of carbon nanotubes is also opening up an important range of applications in ballistics. Carbon fibre shields have been found to be capable of stopping high-speed bullets and thus have great potential applications in making antiballistic equipment including body and head wear (Mylvaganam and Zhang 2007; Grujicic et al., 2008).

Carbon nanotubes also have great potential in biomedical applications. Already, sophisticated biosensing and biocompatible carbon nanotube sensors for early diagnosis of cancer and drug-delivery systems for targeting specific internal tissues including tumor cells are in use. Other uses include autograft bone repair, metal-encapsulating carbon nanocapsules, reinforcement scaffolds for tissue engineering and regenerative medicine, supports for enzyme immobilization to improve biocatalyst performances (Gomez-Gualdron et al., 2011; Wang et al., 2014). Artificial muscles made from yarn woven with carbon nanotubes and filled with wax have been shown to be capable of lifting weights that are 200 times heavier than natural muscles of the same size.

Structural materials used in nuclear applications have short lifetimes because of embrittlement and porosity that occur in the materials under long-term radiation exposure near a reactor core. CNT-reinforced aluminium composite materials hold good promise as a solution to this problem. The dispersion of small quantities of CNT in an aluminium matrix dramatically improves the material's strength and irradiation tolerance, making the composite potentially suitable for applications in nuclear reactors, nuclear waste containers, nuclear batteries, and aerospace components (YuHao Liu, 2016).

The invention of the silicon semiconductor (transistor) in the late 1940s revolutionized electronics and has brought about major improvements in computational power and energy efficiency. The invention paved the way for the miniaturization of modern electronic devices including computers, smart phones, calculators, radios, etc. Although research on silicon-based transistors is ongoing and intensive and major advances with silicon-based electronics continue to be made, the search for possible alternative technologies has focused also on transistors fabricated from carbon nanotubes (CNTs). CNT-based transistors have the potential to outperform silicon in terms of energy efficiency and the prospects for major advances in the near future are bright.

Currently, the smallest features of silicon transistors are about 14 nanometres but research results have shown that nanotube transistors could be as small as 5 nanometres or even smaller. Already, the first computer built entirely using CNT-based transistors was announced in 2013 (Shulaker et al., 2013). Although the computer's capability is very basic, the development is a very significant advance in the search for alternatives and complements to silicon-based transistors. One major problem which needs to be solved is the removal of substantial fundamental imperfections inherent in carbon nanotubes.

11.10   THE FUTURE OF COAL-TO-CHEMICALS
AND PREMIUM MATERIALS

In spite of the collapse of the coal-to-chemicals industry in most regions of the world in the 1950s due to the availability of cheaper and more environment-friendly petroleum feedstock, the process remained alive and virile in South Africa. Presently, SASOL produces over 200 different chemicals from coal. Other countries with vast coal resources, also continued to recover chemicals from coal-fired electric power plants. For example, Australia generates over 90% of electricity requirements from coal and most of the plants have chemicals recovery units. Around 2000, China started an aggressive coal utilization program in order to utilize its vast coal resources and reduce dependence on imported petroleum products. The initial focus was electric power generation but interest soon extended to the production of chemicals.

   The major constraint to the proliferation of coal-based power generation and chemicals production had been the heavy atmospheric pollution but developments in coal gasification, pioneered by South Africa over the last fifty years or so have led to "cleaner coal" technologies which are now being widely deployed, both for power generation and production of valuable chemicals. The momentum generated by the Chinese initiative is spreading rapidly and many countries which have no petroleum resources but are endowed with cheap coal resources are adopting coal-based power generation technologies. Currently, over 600 coal gasifiers are operating in different parts of the world and the number is expected to double by 2019 (Higman, 2014). There are around twenty different technologies available, most of them flexible on feedstocks that include biomass and other carbonaceous materials (Figure 11. 21).

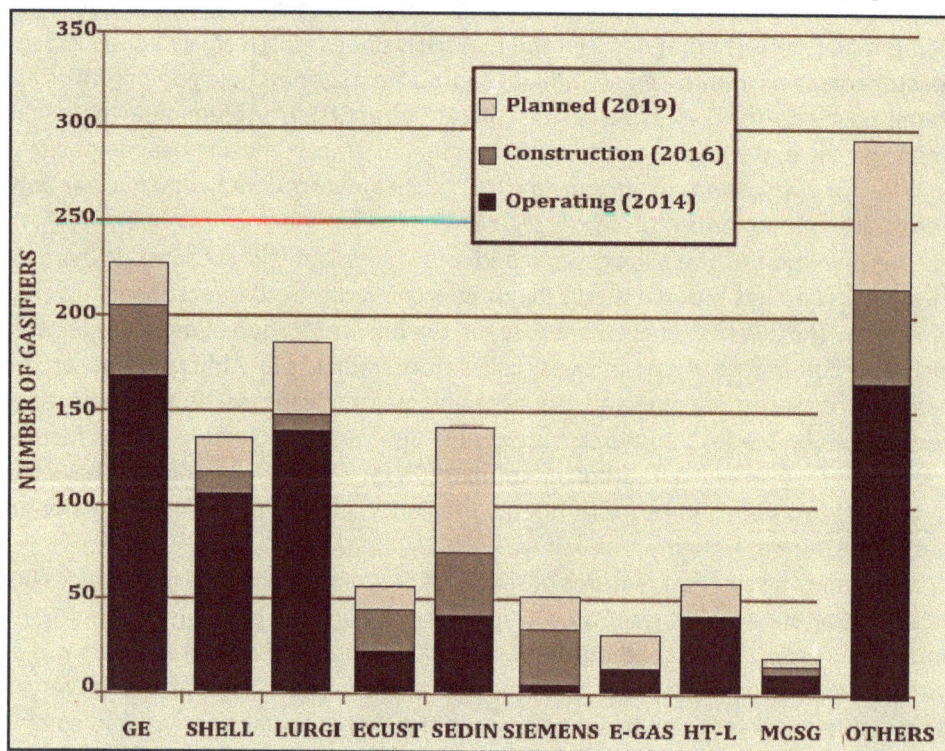

FIGURE 11.21 Global distribution of coal gasifiers. *(Tweddle, 2014).*

The future of coal-to-chemicals will depend on the relative availability and pricing of competing petroleum feedstock, and the progress made in cleaning up coal utilization processes. The relative abundance of coal combined with the progress made in the development of cleaner coal technologies will likely encourage the production of chemicals from coal, particularly in emerging countries , and the opportunities are enormous (Minchener, 2013).

Coal or petroleum derived pitch is the major precursor to synthetic graphite which is the starting material for the production of major high-technology materials including graphite products, industrial diamond, graphene, carbon fibre, carbon nanotubes, nanocarbon products. Since China dominates the world market for both natural and coal-based synthetic graphite supply, the demand for coal tar pitch is likely to grow. Clean coal technologies focus on four main aspects of coal utilization: efficient mining, handling and transportation; efficient coal combustion technologies; coal gasification, carbon capture and sequestration; and deployment of coal gasifiers as part of an energy-efficient Integrated Gasification Combined Cycle (IGCC) system. Significant progress has been made on all fronts, with the exception of carbon sequestration.

Efficient gas cleaning units capable of capturing carbon dioxide and other anthropogenic gases are now part of most modern coal gasification systems. However, progress on management of captured CO_2 is very slow and will likely constitute a significant constraint to clean coal utilization technologies. Furthermore the geopolitics of petroleum may change the dynamics of clean coal technology utilization. For example the recent agreement between China and Russia for the supply of oil and natural gas combined with logistical problems of coal supply and transportation within China may encourage planned and operating coal-based chemical and power plants to convert to natural gas which can be transported more easily. On the other hand, Europe is reducing dependence on Russian oil and gas and reactivating interest in utilization of locally available coal, hence interest in coal gasification is growing.

REFERENCES

Ania, C. (2007). "Pitch-based Carbon materials." *Proceedings, ITA Annual Conference,* Palma de Mallorca, 2007.

Arai, Y. (2001). "Pitch-Based Carbon Fibre with Low Modulus and High Heat Conduction. *Nippon Steel Technical Report* No. 84, 2001.

Ashbury Carbons(2006) "Introduction to Synthetic Graphite." asbury.com/technical-presentations-papers/materials.../synthetic-graphite/

Bae, S. Et al., (2010). "Roll-to-roll production of 30-inch graphene films for transparent electrodes". *Nature Nanotechnology* 5 (8): pp. 574–578.

Bennington-Castro, J. (2014). "Future stretchable electronic devices." Materials360(r), October 17, 2014. Materials Research Society.

CAIB, (2003). "Columbia Accident Investigation Report." Volume 1, August, 2003. Columbia Accident Investigation Board.

CameronCarbon (2006). "Activated Carbon. Manufacture, Structure and Properties." Cameroncarbon.com

Chen, Liming and Warren A. Dick (2011). "Gypsum as an agricultural amendment." *Ohio State University Agricultural Extension* Bulletin 945.

EU (2013). "European Commission Press release Brussels, 28th January 2013. Graphene and Human Brain Project wins largest research excellence award in history, as battle for sustained science funding continues."

Gabrielsen, P. (2013). *Stanford News Service*. Gabrielsen, P (Ed.). http://news.stanford.edu/pr/2013/pr-thin-graphene-sheets-013113.html.

Gibson, P (2007) "Coal to Liquids at SASOL." Kentucky Energy Security Summit, October, 2007.

Gomez-Gualdron, D. A., Burgos, J. C., and P. B. Balbuena (2011). "Carbon nanotubes: engineering biomedical applications." *Prog. Moi Biol Transl. Sci.,*V. 104, pp. 175-245.

Gonalbi, H. (2012). "Carbon nanotube research developments in terms of published papers and patents, synthesis and production." *Scientia Iranica,* 19(6), pp. 2012-2022.

Graphene-Info (2016). *The Graphene Handbook.* www.graphene-info.com/handbook. (Accessed June 2016).

Grujicic, M. et al., (2008). "Ballistic-protection performance of carbon-nanotube-doped poly-vinyl-ester-epoxy matrix composite armor reinforced with E-glass fibre mats." *Materials Science and Engineering: A.* Vol. 479, Issues 1-2, pp 10-22.

Higman, C. (2014). "State of the Gasification Industry: Worldwide Gasification Database 2014 Update." *Gasification Technologies conference,* Washington DC.

IDL (2009). "HPHT synthesis." Internatonal Diamond laboratories, www.diamondlab.org.

IEA (2005*). Premium carbon products and organic chemicals from coal. International Energy Agency,* IEA Report PF 05-05.

Inagaki, M. Yang, Y., and F. Kang (2012). "Carbon nanofibers prepared via electrospinning." *Advanced Materials*, Vol. 24, Issue 19, pp. 2547-2566.

Inagaki, M. Kang, F. Toyoda, M. and H. Konno (2014). *Advanced Materials Science and Engineering of Carbon.* Tsinghua University Press/Elsevier.

Jensen, J. H., Poulsen, J. M., and N. U. Andersen (2011). "From coal to clean energy." Nitrogen+Syngas 310, March-April, 2011. US Department of Energy (DOE). https://www.netl.doe.gov/File%20Library/research/coal/energy%20systems/gasifica-tion/gasifipedia/from-coal-to-clean-energy-march-april-2011.pdf

Johnson, D. (2014). "Graphene transforms itself into a sphere for drug delivery." *IEEE Spectrum,* 06 August, 2014.

Johnson Mathey (2015). *Johnson Matthey Methanation Process.* www.jpmprotech.com

Kang, W. R. and K. B. Lee (2013). "Effect of operating parameters on methanation reaction for the production of synthetic natural gas." *Korean J. Chem. Eng.,* Vol. 30(7), pp. 1386-1394.

Kim, Y et al., (2013). "Strengthening effect of single-atomic-layer graphene in metal-graphene nanolayered composites." *Nature Communications*, Vol. 4, Article No. 2114.

Lalwani, G; Henslee, A. M.; Farshid, B; Lin, L; Kasper, F. K.; Qin, Y. X.; Mikos, A. G.; Sitharaman, B (2013). "Two-dimensional nanostructure-reinforced biodegradable polymeric nanocomposites for bone tissue engineering". Biomacromolecules 14 (3): 900–9. doi:10.1021/bm301995s. PMC 3601907

Lalwani, G; Cai, X; Nie, L; Wang, L. V.; Sitharaman, B (2013). "Graphene-based contrast agents for photoacoustic and thermoacoustic tomography". Photoacoustics 1 (3-4): 62–67. doi:10.1016/j.pacs.2013.10.001. PMC 3904379. PMID 24490141.

Lalwani, G; Sundararaj, J. L.; Schaefer, K; Button, T; Sitharaman, B (2014). "Synthesis, Characterization, *In Vitro* Phantom Imaging, and Cytotoxicity of a Novel Graphene-Based Multimodal Magnetic Resonance Imaging - X-Ray Computed Tomography Contrast Agent". Journal of Materials Chemistry B 2 (22): 3519–3530.

doi:10.1039/C4TB00326H. PMC 4079501. PMID 24999431.

Lee, C. (2008). "Measurement of the elastic properties and intrinsic strength of monolayer graphene." *Science,* Vol. 321, pp. 385-388.

Li, Chunqi (2014). "Current development situation of coal to SNG in China." Datang International Chemical Technology Research Institute, China. www.iea.org/media/workshops/2014.

Liu, Z. Et al., (2012). "Total methanation of syngas to synthetic natural gas over Ni catalyst in a micro-channel reactor. " *Fuel,* Vol. 95, pp. 599-605.

Liu, YuHao (2016). "Energy focus: Carbon nanotubes improve radiation resistance of aluminum." MRS Bulletin, 41, pp. 353-354.

LichaoFeng, NingXie, and Jung Zhong (2014). "Carbon nanofibres and their composites: A review of synthesizing, properties and applications." *Materials,* Vol. 7, pp. 3919-3945.

Littlewood, K. (1977). " Gasification: Theory and Application." *Prog. Energy Combust. Sci.,* Vol. 3, No. 1, pp. 35-71.

Majumder, M., Tkacz, R., Oldenbourg, R., Mehta, S; Miansari, M., and A. Verma (2014). "pH dependent isotropic to nematic phase transitions in graphene oxide dispersions reveal droplet liquid crystalline phases." *Chemical Communications* (Royal Society of Chemistry) 50 (50): 6668–6671.

Minchener, A. (2013). *Challenges and opportunities for coal gasification in developing countries.* IEA Clean Coal Centre, Publ. No. CCC/225.

Mitsubishi Rayon Co. (2016). Carbon Fibers and Composite Materials. Mrc.co.jp/english/products. (Accessed June 2016).

Mukhulyonov, I. U. et al. (2013). *Chemical Technology.* Mir Publishers, Moscow, 1974.

Mylvaganam, K. and L. C. Zhang (2007). "Ballistic resistance capacity of carbon nanotubes." *Nanotechnology,* Vol. 18, No. 47.

NETL.DOE (2017). "Coal-derived chemicals." Nnetl.doe.gov. Accessed 2017.

Nexant (2006). "Coal to Chemicals: Is it Coal's Time Again?" *Nexant/ChemSystems MultiClient Report,* www.nexant.com.

Nippon Steel (2001). "Pitch-based Carbon Fibre with Low Modulus and High Heat Conduction." *Nippon Steel Technical Report* No. 84, July, 2001.

Nippon Steel (2011). "Production of Differentiated Carbon Fibre Using High-quality Impregnated pitch." Nippon Steel News No. 383, 2011.

Ngoroje, L. (2014). "High conductivity suppercapacitors achieved with graphene nanocomposites." Materials360(r), July, 17, 2014. Materials Research Society.

Nobel Foundation (2010) "The Nobel Foundation Prize in Physics, 2010: for groundbreaking experiments regarding the two-dimensional material graphene."

Nuwer, R. (2014). "A new method for mass producing high quality graphene." Materials360(r), July, 17, 2014. Materials Research Society.

OSU (2005). *Gypsum for agricultural use in Ohio – sources and quality of available products.* Ohio State University Extension fact sheet ANR-20. www.osu.edu.

Palucka, T. (2013). "Honeycombe-like 3D graphene excels as a counter-electrode catalyst in solar cells." Materials360(r), August, 27, 2013. Materials Research Society.

Palucka, T. (2013). "Graphene-coated porous silicon opens path to integrated energy storage." Materials360(r), November, 15, 2013. Materials Research Society.

Patel, P. (15 January 2009). "Bigger, Stretchier Graphene". *MIT Technology Review.*

Park, S. (2015). *Carbon Fibers.* Springer.

Plotkin, J. (2012). "China invents and reinvents coal to chemicals." http://www.icis.com/resources/news/2012/04/16/9549986/china-invents-and-reinvents-coal-to-chemicals/

Ronsch, S. et al., (2016) "Review on methanation – from fundamentals to current projects." Fuel, Vol. 166, pp 276-296.

Sheldon, R. A. (1983). "Chemicals from synthetic gas." Reidel Pub. Co., Boston.

Shulaker, M. M. et al. (2013). "Carbon nanotube computer." Nature, Vol. 501, pp. 526-530.

Sudiro, M. and B. Alberto (2010). "Synthetic natural gas (SNG) from coal abd biomass: a survey of existing process technologies, open issues and perspectives." www.cdn.intechopen.com.

Tweddle, G. (2014). "Coal and Natural Gas Chemicals". Technol. OrbiChem, APIC 2014. www.orbichem.com

Udengaard, N. R., Olsen, A. and Wix-Neilsen (2006). "High temperature methanation process – revisited." Pittsburg Coal Conf., 2006.

University of Manchester (2014). "The Story of Graphene." http://www.graphene.manchester.ac.uk/explore/the-story-of-graphene/

Wagner, S. C.. (2011) Biological Nitrogen Fixation. Nature Education Knowledge 3(10):15.

Wang, W., Y. Zhu, S. Liao, and J. Li, (2014). "Carbon nanotubes reinforced composites for biomedical applications." BioMed Research International, Vol. 2014, Article ID 518609.

Weis, M., Walter, S. and U. Berger (2008). "Lurgi's methanation technology for production of SNG from coal." Gasification Technologies Conference 2008.

Wittcoff, H., Reuben, B. G., and J. S. Plotkin (2013). Chemicals from Coal-Industrial Organic Chemicals.

Wix, C. and Ib Dybkjoer (2007). "Coal to SNG – The methanation process." 24th Annual International Pittsburg Coal Conference, Johannesburg, South Africa. Third Edition.onlinelibrary.wiley.com.

Yang, C. and R. Jackson (2012). "China's growing methanol economy and its implications for energy and the environment." www.biology.duke.edu/jackson.

Zhang, X. Et al., (2007). "Ultrastrong, Stiff, and Lightweight Carbon-Nanotube Fibers." Advanced Materials, Vol. 10, Issue 23, pp 4198-4201.

Zhu, Y. Et al., (2012). "Deposition of TiC film on titanium for abrasion resistant Implant material by iron-enhanced triode plasma CVD." Applied Surface Science, Vol. 262, pp. 156-158.

12   Coal utilization and the environment

12.1   INTRODUCTION

Energy is a primary input into most human activities, the fundamental driver of economic development, and all forms of energy production have negative as well as positive impact on the environment. Over 90% of the global anthropogenic (environmental pollution) emissions in 2013 emanated from primary energy production and utilization (IEA, 2015). Virtually all energy production and utilization processes generate pollution gases. Oil drilling and production, gas harvesting and coal mining processes have negative effects on the environment because mining and winning degrade landscapes and utilization results in the release of carbon dioxide, carbon monoxide and methane into the atmosphere. However, the greatest intensity arises from energy utilization, in particular electric power generation and transportation.

Atmospheric pollution caused by energy utilization is of major global concern and has been linked to various environmental problems which are believed to have increased significantly with the exponential increase in fossil fuel utilization over the last century or so. Negative consequences of fossil fuel utilization include global warming, extreme weather, ocean flooding, desertification and human ailments. There are no physical atmospheric boundaries between countries and regions and consequences of activities in one location can resonate over long distances, hence the current efforts by the United Nations to galvanize concerted global effort to resolve the problem of environmental pollution.

12.2   THE EARTH-ENVIRONMENT SYSTEM

The Earth-environment system (EES) comprises the Earth and the environment that surrounds it. The system is both the source and sink of all human activities. Oxygen that sustains life is derived from the Earth's atmosphere and the expelled carbon dioxide goes back into the atmosphere. Food, minerals and energy are derived from the Earth's crust and its environment and the waste that results from utilization is recycled through the system.

The Earth's environment is a natural system that includes the Earth and the ecological units, and natural resources that surround it. These include all living (mankind, animals, birds, fish, microorganisms), all non-living (soils, vegetation, oceans, rocks, etc.) and the atmosphere (air, climate, energy and magnetic radiation). The Earth is central to a global system commonly defined in terms of four major spheres: the *lithosphere*, the *hydrosphere*, the *atmosphere*, and the *biosphere.* The lithosphere comprises the Earth's crust (from the surface to a depth of about 100km), and the upper part of the solid mantle that extends to a depth of about 3000km. The hydrosphere comprises water, liquid or frozen, that exists under or over the surface of planet Earth - oceans, lakes, streams, glaciers, ground waters. The atmosphere comprises layers of gases that extend from the Earth's surface to about 20 km into space. The biosphere refers to parts of the land, sea, and atmosphere that host life. The atmosphere sustains life on Earth through the supply of oxygen needed by most organisms for respiration, carbon dioxide required by vegetation (plants, algae etc) for photosynthesis, the regulation of the Earth's temperature, and the control of the potentially damaging effects of the Sun's ultraviolet radiation. The biosphere refers to parts of the land, sea and atmosphere that host all living organisms, microorganisms and plants.

12.2.1 The Earth's natural environment

The four major spheres that make up the ecological system are characterized by intricate intra and interactions, all interdependent and interconnected. The interaction between these spheres to a large extent determines the climate. The functions of the spheres are controlled by natural regulatory processes and the boundaries between them may be clear or ill defined. While the boundaries between organisms and vegetation are fairly well defined there are no clear boundaries between air, water, climate, energy, radiation, etc. The environment that is controlled by natural forces with little or no human interference is known as *natural environment.*

12.2.2 The Anthropogenic Environment

The natural environment is imperfect in many ways. Extreme weather, flooding, volcanoes, earthquakes are natural occurrences which have always been part of the ecosystem. However the natural processes can be influenced or affected by human actions which are also commonly referred to as anthropogenic factors. When the environment is affected significantly by human actions, it is no longer natural but an *anthropogenic environment.*

The environment is both the source and sink for virtually all natural resources which human activities depend upon. Ample evidence from the deep oceans to the stratosphere indicates significant changes in climate due principally to human activity including burning of fossil fuels, destruction of forests, agricultural activity and use of heat-trapping aerosols. These changes have many far-reaching repercussions including significant changes in climate, rising sea waters, more frequent and intense storms, the extinction of animal and plant species, worsening draught and crop failures, acid rains, fouling of the air by smog and particulate matter which are injurious to human health, and exposure of the human race to dangerous radiations.

12.3   THE EARTH'S NATURAL ATMOSPHERE

The Earth's atmosphere comprises gravity-controlled and stratified layers of gases which surround the Earth and create conditions that support life on Earth - temperature, pressure, climate, radiation. For example, the atmosphere provides the oxygen which supports life and regulates the Earth's temperature. Without the control by the atmosphere, temperatures on Earth would be sub-zero, the pressure would be too low to sustain respiration, and the Earth would be uninhabitable. The atmospheric environment also regulates the supply of carbon dioxide that supports terrestrial vegetation and controls the intensity of the Sun's ultraviolet radiation without which living organisms would suffer genetic damage.

The atmospheric air is made up of a mixture of gases in variable proportions, the main ones being those listed in Table 12.1. Although traces of atmospheric gases have been detected well out into space, 99% of the mass of the atmosphere lies below about 25km to 30km altitude, while 50% is concentrated in the lowest 5 km. This gaseous mixture remains remarkably uniform in composition, and is the result of efficient bio-geochemical recycling processes and turbulent mixing in the atmosphere. The two most abundant gases are nitrogen (78% by volume) and oxygen (20% by volume), held in a layer by the Earth's gravitational force, and together they make up over 99% of the lower atmosphere. There is no evidence that the relative levels of these two gases are changing significantly over time.

Despite their relative scarcity, some of the relatively minor gases, in particular,carbon dioxide, methane and ozone, play a vital role in the regulation of the atmosphere's energy balance. These gases are often referred to as *greenhouse gases*. The air density is highest near the Earth's surface and decreases gradually with altitude. There are five main layers primarily defined by temperature (Figure 12.1).

12.3.1   The Troposphere

The gas layer that is in immediate contact with the Earth is known as the troposphere. It extends from the Earth's surface to between 9km near the poles and 17km to 20km near the equator although there may be variations due to changes in the weather. It is slightly lower in winter at mid- and high-altitudes, and slightly higher in the summer.

TABLE **12.1.** Average composition of the atmosphere below 25km.

Component	Volume% (dry air)	Component	Volume% (dry air)
Nitrogen (N_2)	78.08	Hydrogen (H)	0.0006
Oxygen (O_2)	20.98	Krypton (Kr)	0.0011
Argon (Ar)	0.93	Xenon (Xe)	0.00009
Carbon dioxide (CO_2)	0.035	Methane (CH_4)	0.0017
Neon (Ne)	0.0018	Ozone (O_3)	0.00006
Helium (He)	0.0005		

Notes: Concentration of ozone in the air is variable; Helium, Krypton and Xenon are inert gases

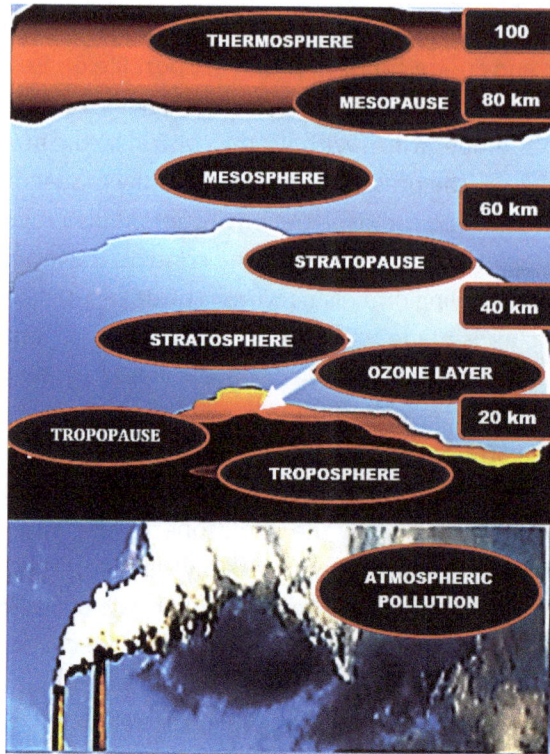

FIGURE 12.1 The Earth's atmosphere.

The troposphere accounts for about 80% of the total mass of the atmosphere, holds nearly all the water vapour and dust particles in the atmosphere and the clouds that control the weather. The layer also controls the reflection of the Sun's energy from the Earth's surface in such a way that a temperature gradient is established, highest at the Earth's surface and decreasing with altitude. The Sun's rays heat up the Earth which in turn radiates the heat into the air immediately above it. The density of the air drops, causing it to rise to a higher altitude. Pressure also decreases with altitude. At the top of the troposphere the temperature is about -55°C. The chemical composition of the troposphere is fairly stable and is made up of mainly nitrogen and oxygen, with about 1% made up of argon, traces of carbon dioxide, hydrogen, ozone, and other gases. Most of the dust and particles in the atmosphere also reside in the troposphere. The troposphere is in a perpetual state of turbulence due to temperature and pressure changes, and plays a prominent role in determining the weather system on the Earth.

The variable and uneven heating of the troposphere by the Sun due to the Earth's rotation around the Sun causes convection currents, turbulence and large scale movement of winds that transport moisture and redistribute the heat from the Sun around the Earth's surface. The troposphere controls the direction of the winds at any given time, and the water cycle which involves evaporation from the Earth's surface, seeding, precipitation and condensation in the troposphere resulting in clouds, rain, snow, slits etc. It also controls the movement of winds which form thunderstorms, hurricanes and tornadoes, heat waves and season patterns around the world.

While the relative proportions of the constituent gases of the troposphere remain fairly constant, the water vapour is variable, depending on evaporation and transpiration processes on the Earth's surface which is the source. The evaporation process plays a prominent role in controlling temperatures on the Earth's surface because the energy that sustains the endothermic processes is sourced from the solar energy and thermal radiation re-emitted or reflected from the Earth's surface.

12.3.2   The Stratosphere

Immediately above the troposphere is the stratosphere. It extends to a height of about 50km and includes the ozone layer, which blocks much of the Sun's harmful ultraviolet rays from reaching the Earth's surface. The stratosphere is warmer than the troposphere because of the energy from the ultraviolet light absorbed by the ozone layer. At its base, the stratosphere is extremely cold, about minus 80°C, but the temperature rises gradually with height and, at its top, the temperature has risen back nearly to freezing.

12.3.3   The Mesosphere

The mesosphere starts at around 50km above the Earth's surface and extends to about 85km although these boundaries vary with latitude and with season. Temperatures continue to fall with rise in altitude and can be as low as minus 140°C at the top of the mesosphere, the coldest naturally occurring place in the Earth's system.

12.3.4   The Thermosphere

The thermosphere layer is directly above the mesosphere. It begins at about 85km and extends to 500 to 1000km above the Earth's surface. The gases, mainly hydrogen mixed with some minor gases, are highly stratified according to their molecular masses and solar radiation controls the activities in this layer. Temperatures rise sharply in the lower thermosphere below 200 to 300km due to the impact of strong solar radiation, then become fairly steady with increase in altitude. Temperatures in the higher layer of the thermosphere are highly variable depending on solar activity and can range from 500 to 2,500°C in the daytime but the air is extremely thin, at near vacuum conditions. The middle part of the layer is fairly stable and hosts many orbital spacecrafts including the international space station which orbits at altitudes between 300 and 400km.

The stratosphere, mesosphere and lowest part of the thermosphere, between 10km and 100km above the Earth's surface are collectively referred to as the middle atmosphere. The mesosphere is the least understood part of the Earth's atmospheric system because it is above the altitude for aircraft and below the minimum altitude for orbital spacecraft. The little that is known has been through the use of probing rockets.

12.3.5 The Exosphere

The exosphere is the Earth's atmospheric boundary with interplanetary space, although the boundary is not clearly defined. For this reason the exosphere is sometimes regarded as part of the outer space. The main gas in the Earth's exosphere is hydrogen, with small quantities of helium, carbon dioxide and atomic oxygen. The exosphere is almost a vacuum, the air becomes extremely thin and air molecules escape into space. Even though the few particles in the zone are very hot, they are too few to transfer much energy.

12.3.6 Atmospheric boundaries and overlaps

The troposphere is commonly referred to as the lower atmosphere of the Earth, separated from the Earth's surface by a very small layer in which mankind functions, known as the boundary layer. As mentioned earlier, the boundaries between the spheres are not constant and vary significantly depending on altitude and season. Between these layers there are boundaries which are characterized by conditions that differ significantly from those in the upper and lower spheres.

12.3.6.1 *The Tropopause*

The tropopause separates the troposphere and the stratosphere. At this boundary there is a dramatic change in the atmospheric variables (mainly temperature) which prevail in the troposphere. This occurrence is also known as environmental lapse rate. From the lowest boundary layer the temperature drops at an average rate of around 6.5°C per kilometer gain in altitude. This gradient becomes inverted at the tropopause and temperature starts to increase with height. The tropopause boundary is between 6 and 1km above the Earth, highest at the equator, and has the lowest temperatures in the troposphere. Above this boundary the temperature starts to rise.

12.3.6.2 *The Stratopause*

The stratopause is the boundary between the lower stratosphere and the higher mesosphere and is located at an altitude of 50 to 55km above the Earth's surface. The continuous rise in temperature from the tropopause through to the stratosphere is due to reflection of energy from the ozone layer but the effect thins out as the ozone density reduces. Temperatures continue to fall with height, through the mesosphere, the top of which is the coldest part of the atmosphere.

12.3.6.3 *The Mesopause*

The mesopause is in the altitude range 85 to 100km where the temperature rise in the stratosphere stops and temperatures begin to fall. It is the coldest part on Earth with temperatures as low as minus 100°C. The drop in temperature from this boundary is believed to be due to an increase in concentration of carbon dioxide and is still a subject on intensive research.

12.3.6.4 *The Ionosphere*

This is not a layer or boundary but a zone located in altitudes between 85 and 600km from the Earth's surface. It is situated mainly in the thermosphere but extends to the exosphere above and the mesosphere below. In the region, the gas density is extremely low and positive and negative ions are bonded by weak electrostatic forces. The shorter wavelengths of solar radiation provide sufficient energy for gas atoms to ionize and stay apart for short periods from time to time before recombination takes place, creating a density of free electrons, hence the name ionosphere.

The processes of ionization and recombination are spontaneous and continuous, the intensity depending on the energy balance between the energy received from the Sun and the energy released in the recombination process. This region is vital to radio propagation known as DX communication between the various regions of the Earth. This communication system transmits high frequency shortwave radio signals over very long distances by using the reflective properties of the ionosphere to bounce a signal from one point source to a very distant destination. Life on Earth depends critically on activities in the first twenty kilometres or so above the surface - the weather, respiration, temperature and radiation control, photosynthesis that sustains vegetation and agriculture, etc.

12.4 THE SUN'S RADIATION

The Sun is the primary source of energy that drives the numerous natural systems on Earth including climate systems, ecosystems, and hydrological systems. The Sun provides the energy required for photosynthesis in plants, the process that sustains the immense diversity of life forms that are found on the Earth. The imbalance between the Sun's energy that hits the Earth and the amount radiated back by the Earth, and the uneven geographical distribution of its intensity on the Earth's surface cause pressure variations which fuel the winds that determine the climate. The Sun radiates energy equally in all directions but the intensity that hits the Earth's surface depends on the angle of the Earth with respect to the Sun as it rotates in an uneven orbit round the Sun. This is variable from point to point on the Earth's surface and determines the Earth's weather pattern. When the Earth tilts the North Pole towards the Sun, the Northern Hemisphere receives direct, intense radiation and experiences Summer while the Southern Hemisphere is in winter and vice versa. The regions of the equator which receive close to perpendicular rays have no major seasons.

The Sun is about 329,000 times larger than the Earth and its core which spans about a quarter of its radius of 1.4 million kilometers is filled with hydrogen. Like the Earth, there is a gravitational pull of the Sun's mass inwards and this creates very high pressures which cause the exothermic fusion of two hydrogen atoms to form helium, with the release of energy and free photons, a process known as nucleo-synthesis. The photons travel a short distance before reacting with other gas molecules to release more energy and more photons. Technically the energy released is considerably more than the amount required to cause the reaction (nuclear fusion). Temperatures at the core are about 15 million degrees Kelvin, sufficiently high to keep the core in a hot plasma phase, made up of a dense, highly conductive population of protons, neutrons and electrons. The high-energy protons are transported by radiation through the plasma towards the surface, each proton taking about 100,000 years to cover the first 70-80% of the distance.

The protons have now lost considerable energy and travel the last 20% of the distance to the Sun's surface by convection over a month or two. A significant number of the photons are trapped within the Sun's internal region but enough reach the surface to form a hot shell of gas known as the photosphere, about 500km thick and nearly 6000°K around the Sun. The photosphere is about 20 times hotter than the Earth. This is the source of the radiant energy that the Sun transmits and radiates in all directions towards the Earth and other planets in the solar system. Electromagnetic radiation travels the 150 million kilometers from the Sun's surface, reaching the Earth in about 8 minutes, and, in the process, different components of the rays are absorbed, scattered or deflected. Only about 40% of the solar energy intercepted at the top of the Earth' atmosphere eventually reaches the Earth's surface. The Sun radiates a mixture of electromagnetic waves (Sunlight) ranging from infrared (IR) through visible light to ultraviolet (UV) (Figure 12.2). The ultraviolet radiation from the Sun may be divided into four categories based on the wavelength as shown in Table 12.2.

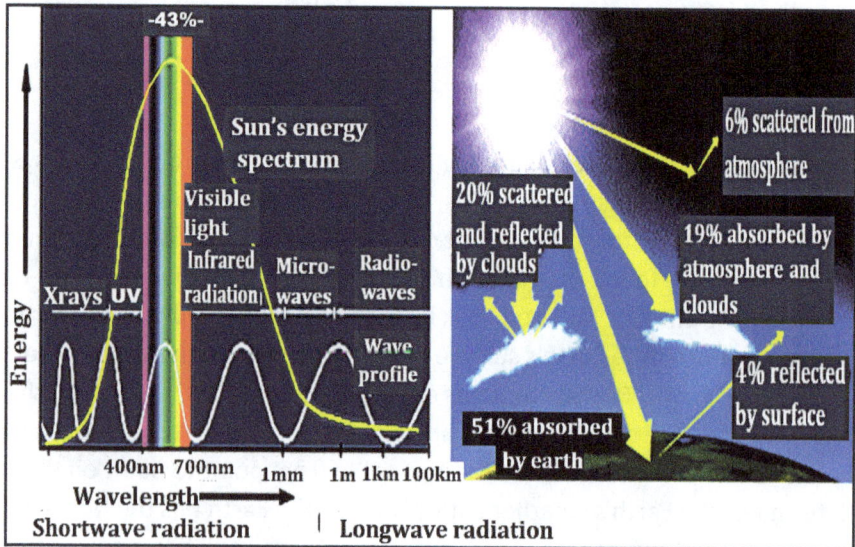

FIGURE 12.2 Spectral analysis of the Sun's radiant energy and passage through the Earth's atmosphere.
(ucar.edu; the Comet Program).

TABLE 12.2 Wavelength analysis of the Sun's ultraviolet radiation

UV WAVELENGTH (NM)	CATEGORY
10 - 100	Vacuum UV
400 -315	UV-A
315 - 280	UV-B
280 - 100	UV-C
700-1400	Infrared-A
400-700	Visible light
1400-3000	Infrared-B
3000-10^6	Infrared-C

About 43% of the radiant energy of the Sun is concentrated in the visible and near visible parts of the spectrum between 400 and 700nm. Although the wavelengths below 400nm account for only 7 to 8% of the total energy radiated by the Sun, the energy intensity is very high. UV-B rays have a longer wavelength and lower energy than UV-C and, like UV-C, are retained by the ozone layer, supplying the energy required for the chemical reactions in the layer. However, some may pass through the ozone layer because the energy is not sufficient to break the ozone bonds. Ozone is transparent to UV-A which has even lower energy and most of the radiation reaches the Earth's surface. It is significantly less harmful than UV-B but excessive exposure is believed to be still capable of genetic damage. The atmosphere is transparent to visible light and most of it reaches the Earth's surface. Part of it is absorbed by land, oceans and vegetation and the balance is radiated back to the atmosphere in the form of invisible infrared radiation.

12.5 THE GREENHOUSE EFFECT

About two-thirds of the solar short wave radiation reaching the Earth is absorbed and the balance is reflected into space as infrared long-wave radiation. A mixture of gases between the Sun and the Earth absorbs this long-wave radiation and reflects it downwards to warm the Earth's surface. Part of the reflected energy passes through the atmosphere back into space while the balance is trapped in the stratosphere, 20 to 40km above the Earth's surface. Some gases present in the stratosphere are capable of absorbing and re-emitting infrared radiation. This natural process ensures that a significant part of the infrared energy reflected from the Earth's surface at night when the Sun goes down is absorbed and retained in the stratosphere. When the Earth's surface temperature drops heat is radiated from the gas layer back to Earth. In effect, the gases form a heat blanket in the stratosphere which helps to control the temperature on the Earth's surface, keeping it at a life-supporting average of 13-15°C (Figure 12.3).

Gases which are capable of absorbing and re-emitting infrared radiation are known as *greenhouse gases* (GHGs) and the natural absorption-emission-absorption cycle is known as *greenhouse effect* (GHE) (Figure 12.3). Without this natural activity, the temperature of the

Earth's surface would be some 33°C cooler and would be uninhabitable. Many gases present in the Earth's atmosphere exhibit these greenhouse properties. Some of them occur naturally in the atmosphere while others are produced by human activities. The main natural GHGs are water vapour, carbon dioxide, methane, and nitrous oxide. Not all of these gases make an equal contribution to the greenhouse effect. For example, one molecule of methane (CH_4) has 20 times the impact of a molecule of carbon dioxide; nitrous oxide (N_2O) 200 times; ground-level ozone 2,000 times; and a chlorofluorocarbon molecule has from 13,000 to 20,000 times the impact of a molecule of carbon dioxide.

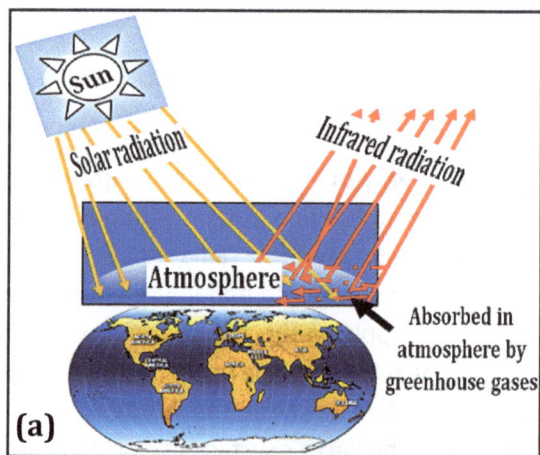

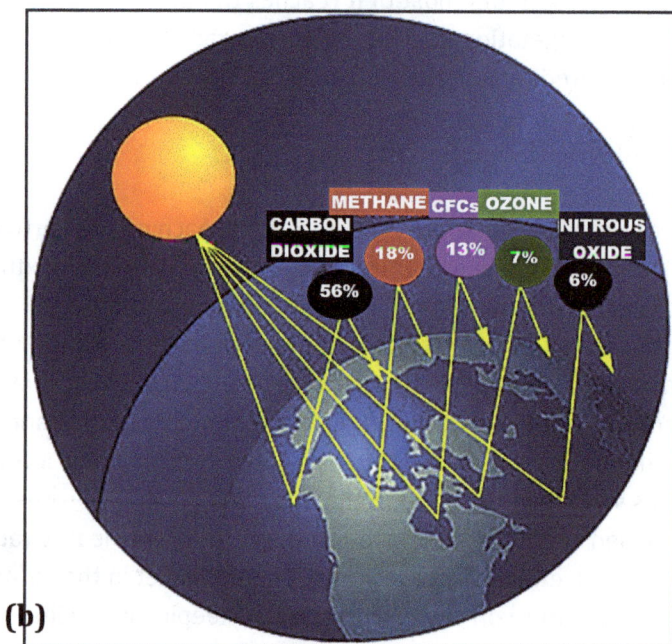

FIGURE 12.3   The Greenhouse Effect (a) the Sun's radiation hitting the Earth. (b) Greenhouse gases. *(Source of (b): www.koshland-science-museum.org).*

Human activities disturb the natural balance by increasing the concentration of these gases in the atmosphere and produce others including ozone, (O_3), halocarbons - chlorofluorocarbon (CFCs), hydro-chloro-fluorocarbons (HCFCs), Hydro-fluoro-carbons (HFCs) per-fluoro-carbons (PFCs), tri-chloroethane, etc., and sulphur hexafluoride (SF_6). Water vapour is the most active component of these gases, responsible for about 67% of the activity. Its concentration in the atmosphere is largely determined by climatic conditions and is relatively unaffected by human activity. Only four (carbon dioxide, methane, nitrous oxide and CFCs) are the most important in the evaluation of the anthropogenic impact of human activity on the environment. Greenhouse gases are made up of tightly bonded atoms but the molecules are capable of absorbing infrared energy which causes the atoms to vibrate and eventually re-emit radiation which will probably be absorbed and discharged by another greenhouse gas molecule thus creating a natural cycle that regulates temperatures on the Earth's surface.

12.5.1 Greenhouse carbon dioxide

Carbon is a naturally occurring element and an essential component of the life sustaining cycle known as the Carbon Cycle. It is present in all living organisms, in oceans, in the air and in the Earth's crust. The element is constantly recycled as shown in Figure 12.4. Nature maintains a delicate balance of carbon between the various units of the Carbon Cycle such that the concentration in the atmosphere is fairly constant. It is present in the atmosphere as carbon dioxide. Plants draw this chemical compound from the atmosphere and, using the Sun's energy through a process called photosynthesis, convert it to carbohydrates which they need to grow. Oxygen and water are by-products of this process and are discharged into the atmosphere from where mankind, animals and some fungi and bacteria draw oxygen which is vital for life sustenance.

The by-product of the human respiration process is carbon dioxide which is returned into the atmosphere. Human beings and animals also accumulate carbon through consumption of plants and when they die, the carbon is returned to the Earth. Coal is formed from plant mass buried in the earth under suitable conditions while oil and gas derive from microorganisms buried in oceans over millions of years. The primary element in all three fossil fuels is carbon.

12.5.2 Greenhouse methane

Methane is a colourless, odourless and highly inflammable gas which gets into the atmosphere through various sources. It is produced primarily by anaerobic (oxygen-deficient) decomposition of organic matter in biological systems, in particular, coal beds and ruminant digestion.

12.5.3 Greenhouse nitrous oxide

Nitrogen and oxygen are normally present in the atmosphere and do not react chemically under normal atmospheric temperatures and pressures. However, lightning can cause very high temperatures and pressures in the atmosphere which provide sufficiently high energy for reaction of nitrogen with oxygen to form nitrogen oxides, including nitrous oxide which remains in the stratosphere. Nitrous oxide gas is also produced and discharged into the atmosphere during the natural nitrogen cycle in which legume plants and algae plants take nitrogen from the atmosphere to produce nutrient nitrate ions in a nitrification process driven by the action of nitrogen-fixing bacteria.

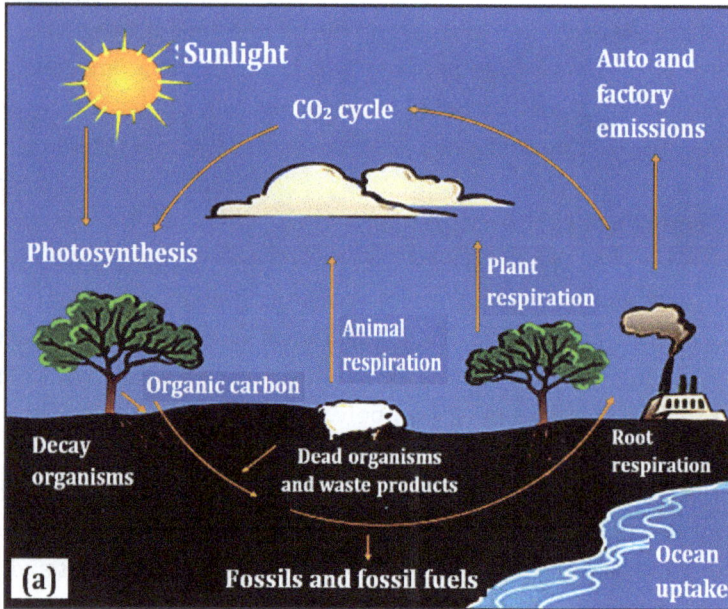

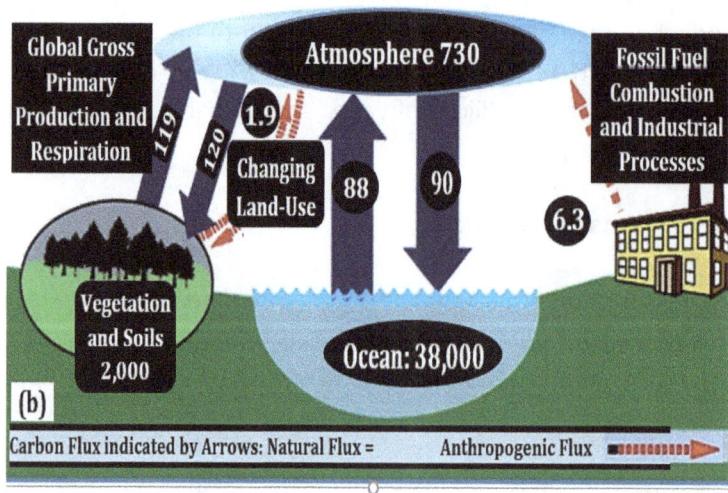

FIGURE 12.4   (a) The Natural Carbon Cycle *(eo.ucar.edu)* (b) Global Anthropogenic Carbon cycle. *(IPCC, 2001).*

Plants and animal waste decay in the soil, discharging nitrates in the soil. Denitirifcation bacteria digest the nitrates to produce nitrogen which goes back into the atmosphere. Nitrous oxide is a by-product of this process and the gas is also discharged into the atmosphere.

12.5.4 Halocarbon gases

Halocarbons are synthetic organic compounds comprising carbon and halogen atoms. Many of these compounds exist naturally in the atmosphere, but extensive use in many technological applications results in an increasing concentration of the gases in the atmosphere, with a negative impact on the environment.

12.6 THE OZONE LAYER

Natural ozone is concentrated in a thin layer in the lower portion of the stratosphere, from approximately 10 to 50km above Earth surface, although the thickness varies seasonally and geographically. The concentration is highest (2-8 parts per million) in the 20-40km altitude range. Ozone breaks down in the lower part of the atmosphere due to ultraviolet radiation and during atmospheric discharges, forming diatomic oxygen and atomic oxygen. Ozone in the Earth's stratosphere is created by the Sun's ultraviolet rays (with wavelengths shorter than 240nm) which supply energy for the dissociation of some of the oxygen molecules into two highly unstable oxygen atoms in accordance with the endothermic reaction shown in Equation 12.1.

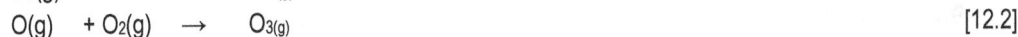

$$O_2(g) \quad \rightarrow \quad 2O_{(g)} \qquad [12.1]$$
$$O(g) \quad + O_2(g) \quad \rightarrow \quad O_{3(g)} \qquad [12.2]$$

The oxygen atoms are extremely unstable and react with oxygen molecules according to the exothermic reaction shown in Equation 12.2, with the release of heat energy. The net effect of the above two reactions is the formation of two molecules of ozone from three molecules of oxygen, accompanied by the conversion of light energy to heat energy according to Equation 12.3. The ozone formed is unstable and absorbs the Sun's ultraviolet rays of up to 290nm which causes it to decompose into oxygen molecules and atoms, with the release of heat energy (Equation 12.4).

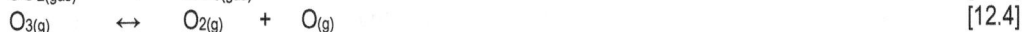

$$3O_{2(gas)} \quad \rightarrow \quad 2O_{3(gas)} \qquad [12.3]$$
$$O_{3(g)} \quad \leftrightarrow \quad O_{2(g)} \quad + \quad O_{(g)} \qquad [12.4]$$

This continuous set of reactions involving the breakdown and formation of ozone with the accompanying absorption and dissipation of energy is known as the ozone-oxygen cycle (Figure 12.5). These reactions play a critical role in screening off potentially harmful radiations from the Sun from the Earth's surface.

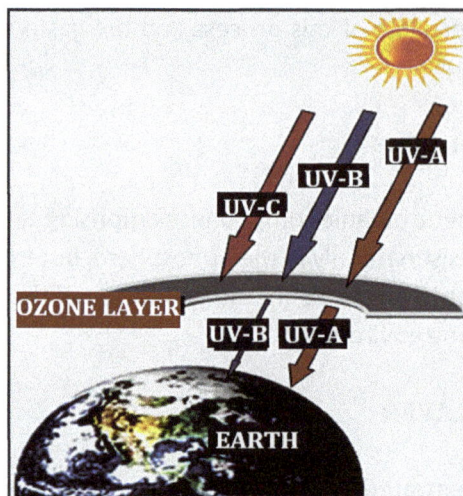

FIGURE 12.5   Role of the ozone layer in controlling the Sun's ultraviolet rays that reach the Earth.

There is no net ozone depletion because the process produces atomic oxygen that reacts with molecular oxygen to form another ozone molecule. However, the natural ozone-oxygen equilibrium is disrupted by the presence of certain radicals that contain chlorine, bromine, nitrogen, hydrogen or oxygen atoms. An ozone layer is thus created in the stratosphere between 10 to 50km above the Earth's surface, although most of the compound (around 90%) is concentrated between 20-40km. Even though the concentration of the ozone in the ozone layer is very small, it is vitally important to life because it absorbs biologically harmful ultraviolet radiation coming from the Sun and prevents it from reaching the Earth's surface by using it to energize the chemical reactions discussed above.

12.7   ANTHROPOGENIC GREENHOUSE GASES

The greenhouse effect and ozone layer are nature's mechanisms for controlling the temperature on the surface of the Earth and screening out components of the Sun's radiation which are potentially injurious to life. Human activities have always had a negative effect on the Earth-Environment system but the rate of damage was such that they were neutralized by natural system processes. However, the Industrial Revolution of the late 19th century greatly accelerated industrial and commercial development, the prime mover of which was energy. Coal was the main source of energy and mines sprang up in many parts of the world. Coal fueled iron and steel production, steam engines which were the prime movers, the electric power generation plants, and the rail system which became the major form of transportation. The development of diesel and petrol engines towards the end of the century was made possible by production of oil and gave birth to auto and air transportation industries in the first two decades of the 20th century. Since then, the demand for coal and other fossil fuels (oil and gas) has grown exponentially, with profoundly negative impact on the Earth-Environment system. Combustion of fossil fuels produces greenhouse gases which have greatly increased the natural concentration in the atmosphere, causing more infrared radiation from the Earth to be trapped in the stratosphere and reflected back to the Earth's surface. As a result, global temperatures have been increasing steadily over the past five decades, with potentially negative effect on the ecosystem. This phenomenon is known as *global warming*. Greenhouse gases in the atmosphere which emanate from human activities are known as *anthropogenic greenhouse gases*.

Human activities are also generating organic compounds known collectively as halocarbons. Although many of these gases are greenhouse gases, increased concentration in the atmosphere disturbs the natural balance and contributes to global warming. Furthermore, the halocarbons react with ozone, another natural greenhouse gas, decreasing its concentration in the atmosphere and therefore reducing the effect of the ozone heat shield which screens out dangerous ultraviolet radiation from the Sun and protects life on the Earth. Potential consequences of anthropogenic degradation of the environment include increased desertification, rise in ocean levels, climate change, extreme weather and human health risks. Coal combustion is the main source of anthropogenic gases and the effects of all other sources are measured relative to coal.

12.7.1   Anthropogenic carbon oxides

Carbon reacts with oxygen present in the air to produce carbon dioxide according to equation 12.5. In effect, every kilogram of carbon burnt produces nearly four kilograms of carbon dioxide (CO_2). When the oxygen supply is insufficient, some carbon monoxide is produced according to Equation 12.6.

$$C + O_2 => CO_2 \quad (12 \text{ kg C}) + (32 \text{ kg O}) => (44 \text{ kg } CO_2) \tag{12.5}$$
$$2C + O_2 => 2CO \tag{12.6}$$

Carbon monoxide usually produced by incomplete combustion of fossil fuel is strictly not a greenhouse gas but its presence in the atmosphere impacts negatively on the greenhouse effect. It elevates concentrations of methane and tropospheric ozone through chemical reactions with other atmospheric constituents, for example hydroxyl radical (OH) that would otherwise assist in destroying methane and tropospheric ozone. Carbon monoxide has a short life span because it is eventually oxidized to carbon dioxide.

12.7.2   Anthropogenic methane (CH_4)

Methane is a colourless, odourless and highly inflammable gas which gets into the atmosphere through various sources. It is present in the natural atmosphere, produced primarily by anaerobic (oxygen-deficient) decomposition of organic matter in biological systems. The compound is also produced by termites which harbour bacteria that are capable of breaking down organic matter in wetlands and swamps. Agricultural activities (biogenic sources) such as wet rice cultivation, soil fertilization and vegetarian livestock production release significant quantities of methane into the atmosphere. For example, the digestive process of a cow emits about a quarter of a kilogram of methane into the atmosphere daily (Figure 12.6). Significant methane emissions emanate from decaying vegetation, especially in hydro dams, artificial lakes and reservoirs.

Human activities such as fossil fuel combustion produce methane but in relatively small quantities compared with carbon dioxide. Other significant non-natural sources are oil an gas production and coal mining. During coalification, large quantities of methane are generated and stored within the coal bed on internal surfaces. Coal beds have large internal surfaces and can store huge amounts of methane-rich gas, about seven times as much as a conventional natural gas reservoir of equal rock volume can hold (USGS, 2017). Methane is explosive and poisonous, and is a major cause of coal mine fires. It is common practice to drill escape holes in coal beds to let the gas into the atmosphere prior to mining, and periodically during mining operations, although it is now being treated as a valuable energy resource which can be utilized for electric power production.

Atmospheric concentrations of methane have increased by about 150% since pre-industrial times and it is estimated that more than half of its concentration in the atmosphere is anthropogenic, from human activities (IPCC, 2001). The life span of the gas in the atmosphere is relatively short because it is destroyed in the lower atmosphere (troposphere) by reactions with free hydroxyl radicals (OH) (Equation 12.7). Some methane is also converted to carbon dioxide.

$$CH_4 + OH = CH_3 + H_2O \qquad\qquad\qquad [12.7]$$

Increasing emissions of methane reduce the concentration of the hydroxyl radicals in the atmosphere, the net effect of which may increase methane's atmospheric lifetime. The total contribution of methane to enhanced GHGs is estimated at about 15%. Although methane concentration in the atmosphere is lower than carbon dioxide and the gas has a short life span of only about ten to twelve years in the atmosphere before it is removed by a natural process, it has about twenty times more potent heat trapping capacity than carbon dioxide.

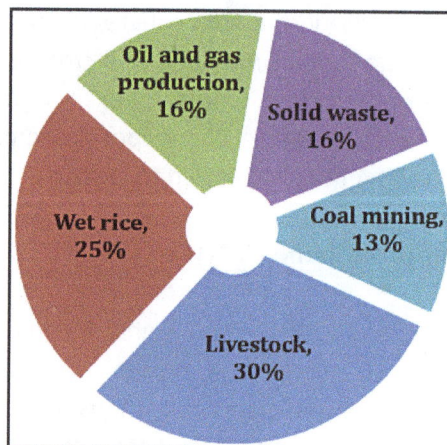

FIGURE 12.6   Sources of methane in the atmosphere. *(epa.gov).*

12.7.3   Anthropogenic nitrogen oxides (NOx)

Nitrogen oxides are chemical compounds formed from reactions between nitrogen and oxygen. Normally both elements coexist in the atmosphere but at high temperatures such as exist in a steam boiler, combustion engine or atmospheric storm, endothermic reactions take place between the two elements as shown in equations 12.8 to 12.10.

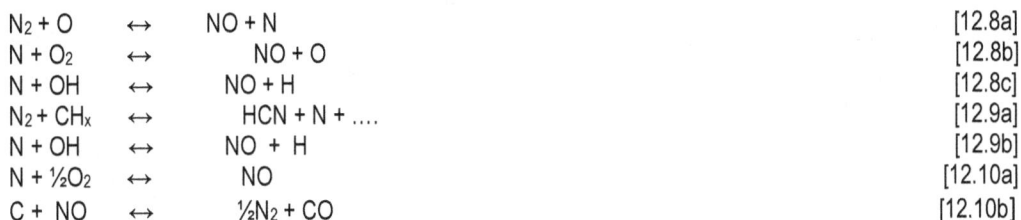

$$N_2 + O \leftrightarrow NO + N \tag{12.8a}$$
$$N + O_2 \leftrightarrow NO + O \tag{12.8b}$$
$$N + OH \leftrightarrow NO + H \tag{12.8c}$$
$$N_2 + CH_x \leftrightarrow HCN + N + \ldots \tag{12.9a}$$
$$N + OH \leftrightarrow NO + H \tag{12.9b}$$
$$N + \tfrac{1}{2}O_2 \leftrightarrow NO \tag{12.10a}$$
$$C + NO \leftrightarrow \tfrac{1}{2}N_2 + CO \tag{12.10b}$$

The reactions in Equations 12.8 a&b also known as Zeldovitch mechanism are strongly endothermic and require temperatures of 1500-1600°C to dissociate nitrogen and oxygen for the reactions to take place. *Thermal NO* is produced in these reactions which are most likely to occur during the combustion of natural gas. More recent work has shown also that atomic nitrogen can react with the hydroxyl radical OH as shown in Equation 12.8c. *Prompt NO* is produced in the combustion of fuels which are rich in CHx compounds such as natural gas, petrol, diesel as shown in equation 12.9a. In effect, gas fired burners and internal combustion engines are the main sources of this type of NOx .

A third type of NOx known as *Fuel NO* is produced mainly when coal is burned. About 75% of the nitrogen oxide results from the reaction between carbon and nitrogen present in the volatiles as shown in Equations 12.10 a&b. Carbon and nitrogen still present in the char can also react to produce some nitrogen oxide accounting for about a quarter of the total. Nitric oxide (NO) is produced at very high temperatures during combustion, the extent depending on availability of oxygen hence production is very limited or absent at lower temperatures or when air supply is limited. Nitric oxide is unstable and rapidly reacts with oxygen to form nitrogen dioxide (NO_2).

Presence of nitric oxide in the stratosphere where there is an abundance of ozone and oxygen molecules, highly reactive oxygen atoms and ultraviolet light energy causes a reaction with ozone according to equations 12.11 a&b. The net effect of these reactions is a depletion of ozone according to equation 12.11c (addition of equations 12.11a &b).

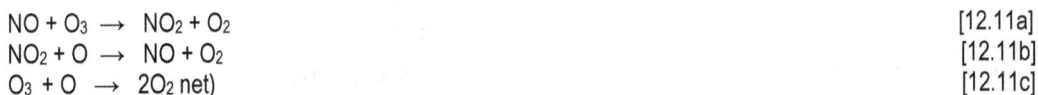

$$NO + O_3 \rightarrow NO_2 + O_2 \tag{12.11a}$$
$$NO_2 + O \rightarrow NO + O_2 \tag{12.11b}$$
$$O_3 + O \rightarrow 2O_2 \text{ net)} \tag{12.11c}$$

Nitrous oxide (N_2O) is a colorless and non-flamable gas with pleasant odour and taste. In induces euphoria on inhalation, hence its common name "laughing gas". About two-thirds of nitrous oxide released into the atmosphere is from natural processes of decay in ocean beds and soil bacterial action (denitrification).

Nitrous oxide is also produced as a result of human activities, mainly agriculture: soil cultivation, use of nitrogen fertilizers, decaying animal waste. It is estimated that livestock farming produces about a third of human-related nitrous oxide. Synthetic nitrous oxide is also produced and used extensively as analgesic in anaesthesia.

Nitrous oxide makes up an extremely small amount of the atmosphere - it is less than one-thousandth as abundant as carbon dioxide. However, it is 200 to 300 times more effective in trapping heat than carbon dioxide. In fact, nitrous oxide is considered to be the largest anthropogenic emission of an ozone-destroying compound currently (Portmann et al., 2012). However, recent research suggests that the hydrogen radical family may be even more important. Nitrous oxide (N_2O) is not often considered a greenhouse gas but it does contribute to environmental degradation as well as to greenhouse gases. Like carbon dioxide it absorbs long wavelength infrared radiation and stores heat, thereby contributing to global warming. It reacts with ozone in both the troposphere and in the stratosphere, although slowly. It also has a very long half-life estimated to be between 100 and 150 years.

12.7.4   Anthropogenic halocarbons (HFCs)

Halocarbons is a general term for any group of synthetic organic compounds that contain carbon and halogen atoms such as fluorine, chlorine, bromine. Like nitrogen oxides, many of these compounds, such as chloro-fluoro-carbons (CFCs), hydro-chloro-fluoro-carbons (HCFCs), hydro-bromo-fluoro-carbons (HBFCs), methyl chloroform, and carbon tetrachloride, react with ozone in the stratosphere resulting in ozone depletion. Because they can be easily converted from gas to liquid or liquid to gas, CFC's and HCFCs can be used in aerosol cans, refrigerators, air conditioners, foam manufacture, suppression, solvent degreasing, sterilization, pest control and laboratories etc. The use in these various applications has greatly increased the atmospheric concentration of these gases which are known to react with and break down ozone in the stratosphere.

Although many of the halocarbons are also powerful greenhouse gases, their net radiative forcing effect on the atmosphere is reduced because they cause ozone depletion which in itself is a potent greenhouse gas in addition to shielding the Earth from harmful levels of ultraviolet radiation. Halocarbon gases are very stable and are believed to be capable of remaining in the atmosphere for hundreds of years. In the last six decades or so gases used in refrigeration and air conditioning have become a potent contributor to enhanced GHGs, accounting for over 20% of the total.

Worldwide action against the use of halocarbons gained momentum in the last three decades, leading to the Montreal Protocol of 1987 which sought to restrict the use of ozone depleting substances. This action has been very effective and use is now restricted mainly to developing countries. The substitutes for CFCs [hydro-fluoro-carbons (HFC's), per-fluoro-carbons (PFCs), and sulphur hexafluoride (SF_6)] do not harm or breakdown the ozone molecule and currently have a small aggregate radiative forcing impact. However they contribute significantly to greenhouse effect.

12.7.5 Anthropogenic sulphur Oxides (SOx)

SOx refers to chemical compounds formed from the combination of elemental sulphur and oxygen, the most important being sulphur dioxide (SO_2) and sulphur trioxide (SO_3). The dioxide is the acid anhydride (a compound that combines with water to form an acid) of sulphurous acid; the trioxide is the acid anhydride of sulphuric acid. Sulphur is present in many coals in varying quantities up to about 12%. It may be part of the organic structure of the coal or may be present combined with iron as pyrite or with copper, also as pyrite. The element reacts with oxygen during combustion to form sulphur dioxide as shown in Equation 12.12. Every kilogram of sulphur forms 2 kilograms of sulphur oxide.

$$S + O_2 => SO_2 \ (32 \ kg \ S) + (32 \ kg \ O) => (64 \ kg \ SO_2) \qquad\qquad [12.12]$$

Sulphur dioxide also occurs in nature in volcanic gases and in solution in warm springs. The compound is also formed by the oxidation of hydrogen sulphide (H_2S) by atmospheric oxygen. Hydrogen sulphide is released into the atmosphere from biological decays in marshes and oceans, and is frequently found in combination with natural gas. The gas is also produced when high sulphur coal is burnt. The transportation sector is the major contributor to the emissions of sulphur oxides but a significant proportion also comes from coal-fired electric power plants. The oxides are products of reactions between sulphur constituents of fossil fuels and oxygen. When ejected into the atmosphere from combustion chamber exhaust systems, they react with atmospheric moisture to form corrosive acids commonly referred to as acid rain which not only attacks metals especially steel, but also causes considerable deforestation. Acid rain damages forests and agricultural crops and acidifies lakes to the detriment of aquatic life. Sulphur dioxide, especially at high concentrations can cause significant human health problems including irritation of skin, eye, throat and lungs, inflammation of the respiratory system. Long-term exposures to sulphur dioxide and sulphate particles can cause respiratory illness, aggravate existing cardiovascular diseases and worsen asthma attacks.

12.8 ENERGY UTILIZATION AND THE ENVIRONMENT

Human activities especially since the beginning of the Industrial Revolution have consistently undermined the natural balance in the carbon cycle as designed by nature, with various activities that produce anthropogenic gases, dust and smog which end up in the atmosphere. Increases in the natural concentration of gases greatly disturb nature's control system which enables the sustenance of life on the planet and could have a profound effect on climate systems. Increased concentration of dust and smog alter the natural response of clouds to the Sun's radiation and also cause respiratory problems for human beings. Fossil fuels (oil, natural gas and coal) filled about 80% of total world energy demand in 2013 with oil contributing 31.5%, (Figure 12.7), followed by coal (28.8%), and natural gas (21.3%) (IEA, 2015). Although all forms of energy production have direct or indirect negative effect on the environment, fossil fuel combustion (oil, gas, coal) contributes by far the largest proportion.

Over 80% of the total global emissions of GHG gases in 2010 were from the energy sector (Figure 12.8). The United States is by far the largest consumer of fossil fuels. With only 5% of the world population, the country uses over a quarter of the global supply of fossil fuels. Over 90% of the country's contribution to the total global GHG emissions come from the combustion of fossil fuels. Carbon dioxide accounts for about 76% of GHGs and is mostly a by-product of the combustion of oil, natural gas and coal Figure (12.8). In view of the dominance of the gas in atmospheric pollution, the intensities of other gaseous pollutants are expressed in terms of CO_2 equivalent (tCO_2) and their effects on the environment expressed relative to carbon dioxide.

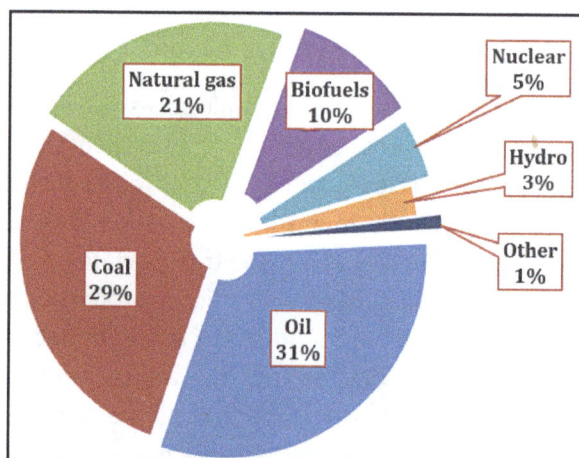

FIGURE 12.7   Fuel shares of total global primary energy supplies in 2013. *(IEA, 2015).*

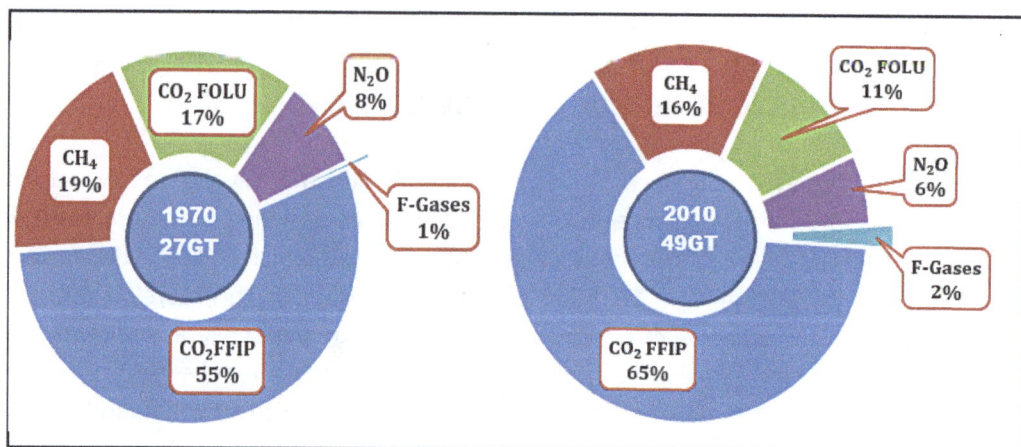

FIGURE 12.8   Global anthropogenic GHG emissions by groups of gases, 1970 & 2010. Greenhouse gases (GHG); Fossil fuel and industrial processes (FFIP); Forestry and other land uses (FOLU); carbon dioxide (CO_2); methane (CH_4); nitrous oxide (N_2O); Fluorinated gases; (F-Gases). *(IPCC, 2014).*

Several other parameters for assessing environmental damage caused by greenhouse gases have also been developed. These include, carbon intensity and carbon footprint.

12.9   COAL UTILIZATION AND THE ENVIRONMENT

The production of coal causes severe degradation of the environment. Coal mining, especially by opencast method degrades vast expanses of land and generates dust, mining releases methane into the atmosphere, oil and gas production cause severe pollution and ecological problems. Furthermore, combustion of all fossil fuels generate anthropogenic greenhouse gases which cause significant changes in the natural Earth-atmosphere system that can impact negatively on global environment and undermine nature's protection system for life on Earth.

The bulk of the world's annual consumption of coal is for electric power generation and metallurgical coke production. such as town gas, synthetic natural gas, petroleum products, activated carbon and chemicals (Figure 12.9). Up to the 1950s, many important industrial chemicals were synthesized from coal-derived feedstock, mainly by-products of metallurgical coke or town gas production, until petroleum became the preferred source. Currently, coal-to-chemicals technologies are being actively revived.

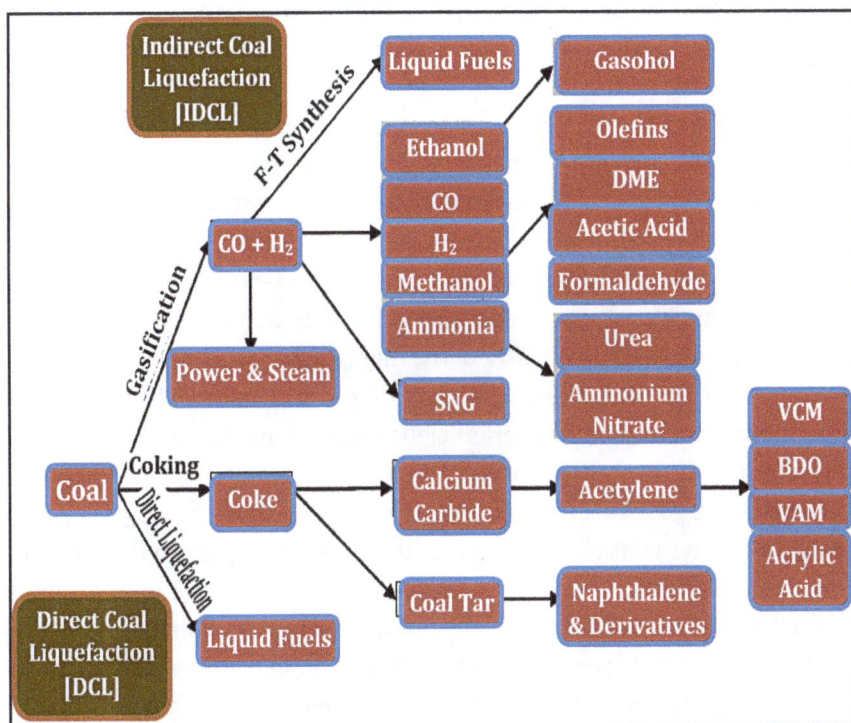

FIGURE 12.9 Main coal utilization processes.
(Adapted from www.chemsystems.com, 2011).

Coal production involves prospecting, mining, washing and transportation to consumers. Prospecting is by geological mapping and core drilling, and does not pose any significant danger to the environment. However, production which is either by underground or surface mining causes environmental problems. While underground mining affects very small areas per tonnage of coal produced, significant amounts of anthropogenic methane are vented out of mines into the atmosphere to reduce the risks of explosion and fires. Mechanized recovery removes large amounts of shale and soil with coal which make washing necessary. In pre-wash grinding particulate carbon is released into the atmosphere. Carbon dust is injurious to human health and the washery waste contains acidic sulphates which damage the ecosystem.

Abandoned underground mines often fill up with water which may become acidic and drain out, polluting streams, rivers and soils. Surface mining involves the removal of the soil and rock overburden over large surface areas to reach the coal bed, using excavators and explosives. Large areas of the landscape are disturbed and, in many cases destroyed. Methane is also released into the atmosphere during mining operations. While some countries have strict regulations about reclamation and rehabilitation of damaged mining areas, many, especially emerging countries do not and disused opencast mines cause significant damage to the landscape, creating barren land and polluted artificial lakes.

Although coal produced by opencast mining often does not need to be washed, water draining from mines may contain acidic compounds which pollute rivers and streams, degrade the fertility of the land, and pose danger to aquatic wildlife. When coal is burnt, whether for cooking, heating or steam generation, carbon particles and gases including carbon dioxide, carbon monoxide, nitrogen oxides, sulphur oxides, and methane are released into the atmosphere. Fly ash, toxic mercury and other heavy metals are also released into the atmosphere and pollution leaching from ash residues which are often dumped may contaminate groundwater and cause other environmental concerns.

Compared with natural gas fired power generation, coal-fired power units produce around a hundred times as much sulphur dioxide (depending on the sulphur content of the coal), twice as much carbon dioxide and five times as much nitrogen oxides per unit of electricity generated (USGAO, 2012). The high disparity is due to several reasons: coal contains more carbon and sulphur than natural gas and many natural gas-fired power plants are more modern, operate at higher efficiencies, and have more efficient installed emissions control systems. The intensity, concentration, and life span in the atmosphere and anthropogenic effects of the gases vary significantly. The release of anthropogenic gases, in particular, carbon dioxide into the atmosphere poses the greatest danger to the environment.

Production of chemicals from coal declined when alternative petroleum sources became widely available in the 1950s. However, many factors have combined to bring coal-derived chemicals back to the forefront in the last decade or so, the most important being the wide availability and geographical distribution of coal relative to oil and gas. The prime mover of this development is China which now produces a wide range of chemicals from coke oven byproducts and coal gasification. Coal gasification based systems offer significant environmental advantages over competing technologies, particularly coal-to-electricity. The ability to capture particulate, CO, CO_2, SOx, NOx and mercury emissions at the gasification stage is of major environmental benefit. Processes for multi-contaminant control, such as selective catalytic reduction are available already or under development. The relatively higher thermodynamic efficiency of Integrated Gasification Combined Cycle (IGCC) systems minimizes carbon dioxide emissions. Typical system configurations featuring coal gasification include carbon capture and sequestration and various studies have shown that the cost of capture in an IGCC system is around half the cost for a pulverized coal power plant, because the gas is in a compressed state and nearly pure.

Underground coal gasification produces even lower pollution because a significant proportion of the potential pollutants remains trapped underground. Other coal gasification-based systems for production of liquid fuels and chemicals have even lower carbon dioxide capture costs. Virtually all the carbon in coal is converted to energy in a gasification system and the mineral matter in the feedstock ends up in ash and char, both of which are useful in construction as aggregates and fills. Water requirements for the production of a unit power is around 20% less than other coal-based technologies. The amount of CO_2 produced per barrel of coal-derived fuel (CTL) is still much higher compared with gasoline from petroleum crude oil (Figure 12.10). However, the processes greatly facilitate carbon dioxide capture and sequestration (CCS) and if a CTL plant includes a CCS unit, CO_2 production levels can be reduced to the same level as for gasoline.

Hydraulic fracturing (fracking) technology has been used for decades to access unconventional gas deposits locked up in rocks and shales at great depths of up to 4000 metres. However, it is only in recent times that the potential negative impacts on the environment have become subjects of intensive discussion. Of particular concern is the increasing evidence of leakages in production wells which can lead to undesired entry of potentially carcinogenic fracking fluids into the neighbourhood rocks and water systems. Also, many potentially toxic compounds like methane can contaminate both the air and ground water. It is also speculated that seismic disturbance caused by fracking can trigger earth quakes.

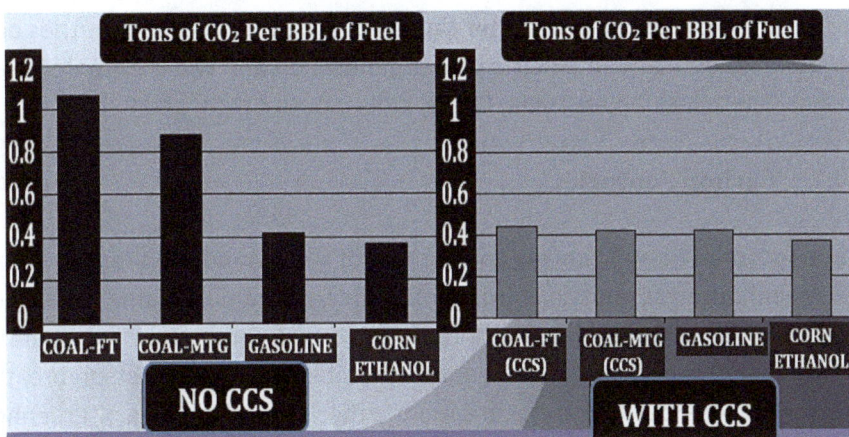

FIGURE 12.10 Carbon dioxide footprint of coal-derived liquid fuels (tons/bbl of fuel) (COAL-FT = Fischer-Troopsh coal gasification, COAL-MTG = gasification-to-methanol, CCS = carbon capture and sequestration). *(ExxonMobil Research & Engineering, 2009).*

12.10   QUANTIFICATION OF ANTHROPOGENIC POLLUTION

Various measures have been devised for quantifying the contribution of anthropogenic gases and related human activities to the overall environmental degradation. These include *carbon intensity, carbon footprint, life cycle analysis*, etc.

12.10.1   Carbon Intensity

Carbon Intensity (also called Carbon Intensity Index or emission intensity) is the amount of tonnes of carbon dioxide or equivalent produced per unit of output of an activity, for example per Mega joule of energy produced, per capita or per GDP. Population and GDP are the major determinants of a country's emissions and is a composite of two other major factors, namely energy intensity and fuel mix. While energy intensity gives some indication of a country's level of economic development, there are many exceptions. For example, developed countries tend to higher efficiency in energy utilization and may have lower energy intensities than emerging countries with energy-intensive primary industries such as primary metal production. Emissions intensity also depends on fuel mix (carbon content of energy consumed). Coal has the highest carbon content, followed by oil. Natural gas has the lowest carbon content of fossil fuels.

Carbon intensity is useful in comparing the effect of greenhouse gases from various sources on the environment. Values of carbon intensity can be calculated for electric power plants per unit of energy produced using different fuels, for heat generating units per joule of heat. Typical carbon intensities of electric power plants fired by different fossil fuels are shown in Table 12.3. The carbon intensity of total emissions has more than doubled over the period 1971 to 2011 (Figure 12.11) and varies widely between regions, with North America leading the world (Figure 12.12). The main sources of carbon pollution are electric power generation and transportation (Figure 12.13). Only ten countries accounted for two-thirds of the total world GHG emissions in 2010 and China leads the world, but the United States of America has the highest emissions per capita (Figure 12.14). Some countries depend on coal for between 65 and 94% of electric power generation and will inevitably have high carbon intensities. These include Australia, China, India, South Africa, and Poland.

12.10.2   Carbon Footprint

A carbon footprint literally means an estimate of the amount of direct or indirect contribution of greenhouse gas emissions from any source over a time frame (usually one year), for example, events, products, organizations or persons. A comprehensive, quantitative assessment based on this definition would not be possible since every person breathes oxygen in and carbon dioxide out and there are many other natural sources of greenhouse gases. A carbon footprint is measured in tonnes of carbon dioxide equivalent (tCO_2e). The carbon dioxide equivalent allows the different greenhouse gases to be compared on a common basis, rated compared to carbon dioxide which is assigned one unit. CO_2e is calculated by multiplying the emissions of each of the six greenhouse gases by its 100 year global warming potential (GWP). A carbon footprint considers all six of the Kyoto Protocol greenhouse gases: Carbon dioxide (CO_2), Methane (CH_4), Nitrous oxide (N_2O), Hydrofluorocarbons (HFCs), Perfluorocarbons (PFCs) and Sulphur hexafluoride (SF_6). Carbon footprint calculation has become a powerful tool for assessing the impact of human (including personal) behavior on global warming. It should be noted that carbon represents all greenhouse gases which are now added as carbon dioxide equivalents. The foods and goods that humans buy everyday, polymer shopping bags, travels have carbon footprints which must be accounted for in calculating personal contribution. For example, ten litres of petrol or diesel burnt in a car, or of oil used in home heating contribute 23-27kg of carbon dioxide to the atmosphere. The true assessment of a product must not only take account of direct emissions from the manufacturing and transportation to retailers but must also take account of a host of indirect emissions such as those caused by the production of the raw materials used in the production of the good.

TABLE 12.3 Typical emissions from electric power plants by fuel. *(EEA, 2008).*

POLLUTANT	HARD COAL	BROWN COAL	FUEL OIL	OTHER OIL	GAS
CO_2 (g/GJ)	94,600	101,000	77,400	74,100	56,100
SO_2 (g/GJ)	765	1,361	1,350	228	0.68
NO_2 (g/GJ)	282	183	195	129	93.3
CO (g/GJ)	89.1	89.1	15.7	15.7	14.5
CO_2 equivalent (g/GJ)	95,746	102,633	78,961	74,473	56,209
Particulate matter (g/GJ)	1,203	3,254	16	1.91	0.1
Flue gas (m^3/GJ)	360	444	279	276	272

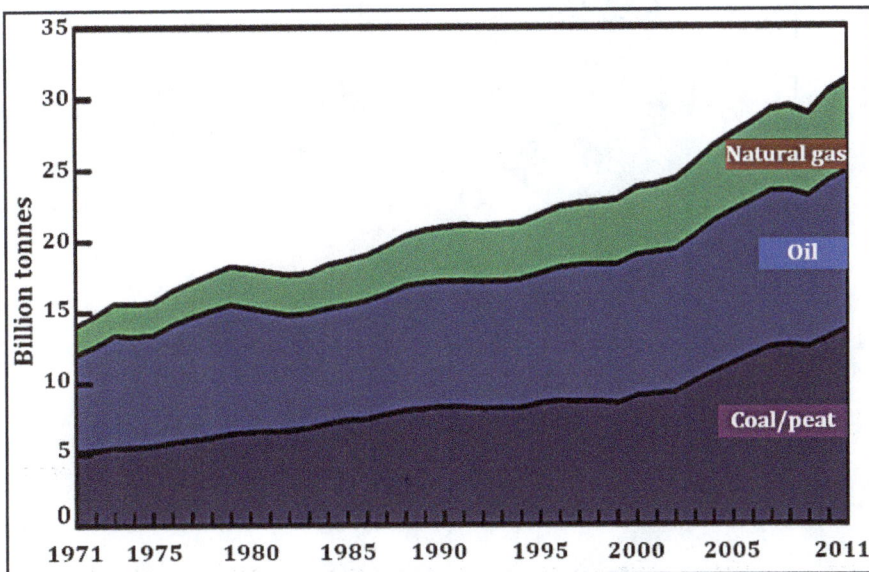

FIGURE 12.11 World CO_2 emissions by fuel, 1971-2011. *(IEA, 2013).*

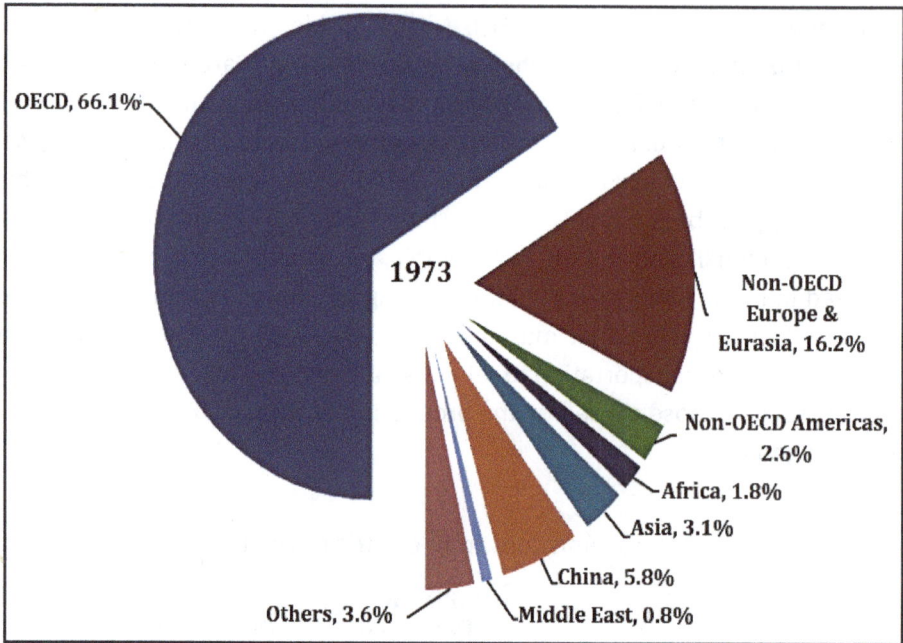

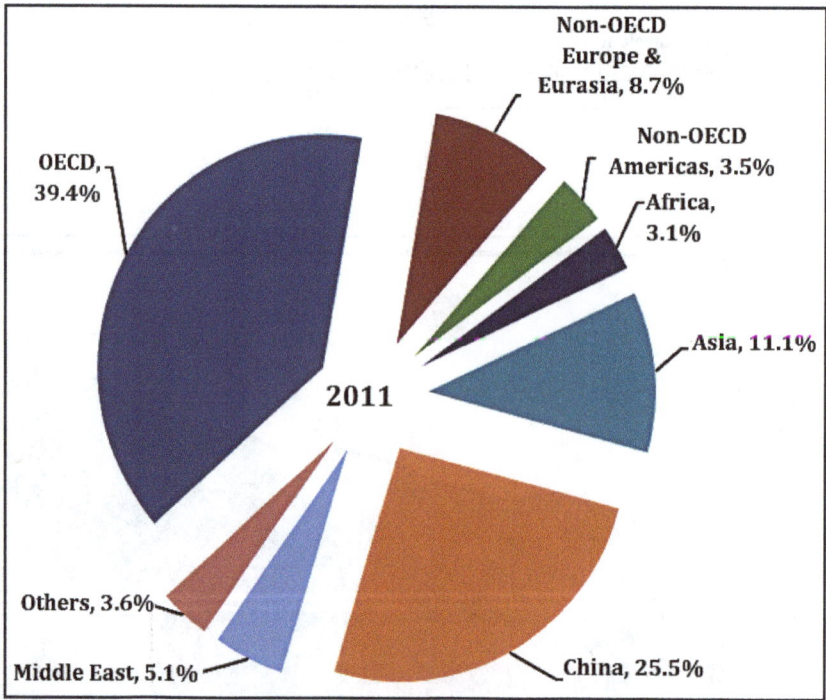

FIGURE 12.12 Share of CO$_2$ emissions by region in 1973 and 2011. *(IEA, 2013).*

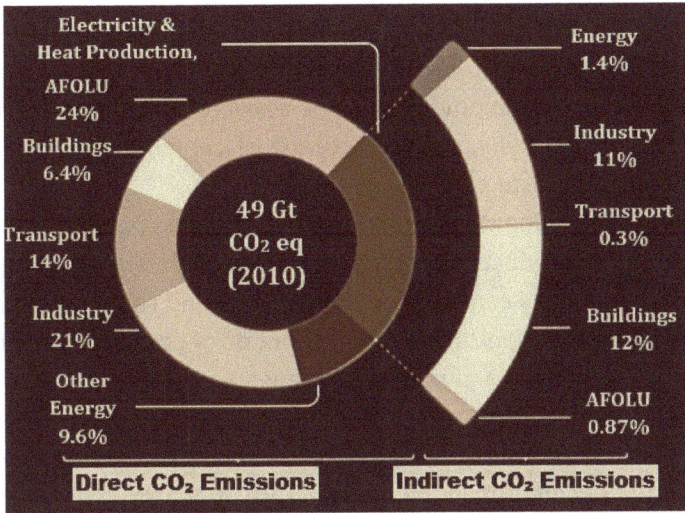

FIGURE 12.13 Greenhouse gas emissions by economic sector. Pull-out shows how indirect CO_2 emission shares (in % of total anthropogenic GHG) from electricity and heat production are attributed to sectors of final energy use. The emissions from Agriculture, Forestry and Other Land Use (AFOLU). *(IPCC, 2014).*

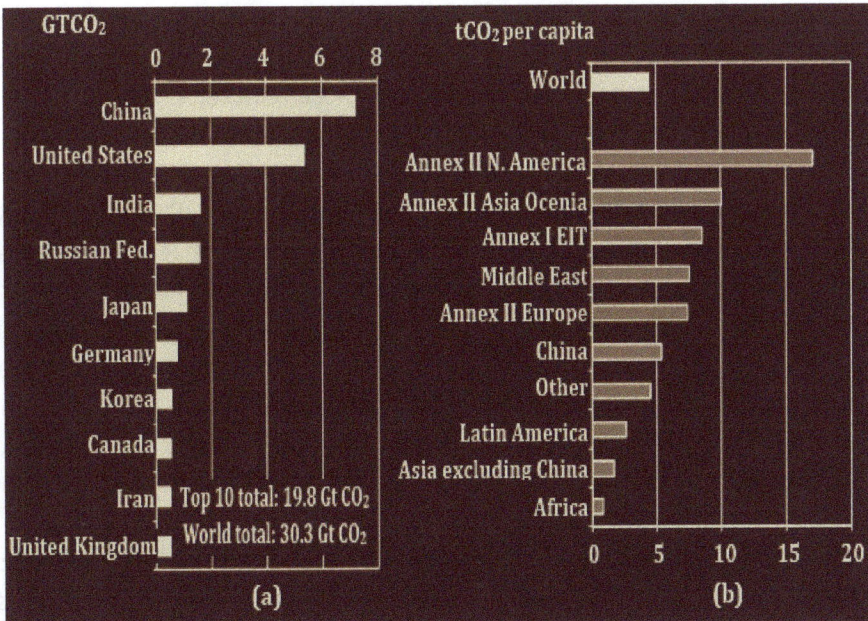

FIGURE 12.14 (a) Top ten emitting countries in 2010 (b) CO_2 emissions by major world regions in 2010. *(IEA CO_2 Emissions Highlights, 2012).*

12.10.3   Life cycle analysis of greenhouse gases

Total Life Cycle Analysis (LCA) is an accounting process that considers the environmental impact of all stages of a particular process or product chain. LCA for an energy resource for example involves consideration of emissions at every stage from production to utilization, from material and fuel mining, manufacturing of components, construction, installation, de-commissioning, to waste management.

Emissions at the mining stage, transportation, processing, ultimate utilization and recycled or discarded by-products are taken into account, so also are the emissions from the production of materials of construction such as steel and cement used in dam construction, materials used in the transportation and distribution of natural gas and fuel oil. For example, solar energy is considered a zero emission energy resource in the traditional evaluation process. However, Life-Cycle-Analysis presents a different picture. The overall carbon footprint is high because enormous energy (usually supplied by coal-fired plants) is required for the production of silicon used in the manufacture of photovoltaic solar cells, collectors and other equipment. In effect, there is nothing like near-total generation of electricity from zero carbon sources (solar, hydroelectric, nuclear, wind) as claimed by some developed countries.

Another example is the wide assumption that gas-fired power generating plants emit around 50% lower environmental pollutants compared with coal. However, total life-cycle analysis shows that, unless methane emissions can at the gas production stage can be kept below 2-3%, there is no relative advantage, since methane is a significantly more potent anthropogenic gas compared with carbon dioxide which is the predominant pollutant in coal-fired plants (Nalbandian-Sugden, 2015). Life cycle analysis also involves the quantification of all emissions in the chain, including emissions saved (displacement emission) where for example the by-products of one process are used to replace products of another process, thereby avoiding the emissions associated with those replaced products.

Steel, aluminium, copper and indeed most metals are 100% recyclable. The energy required in processing recycled steel is about 30% less than required for processing virgin steel raw materials. Recycled aluminium only consumes about 10% of the energy required to produce the material from bauxite. It is commonly assumed that replacement of steel with lighter materials such as plastic or aluminium makes an automobile lighter and therefore more fuel efficient. While this may be true, the analysis fails to take into effect the fact that most of these lighter materials require more energy for production than steel. In effect, in order to provide a true picture about the contribution of any process to environmental degradation the total life cycle "from cradle to grave" should be evaluated. This would enable society to make informed choices on the environmental impact of products and processes. Fly ash, usually a waste product from coal combustion in now increasingly used to replace cement in concrete mixtures, thereby reducing the cement required and saving the emissions that would have been discharged into the atmosphere during its production and the emissions associated with the energy saved.

Life cycle analysis for conventional steelmaking comprises emissions from the coke oven, sinter plant, blast furnace, oxygen converter, slab caster, rolling mills, oxygen and electricity production and utilization. Utilizing waste process energy from a typical integrated steel plant to generate electricity can displace about 8 GJ of grid electricity which in turn displaces the combustion of the equivalent of 22 GJ of feed coal in a coal-fired electricity power plant. These should be taken into account in a comprehensive LCA. Life cycle analysis can significantly change the widely held views about the relative emissions from fossil energy resources. About 97% of coal emissions are generated at the combustion stage, the balance at the production and cleaning stages. By comparison, only about 60% of emissions from natural gas occur at the burner tip. However, considerable emissions, mainly methane occur at the well head (about 15%), pipe line leakages during transportation, and at other points in the cycle. Conventional emission analysis indicates that the contribution of coal is about 30% higher than that of natural gas. However, life cycle analysis shows no significant difference.

Calculation of LCAs is extremely difficult and the results depend on the assumptions made in estimating energies associated with the various stages of the total life cycle, and the methodologies adopted. Another complication is the effect of the efficiency of each stage in the life cycle which may vary significantly between installations in different parts of the world. Improved efficiency of coal-fired generating plants results in lower output of anthropogenic gases. Hundreds of studies on electric power generation have been published, showing very divergent results (Table 12.4). Nevertheless, the figures still show the predominance of fossil fuels as environmental pollutants.

12.11 CONSEQUENCES OF ENVIRONMENTAL POLLUTION

The consequences of anthropogenic interference with the natural equilibrium of greenhouse gases and the ozone layer are multi-dimensional. Climate change through global warming is one of the major negative effects. Others are solar radiation damage, acid rain and atmospheric pollution. All types of pollution (air, water and soil) can cause serious damage to human and animal health, marine life, and other living organisms, plants, vegetation, etc.

TABLE 12.4 Total life cycle analysis (LCA) of some electric power generation fuels. *(Compiled by averaging values from 25 different publications).*

Fuel	Emission (g/KWh)
Coal	950-1050
Gas	400-1000
Diesel/Heavy oil	620-800
Nuclear	3-130
Biomass	10-40
Solar thermal	10-20
Solar PV	30-220
Hydro	5-30
Wind	4-80
Geothermal	30-50

12.11.1   Ozone Layer Depletion and enhanced concentration of greenhouse gases

Ultra-violet rays originate from the Sun and pass through the various atmospheric layers to reach the Earth. Ozone (O_3) is an allotrope of oxygen that contains three atoms of elemental oxygen compared with two for the (diatomic) oxygen (O_2) which makes up about 22% of the lower atmosphere and sustains human life on Earth. Unlike diatomic oxygen which is stable, colourless and odourless, ozone is unstable, pale blue in colour, with an extremely sharp and pungent smell. Ozone constitutes only about 0.6 parts per million of all atmospheric gases, insignificant compared with oxygen concentration. It is clear from the discussion earlier in this chapter that nature has in place a process for regulating temperatures and solar radiation damage on the Earth's surface through the ozone layer reactions in the stratosphere. Without this, life on earth as we know it cannot be sustained. However, human activities which release ozone-depleting nitrogen oxides and chlorofluorocarbons (CFC) into the stratosphere cause more harmful ultraviolet radiation to reach the Earth's surface. Apart from the potentially harmful effect on humans, depletion of the ozone layer also causes the Earth's surface temperatures to rise.

Natural greenhouse gases regulate temperatures on the Earth's surface. Increases in the concentration, in particular, carbon dioxide, methane and nitrous oxide cause more of the infrared radiation that is being reflected into the atmosphere from the Earth to be trapped in the stratosphere and reflected back to the Earth, causing increased temperatures.

The combined effects of ozone depletion and enhanced concentration of greenhouse gases in the atmosphere are believed to be largely responsible for the observed increase in the Earth's average temperature over the last few decades. Numerous studies have shown that an increase of just half a degree can have profound negative climate impact (Jackson, 2016). This phenomenon known as global warming has been a subject of intensive research and debate in the last two decades or so. While there is strong scientific evidence that the Earth's temperature is increasing, there is no consensus on its attribution entirely to human anthropogenic activities.

12.11.2   Global Warming and Radiative Climate Forcing

The overall state of the global climate is determined by the balance of solar and terrestrial radiation activities. How this energy balance is regulated depends upon the fluxes of energy, moisture, mass and momentum within the global climate system, made up of five components: the atmosphere, the oceans, the cryosphere, the biosphere and the geosphere. A sixth component, mankind (an anthropogenic system) has emerged in the last two centuries due to increased utilization of the global natural resources. Many human activities are having significant, mostly negative effect on the global climate and environment. Release of greenhouse gases into the atmosphere alters the natural balance and causes disequilibrium in the atmospheric gas mix, thus enhancing greenhouse effect and increasing the Earth's surface and sub-surface ocean temperatures, with potentially negative consequences.

Natural or anthropogenic activities, for example changes in the greenhouse gas concentration or composition, can shift the balance of energy transfers between the atmosphere, space, land, and the oceans, thus causing a change in the energy available to the Earth-atmosphere system. The shift could be in such a way as to cause the Earth's lower atmospheric temperature to rise, either by increasing the Sun's energy that reaches the Earth or decreasing the amount that is reflected back to space. On the other hand, the reverse could be the case and the Earth's surface cools. When changes in the environmental energy balance persist over very long periods, the global climate could be affected. The quantified contribution of these factors to climate change is known as *radiative climate forcing* (RCF). Positive or high RCF indicates a heating effect while negative (low) value means that the Earth cools as a result (IPCC 1996).

There is ample evidence (see earlier sections of this chapter) that human activities can affect concentrations, distribution and life cycles of atmospheric gases. Increases in the concentration of greenhouse gases enhance the heat-trapping ability of the Earth's atmosphere resulting in a gradual increase in the global mean surface temperatures. The ten hottest years ever have been recorded in the last fourteen years. The average temperature across global land and ocean in 2015 was the hottest on record, surpassing 2014 which had been the highest (Howard, 2016). Record temperatures are causing melting of arctic ice and rising seas. Scientists predict that the average global surface temperature could increase by 1 to 3°C in the next fifty years and up to 6°C by the end of the century, although there may be significant regional variations.

12.11.2.1 *Global Warming Potentials (GWPs)*

Greenhouse gases may remain in the atmosphere for very short periods or for as long as 150 years. *Global Warming Potentials* are a quantified measure of the globally averaged relative radiative forcing impacts of a particular greenhouse gas (Table 12.5). It is a cumulative radiative forcing, (both direct and indirect effects) integrated over a period of time from the emission of a unit mass of gas relative to some reference gas which is carbon dioxide (IPCC 1996). Carbon dioxide is assigned a unit global warming potential (GWP) over all time. In effect, a gas which has a high radiative forcing but a short life compared with carbon dioxide could have a high GWP on a 20-year time scale or a low value on a 100-year time scale. On the other hand, a gas that has longer atmospheric life than the reference gas will have higher GWP with the time scale.

Direct effects known as *direct radiative forcing* occur when the gas itself is a greenhouse gas and *indirect radiative forcing* refers to situations when chemical transformations involving the original gas produce a gas or gases that are greenhouse gases, or when a gas influences other radiatively important processes such as the atmospheric lifetimes of other gases. For example, methane has both direct and indirect effects, an atmospheric lifetime of 12 ± 3 years and a GWP of 56 over 20 years, 21 over 100 years and 7.6 over 500 years. The decrease in GWP at longer times is because methane is degraded to water and carbon dioxide through chemical reactions in the atmosphere.

TABLE 12.5 Global warming potentials (GWP) and atmospheric lifetimes (yrs) of select greenhouse gases. *(IPCC, 1996).*

Gas		Atmospheric Lifetime (yrs)	GWP (100 yrs)	GWP (20 yrs)
Carbon dioxide (CO_2)		50-200	1	1
Methane (CH_4)*		12±3	21	56
Nitrous oxide (N_2O)		120	310	280
Hydrofluorocarbons				
	HFC-23	264	11,700	9,100
	HFC-125	32.6	2,800	4,600
	HFC-134a	14.6	1,300	3,400
	HFC-143a	48.3	3,800	5,000
	HFC-152a	1.5	140	460
	HFC-227a	36.5	2,900	4,300
	HFC-232fa	209	6,300	5,100
HFC4310mee		17.1	1,300	3,000
CF_4		50,000	6,500	4,400
C_2F_6		10,000	9,200	6,200
C_4F_{10}		2,600	7,000	4,800
C_6F_{14}		3,200	7,400	5,000
Sluphur hexafluoride (SF_6)		3,200	23,900	16,300
* The methane GWP includes the direct effects and those indirect effects due to the production of tropospheric ozone and stratospheric water vapour				

The relationship between gigagrams (Gg) of a gas and TgCO₂Eq can be expressed as in equation 12.13 (UGGGIP 2002). Global Warming Potentials and Atmospheric Lifetimes of greenhouse gases are shown in Table 12.5.

$$TgCO_2\ Eq\ =\ (Gg\ of\ gas)\ x\ (GWP)\ \ x\ \ (Tg\ /\ 1,000Gg) \qquad [12.13]$$

where

$TgCO_2$ Eq = Telegrams of Carbon Dioxide Equivalents, Gg = Gigagrams (equivalent to a thousand metric tons), GWP = Global Warming Potential, Tg = Teragrams

12.11.2.2 *Consequences of Global Warming*

Global warming will likely increase evaporation which will in turn increase average global precipitation. Solid moisture is likely to decline in many regions as a consequence of global warming, causing severe draught and poor agricultural yield. Unusual weather patterns are becoming increasingly evident and land use patterns are changing. Rainstorms are becoming more frequent and intense and sea levels are rising along many coasts. Ice glaciers are melting, causing severe flooding. Tropical storms nucleate on the sea surface at about 26°C, hence, a rise in average global temperature may increase the frequency and intensity of storms and hurricanes. Increases in global temperatures are causing melting of arctic ice. In the 1980s, old ice (ice which has survived at least four hot seasons), was about 4 metres thick and covered nearly 2 million square kilometres. However, in 2016, the average thickness is only about one metre and covers only about 100,000 square kilometres (NASA, 2016). The older ice is thicker and more resistant to melt than new ice, hence it protects the new ice cap from melting during warm summers. The melting rate of Greenland's ice sheet has increased threefold in the last two years. Other changes attributable to global warming are disappearing mountain glaciers, earlier springs and hotter summers in temperate regions. While there is ample evidence to support these observed changes, it is not clear whether they can all be attributed to global warming.

Some of the climate changes triggered by global warming can create a *'feedback effect'* which can cause the warming trend to accelerate. For example, the melting of the arctic ice creates a magnifier effect by increasing global temperature since bright white ice normally reflects back much of the Sun's rays, thus helping to keep the planet cool. Melting exposes a darker sea surface which absorbs more heat which in turn causes more ice to melt, exposing more dark sea surface. This *'feedback loop'* has the potential to accelerate global warming and is also evident in some other climate change phenomena.

Long-term fluctuations in temperature, precipitation, wind and other elements of the Earth's climate system are commonly referred to as 'climate change'. However, greenhouse gases are not the only causes of climate change, other natural processes such as solar radiation variations, variations in the Earth's orbital parameters, and volcanic activity can also produce climate change.

12.11.3   Environmental pollution and Climate change

Climate is determined by a complex mix of natural processes, such as the rotation of the Earth around the Sun, changes in the equilibrium of energy transfer between the Sun and the Earth, the pattern of distribution of surface moisture by winds, ocean currents and other mechanisms. Short-term changes in climate are known as weather changes but climate change refers to changes in the statistical properties of the weather pattern over of long period of time, covering several decades. Climate change poses risks for human and natural systems, the later being the much more vulnerable. The report of the Intergovernmental Panel on Climate Change (IPCC, 2014) presents evidence of change in climate in recent decades, and the impact on natural and human systems on all continents and across the oceans, the strongest and most comprehensive impact being on natural systems. The changes observed in the report are summarized below.

12.11.3.1   *Changes in weather pattern*

Changes have been observed in weather patterns, geophysical systems such as floods, draughts, desertification, sea level rise, wild fires, extreme weather, geographical ranges. Negative effects on seasonal activities and migration patterns, crop yields across regions have also been observed. What is not quite clear (and is being widely disputed) is the extent to which the observed changes can be attributed to human anthropogenic activities, since natural climate change can cause significant shifts in the ecosystem and species extinction.

12.11.3.2   *Acid rain*

Acid deposition is a general term that includes more than simply acid rain. Acid deposition primarily results from the transformation of sulphur dioxide (SO_2) and nitrogen oxides into dry or moist secondary pollutants such as sulphuric acid (H_2SO_4), ammonium nitrate (NH_4NO_3) and nitric acid (HNO_3). The transformation of SO_2 and NOx to acidic particles and vapours occurs as these pollutants are transported in the atmosphere over distances of hundreds to thousands of kilometres.

Acidic particles and vapours are deposited via two processes - wet and dry deposition. Wet deposition is *acid rain*, the process by which acids with a pH normally below 5.6 are removed from the atmosphere in rain, snow, sleet or hail. Dry deposition takes place when particles such as fly ash, sulphates, nitrates, and gases (such as SO_2 and NOx), are deposited on, or absorbed onto, surfaces. The gases can then be converted into acids when they contact water.

Sulphur dioxide emissions come from a variety of sources including coal-fired power generation, transportation, industrial processes, fossil fuel combustion, ore smelting and natural gas processing. The main source of NOx emissions is the combustion of fuels in motor vehicles, residential and commercial furnaces, industrial and electrical utility boilers, engines, and other equipment. Acid rain is a problem in many parts of the world because many of the water and soil systems lack natural alkalinity such as a lime base and therefore cannot neutralize acid naturally. Increased acidity of rivers, lakes and streams destroys aquatic life and acidified soils harm crops and forests. Acid rain also causes corrosion of road and building metal fixtures and automotive metal components.

12.11.3.3   *Effect on human health and safety*

The impact of environmental pollution and climate change on the human system is not very clear. While increases have been observed in human ill-health, heat-related mortality, distribution of water-borne illnesses, much of it could have been the result of other stressors some of which are still not well understood. It is accepted however that climate-related hazards can exacerbate the impact of other stressors. Mining operations, in particular, underground mining are hazardous and major mine disasters occur frequently. Hazards include roof falls, rock bursts, fires and explosions due to accidental ignition of flammable gases such as methane which are trapped in the coal beds and are released in mining operations. Other health hazards include exposure to methane, carbon monoxide, and mercury all of which are toxic.

Particulate matter, mainly soot and fly ash can contribute to long-term respiratory problems, lung cancer, chronic obstructive pulmonary disease, and heart disease. Inhalation of these extremely small particles over a long period of time reduces the functional capacity of lungs, causing severe breathing and respiratory problems known as pneumoconiosis (black lung). Chronic asthma or emphysema can also result. The particles can also carry along small amounts of toxic trace elements including mercury, and potentially carcinogenic organic molecules. The World health Organization estimates that more than one-in-every-nine premature deaths worldwide result from pollution (both household and outdoor).

About one million children under five years die yearly from pneumonia, with more than half being caused by exposure to household air pollution (WHO, 2016 a&b).

Ozone, nitrogen dioxide and sulphur dioxide are linked to asthma, bronchial disease, reduced lung function and lung disease. Sulphur trioxide causes irritation of the mucous membranes of the respiratory tract, causing coughing and choking. Inhalation of droplets of acid rain formed by the reaction between sulphur trioxide and atmospheric moisture can also harm the respiratory system. High levels of carbon monoxide are also believed to cause problems in the nervous system. Sulphur and nitrogen oxides in gaseous products of coal combustion often react with moisture, soot and fly ash particles to form a dense smog which not only obscures sunlight but can cause eye irritation and breathing problems. Long-term exposure can cause chronic pulmonary diseases such as asthma, bronchitis, lung cancer, and cardiac problems, including arrhythmias and heart attacks (USEPA, 2002).

Depletion of the Earth's protective heat shield formed by the stratospheric ozone layer enables potentially dangerous ultraviolet radiation to reach the Earth's surface, causing skin burns and skin cancer. Incidences of fluorosis, arsenism, selenium intoxication, pneumoconiosis and silicosis are high among rural communities that live near coal mines or burn coals with high fluorine, arsenic, selenuim, iron and silicon contents in poorly ventilated environment (Zheng et. al., 1999; Huang et al. 2005). Global warming and consequent climate change may have wide ranging and mostly negative direct and indirect impacts on human health. Increasing heat waves can cause heat strokes and deaths. Increasing temperatures can enhance increased production and migration of some disease-carrying vectors such as mosquitoes which spread malaria and zika virus. It is difficult to quantify the effects of human anthropogenic activities on human health because most diseases are multi-factorial and could be caused by other risk factors. However, there is fairly strong scientific evidence linking respiratory diseases and exposure to anthropogenic pollution. Even in such cases, the contribution of other factors that may cause or aggravate respiratory diseases is unclear.

12.12   BETWEEN GLOBAL ENERGY SECURITY AND CLIMATE CHANGE

Energy is an indispensable requirement for many human activities and there is a symbiotic relationship with human development. Energy availability is a prime mover of human development. On the other hand, human development promotes increased demand for energy. This mutual dependence has been a major impediment to global efforts to mitigate anthropogenic activities. The key challenge is how to decarbonize electricity production while still providing a high standard of living (zero-carbon, zero-poverty). The primary concerns of most countries is to meet the energy needs of its citizens, hence global concerns on climate change are not considered a priority.

Fossil fuels still supply the bulk of the global primary energy needs and this is unlikely to change very much in the foreseeable future. It is inevitable therefore that most countries will rely on local resources which in many cases are dominated by coal. Coal is available in many countries, in contrast with oil and gas, and will be the primary source of energy, in particular, for power generation. While the bulk of the increase in coal utilization for power generation will be in emerging countries, many developed countries also still rely heavily on coal. Coal still accounts for about 40% of power generation in the USA (USEIA, 2015), although the proportion is much higher, up to 60% in some states. Some other countries depend on coal for over 90% of primary power supply. The prime role that coal and other fossil fuels continue to play in the primary energy mix of many countries accounts for the ineffectiveness of global agreements on climate change.

All available projections indicate that global primary energy demand will continue to grow steadily and fossil fuels will remain dominant in the foreseeable future. The most recent projection by the International Energy Agency (IEA, 2016) estimates that demand in 2040 will be 48% higher than the level in 2012 and most of the energy demand growth will be in emerging non-OECD countries. The demand in OECD countries will increase by about 28% compared with 57% in the non-OECD countries. Fossil fuels will still account for 78% of primary energy use. Although coal share of global primary energy use will decline from 28% in 2010 to 22% in 2040, use for electric power generation which was nearly 40% in 2012 will remain virtually unchanged in 2040, and industrial use will grow by about 2% to 38%. Three countries (United States, China and India) will account for more than 70% of world coal use in 2040.

The implication of the above projections is that anthropogenic emissions which result primarily from the use of fossil fuels will continue to increase over the coming years and will be 34% higher in 2040 compared with the level in 2012. Also, there will be significant growth disparity between developed and emerging countries. While the projected growth in OECD countries in only 8% , emission levels in non-OECD countries will grow by 51%. Many of these countries simply do not have the competence or funding to take the actions required to mitigate environmental pollution.

In 1992, the United Nations convened an Earth Summit meeting in Rio de Janeiro, the outcome being an International Environmental Treaty - United Framework Convention on Climate Change (UNFCCC). The thrust of the Treaty was the stabilization of greenhouse gas concentrations in the atmosphere at a level that would prevent dangerous anthropogenic interference with the climate system (UNFCCC, 1992). The parties to UNFCCC have met annually since 1995 as Conference of the Parties (COP) to assess progress in dealing with climate change. The meeting in Kyoto in 1997 developed a Kyoto Protocol which established legally binding obligations for developed countries to reduce their greenhouse gas emissions. Six greenhouse gases were identified which, if reduced could reduce global warming significantly. From 2005 the conferences have also served as the Meetings of Parties of the Kyoto Protocol (CMP).

The Kyoto Treaty was not ratified by many countries. The USA, the world's second largest emitter did not ratify the Treaty because it would cause serious harm to the US economy, and also because it did not cover developing countries (80% of the world) including China, the leading emitter in the world. Some countries which ratified the Treaty initially, for example, Canada, also withdrew in 2012 because the country was unable to achieve the 6% reduction in emission from the 1990 level. Rather, emission was 17% higher in 2012. Nevertheless, global emissions in 2012 were nearly 23% lower than 1990 levels compared with the set target of 5%, probably not because of the Treaty but because many developed countries including the United States already had policies in place to decarbonize, largely by substituting gas for coal. Also, many developed countries adopted the 1987 Montreal Protocol to eliminate ozone-depleting gases. For example, use of fluorocarbons in refrigeration and air conditioning has been widely discontinued.

The United Nations Conference on Climate Change comprising Conference of Parties (COP) and the Meeting of Parties to the Kyoto Protocol (CMP) was held in Paris in November 2015. The meeting was attended by 197 parties and the Paris Agreement, a global agreement on the reduction anthropogenic emissions was adopted. Signatories were obligated to establish National Greenhouse Gas Inventories and develop strategies for control and removal by 2020. In spite of the concerted effort of the United Nations to stabilize and control greenhouse gas concentrations in the atmosphere, positive results have been minimal, due mainly to the fact the agreements contained no enforcement mechanisms. For example, the Paris Protocol of 2015 set a goal of limiting the increase of the world's average temperature to no more than 1.5-2 degrees centigrade above pre-industrial levels. While this landmark accord was signed by all 196 countries that adopted the original United Nations Framework Convention on Climate Change (UNFCCC), it was non-binding. Also, the recent withdrawal of the United State of America from the protocol is a major setback. However, all other major countries and many states within the U.S.A. have expressed the determination to move forward with the implementation of the agreement.

The major problem with the international protocols and treaties on climate change has been the lack of legal instruments for verification and enforcement. The 2015 conference agreed on a goal of achieving zero net anthropogenic greenhouse gas emissions by the second half of the 21st century. The Paris treaty also introduced a legal framework to compel signatories to the Agreement to adopt and domesticate the Treaty within their own legal systems. However, many sections are promises, aims, goals and indefinite time frames which are not enforceable. For example, each country that ratifies the Agreement will be required to set a voluntary target for emission reduction. However there is no target date, nor a mechanism for enforcement. Furthermore, the Agreement becomes legally binding only after it has been ratified by 55 countries that generate 55% of the global greenhouse gases. It is doubtful if enough countries in this group (including developed countries) will do so. Furthermore, as discussed earlier, the priority of most developing countries is to provide primary energy to their growing population and most lack the wherewithal to deal effectively with pollution control. However, in order to help mitigate this problem, the Paris protocol established a revolving fund to help developing countries adopt non-greenhouse gas technologies.

One major clause in the Paris Agreement calls for zero net anthropogenic greenhouse gas emissions to be achieved globally during the second half of the 21st century. Zero net carbon means carbon neutrality and requires balancing the carbon released into the atmosphere with an equivalent amount sequestered or offset with the use of zero carbon energy technologies. Any shortfall could be compensated for by buying carbon credits. However, as discussed earlier, the most recent projection by the International Energy Agency shows that fossil fuel will fill meet 78% of the global primary energy needs in 2040. Furthermore, anthropogenic emissions will increase by 34% above the 2012 level.

Carbon sequestration technologies are still in the early stages of development and proliferation, there is no zero carbon energy technology when assessed on life cycle basis, and carbon trading is a very complex accounting process. It seems therefore that the target set by the Paris Agreement is unrealistic. Although renewable energy is projected to grow by around 2.6% per year over the period (the fastest-growing energy source), the contribution to global primary energy use in 2040 will still be less than 20%. In effect, fossil fuels will remain the major energy sources well into the second half of the 21st century. Another problem is the fact that some countries (Australia, India, South Africa, China) depend on coal for 60-90% of primary energy supply. It is unlikely that such countries can make a significant departure from fossil fuels in the foreseeable future.

REFERENCES

EEA (2008). "Air pollution from electricity-generating large combustion plants." *European Environmental Agency*, Copenhagen.

ExxonMobil (2009). "An alternative for Liquid Fuel Production." ExxonMobil Research and Engineering (EMRE). *Gasification Technology Conference,* October, 2009.

Howard, B. C. (2016). "2015 was the hottest year on record." www.nationalgeographic.com (Accessed September, 2016).

Huang, Xi, et al., (2005). "Mapping and prediction of coal workers' pneumoconiosis with bioavailable iron content in the bituminous coals." *Environ. Health Perspect.,* Vol. 113 (8), pp. 964-968.

IEA (2012). CO_2 *Emissions Highlights.* International Energy Agency. www.iea.org.

IEA (2015*). Key Energy Statistics.* International Energy Agency. www.iea.org.

IEA (2016*). International Energy Outlook, 2016.* International Energy Agency. www.iea.org.

IPCC (1996). *Climate Change 1995: The Science of Climate Change. Intergovernmental Panel of Climate Change.* J. T. Houghton et al., Eds; Cambridge University Press, Cambridge, U.K.

IPCC (2001). *Climate Change 2001: The Scientific Basis.* Intergovernmental Panel on Climate Change Cambridge University Press, Cambridge, U.K.

IPCC (2007). *Climate Change 2006: The Science of Climate Change. Intergovernmental Panel of Climate Change.* J. T. Houghton et al, Eds; Cambridge University Press, Cambridge, U.K.

IPCC (2007). *IPCC Special Report on Emissions Scenarios. Intergovernmental Panel of Climate Change.*

IPCC (2014). *Intergovernmental Panel of Climate Change, Fifth Assessment Report: Summary for Policymakers.*

Jackson, R. (2016). " Half a degree C may not sound like much, but it makes a world of difference in climate impacts". Nationalgeographic.com. Accessed September, 2016.

Nalbandian-Sugden, H. (2015). *Climate implications of coal-to-gas substitution in power generation.* International Energy Agency (IEA) Clean Coal Centre, London.

NASA (2016) "Arctic Sea Ice is Losing Its Bulwark." http://earthobservatory.nasa.gov/IOTD/view.php?id=89038.

Portman, R. W., J. S. Daniel, A. R. Ravishankara (2012). "Stratospheric ozone depletion due to nitrous oxide: influences of other gases." *Philosophical Transactions of the Royal Society B.*

Saikat Manna, (2014). www.studies-in-botany.blogspot.com.

UNFCCC (1992). United Nations Framework Convention on Climate Change. https://unfccc.int/resource/docs/convkp/conveng.pdf

USEIA (2015). *International Energy Outlook 2014.* www.useia.gov.

USEPA, (2002). *Latest findings on National Air Quality*, September, 2002. United States Environmental Protection Agency.

USGGIP (2002). *Greenhouse Gases and Global Warming Potential Values.* U.S. Greenhouse Gas Inventory Program U.S. Environmental Protection Agency.

USGAO (2012). *Air Emissions and Electricity Generation at U.S. Power Plants.* United States Government Accountability Office, www.gao.gov.

USGS (2017). "Coal-Be Methane: Potential an Concerns." United States Geological Survey, MS 939. usgs.gov. Accessed February 2, 2017.

WHO (2016a). *World Health Statistics, 2016.* World Health Organization, Geneva.

WHO (2016b). *Air Pollution: A Global Assessment of Exposure and Burden of Disease.* World Health Organization, Geneva.

Zheng, B. Et al., (1999). "Issues of health disease relating to coal use in southwestern China." *International Journal of Coal Geology,* Vol. 40(2), pp. 119-132.

Fisk coal-fired Generating Station, Chicago, USA
(Chicagoreader.com)

Ultrasupercritical coal-fired power plant ,Isogo, Japan
(powerengineeringint.com)

13 Cleaner coal technologies

13.1 INTRODUCTION

It is now widely agreed that fossil fuels will remain the main global primary energy resources in the foreseeable future and coal, the most anthropogenic, will continue to play a prominent role, particularly in the emerging economies. Coal use will continue to dominate the world's primary energy mix, particularly in electricity generation as it has done for decades. Since 2010, the growth in power generation from coal has been greater than that of all non-fossil sources of power generation combined (IEA, 2014). The recent trend of switching to natural gas for power generation to reduce anthropogenic emission from coal appears to have stalled due to increasing costs of producing and transporting gas. Global nuclear capacity is stagnating, and, although renewable energy, especially hydropower is becoming increasingly cost competitive, global investments have slowed down due apparently to policy uncertainties and changes. Also, current depression in the global petroleum market has slowed down interest in renewable energy. A projection by the International Energy Agency (IEA, 2012) shows that fossil fuels will still account for about 75% of the total primary energy supply in 2035 (Figure 13.1).

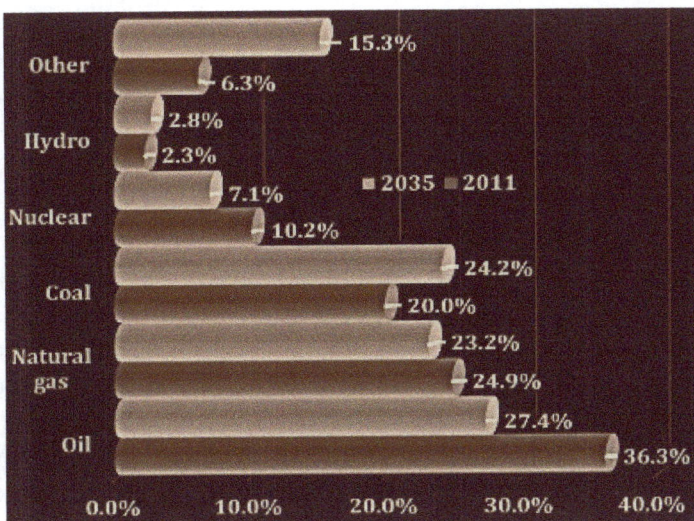

FIGURE 13.1 Projected global total primary energy supply in 2035. *(IEA, 2012).*

It seems inevitable that the carbon intensity of global primary energy use will continue to rise, with undesirable consequences for the environment. Coal is still the dirtiest of all primary energy sources in spite of recent major advances in the development of coal cleanup technologies. The common term "clean coal" used for many new pollution reduction technologies is therefore a misnomer. There is little doubt that major developments in coal mining, cleaning, transportation and utilization technologies in recent years have resulted in a significant reduction (not elimination) of pollution emissions. A more appropriate term therefore would be 'cleaner coal technologies'. The world is beginning to come to terms with the fact that coal will remain dominant in the global primary energy mix for many years and current efforts are focused on how to mitigate the negative impact of coal utilization. The fact that fossil fuels, particularly coal have continued to dominate the global primary energy mix has been the main reason for the slow pace of environmental mitigation through reduction in pollutants from energy use, in spite of positive and substantial developments in cleaner energy technologies.

Progress towards clean coal technologies has been slow. In 2011, the IEA Energy Sector Carbon Intensity Index (ESCII) was 55.9 tCO_2/TJ (2.34 tCO_2/toe), compared with 57.1 tCO_2/TJ (2.39 tCO_2/toe) in 1990. Another factor deterring progress is the fact that new generating plants are not taking advantage of best available technologies (BATs). About half of coal-fired plants built in 2011 worldwide use inefficient sub-critical technologies. However, increasingly stringent environmental regulations, particularly in the industrialized countries have provided the impetus for research and development of advanced cleaner coal technologies that can be deployed in existing technologies to increase efficiency and reduce the pollution intensities of coal utilization processes, in particular, existing as well as future generating plants. New technologies range from those that seek to improve efficiencies of utilization, to capture of anthropogenic combustion by-products, combined fuel processes, coal gasification and liquefaction technologies. The International Energy Agency (IEA, 2008) has proposed four technology options for decarbonizing fossil fuel utilization and achieving near-zero emissions from coal powered processes (Figure 13.2) and global research efforts focused on various aspects of the plan are beginning to yield positive results.

The three major areas that have received the greatest attention are, improvement of efficiencies of existing and new coal-fired power plants from production to end-use, capturing and processing the potential pollutants and development of advanced cleaner coal technologies which deliver higher efficiencies and reduce pollutants to near-zero levels. Efficient coal use is currently the primary means of reducing coal's anthropogenic impact on the environment, and the most pragmatic in the short term. Longer-term focus is on the development of innovative advanced technologies which not only de-emphasize the over-reliance on coal but also feature effective methods of capturing and managing combustion products emitted from coal-fired power plants.

Most existing coal-fired power plants use the sub-critical pulverized coal technology which has been in use for nearly a century. Pulverized coal carried in pressurized air is introduced into coal burners. The pulverization causes the coal to burn more efficiently and the combustion process can be better controlled. Higher boiler steam temperatures and pressures are achievable and boiler sizes have increased as a result of moving from lump coal to pulverized coal technology. The average life span of a coal-fired power plant is 40 years to 60 years, hence many plants in operation currently are old and inefficient. The average operating efficiency is 25% to 35% and many of the plants have no facilities for capturing and

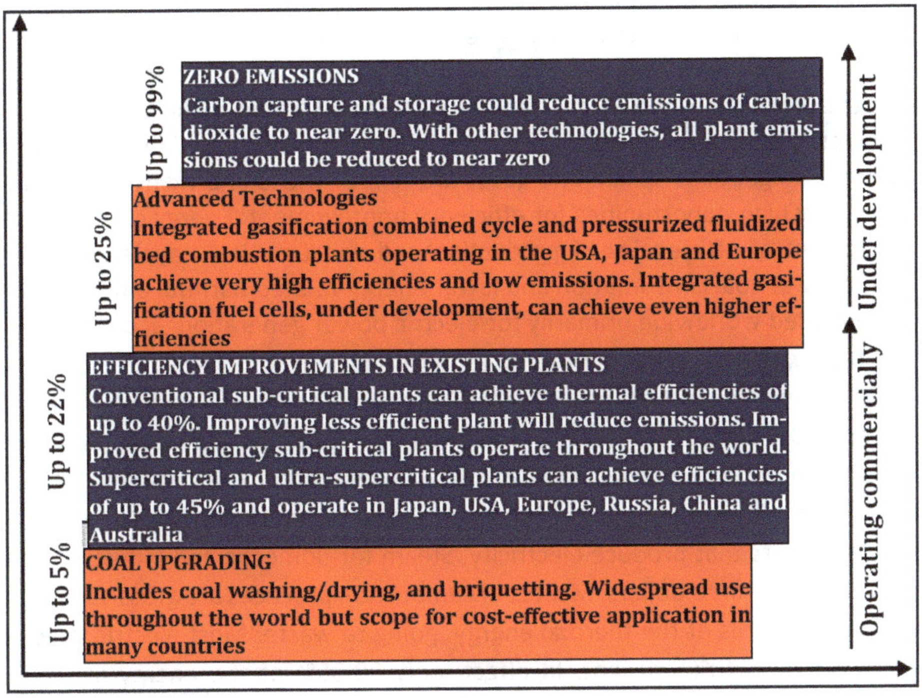
ZERO EMISSIONS
Carbon capture and storage could reduce emissions of carbon dioxide to near zero. With other technologies, all plant emissions could be reduced to near zero
Up to 99%
Advanced Technologies
Integrated gasification combined cycle and pressurized fluidized bed combustion plants operating in the USA, Japan and Europe achieve very high efficiencies and low emissions. Integrated gasification fuel cells, under development, can achieve even higher efficiencies
Up to 25%
EFFICIENCY IMPROVEMENTS IN EXISTING PLANTS
Conventional sub-critical plants can achieve thermal efficiencies of up to 40%. Improving less efficient plant will reduce emissions. Improved efficiency sub-critical plants operate throughout the world. Supercritical and ultra-supercritical plants can achieve efficiencies of up to 45% and operate in Japan, USA, Europe, Russia, China and Australia
Up to 22%
COAL UPGRADING
Includes coal washing/drying, and briquetting. Widespread use throughout the world but scope for cost-effective application in many countries
Up to 5%
Under development
Operating commercially

FIGURE 13.2 Reduction in emissions of CO_2 through clean coal technological Innovation. *(IEA, 2008).*

processing the flue gases and particles which are causes of environmental degradation. Although new, advanced and more efficient technologies have been available for the last decade, around 50% of new power plants coming on stream worldwide still use the outdated sub-critical pulverized coal technology. State and regional policy instruments are now emerging to promote the adoption of emerging cleaner coal technologies for power generation. New technologies which are already featured in many coal-fired generating plants, notably in North America, Europe and China include carbon capture and storage, supercritical pulverized fuel technologies (SCPF), pressurized fluidized bed combustion (PFBC), and Integrated gasification combined cycle technologies (IGCC).

13.2 POLLUTION REDUCTION THROUGH IMPROVED EFFICIENCY OF COAL UTILIZATION

Improved energy efficiency over the whole value chain is now regarded as an energy resource and the easiest and least expensive way of conserving primary energy resources. Projections indicate that up to 40% of future global primary energy requirements could come from improved efficiencies in production and utilization (IEA, 2008; IEA, 2011; WEC, 2013; IEA, 2014). It is estimated also that energy efficiency represents about 40% of the reduction potential of anthropogenic greenhouse gases (McKinsey, 2010). All sectors of the primary energy value chain have efficiency improvement potential: coal, oil and gas exploration, production and transportation, power generation, energy transmission and distribution, domestic and commercial utilization. For example, electric motors, lighting and home appliances account for approximately 70% of the total electricity consumption in the industrialized nations but efficiency is between 30 and 50%. Estimates vary widely but it is believed that between 30% and 50% of primary energy produced worldwide is wasted through inefficient production processes, transmission and utilization. The recovery efficiencies of coal mines, oil and gas deposits are currently low, limited by available technologies. At the production stage, oil is spilled and gases are flared. These and many other energy sectors present very high opportunities for efficiency improvement. Energy conserved through the improvement of efficiencies of production and consumption translates to lower energy demand, lower coal consumption and lower energy-related harmful emissions.

Coal is used worldwide primarily for electric power generation, accounting for over 40% of the total world capacity, more than twice the proportion of any other fuel. The share of coal for power generation has increased steadily for many years, with a 52% increase between 2000 and 2011. A recent projection by IEA (2016) estimates that coal share will decline to about 30% in 2040, although power generation from coal will increase by 23%. Coal-based electric power plants burn coal to raise steam that drives turbines which in turn drive alternators that produce electricity. Steam for domestic and industrial heating is also produced in coal-fired boilers. Most coal combustion practices are grossly inefficient, with as much as two-thirds of the thermal energy going to waste. Most electric generation plants are only 30% to 40% efficient. In effect, only 30-35% of the energy in coal ends up as electricity at the pre-transmission end of the plant and around 10% of primary electricity generated is lost through transmission and delivery. Petrol and diesel engine efficiencies are between 30% and 40%, and numerous human activities waste power.

13.2.1 Improved efficiency of coal mining, preparation and transportation

Currently available technologies limit the amount of coal recovered from underground mines to as low as 30%. More advanced mining technologies are being developed and deployed and previously abandoned mines are being reopened. Energy efficiency in coal, oil and gas exploration is around 20%. New technologies which reduce the energy intensity, increase efficiency of coal extraction ultimately to around 50% and minimize environmental pollution are being developed and deployed. Coal-bed methane, a potent pollutant which is currently vented into the atmosphere in many coal mines is a potentially valuable energy resource, especially for power generation.

Coal is washed to reduce ash in raw coal to facilitate combustion and increase the energy content per tonne of fuel. Sulphur content which is undesirable in coal combustion may also be reduced. There is a high potential for improvement of efficiencies of the unit cleaning processes, from grinding to cleaning, and the recovery and utilization of coal fines. Cleaned coal is transported to consumers by road, rail or ship. It is also commonly transported through slurry pipelines. Many power plants now use coal slurry feed to reduce coal dust. Coal stored for extensive periods during transportation or in stockyards tends to lose heating value, which translates to more coal requirement per unit of energy produced. There is considerable scope for improvement of transportation and storage conditions to minimize weathering. Increased pipeline transportation will not only cut costs but reduce coal weathering. Washed coal can also be dried and processed into smokeless briquettes using established commercial technologies which have been in use in many developed countries for decades. It is estimated that adoption of this process could reduce CO_2 emissions by as much as 5% (IEA, 2008). Briquetting of coal and processing it into smokeless fuel will also make low carbon footprint fuel available for a wider range of applications, including domestic heating and cooking, especially in developing countries. Furthermore, ultrafine coal which would have been discarded can be compacted into valuable smokeless fuel.

13.2.2   Efficient coal-fired power generation

Electric power accounts for about 40% of the total global final energy use and coal-fired plants produce over 40% of the total world electric power. More than half of the fuel energy in a coal-fired power generating plant is lost as waste heat (Figure 13.3). The more inefficient the coal combustion process, the more coal has to be fed for the same power output and the higher the level of carbon dioxide emission. Efficiencies of most power generating plants worldwide are low (30-35%). In effect, only about 35% of the potential energy of the coal fed to combustors actually gets converted to electricity. Raising the thermal efficiency to 50% would emit 40% less carbon dioxide. Production efficiencies are being improved through the design of more efficient, low nitrogen burners, deployment of efficient pulverized coal combustion systems and fluidized bed combustors that allow coal-fueled electricity generating plants to produce more power with less coal and fewer emissions.

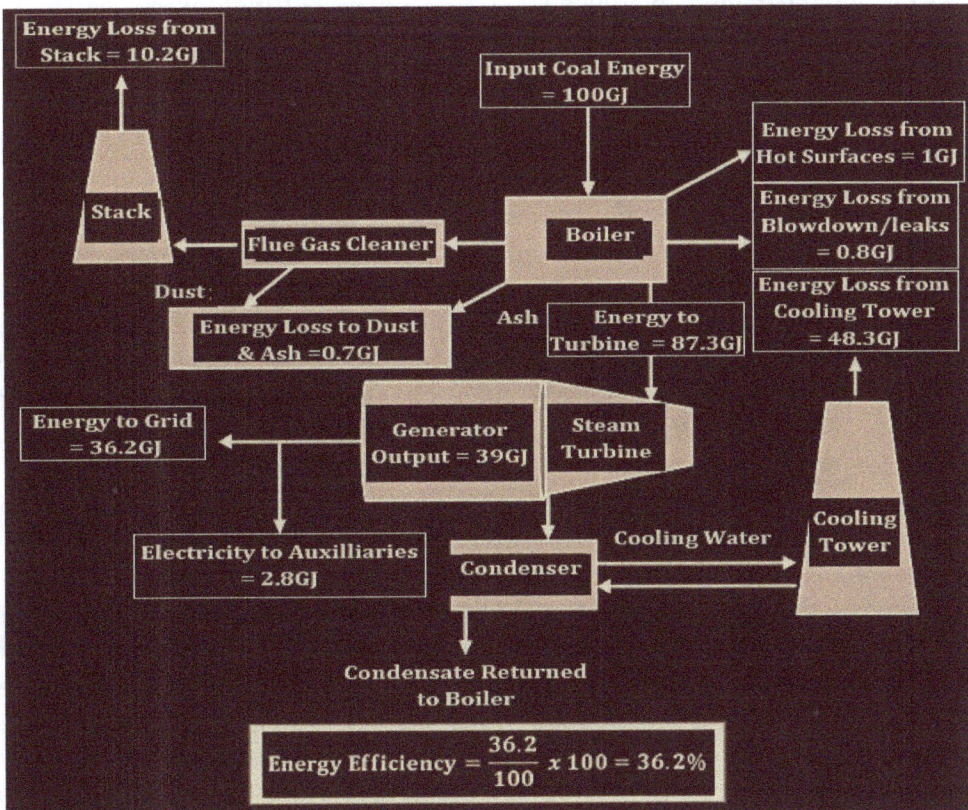

FIGURE 13.3 Energy balance of a typical conventional coal-fired power plant.

Existing plants are being retrofitted and new installations are deploying burners with preci-sion controlled air-fuel ratios for high efficiency combustion. Low nitrogen oxide burners which are capable of reducing NOx emissions by as much as 70% have also been developed and are being deployed in coal-fired power generating plant retrofits.

Residual energy is recycled to improve overall thermal efficiency of power plants. The design of turbine blades has greatly improved in the last decade or so through computer-aided design and optimization of blade profiles, hence efficiencies of both gas and steam turbines generating plants have been rising. Furthermore various energy systems are being integrated to take advantage of their individual merits. Newer plants and many retrofitted installations now operate at around 40% efficiency and adoption of pre-gasification of coal could raise efficiency to around 60%. Even at this level of efficiency there is still a tremendous amount of energy lost in the generation process which could be reduced through the adoption of more advanced combustion optimization technologies and control systems. The economic and ecological benefits of recovering and utilizing more of the energy in the input coal are also enormous.

Efficiency improvement translates to lower coal consumption per kilowatt-hour of generated power and lower emission. One percent increase in efficiency of conventional pulverized coal-fired plants results in 2 to 3% reduction in CO_2 emissions. Highly efficient modern coal plants emit up to 40% less CO_2 than the older plants. The temperature and pressure of the steam are critical factors that determine efficiency, and availability of materials capable of withstanding such extreme conditions is a major constraint limiting current highest operating temperatures to around 600°C to 620°C at 250 atmospheres. Nickel superalloys currently under development may allow temperatures as high as 700°C, with consequent increases in efficiency above 50%.

13.2.3   Cogeneration power plants and Energyplexes

Another innovation which has greatly improved the thermal efficiencies of coal-fired power plants is the development of combined heat and power plants, also known as cogeneration power plants. A cogeneration system is a single, integrated system that produces multiple forms of energy from the same fuel feed. The different forms of energy may be produced sequentially or simultaneously. Some systems produce electrical power and heat while others produce chilled water as well. Typically there is a combustion unit which is capable of burning a variety of fuels, including natural gas, coal, oil, biomass fuels and refinery waste gases.

Steam or hot gases are produced and fed to a prime mover (steam or gas turbine, reciprocating engine, fuel cell) which turns the heat into mechanical energy that can be used to drive an electrical generator or any other load including compressors, pumps and fans. A cogeneration system is designed to recover and utilize as much of the 60% or more of the coal energy that is wasted in coal and other fossil fuel-fired plants. The recovered heat is either used to generate more electricity in a secondary unit or supplied for use in industrial and domestic heating. A typical cogeneration system can have a theoretical thermal efficiency as high as 90% and requires about 25% less primary fuel energy compared with separate power and heat systems which have the same electric power and heat energy output (Figure 13.4). Furthermore, a cogeneration plant is considerably cheaper and emits significantly less pollutants per unit of power and heat energy output compared with electric power and heat supplied from separate power plant and boiler installations. There are many different configurations for combined cycle gas turbine (CCGT) power plants which are perhaps the most advanced and most widely deployed cogeneration systems. A typical combined cycle gas turbine plant would feature a steam turbine and a gas turbine in parallel, each driving an electric generator as shown in Figure 13.5a. In some configurations both the gas and steam turbines are mounted in tandem on a single shaft. Compressed air is heated by combustion of the injected fossil, biofuel, biomass or a combination of fuels. The hot gases are passed into the gas turbine through an expander, turning the rotor which drives the compressor and the generator. Exhaust gases leaving the gas turbine are typically at a temperature of 550 to 600°C and are passed through a heat exchanger (boiler) to recover the heat and raise steam which drives the steam turbine.

Both generators in a CCGT feed into a common grid. Simple cycle turbine power plants average around 30% efficiency or less running on gas and around 25% on oil. Much of the wasted energy ends up as thermal energy in the hot exhaust gases. Because efficiencies of turbines are so low, the electrical power output of the unit accounts for about half of the CCGT plant output. In a combined heat and power plant the waste heat from the turbine is recovered and used to generate more electrical power. Some of the heat may also be used for industrial, commercial or domestic heating. The most efficient conventional coal-fired plants currently in operation have efficiencies in the range 40-45%. This is only half of efficiencies currently achievable with cogeneration plants.

Existing power plants have been reconfigured as cogeneration plants in many countries and are operating at over 70% efficiencies. Cogeneration plants are usually in close proximity to the fuel source and the energy consumer, hence energy losses from long distance transmission systems are eliminated. A typical plant located near a community such as college campuses not only supplies electric power but also meets the heating and cooling require-ments of the community (Figure 13.5b).

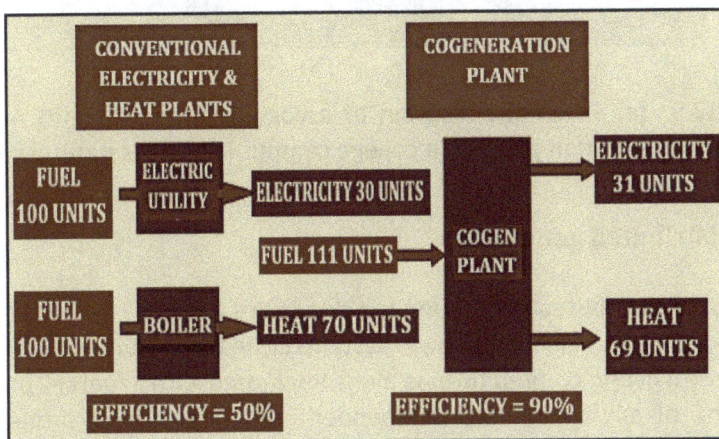

FIGURE 13.4 Comparison between separate and combined heat and power systems.

Gas Turbine with Heat Recovery

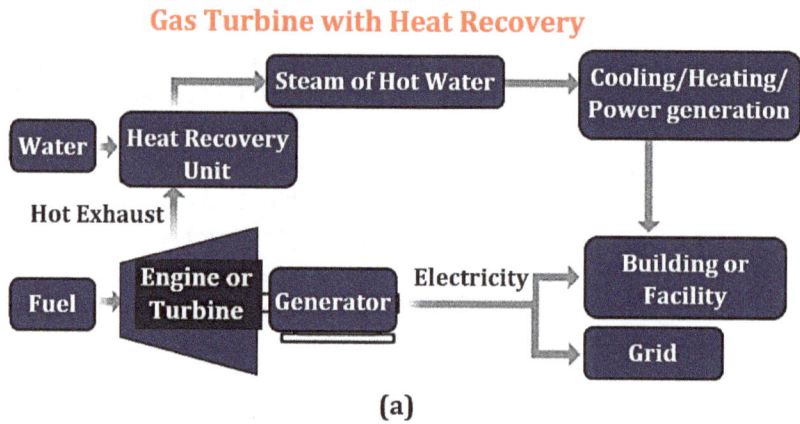

(a)

(b)

FIGURE 13.5 (a) Schematic diagram of a cogeneration plant (b) A clean, coal-fired multi-fuel cogeneration plant on a college campus (also uses natural gas and biofuel).

13.2.4 Distributed generation

The concept of distributed generation is a departure from the grid system and involves the strategic location of small hybrid power systems close to consumer clusters. A typical system may comprise a hybrid co-fired turbine plant integrated with a fuel cell power plant or a solar power plant, or any other nature-dependent and intermittent-prone renewable power systems, feeding consumer demand centers. A typical demand center may comprise a cluster of industrial plants, commercial and domestic consumers utilizing the electric power produced as well as the waste heat from the system. Potential advantages include much higher reliability of energy supply, higher energy utilization efficiencies, lower costs, and uninterruptible power supply for critical units. The need for expensive long distance transmission and the associated, substantial transmission power losses is also eliminated.

13.2.5 Efficient power transmission and distribution

Generated electricity is fed through a transmission and distribution system (T&D) comprising high voltage step-up transformers, high voltage transmission lines, substations and consumer feeder lines. A significant amount of energy, between 10 and 12% is lost in the transformers, distribution lines, and sub-stations. The extent of losses varies depending on the design of the T&D system, the transformers and other equipment, and the operation management of the system. Another source of losses arises from mismatch between electricity supply and demand.

Electricity that is generated but not utilized instantly is wasted energy hence many T&D systems have complex control systems which can tap power from different generating plants and deliver to a network of consumers with differing requirements in terms of power and timing. Also, providers offer attractive incentives to encourage more utilization of energy at low demand periods (early mornings, late evenings and weekends). Areas of efficiency improvement in power transmission include more efficient transformers, and the replacement of copper cables used currently in transformer windings and transmission lines with high-temperature superconducting cables (HTS) which are currently under development. Losses in HTS are only about 0.5% compared with 5-8% in copper cables. Automation in the electric power generation and distribution has been slow and much of the work currently being done manually, like reading meters, tracing broken equipment, synchronizing and distributing power from various generating plants can be automated to improve efficiency.

A new technology known as smart grid technology (SGT) which helps to eliminate or minimize transmission losses is being developed and deployed in some countries. Using computer-based remote control, two-way digital communications technology and automation, it is becoming increasingly possible to remotely monitor and control the whole electricity distribution system, from networks of generation sources, through transmission systems, to the consumer, thus greatly improving efficiency and reliability. Consumer meters can now be read remotely and consumers can control their energy use, like using more energy in the night when it is cheaper or free. Another area of potential efficiency improvement is wider adoption of the High voltage direct current (HVDC) in preference to the widely used high voltage alternating current (HVAC) system for high tension power grids. HVDC systems have 25% lower energy loss and two to five times the capacity of HVAC systems at similar voltage. Hitherto, the use of HVDC systems has been restricted to high voltage transmissions over very long distances because it is relatively very expensive. However, recent developments indicate that DC transmission can now be deployed economically over much shorter distances (ABB, 2014).

13.3 POLLUTION REDUCTION THROUGH IMPROVED END-USE EFFICIENCY

Over half of the global final energy consumption in 2011 was by industry (Figure 13.6). For over a decade, there has been an energy demand growth shift from the developed countries to the emerging regions (Figure 13.7), reflecting a major shift in industrial manufacture in the same direction. Another reason for the drop in industrial energy use by the OECD countries is the relatively high infusion of high-technology, highly efficient manufacturing processes which require less energy use per unit output compared with the developing countries. All sectors of utilization offer a high potential for improved efficiencies and significant reduction in anthropogenic emissions (Table 13.1).

13.3.1   Energy-efficient electric motors

About 40% of the global electric power generation and 65% of industrial power is used to run electric motors in a very wide variety of applications ranging from driving industrial equipment to domestic fans, microwaves and washing machines (Figure 13.8).

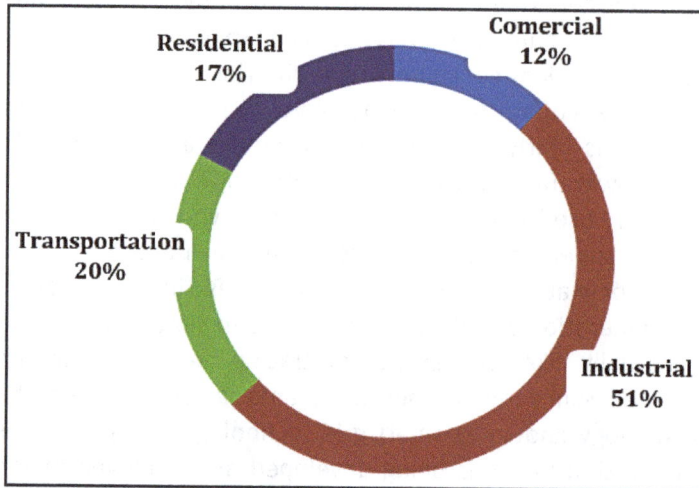

FIGURE 13.6   World final energy use in 2011 by sector. *(EIA,2014).*

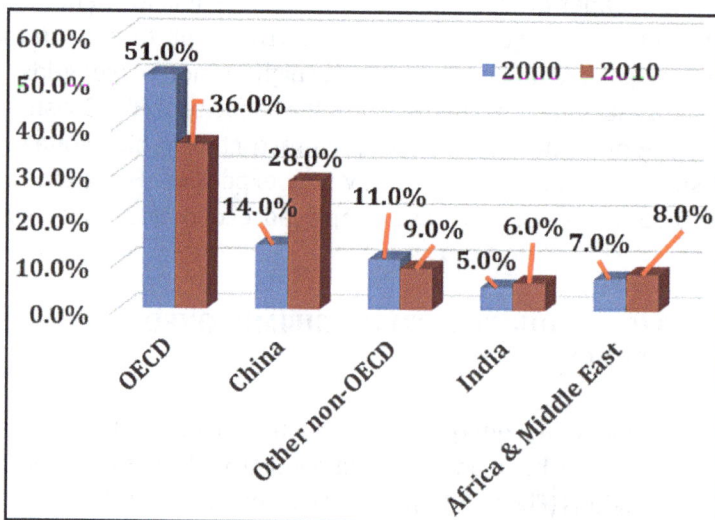

FIGURE 13.7   Global industrial energy consumption by region. *(IEA, 2013).*

TABLE 13.1 Sector contribution to potential emission reduction by 2025. *(IEA, 2014).*

Sector	Potential emission reduction (%)
Power	34
Buildings	25
Transport	18
Industry	16
Other	8

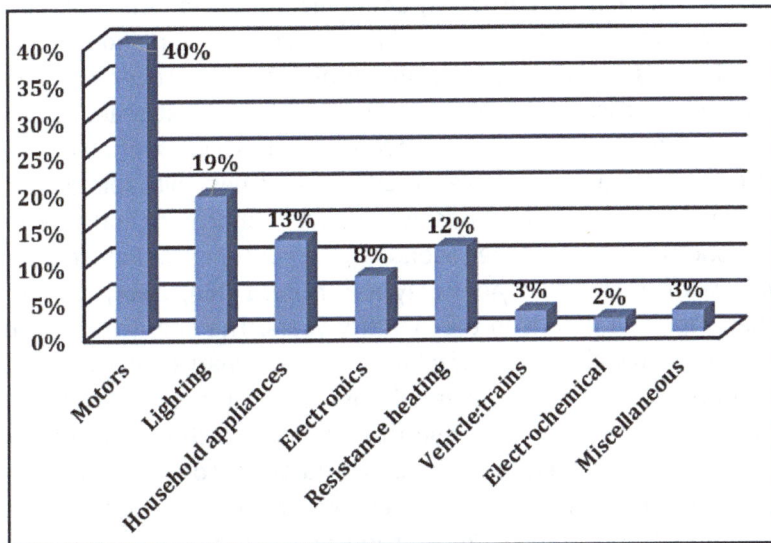

FIGURE 13.8 Global electricity demand by sector. *(WEC, 2013).*

Most electric motors deployed in industry and incorporated in domestic appliances are fixed speed motors which consume substantially more energy than required at light load. Variable speeds are obtained through gear reduction units. Variable speed motors are much more efficient than fixed speed motors and could reduce the world's demand for electricity substantially. However, the relatively high cost has restricted use even in industry. Virtually every industrial process that utilizes electrical motors has a potential for improved efficiency. These include food and consumer product manufacture, machinery and automotive industries, pulp and paper, metallurgical, chemical and petrochemical industries and service industries.

13.3.2 Energy-efficient Industry

Industry accounts for over 50% of the global total final energy use and the demand is growing, up 36% between 2000 and 2011. However industrial energy intensity (amount of energy required per unit product) has been declining in recent years (about 10% since 2000) due to wider adoption of newer, more energy efficient technologies. There has also been a notable shift in the regional pattern of industrial energy demand in favour of emerging regions which now use 66% of global industrial energy. Five energy intensive sectors - chemical and petrochemical, iron and steel, aluminium, cement and pulp and paper - accounted for 67% of the world's total industrial energy consumption in 2011 and many are located in emerging countries. Very few innovative energy-saving technologies have emerged in these sectors in recent years but application of current best available technologies (BATs) could reduce energy use by 10% to 25%. Options include process and plant modification, waste and energy recycling, more efficient process control, and adoption of proven but under-utilized technologies.

The chemical and petrochemical sector accounted for 28% of the global industrial energy use in 2011 and, with the projected strong growth in demand for high-value chemical and petrochemical materials (in particular, ammonia, methanol, ethanol, ethylene, propylene, benzene, toluene and xylene) in the coming years, energy demand by the sector is expected to increase. There is a considerable scope for efficiency improvement in the sector through the adoption of best available technologies (BATs) and best practice technologies (BPTs). Modification and optimization of process routes, optimal integration of processes, refurbishment and retrofitting of existing plants, adoption of best available technologies (BATs) for new plants can make a significant positive impact on efficiency in the sector.

The iron and steel sector is the second largest consumer of industrial energy, accounting for about 22% of total industrial energy use in 2011 (IEA, 2014). Energy efficiency in the industry has improved greatly in the last fifty years. The energy requirement per tonne of steel in 2015 was only about 15% of what it was fifty years earlier. Nearly 70% of the world production of steel is by the coal-powered blast furnace route which utilizes about 700 million metric tonnes of coal annually. Many of the direct reduction processes which account for about 20% of the world steel requirements also utilize coal or natural gas. About 600 kilograms of high grade coking coal and an additional 200 kilograms of low-grade coal are required to produce 1 metric ton of steel. Coke is produced from the coking coals by carbonization in coke oven batteries at about 1000°C in minimum air. This process is a major source of atmospheric pollutants including carbon dioxide, sulphur and particulate carbon. The coke produced is charged with iron ore or sinter or a mixture of both, as well as limestone into the blast furnace.

Coke supplies the carbon monoxide for the iron ore reduction, provides the heat for the reactions and also acts as a porous yet strong support bed for the charge burden while allowing the reducing gases to pass through. The coke oven is not only a major source of environmental pollutants, coke is also the most expensive charge input into iron and steelmaking hence there is considerable interest in reducing the coke intensity of the blast furnace process. Technologies already in commercial operation include installation of flue gas processing and waste heat recycling systems for coke ovens and blast furnaces, preheating blast furnace tuyere air, and injection of supplementary fuel, mainly recycled coke oven breeze, pulverized low grade coal or natural gas. Integration of steel plants with co-generation power units can double the overall thermal efficiency and greatly minimize the amount of pollutants that are ultimately released into the atmosphere.

The direct reduction (DR) steelmaking route as an alternative to the traditional blast furnace-oxygen converter (BFOC) route has been commercially available for decades. The process involves a direct reaction between iron oxide and carbon to produce iron which is refined in an electric arc furnace to produce steel. The direct reduction process option has numerous advantages over the blast furnace route: the carbon can be sourced from low-grade coals, gasified coal (syngas), natural gas, or biogas; the process is less energy intensive; investment per tonne of steel output is lower; process efficiency is higher; and any production capacity can be achieved through integration of a series of small modular units with CO_2 capture (Santos, 2012). Also, the process can be shut down and restarted in response to steel demand, unlike the blast furnace.

In spite of the advantages of DRI process over the BFOC route, adoption has been very slow, due mainly to the inertia in the sector. Most existing blast furnaces are decades old, already have established sources of raw materials (often owned by the steel companies), and process technologies are well established (even if inefficient). Investments required for new blast furnace steel plants are very high and there have not been many lately, given the persisting uncertainties in steel supply and demand, driven by the ups and downs of the global economy. The current depression (2016) in the global steel market and aggressive marketing by China have resulted in the closure of many European steel plants.

In contrary, coal-based direct reduction steel plants (COREX) are operating in South Africa, India, China and Italy and a natural gas based plant (MIDREX) in Quatar has been in commercial operation for over forty years. Also, new plants are running in twenty other countries around the world. Direct reduction plants with capacities as high as 1.8 million tonnes/yr (comparable to a small blast furnace) are now available. A new energy-efficient direct reduction process known as HISmelt produces pig iron directly from ore using a smelt reduction vessel. Iron ore fines, coal, flux and hot air enriched with oxygen are injected deep into a molten metal bath in a smelt reduction vessel. The reactions produce iron and carbon monoxide which is further burnt within the melt to release more energy.

The main difference between HISmelt process and other DR processes is that the product is molten iron while other DR processes produce solid sponge iron. While cold sponge iron must be refined in electric arc furnaces, HISmelt molten iron can be refined in relatively low-cost oxygen converters or electric arc furnaces. A commercial plant has been in operation in Australia and other plants are planned to become operational in China and India in the next couple of years.

In summary, wider adoption of direct reduction technology for steel production could lead to greater efficiencies and flexibility in the industry. Infusion of new, more efficient technologies has been slow mainly because of the slow pace of proliferation of the technologies but there is a high potential for improved efficiencies and lower pollution through more precise process control, use of coal-derived syngas in place of coal and deployment of more efficient burners.

Recycling of materials can lead to substantial reduction in global primary energy demand and therefore reduce atmospheric pollution. Many materials can be recycled - ferrous and non-ferrous metals, lead-acid batteries, polymers, rubber, paper products, glass, electronics, etc. Apart from substantial energy savings, recycling helps to conserve virgin minerals and raw materials as well as reduce environmental pollution by discarded refuse. Energy required to produce copper or aluminium from recycles is only about 10% of that required for production from virgin ores. Steel produced from scrap requires 30% less energy than production from iron ore. Substantial reduction in global energy demand is possible through recycling but there are formidable obstacles, notably, collection strategies and public sensitization.

13.3.3   Energy-efficient transportation

Electric vehicles have been in use primarily by industry for over a hundred years, even before petrol engine-driven cars became popular. They are considered to be more energy efficient with lower pollution generation than gasoline and diesel powered vehicles. The core unit is the electric motor which draws power directly from grids as in trains and trams, or from on-board diesel engines as in ships. Plug-in electric forklift trucks and delivery vans have been in use for decades but the technology only became popular in the consumer auto industry from the early 1990s. Initial development focused on hybrid vehicles which draw power from two distinct sources. A hybrid petrol-electric vehicle draws traction power either from an on-board charged battery or a conventional internal combustion engine. An on-board computer determines which of the two power sources is more suitable for a given driving condition. Electric motors are more efficient in town driving while conventional fuel engines perform better in country driving. Fully electric vehicles (EV) which run entirely on charged batteries are also becoming increasingly popular. A major advantage of electric vehicles is that the electricity they consume can be generated from a wide range of sources, including fossil fuels, nuclear power, solar power, renewable energy, hydropower and wind power. Another major advantage is the fact that electric vehicles have low carbon footprint since most of the associated pollution (on a life cycle basis) is restricted to and can be contained at the power plants, unlike gasoline and diesel vehicles distribute pollution widely. Furthermore, batteries can be charged from electricity derived from sources other than fossil fuels.

In spite of the numerous advantages, sale of electric and hybrid cars has failed to expand beyond core markets, with Japan and the United States accounting for over 90% of total global sales in 2013 (IEA, 2014). The main constraints to the proliferation of electric vehicles have been the prohibitive cost and heavy weight of batteries, lack of networks of charging bays, and the time required to recharge batteries. Batteries are becoming cheaper and more efficient and some countries, notably USA, Japan and Israel are developing charging infrastructure equipped with fast chargers. A system under development is based on the cell phone provider model. Customers would buy electric cars but the battery would be on lease from companies that operate swap bays. Consumers could charge their batteries at home or drive into swap bays and get them replaced with freshly charged batteries within a few minutes. Proliferation of swop bays in cities and along freeways will definitely improve the market for electric cars. Other developments include on-board internal small combustion engines fueled by petrol, diesel or biofuels which will charge the battery continuously, eliminating the need for access to charging bays.

The use of solid oxide fuel cells (SOFC) in place of batteries is also under development. Although fuel cells are less efficient than batteries, they generate electric power on board and can use oil, natural gas syngas or biogas. These technologies also have advantages in heavy, long-haul transportation. The impetus driving the development of electric cars is the assumption that electricity is clean energy and pollution from hybrid electric cars is greatly reduced. This is however not true since, based on a total life cycle analysis discussed in Chapter 12, electricity has a carbon footprint irrespective of the source and the more electric cars are deployed the greater the demand for electricity and coal. Wider use of electric cars will no doubt lead to greater efficiencies and lower pollution in the transportation industry, but the slow pace of proliferation of battery charging/swopping bays remains a formidable obstacle and is probably the main reason why interest is low worldwide.

13.3.4   Energy-efficient commercial and residential buildings

Buildings are the largest energy-consuming sector, accounting for 31% of final energy consumption globally, rising by around 20% between 2000 and 2011. The trend is expected to continue, driven by rising population, urbanization and increasing wealth in some regions. Heating/cooling per unit floor area is the largest end use. Substantial potential exists (up to 25%) for improvement of the energy efficiency of buildings, particularly in heating and cooling which account for one-third (60% in cold climates) of all energy consumed in buildings (IEA, 2013). A major area of efficiency improvement is the upgrade of the building thermal envelope through design and use of best available technologies (BATs). Potential areas of innovation include insulated walls, windows, cool roofs, improved natural lighting, and efficient lighting. The energy utilization efficiency of buildings also known as Energy star rating (ESR) is becoming a critical variable for both home builders and buyers in many countries. New buildings feature effective wall and roof insulation, double glazed windows, compact fluorescent lighting (CFL), which can provide the same light power for 20% of the power consumption of a conventional resistance bulb, control systems which switch off lights in unoccupied rooms, energy-efficient heating and air conditioning systems, and energy star-rated appliances. Also, increasing ability to control home appliances remotely on cell phones means home heating can be switched off when every one is out and reactivated shortly before return. Also, temperatures can be monitored and controlled remotely. There is also a need to sensitize consumers on energy conservation and potential cost savings. Switching off idle equipment and the use of energy efficient lighting could reduce energy cost substantially and reduce coal combustion. Also, energy marketing companies in some developed countries are offering lower unit pricing and bonuses for low energy consumption.

13.3.5   Energy-efficient ICT

The information and communications sector (ICT) has literally exploded in the past decade or so. Internet connectivity is growing at an exponential rate. There were 2.7 billion internet users worldwide in 2013 and internet traffic could double in the next few years. Deployment of network-enabled devices in offices and homes has increased exponentially - broadband connectivity, wireless mobility, cloud computing, e-commerce, social media, video-enabled devices, smart devices , smart systems and sensors, personal computers, servers, laptops, game consoles, printers, etc. Many home appliances – TVs, microwave machines, washing machines, home security surveillance systems – now have internet connectivity. Currently, around 14 billion devices are connected and the number could double by 2020. Projections indicate that network-enabled devices will be around 500 billion by 2030 (WEF, 2012; OECD, 2012). Edge devices – routers, routing switches, integrated access devices, multiplexers, metropolitan area networks (MAN), wide area networks (WAN) – all of which provide authenticated access to enterprise or provider core networks are now indispensable in industry, commerce and homes.

There has been a surge in energy demand to power ICT systems, accounting for more than 8% of total final global energy consumption in 2013. The energy required to meet the projected growth in network-enabled devices is expected to double in the next decade. Most of the energy (up to 80%) is consumed when devices are in standby modes (ready and waiting but not performing the main function) (NRDC, 2011, Hittinger, 2011). The aggregate electronic devices in offices and homes -   modems, routers, TVs, PCs, servers, set-top boxes, games consoles, printers, security equipment, etc - constitute more than 40% of ICT electricity demand (Figure 13.9). Home appliances on standby consume substantial energy. Standby energy accounts for 1 to 2% of global electricity consumption and approximately 10% of residential electricity use. Heating food in a microwave requires 100 times more power than running the clock displayed on the unit but microwaves are in standby modes more than 99% of the time over their life cycle and more energy is used to run the clock than to cook the food (EES, 2011). There is already a strong, consumer-driven industry effort to reduce ICT-related energy demand. Consumers now demand smart phone and other mobile device batteries that last longer between charges. Also, consumers can control their home appliances on their cell phones from almost anywhere. Electronic and non-electronic home appliances: home (kitchen and laundry) appliances, heating, ventilation and air-conditioning (HVAC) systems, lighting etc are already featuring smart control systems that could improve energy utilization efficiencies.

However, in spite of the above improvements, the potential areas of energy saving are enormous. Implementation of best available technologies (BATs) and solutions could reduce IT energy demand by up to 65%, for example, high efficiency mobile devices that can maintain network connections at very low energy consumption are coming on stream but this development is not penetrating other network-enabled devices. There are also initiatives by world and regional bodies notably International Energy Agency (IEA), European Union (EU) to establish industry standards that specify minimum energy performance requirements to reduce standby energy consumption in network-enabled devices. This needs policy intervention such as specifications/minimum energy performance requirements or standards, and consumer awareness campaigns.

Electronic devices consume very low energy compared with electrical devices but the total life cycle energy input is very high. For example, enormous energy is required to manufacture silicon and other materials used in the production of electronic devices. Presently there is very limited facility for recycling electronic devices, even in the developed countries whereas greater recycling offers high potential for overall energy efficiency in the industry, apart from the positive effect on the environment.

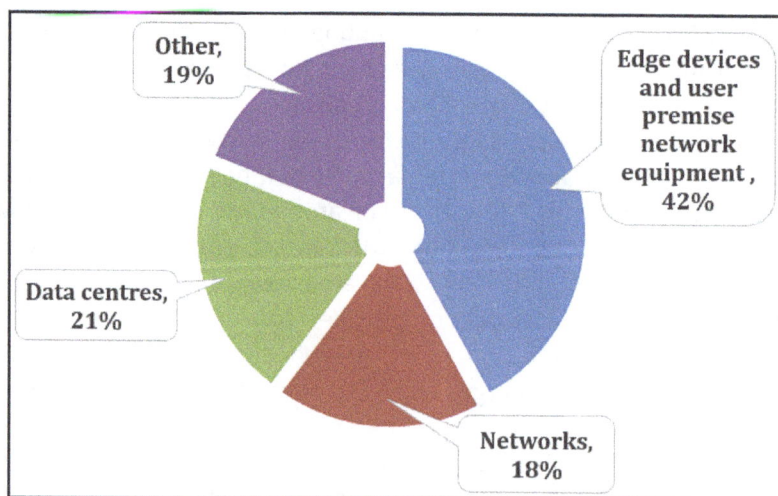

FIGURE 13.9 ICT energy use by sub-segment. *(IRA, 2014).*

13.4 CAPTURE AND UTILIZATION OF POLLUTANTS

Coal carbonization and combustion processes generate a wide range of pollutants, including tar, ammonia, carbon dioxide, carbon monoxide, methane, sulphur oxides, nitrogen oxides and mercury. Many countries, in particular the industrialized world have put in place laws which control the discharge of potentially anthropogenic pollutants from industrial processes into the environment. This has prompted industry to adopt retrofitting or new technologies in order to meet the stringent requirements. Options for reduction of pollutants emanating from coal utilization include retrofitting existing plants with cleaning and capture systems for solid, liquid and gaseous by-products, and replacing ageing plants with new, more efficient and cleaner plants. Some new power plants that have been commissioned in recent times have efficiencies of around 40% compared with about 30% for older plants, and emit up to 90% less pollutants than the plants they replace. Multi-stage capture systems strip flue gases of tar and ammonia in condensers, particulate carbon in electrostatic precipitators, sulphur oxides, particulate matter and other impurities in scrubbers. Selective catalytic reduction units which are capable of removing up to 90% of nitrogen oxides from flue gases are now in common use. Considerable progress has been made in the development of processes for capture of as much as possible of the pollutants produced during fossil fuel utilization, thus reducing the negative effect on the environment.

13.4.1 By-product gas processing

In response to the increasingly strict control of discharge of environmental pollutants which is being enforced in many developed countries, industrial plants are either being retrofitted with capture systems or replaced with new, more efficient plants which produce lower pollutants and feature effluent processing units. Effluent gases from industrial units are processed to separate solids, condense out liquids and recover useful heat. Most of the captured by-products are converted to useful products and the recovered heat is recycled (Figure 13.10). Some of the most important ones are discussed briefly in the following sections.

13.4.1.1 *Particulate filtration.*

Carbon particles produced as a result of incomplete combustion of fossil fuels are emitted into the atmosphere in varying degrees from all combustion processes. The emissions may be in form of soot (lamp black, carbon black) such as in transportation vehicles, or heavier particles from most industrial fossil fuel combustion processes. Cyclones, electrostatic precipitators, fabric filters, wet particulate scrubbers and hot filtration systems are now standard particle capture equipment featured in most industrial combustors. The captured material is either recycled or used as feedstock for the production of other useful premium carbon materials (see Chapter 11).

13.4.1.2 *Removal of chemicals from flue gases.*

Carbon dioxide, sulphur oxides and nitrogen oxides are the most potentially damaging gases emitted as by-products of fossil fuel combustion. The most common method is to pass the post-combustion gases through a series of processes to remove the undesirable contents.

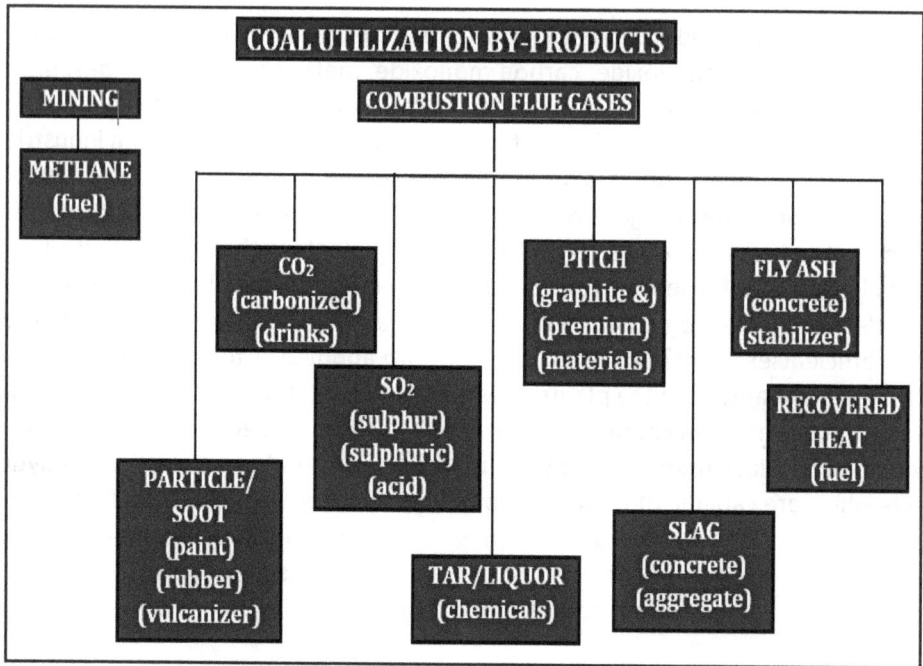

FIGURE 13.10 Uses for coal utilization by-products.

The technology for capturing carbon dioxide is well established and routinely used in separating the gas from combustion flue gases or natural gas at the well head. The CO_2 capture process involves filtration through fabric filters and wet particulate scrubbers to remove up to 99% of particulate matter, followed by solvent scrubbing to remove sulphur oxides and selective catalytic reduction to remove up to 99% of nitrogen oxides. The residual gas is mainly carbon dioxide. The nearly pure carbon dioxide is sold to food processing and chemical industries. However, the carbon dioxide concentration in most fossil fuel combustors is usually low and needs to be upgraded by oxygenation to convert the carbon monoxide content. Flue gas rich in carbon dioxide can also be produced by burning coal in oxygen-rich atmosphere to produce carbon dioxide rather than carbon monoxide.

NOx gases are difficult to remove from flue gases released by conventional coal burners but the quantity produced can be reduced significantly by the use of low NOx burners capable of reducing emissions by up to 70% . These burners are now commercially available and are being deployed in fossil fuel-fired plant retrofits as well as new plants. Fluidized-bed combustors are particularly efficient in removing sulphur and nitrogen oxides from coal because of the high operating efficiency which reduces production of the gases, but also because limestone can be injected into the bed to remove the sulphur oxides. The removal efficiency of NOx gases is over 90% and most of the sulphur (up to 99%) can also be removed. The integrated gasification combined cycle process (IGCC) offers a promising option for the removal of carbon dioxide because the gaseous products of combustion of the syngas in the turbine have much higher concentrations of carbon dioxide and no further oxygenation is required to enhance capture. Furthermore, the gas is at high pressure which makes capture and sequestration relatively easy.

13.4.2   Recycling and utilization of captured waste and pollutants

Virtually all the processes developed so far to clean up flue gases from coal carbonization combustion yield products that have found uses in industry, including food and chemicals industries, building and construction industries, and premium materials industries (Figure 13.10).

13.4.2.1   *Particulate carbon and soot.*

Fine carbon particles recovered from flue gases are products of incomplete combustion and because of their high calorific value they are usually recycled into the furnace. Soot, also called lampblack, pigment black, carbon black or black carbon, is a dark powdery deposit of unburned fuel residues, usually composed mainly of amorphous carbon, that accumulates in chimneys, automobile mufflers and other surfaces exposed to smoke. Soot is produced by incomplete combustion of carbon-rich fuels due to insufficient oxygen supply. Lampblack has been used as the black pigment in paints and inks since prehistoric times, and is still widely used in printing inks, toners for xerography, laser printers, and in the chemical industry. It is also used as food coloring, for example, in sweets. The black color of rubber tires is due to the use of lampblack as an ingredient in the vulcanization process. This use accounts for around 85% of the market use of carbon black.

13.4.2.2   *Flue gas desulphurization product.*

The process of removal of sulphur dioxide from flue gases typically involves the injection of a reagent usually hydrated quicklime into the flue gases to form a solid by-product commonly referred to as FGD filter cake, comprising mainly   varying amounts of 10–13% calcium sulphate, 60–70% calcium sulphite, unreacted lime, fly ash and water. Traditionally, the cake is used for land fill but new uses have now been developed, notably its use in highway pavement, as low-toxicity, low-permeability liners for water ponds, and as hydro dam liners and seepage cut-off walls. Many coal-fired plants have systems that can remove up to 95% sulphur from flue gas.

The wet sulphuric acid process (WSA) is a very efficient process for recovering sulphur from flue gases in the form of commercial quality sulphuric acid. Sulphur dioxide (SO_2) is converted to trioxide (SO_3) by oxidation. Hydration produces commercial sulphuric acid. The SNOX process designed specifically for power and steam generation plants can remove sulphur dioxide and nitrogen oxides simultaneously from flue gases. The sulphur dioxide is converted to sulphuric acid by catalytic oxidation to trioxide followed by hydration. An SNOX recovery plant can be configured to produce sulphuric acid or elemental sulphur. The nitrogen oxide is converted to environmentally benign nitrogen by spraying with ammonia.

Gypsum (hydrated calcium sulphate, ($CaSO_4 \cdot 2H_2O$) is a naturally occurring mineral used extensively to make wallboard (drywall) for homes, offices, and commercial buildings. It is also used worldwide in concrete set control for highways, bridges, buildings and many other structures, cement production, water treatment, subsidence control and barrier to acid mine drainage formation in underground mines, and in glass making. The use of gypsum as a soil conditioner and acidity control on large tracts of land in suburban areas, as well as fertilizer in agriculture is also widespread. Although the bulk of gypsum in use worldwide occurs naturally, synthetic flue gas desulphurization (FGD) gypsum can be produced by passing the flue gas from coal combustion and carbonization through a slurry of calcium carbonate (limestone). Sulfur dioxide in the flue gases reacts with the calcium carbonate to produce calcium sulphite ($CaSO_3$), which is converted into gypsum by oxidizing it with water. Chemically, the synthetic gypsum is nearly identical to virgin gypsum and is often of higher purity. About 30% of drywall is now produced from synthetic gypsum.

13.4.2.3   *Tar, liquor and pitch.*

Tar is a primary condensate of coal carbonization or combustion. Distillation yields a wide variety of aromatic and phenolic chemicals. Amoniacal liquor is also a by-product of coal combustion or carbonization and is a feedstock for the production of ammonia and ammonium sulphate. The solid residue from tar distillation is pitch which is the feedstock for the production of high-premium industrial materials including graphite, carbon fibre, industrial diamond, fullerene, carbon nanotubes (see Chapter 11). Pitch is also used as a binder in coal briquetting. It is used extensively as base for coatings and paint, in roofing and paving, and as a binder in asphalt products.

13.4.2.4   *Fly ash.*

Ash is the residue of coal combustion processes. The bulk of ash produced settles at the bottom of the combustion chamber (bottom ash). However, a significant amount is carried in the flue gases and captured in electrostatic and other particulate filters. This ash is glass-like and fine, and is known as fly (or flue) ash. The chemical composition of fly ash varies depending on the type of coal and the geological setting in which the coal matured. However, most contain substantial amounts of silica (SiO_2), alumina (Al_2O_3), iron oxides (Fe_2O_3, Fe_3O_4, etc.), and lime (CaO) all of which are present in the coal-forming precursors as well as coal seam intrusions and overburden.

The proportions of the various compounds in coal vary widely depending on the type and rank of coal and could range from 1% for alumina to 60% for silica. A large number of other elements and compounds may also be present and most end up in the ash. These include chromium, cadmium, boron, beryllium, selenium, strontium, molybdenum and vanadium. Fly ash produced in coal combustion had traditionally been released into the atmosphere. However, increasingly stringent pollution control regulations now require that the product be captured and safely disposed of. Although the bulk of the captured fly ash is either dumped as waste or used as land fills, recent research has established its potential for other uses. For example, because of the high content of silica, alumina and lime, fly ash reacts with water to produce a cement-like compound known as pozzolan, an inexpensive replacement for up to 35% of portland cement used in concrete. Also, because of the spherical shape of fly ash particles, pozzolan has been found to actually reduce segregation and improve strength, density, workability and ease of pumping of the concrete. Concrete containing pozzolan produces a smoother surface and sharper detail. Other enhancements of concrete quality include greater resistance to expansion due to sulphate attack in fertilized soils or coastal area soils, and lower heat of hydration. Improvement in concrete plasticity and moderation of solidification rate reduce surface cracks which facilitate the ingress of reinforcement-damaging corrosive fluids into concrete, especially when immersed in sea water (Wischel,

1991, 2005; Jones et al.,2003; McCarthy and Dihr, (2005; Blissett and Rowson, 2012; Arya and Hooda, 2014). Fly ash is also used as an ingredient in brick, block, paving, and structural fills.

13.4.2.5 *Slags.*

Slags are complex, non-metallic by-products of many metallurgical operations and are products of the flushing process for the removal of undesirable elements and compounds from the final products. They consist primarily of calcium, magnesium, and aluminium silicates in various combinations. Iron and steel slags are co-products of iron and steel manufacturing. In the production of iron, the blast furnace is charged with iron ore, fluxing agents, usually limestone and dolomite, and coke as fuel and the reducing agent. The iron ore is a mixture of iron oxides, silica, and alumina. From this and the added fluxing agents molten slag and iron are formed. The slag which is lighter than molten iron is tapped separately and processed. The physical characteristics of slags such as density, porosity, and particle size, are affected by the chemical composition and processing conditions, in particular, cooling rates. Depending on the cooling method, three types of iron slag are produced. Slag that is air-cooled relatively slowly under ambient conditions is crushed, screened, and used principally as aggregate for road base and surface construction, concrete, and also for the production of asphaltic concrete. The second type, granulated slag is used mainly for cement production. The third type of slag is produced when the molten slag is quenched in water. It forms fine granular aggregate which has found use in concrete mixes. The material reacts with free lime to develop strong, cement type properties. Expanded or pelletized slag is produced by quenching in a jet of water. The steam generated transforms the material into a spongy, low-density, porous structure capable of forming strong bonds with hydraulic cement paste.

13.4.2.6 *Methane.*

Methane, the main component of natural gas is a by-product of coal combustion, although in much smaller quantities compared with carbon dioxide. Coal mining, oil and natural gas production are significant sources emission of the gas in the atmosphere, contributing about 30% while other human activities such as livestock production, rice cultivation and waste treatment account for the balance. Apart from its very high greenhouse gas potential, methane is a poisonous, highly explosive and highly inflammable gas. Large quantities of methane-rich gas are formed during the coalification process and stored within the coal bed on internal surfaces. Because of the relatively shallow depths of coal compared with conventional natural gas reservoir, recovery wells are easier and less expensive to complete. Also, methane is anthropogenic and recovering coal bed methane prior to mining coal may reduce significantly methane emissions to the atmosphere during coal mining.

Coal bed methane (CBM/CMM) and oil-gas associated methane are valuable energy resources and are being used for on-site electric power generation or the production of methanol, also a valuable fuel. Although there are plentiful resources of coal bed methane in many countries such as the USA, China, Russia, India, many countries of Africa and Europe, the only countries with significant coal bed methane utilization projects in progress are China, Australia, Germany, the USA and the UK. Despite the potential for using CBM/CMM as an inexpensive and clean source of energy, the investment in new CBM projects has been somewhat limited due to economic, physical, social barriers and technological limitations

(IEA, 2005). One major problem is the disposal of large quantities of contaminated water that must be pumped out prior to coal bed methane recovery.

13.5   CARBON CAPTURE AND STORAGE (CCS)

Carbon dioxide is the main constituent of coal carbonization and combustion flue gases and the most prominent of the six major anthropogenic greenhouse gases. The gas is the benchmark for assessing the contribution of all the other gases to atmospheric pollution. Carbon dioxide emission intensities of some carbonaceous fuels are shown in Table 13.2. Anthropogenic carbon dioxide emissions result primarily from the combustion of fossil fuels. According to the International Energy Agency projection, emissions will increase by around 34% in 2040 compared with the 2012 emissions level (EIA, 2016). Much of the growth in emissions will come from developing economies. While the growth projection for Organization for Economic Cooperation and Development (OECD) countries will be around 8%, non-OECD emissions will grow by about 51%. There are many international agreements and treaties on limiting discharge of anthropogenic greenhouse gases into the atmosphere to mitigate environmental pollution and climate change. This has accelerated the development of technologies for the capture of CO_2 from coal combustion and carbonization flue gases in recent years.

13.5.1   Carbon dioxide capture technologies

Carbon dioxide capture technology involves the separation of the gas from other by-product gases, preferably in concentrated form that can be readily transported to a storage site. While the gas can also be transported in lower concentrations, energy and other associated costs would be relatively high and make this method impractical. CO_2 capture practice is well established for some industrial processes, for example, the gas is captured in the purification of natural gas, in coal and other fossil fuel combustion processes, and from the production of hydrogen-containing syngas in the manufacture of ammonia, alcohols and synthetic liquid fuels. There are three main process routes for carbon capture, but suitability depends on the process or power plant application (Figure 13.11). Post-combustion and pre-combustion systems are already in commercial operation and a third (oxyfuel) is under development.

13.5.1.1   *Post-combustion capture technology*

The flue gases produced by the combustion of coal and other carbonaceous primary fuels in air comprise mainly nitrogen, with a small fraction of carbon dioxide, typically 3-15% by volume. Several capture technologies have been developed but have not been tested extensively. Some involve chemical absorption using aqueous solutions while others use adsorption and membrane technologies. The capture technology using chemical solvents is considered the more promising since it offers the higher capture efficiency, with the lowest energy use and cost. Furthermore, units can be more readily installed in existing power plants as retrofits. Most plants fitted with capture systems employ an organic solvent such as monoethanolamine (MEA) but many have technological and operational problems. What to do with the captured carbon dioxide poses an even more formidable challenge. The current status of capture technologies and potential uses of captured carbon dioxide have been reviewed recently by several authors (Joao Gomes, 2013; Leung et al., 2014).

TABLE 13.2 Direct CO_2 emission factors for some carbonaceous fuels. *(IPCC, 2005).*

Carbonaceous fuel	CO_2 Emission Factor ($gCO_2 MJ^{-1}$)
Coal	
Anthracite	96.8
Bituminous	87.3
Sub-bituminous	90.3
Lignite	91.6
Petroleum fuel	
Motor Gasoline	69.3
Distillate Fuel Oil	68.6
Liquefied Petroleum Gas	59.1
Kerosene	67.8
Other fuel	
Biofuel	78.4
Natural gas	50

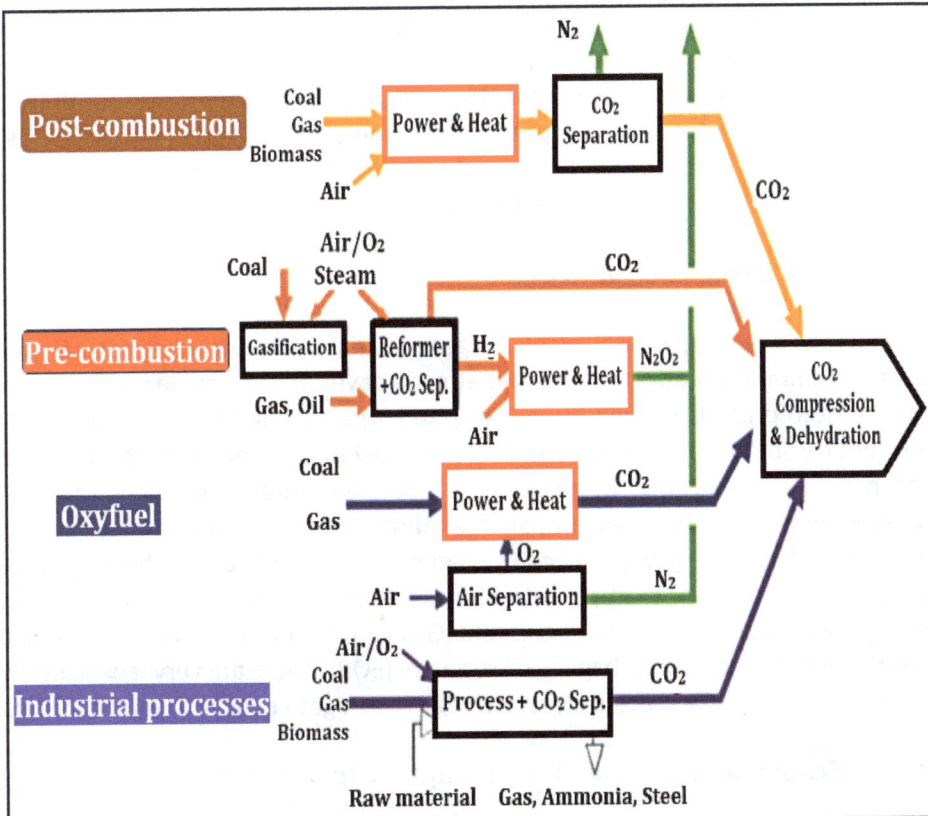

FIGURE 13.11 Technology options for CO_2 capture. *(IPCC, 2005).*

13.5.1.2 *Pre-combustion capture technology*

Modern coal gasification systems produce synthesis gas, a mixture of mainly carbon monoxide and hydrogen. Additional hydrogen is produced by treating the synthetic gas with steam (shift reaction) to convert the carbon monoxide to hydrogen. High-concentration carbon dioxide (typically 15 to 60% by volume on a dry basis) is the main by-product of this process. The gas is separated from hydrogen which is used as fuel to generate power and heat. Although pre-combustion capture technology is more elaborate than post-capture systems and costlier, the high-concentration carbon dioxide produced by the shift reactor is at high pressure, a form that facilitates relatively easy separation, transportation and storage. Furthermore, pre-combustion capture is more suitable for power plants that feature coal gasification units, for example, integrated gasification combined cycle (IGCC) technology.

13.5.1.3 *Oxyfuel combustion capture technology*

Oxy-fuel combustion features an air separation unit which provides oxygen stream (90-99% purity) for combustion instead of air to improve efficiency. The flue gas which is primarily carbon dioxide (and steam vapour) is compressed without further chemical capture or separation. Typical concentrations of carbon dioxide in oxyfuel system flue gases are higher than 80% by volume, which makes the gas easier to capture. The water vapour is removed by cooling and compressing the gas stream. Oxy-fuel systems are capable of capturing up to 90% of the plant's carbon emissions. However the captured gas may contain some impurities, in particular, SOx and NOx gases and may not meet sequestration standards which require a very clean gas. Further treatment of the flue gas may be necessary to remove these pollutants. However, the technology has high potential for new generating plant installations especially boilers and possibly gas turbine systems. Research efforts are currently focused on purification of captured gas to meet required standards.

13.5.1.4 *Carbon dioxide utilization*

Many existing power generating plants, large industrial plants, including natural gas treatment plants and ammonia production facilities already have carbon dioxide separation units. Some of the captured gas is utilized by industry and the unutilized gas is discharged into the atmosphere. Some plants capture the carbon dioxide for useful industrial applications and uses include pressurization of gas wells to increase production, chemical and biological processes where CO_2 is a reactant, such as those used in urea and methanol production, purification of gases (e.g. oxygen and hydrogen) as well as various technological applications that use CO_2 directly, for example in the horticulture industry, refrigeration, food packaging, welding, beverages, and fire extinguishers. Gas captured from oil and gas fields is pumped back into wells to facilitate enhanced oil recovery (EOR). There are very few plants that feature carbon dioxide capture specifically for storage to mitigate environmental degradation.

13.5.2 Economics of carbon dioxide capture technologies

Carbon dioxide capture for the purpose of protecting the environment can be very expensive because it must be in concentrated, highly compressed state that can readily be transported to a suitable storage site. Furthermore, impurities such as SOx and NOx gases must be removed. Studies to determine impact of capture on the economics of various fuel combustion processes have yielded wide variations, even for the same process because of differences in fuel types and the capture technology adopted, plant size, location, efficiency, cost of capital, etc. Furthermore, there are not many existing plants (small or large) fitted with CO_2 capture systems and operating experience is limited, especially for new generation natural gas combined cycle (NGCC) and integrated gasification combine cycle (IGCC) plants. In effect, capture cost estimates are based on assumed design and operation models and sets of assumptions which do not necessarily apply to all situations or all parts of the world. The vast majority of emission sources have concentrations in the range of 5 to 15% and expensive, energy-intensive processing is required for concentration, purification, dehydration and compression to achieve the specifications required for transportation and sequestration. Comparative cost estimates compiled from literature are presented in Table 13.3.

Investment costs and energy consumption of power plants fitted with carbon capture are increased significantly and plant efficiency is reduced. However, CO_2 emission is reduced by nearly 90%. It would be difficult for many power plants in the developed countries to absorb the potentially heavy increases in both capital and operating costs resulting from the adoption of carbon capture and sequestration technologies. In fact, most emerging countries would consider it unacceptable. While many developed countries have the wherewithal to enforce compliance with environmental control regulations, for many developing and emerging countries, it is neither urgent nor affordable and most do not have the wherewithal to enforce compliance.

13.5.3   Carbon dioxide transportation

Unless a plant with CO_2 capture is close to a suitable storage site, the capture products must be transported to a suitable location and sequestered (permanently stored) in environmentally benign state or utilized for the production of environmentally friendly products. Carbon dioxide may be transported in compressed state by pipeline. The gas can also be transported as a liquid in ships, road or rail tankers. Transportation adds significantly to capture costs and cost increase depends on the mode of transportation, the distance and the quantity transported. Most of the CO_2 being transported currently is through pipelines. There is also some transportation by ship but on a small scale because of limited demand.

TABLE 13.3 CO_2 capture costs for new power plants. *(IPCC, 2005)*. (NGGC = Natural Gas Combined Cycle, IGCC = Integrated Gasification Combined Cycle, PC = Pulverized Coal power generation).

Performance and cost measures	NGCC	IGCC	PC
% increase in cost of energy.	46	57	33
% increase in capital investment with capture.	76	63	37
% CO_2 emission reduction per kWh.	86	85	86

The costs of pipeline transportation of captured carbon dioxide depend on whether the pipeline is onshore or offshore, the terrain, or whether the pipeline passes through heavily congested areas. All these factors determine the cost of laying the pipeline, the number of recompression booster stations, the cost of pipeline operation, and could double the cost per unit length.

13.5.4   Carbon dioxide sequestration

Extensive research is ongoing on suitable sites for permanent storage of captured carbon dioxide and current focus is on geological storage. Three types of geological formations are currently under development: depleted oil and gas reservoirs in onshore and offshore sedimentary basins; deep saline formations, shale and basal formations; and porous coal beds that are unlikely to be exploited. In each case, geological storage of CO_2 is achieved by injecting it in dense form into a rock formation below the Earth's surface. Other potential sites include porous rock formations that hold or had previously held gas, oil or brine.

Progress in developing carbon storage technologies has been slow because of the potentially prohibitive costs and lack of consistent policies and framework for the enforcement of strict environmental pollution regulations at national and international levels. Carbon capture and storage is already being practiced on a limited scale in some countries but issues of ownership and long-term monitoring of storage sites remain unresolved. It is estimated that two thirds of the growth in energy consumption in the next twenty to thirty years will be in developing countries and most of it will be filled by fossil fuel plants. The capacity of these countries to adopt emission reduction and control measures is doubtful. In contrast, many developed countries will replace old plants with new low-emission technologies and a significant drop in greenhouse gas emissions is anticipated. In effect, the impact of new carbon capture and sequestration (CCS) technologies on the global emissions may be minimal in the foreseeable future.

13.5.4.1   *Carbon dioxide sequestration in gas wells, coal deposits, and oceans.*

The practice of pumping carbon dioxide back into oil and gas reservoirs to enhance recovery (EOR) is well established and the gas stays underground as long as the original pressure of the reservoir is not exceeded. Another technology under development is the injection of captured carbon dioxide to displace methane from coal beds. Carbon dioxide is readily adsorbed on internal coal surfaces and has twice the adsorptivity of methane. The more valuable coal bed methane is easily displaced from the coal bed and recovered for the production electric power or ammonia. This process is known as Enhanced Coal-Bed Methane Recovery (ECBM).

Considering that about 90% of the world coal reserves cannot be mined by existing technology due to sporadic seam depths, low seam thickness and structural integrity of the geological setting, this storage technology has immense potential. Research is also ongoing on the feasibility of storing carbon dioxide in disused salt mines, aquifers, and depleted oil and gas fields which are very common in many countries. Storage in deep oceans is also under active investigation. Estimates indicate that the oceans of the world are capable of storing all the carbon dioxide produced by human activity for centuries. It is not clear however whether the storage will be permanent or the gas will eventually escape into the atmosphere, or how the marine ecosystem will be affected. The effect on aquatic life is also unknown.

13.5.4.2 *Terrestrial sequestration of carbon dioxide.*

The terrestrial biosphere sequesters large amounts of carbon estimated at about 2 billion tons a year. Current research on carbon storage includes measures to protect ecosystems that store carbon so that sequestration can be maintained and methods of manipulating the ecosystems through genetic engineering to increase carbon storage beyond natural levels. Ecosystems under investigation include forest lands, agricultural lands, use of photosynthetic micro and macro organisms capable of converting carbon dioxide directly to methane and acetate, and micro-algae capable of increasing the rate of photosynthesis and carbon dioxide conversion.

13.5.4.3 *Chemical sequestration of carbon dioxide.*

Carbon dioxide can be caused to react with various chemical compounds to form solid, environmentally benign compounds. One technique under development is mineralization by reaction with calcium and magnesium silicates to form carbonates which are thermodynamically stable and the carbon remains stored permanently. The potential of chemical sequestration technology is rated high since the entire world carbon dioxide emission for one year can be stored in a 5 sq.km land space. Furthermore, storage sites can be located close to recovery sites and the need for transportation over long distances would be eliminated. This sequestration method is probably the most promising. Another current line of research is the development of methods of accelerating natural weathering of alkaline rock which is a natural carbon sequestration process that occurs over long periods of time.

13.6 ADVANCED CLEANER COAL POWER GENERATION TECHNOLOGIES

Much has been achieved globally in improving efficiencies along the value chain of coal utilization, from exploration to utilization. In electric power generation, efficiencies have been raised from the world average of around 30% to as high as 40% mainly through retrofits of existing plants. New plants being built feature more advanced technologies which have raised efficiencies to as high as 60% and halved the pollution emissions. The most promising include advanced coal combustion systems, coal gasification, and integrated gas combined cycle plants (IEA, 2008; IEA, 2014; Reddy, 2014).

13.6.1 Supercritical pulverized fuel technologies (SCPF)

The basic concept of advanced coal combustion systems is to burn coal in a more efficient way to maximize the use of the energy content and minimize pollution emissions. Many concepts are under development but the most promising processes involve coal gasification. While coal gasification has been a commercial technology for centuries, its integration with combined cycle gas turbines for electricity generation is a recent technological innovation that has triggered intensive research and development of many new power generating processes, some of which have been commercialized. The advanced power generation coal combustion technologies already in commercial operation may be classified into two categories: Supercritical (SC) and ultra-supercritical (USC) technologies. The main focus is how to achieve higher outlet steam temperatures. Supercritical technology for hard coal-fired plants achieves outlet steam temperatures of 540-566°C at an operating pressure of 250 bar. Power plants which adopt this technology operate at efficiencies of around 45%. Ultra-supercritical technologies capable of achieving temperatures up to 700°C are under development and efficiencies of around 50% are achievable.

13.6.2   Pressurized fluidized-bed combustion (PBFC)

Coal combustion fluidized-beds are pressure vessels in which fine granular coal is held in suspension in a vessel by pre-heated primary or secondary combustion air blown through uniformly distributed nozzles at the bottom of the vessel. The coal is mixed with fine, uniformly sized sand or limestone and the upward flow of the air keeps the bed in constant motion thereby promoting uniform mixing and heat transfer. Combustion of the coal in fluidized-beds occurs at lower temperatures than usual, resulting in lower production of NOx gases. The use of limestone in the bed promotes the removal of sulphur as a result of direct reaction between the limestone and sulphur dioxide gas produced in the bed. The process is a versatile option for converting various other fuels including coal rejects, biomass, refuse or mixtures of these into steam for electric power generation, industrial or home heating, or to hot gases for driving gas turbines.

Typical thermal efficiencies of circulating (sub-critical) fluidized- bed systems are of the order of 30% but pressurized (supercritical) combustors under development, operating at up to 30 atmospheres and incorporating both gas and steam turbines have achieved thermal efficiencies of around 45% (Figure 13.12). Over four hundred supercritical fluidized-bed plants are in commercial operation worldwide and projections indicate that this plant will dominate the future electric power generation scene. Ultra-supercritical (USC) plants have the supercritical plant configuration but operate at higher temperatures and pressures and are capable of efficiencies of about 50%. Some plants are already in commercial operation. Ultra-supercritical fluidized bed plants have great potential for reducing carbon dioxide emissions by up to 25% since improved efficiencies translate to reductions in fuel consumption and greenhouse gas emission. Increase in efficiency from 35% to 50% is estimated to reduce anthropogenic emissions by around 30%.

13.6.3   Integrated gasification combined cycle (IGCC)

The integrated gasification combined cycle (IGCC) technology allows the use of solid or liquid fuels which are gasified to achieve low carbon footprints similar to natural gas. In the integrated gasification combined cycle systems, coal, oil or biomass is not burned directly but reacted with oxygen and steam to produce a synthetic gas (syngas) composed mainly of hydrogen and carbon monoxide. The syngas is cooled, cleaned of particulate matter and sulphur oxides and then burned in a gas turbine to generate electricity and produce steam for a steam turbine power cycle. The removal of pollution constituents from the syngas under pressure prior to combustion greatly reduces the carbon footprint, and cogeneration using gas and steam turbines in parallel raises efficiencies significantly.

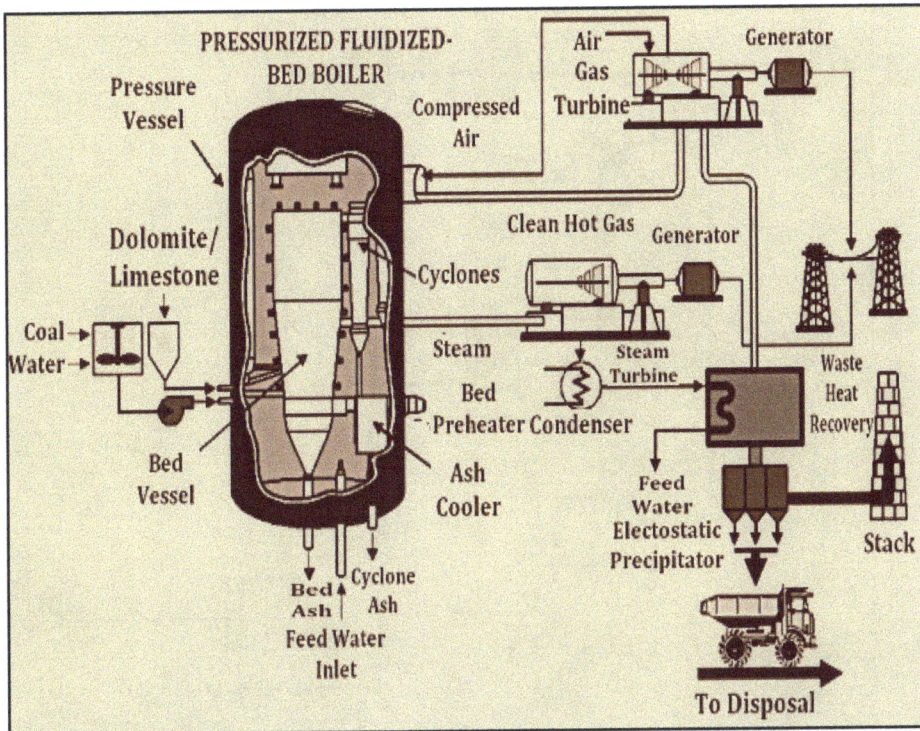

FIGURE 13.12 PFBC power generation process. *(NETL 2010b).*

There are many variations in commercial operation, with different gasification technologies, gas cleanup arrangement and integrated configurations. IGCC plants with thermal efficiencies of around 45% are in operation in many parts of the world. Current developments in gas turbine technology and utilization of the gas turbine waste heat to generate steam for driving an integrated steam turbine will raise efficiencies to around 60% in the near future if the materials problems can be resolved. A typical IGCC flow chart is shown in Figure 13.13. Another major asset of the IGCC technology is the relative ease with which up to 99% of the NOx and SOx gases and over 80% of carbon dioxide can be removed from the syngas. The technology also has high potential as a ready source of cheap, clean hydrogen for future hydrogen power systems. Wider deployment of IGCC plants will no doubt lead to cleaner environment as shown in Figure 13.14.

13.6.4 Hybrid power systems

Co-fired power plants are capable of using a combination of a wide variety of fuel feedstocks including coal, oil, gas, biomass, municipal and agricultural waste to produce some combination of electricity, process heat, heat for domestic heating, chemicals, transportation fuels. Synergies between coal and biomass significantly increase operating thermal efficiency, reduce greenhouse gas emissions, and provide economic use for agricultural and municipal waste.

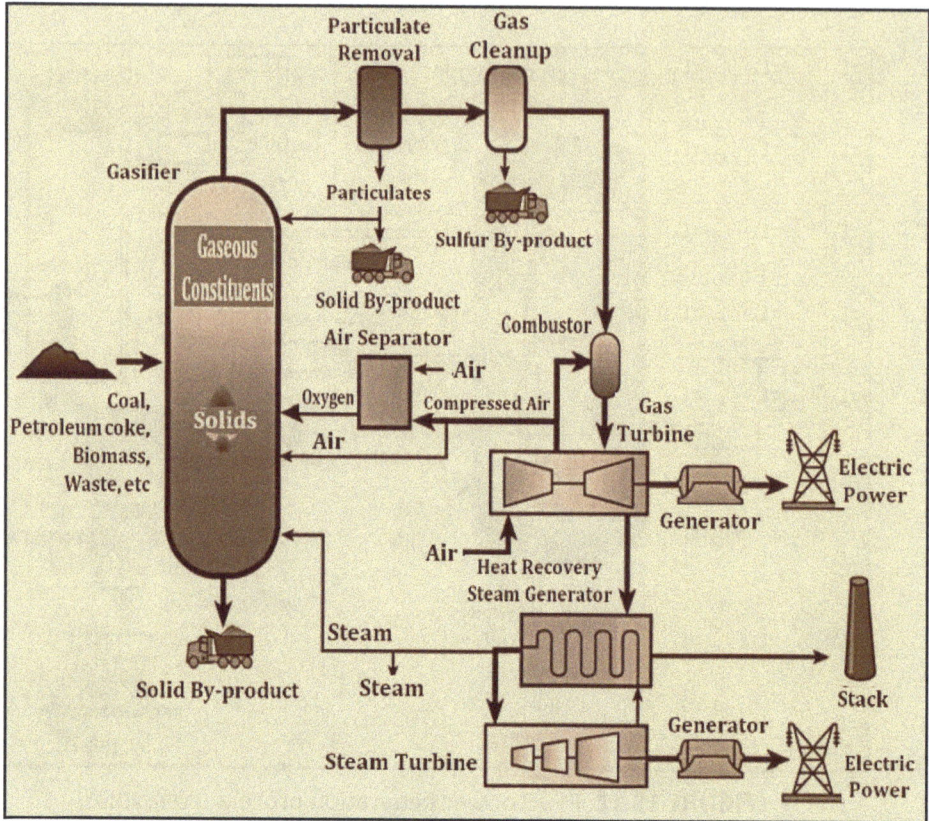

FIGURE 13.13 IGCC power generation process. *(NETL 2010a).*

(a)

(b)

FIGURE 13.14 (a) A typical coal-fired power generating plant (b) A cleaner coal-to-syngas power plant with carbon capture and storage. *[(a) US Department of Energy, b) energy.gov;abb.com].*

Renewable energy technologies are becoming increasingly important but are not expected to contribute more than about 10 to 15% to the world's primary energy requirements in 2035 because of problems of development and sustainable deployment. The relatively high cost of renewable energy technologies compared with fossil systems and lack of global investment are the most formidable barriers but there are several other problems. Most of the technologies are dependent on natural forces and therefore unpredictable. Furthermore, they are site-specific and are only available at certain locations. Another problem with renewable energy is the intermittent nature of availability which means that plants cannot respond to continuous and often variable load demands. For example, even though onshore wind energy is relatively cheap (perhaps the cheapest of the renewable energy technologies), the annual load factors of onshore wind electricity generation plants are low, typically 18-20% and wind peaks seldom match electricity demand peaks. Furthermore, wind energy depends on the weather and the wind intensity and wind energy-based plants located at very good sites do not normally operate more than one third of the time.

Solar energy depends on the intensity of the Sun and the duration in which it is available at a particular site. Hydropower depends on the right sort of geographic conditions and rainfall and power plants often shut down because of low water levels. Furthermore, there is a growing opposition to new hydropower plants, particularly in developed countries because of environmental concerns. Geothermal energy is only available in certain specific sites. Biomass is seasonal, weather sensitive and difficult to transport. Intensive research is ongoing on the various problems that are restricting the rapid growth of renewable power utilization. One of the major solutions under development is co-configuration of renewable energy and fossil energy technologies, in particular coal-based systems, in order to smoothen ripples and ensure sustainable and continuous power supply (fuel-switch).

A hybrid system which is already in commercial operation in many locations in some developed countries combines solar generation with fossil fuel-fired power generation. Solar power is used in day time and fossil power at night time. Hybrid plants have a bright future but they require complex grid systems and controls in order to effectively synchronize electricity supply with demand. Hybrid systems already in operation in many parts of the world include multi-fuel fired single plants (coal-biomass, coal-gas, coal-fuel oil) and hybrid co-plants operating on the same site (coal-solar, coal-hydro, coal-wind, coal-geothermal, gas-solar). Combined solar and coal or gas generation plants are particularly popular and are feeding power to many grids in North America, and many other sunshine states all over the world.

13.6.5   New solar technologies

Wide adoption of solar technology has been stunted for two reasons: the need for good sunshine over extensive periods, and the prohibitive cost of photovoltaic cells, collectors, and other ancillary equipment. Apart from cost, photovoltaic collectors are non-concentrating and the maximum temperature achievable is around 200°C, too low for generating enough steam at the right pressure for large-scale thermal generating plants. A new generation of cost-effective concentrating collectors is now available and are deployed in many solar plants commissioned in the last two decades (ISES, 2003). Basically, a concentrating collector is a parabolic reflector which concentrates the sunlight to a focal line or focal point. Most parabolic collectors are capable of tracking the Sun's rays all day and some are capable of collecting significant amounts of the Sun's energy on cloudy days and at night.

There are two variants of the concentrating collector: one-axis and two-axis systems. One-axis tracking systems concentrate the sunlight onto an absorber tube in the focal line while two-axis systems focus on a relatively small absorber surface near the focal point. Concentrating collectors are large curved mirrors (parabolic collectors) or parallel rows of mirrors up to 600 metres long (parallel collectors). The absorber may be an oil-filled heat exchanger which may be a pipeline, an oil-filled tower or a molten salt tank. The heat is transferred from the hot oil (around 400°C) to a boiler that generates high pressure steam for driving the turbine-generator system. Parabolic concentrators are the most efficient and can usefully convert up to 80% of the Sun's energy collected into heat. There are other less efficient concentrators and different configurations of solar collector fields (Quaschning, 2003). Collected solar energy that cannot be used immediately is stored in molten salt and can be used to fill the periods when the Sun's energy is low (cloudy days and night). This resolves the problem of intermittent power generation (Figure 13. 15).

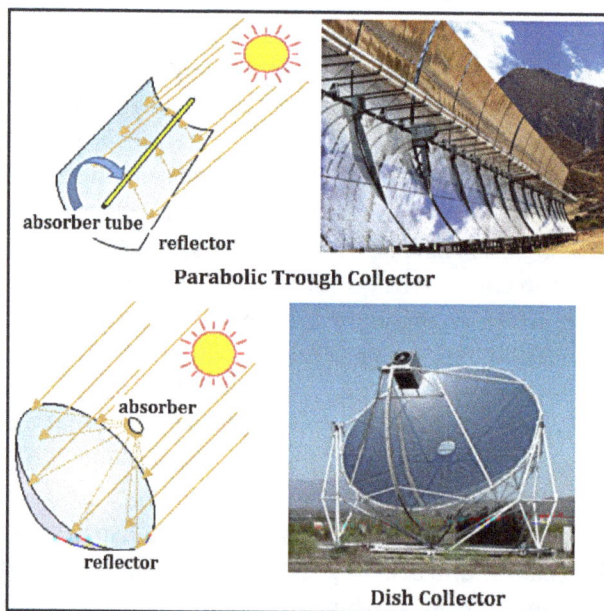

Parabolic Trough Collector

Dish Collector

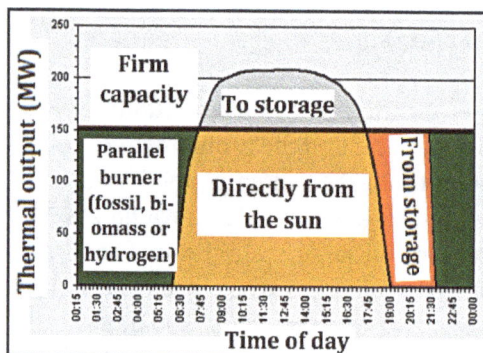

FIGURE 13.15 Solar power generating systems. *(Quaschning, 2003)*

The overall efficiency of a solar plant is still very low, 15-20%. While solar energy collection can be up to 80% efficient, the thermal steam cycle is 35% or less and there are also field losses. In spite of the low efficiency, solar thermal power plants configured for 24-hour operation using supplementary gas burners and thermal storage molten salt tanks are becoming increasingly widespread. Solar plants with capacities ranging from 100 to 350 MW are operating in many countries including the USA, India, Spain, China, and South Africa. The largest concentrated solar plant (CSP) plant in the world (580 MW) funded by the European Union' World Bank and African Development Bank is under construction, located at Quarzazate in the Moroccan desert. When completed (in four phases, 2015-2019) it will have half a million crescent-shaped solar concentrator mirrors.

13.6.6 Coal-to-gas, coal-to-liquid fuel technologies

Converting coal to gas before use as fuel has several potential advantages. Conversion can be done at power generation plants or in-situ in underground coal mines thereby facilitating efficient carbon capture and producing gas which has significantly (about 40%) lower carbon footprint than coal. When the power plant is located at or close to the gas source, pollution from pipeline explosions and gas leaks is eliminated. Syngas is also useful as feedstock for the production chemicals and hydrogen. Hydrogen is considered the ultimate zero-emission fuel of the future and various studies have shown that gasified coal (syngas) is the cheapest source of hydrogen (EPRI, 2006). Coal-based hydrogen production with carbon capture and storage allows for large-scale CO_2 removal and storage while producing automotive fuel that has no carbon dioxide emissions. Syngas is also a potential feedstock for coal-to-methanol, coal-to-dimethyl ether, and other products.

Coal has been converted to automotive and aviation fuels for decades and the technology is well established. The technology was pioneered in Germany and provided fuel for the second world war effort. Development was taken to higher levels in South Africa during the apartheid years (See Chapter 10). The technology becomes cost competitive at crude oil prices of around USD 50/barrel or above but, without comprehensive CCS, the carbon footprint is at least 150% to 175% higher than that of conventional petrol and diesel production.

13.6.7 Coal-fueled solid oxide fuel cell technologies

A fuel cell is an electrochemical device that can produce direct current electricity by oxidizing a fuel in a chemical reaction. A fuel cell is made up of two electrodes: positive anode and negative cathode, and an electrolyte which facilitates the movement of electrical charges between the two electrodes. The basic reaction is as shown in Equation 13.1.

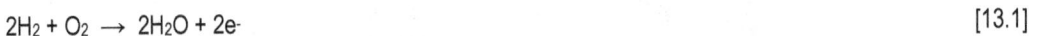

$$2H_2 + O_2 \rightarrow 2H_2O + 2e^- \tag{13.1}$$

A fuel cell is characterized primarily by the electrolyte which may be liquid or solid. A fuel cell has many advantages over other power generation devices. The operating efficiency is high, between 40-60% and could be up to 85% if deployed in combined heat and power (CHP) cogeneration systems. This level of efficiency is unmatched by any other energy conversion technology. Fuel cells can operate on a wide variety of hydrocarbon fuels including natural gas, syngas, diesel oil, alcohols and hydrogen. Furthermore, fuel cells are stationary and have no moving components hence maintenance is relatively easy. Emission of anthropogenic pollutants from fuel cells is relatively low and can be easily captured and processed. Capital and operating costs are relatively low and installations have high long-term stability. Modular units can be stacked in planar or tubular configurations and coupled to form a large power generating unit (Figure 13.16). Fuel cells need to operate at high temperatures to achieve high efficiency but the non-availability of suitable high-temperature, corrosion-resistant materials has been a major limitation.

Fuel cell was invented in 1838 but the first major application was in the NASA space programme in the early 1960s as a power source for probes and satellites. Since then the technology has found wide applications as primary sources of power and power backups for boats, submarines, forklift trucks, automobiles, buses, and in many other industrial, commercial and residential applications. There are many varieties, the most widely used being the polymer-based proton exchange membrane fuel cell (PEMFC). They have been the most extensively studied, due in large part to their lower operating temperature of around 80°C. This factor and the fast start-up time of around 1 second make them more amenable to transportation and other applications where rapid start-up from ambient temperature and portability are desired. High-purity hydrogen is required as fuel and supply requires expensive production infrastructure. Natural gas is currently the main source of hydrogen and the need for extensive and expensive purification increases manufacturing costs significantly. Coal-derived syngas also has great potential as a source of hydrogen. Systems that have been deployed commercially have efficiencies of around 45%.

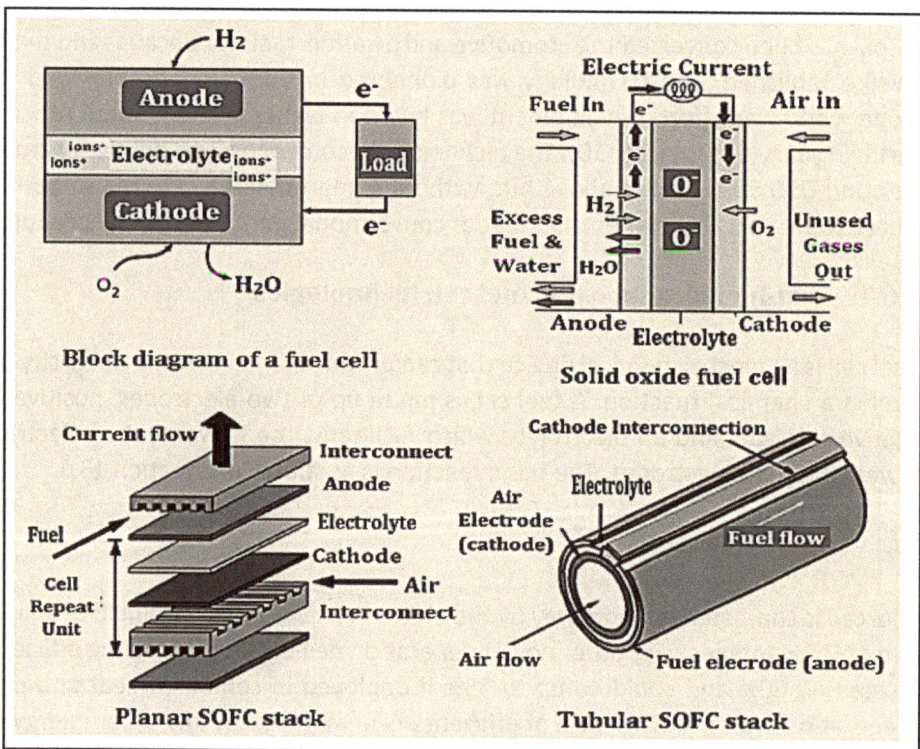

FIGURE 13.16 Fuel cells. *(Wikipedia.org, proquest.com)*.

Solid oxide fuel cells (SOFC) use an oxygen-ion conducting electrolyte instead of proton-conducting medium. The ions transport oxygen from the cathode to oxidize fuel at the anode, thus providing the ability to directly use a variety of hydrocarbon fuels, with high fuel to electricity conversion efficiency of above 55%. The by-products of the reaction are carbon dioxide, water and heat. SOFCs use solid electrolytes and can operate at higher temperatures than liquid-electrolyte systems (800°C-1000°C). They are fuel-flexible and can tolerate higher levels of impurities. However, material issues have been the most severe limitation to wide commercialization of the system.

Intensive materials research over the last decade or so has now extended the range of available suitable, cost-effective materials, including nanomaterials. Also, development of advanced micro-elecromechanical systems-based SOFCs capable of operating at lower temperatures (< 650°C) as well as new designs which are more materials-flexible have greatly enhanced performance, lowered cost, and dramatically increased the number of potential applications of the technology. Developments in coal gasification have also produced advanced combined coal-fired and SOFC cogeneration power systems. A typical SOFC cell consists of an anode (fuel electrode) and a cathode (air electrode) and an electrolyte in between the electrodes. Cells are stacked and interconnected to form large power units (Figure 13.16). Development of a suitable solid electrolyte has been a major subject of intensive research because of the need to find the best compromise between conductivity, stability and cost. Development of thin-film electrolytes and more active anode and cathode materials in tailored composite structures has fast-tracked the development and commercial deployment of intermediate temperature (IT-SOFC) systems operating around 800°C in recent years. The ultimate target is low-temperature systems (LT-SOFC) operating at temperatures around 650°C or below but this will require the development of ultra-thin, sub-micron conductive electrolytes (Wachsman et al., 2014).

13.7 COAL-BASED HYDROGEN

The problem of supply of high-purity hydrogen supply infrastructure has been a major constraint to the wider commercial deployment of solid oxide fuel cells, particularly in automobiles. Research is currently focused on the development of cost-effective and reliable pure hydrogen production processes. Candidate fuel sources are natural gas, coal and biomass. Hydrogen can be produced from coal by gasification which involves partial oxidation of coal under high temperatures and pressures to form synthesis gas which is a mixture of carbon monoxide and hydrogen. The synthesis gas is cleaned to remove impurities and the carbon monoxide is reacted with steam to produce additional hydrogen and carbon dioxide. Hydrogen is removed by a separation system and the highly concentrated carbon dioxide can be captured in an integral unit and sequestered. The production of hydrogen from coal takes place in two stages as shown in the reactions shown in equations 13.2 and 13.3.

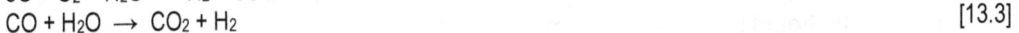

$$3C + O_2 + H_2O \rightarrow H_2 + 3CO \qquad [13.2]$$
$$CO + H_2O \rightarrow CO_2 + H_2 \qquad [13.3]$$

Hydrogen is particularly attractive as a fuel of the future because it has a zero carbon footprint at point of use. Of course this is not true when evaluated on a total life cycle basis, However, pollution emanating from hydrogen production can be contained more easily at the production site compared with other fuels which generate high pollution at virtually every point of use. The potential of coal gasification process is enormous since it makes pure hydrogen available for a wide variety of applications including solid-oxide fuel cells, cogenerators and as feedstock in chemical processes. High-temperature fuel cells typically consume only 70% to 80% of the feed fuel and the unconsumed hydrogen usually goes to waste. Technologies are under development to capture this waste stream and utilize it for the production of by-product hydrogen, reducing gas (hydrogen-carbon dioxide mixture) for the bright annealing industry, and additional power and heat. Advanced combined power heat, hydrogen and power systems which feature carbon capture are also under development.

13.8   FUTURE PROSPECTS OF COAL AND OTHER FOSSIL FUELS

One of the key decisions at the Paris Conference on climate change in 2015 set a target of second half of the 21st century for zero net anthropogenic emissions in order to limit global warming to less than 2°C compared with pre-industrial levels. The implication is that the world should be able to sequester, utilize or compensate for as much anthropogenic gases as are generated. This goal is only achievable if the amount of anthropogenic gases generated can be reduced and any amount generated could be captured, utilized or safely disposed of. Any difference between amount generated and captured should be offset by buying carbon credits. It is useful to assess the feasibility of achieving these targets.

13.8.1   Reduction of anthropogenic gases generated

In 2012, fossil fuels accounted for 84% of worldwide energy consumption. Several recent projections indicate that global primary fuel consumption will increase by 51% by 2040 compared with 2011 level but the the contribution of fossil fuels will drop by only 6% to around 78%, although there will be a significant change in mix (Figure 13.17). Coal use will drop from 28% to 22%, liquid fuels from 33% to 30%, while natural gas share will rise from 23% to 26%. Also, there will be a significant change in regional share (Figure 13.18). Without this shift in fossil fuel mix, the increase in emissions (assuming planned policies come into effect) would be about 23%. Currently just two countries, the United States and China account for about 40% of the global anthropogenic emissions. Both countries have extensive plans for emission reduction but it remains to be seen to what extent they can be enforced.

On the basis of the projection by the International Energy Agency (IEA, 2016), generation of anthropogenic gases from combustion in 2040 should be 40% lower than the level in 2014 if the goal of zero net anthropogenic emissions by the second half of the 21st century is to be achieved (Figure 13.19 to 13.21). However, the same projection shows that the level in 2040 will increase by 13%, even if all policies for reduction pollution currently under consideration are activated successfully. It remains unclear whether most of the policy goals can be met and by which countries. In view of the fact that, even in 2040, renewable and nuclear energy will only contribute about 20% to the global primary energy mix, the policy targets are rather optimistic. It should be noted also that most of the rise in energy consumption over the period will be in emerging countries and the greatest increase will be in the power sector. Most of these countries are unlikely to be able to enforce strict environmental control policies. Furthermore, coal, the most potent pollutant is probably the only locally available major primary energy resource.

The primary energy needs of many developed countries have been flat or falling for some time due mainly to improved efficiencies of use, and many are working to replace coal with lower carbon alternatives. For these reasons the demand for coal in the European Union and the United States (currently nearly 20% of the global total) is projected to fall by 60% and 40% respectively over the period to 2040 (EIA, 2016). On the contrary, the demand for coal will rise significantly in emerging countries - India, Southeast Asia, Africa - all of which are exploiting multiple sources of energy to meet the fast growth in demand. Many of these countries have substantial coal resources and very few other sustainable options. It seems inevitable that carbon dioxide emissions will continue to rise globally in the foreseeable future. Since there is no physical atmospheric partition between countries, global success in decarbonizing primary energy will very much depend on the extent to which the developed economies can help the emerging countries in the adoption of lower carbon, cleaner coal technologies.

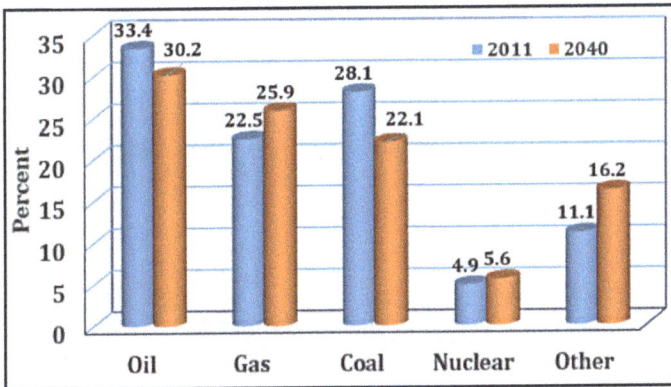

FIGURE 13.17 Projected global primary energy consumption, 2011-2040. *(Eia.gov).*

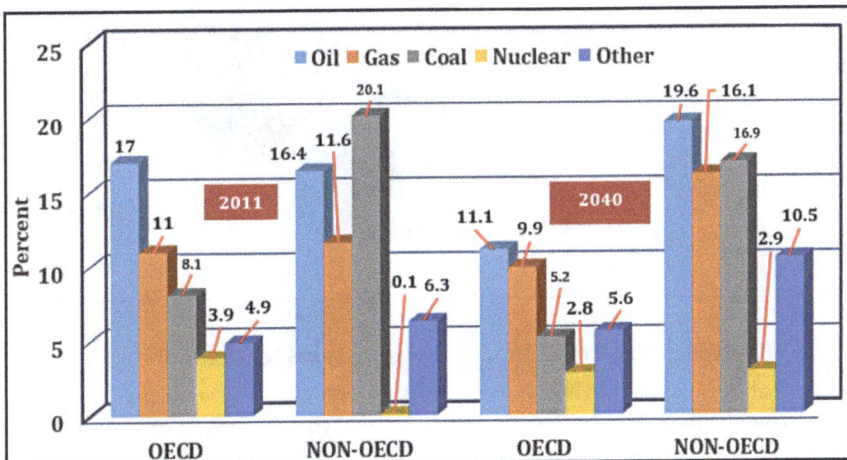

FIGURE 13.18 Projected regional share of global primary energy consumption, 2011-2040. (*eia.gov, 2016).*

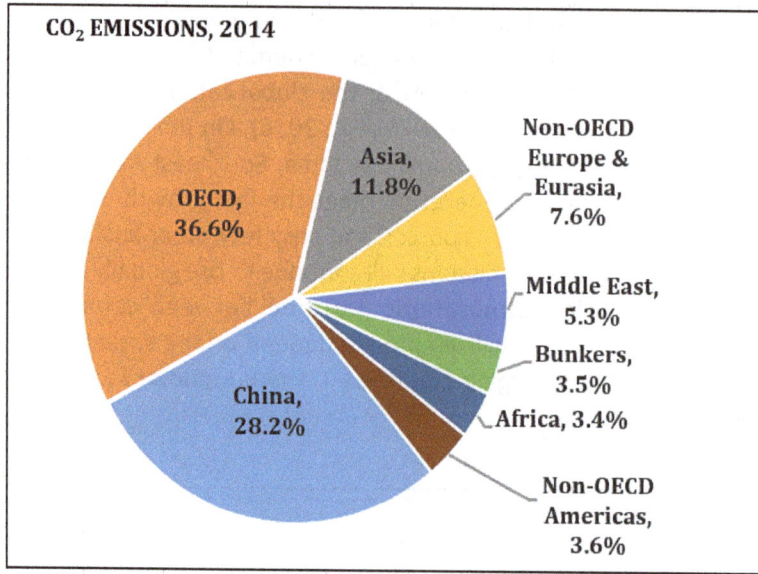

FIGURE 13.19   Actual global CO$_2$ emissions by region in 2014. *(eia.org,*

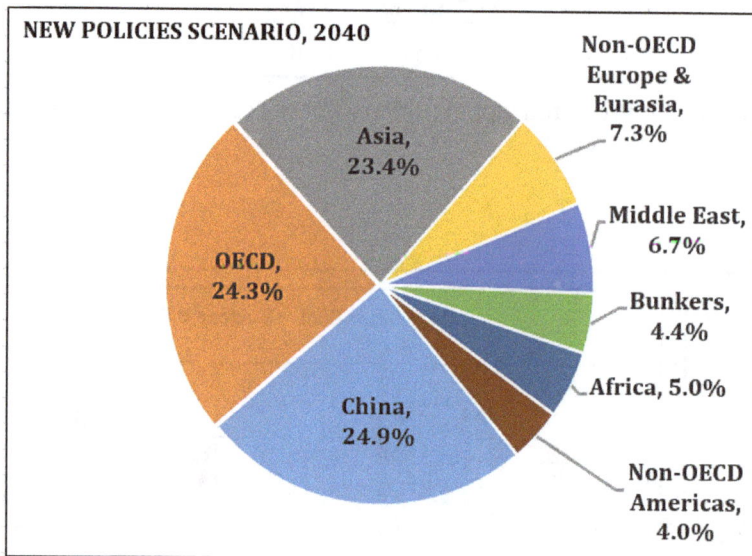

FIGURE 13.20   Projection of CO$_2$ emissions to 2040 by region, New Policies Scenario (NPS). *(eia.org, 2016).* (NPS assumes that all policies under consideration globally for reducing energy-sourced pollution will be effected)

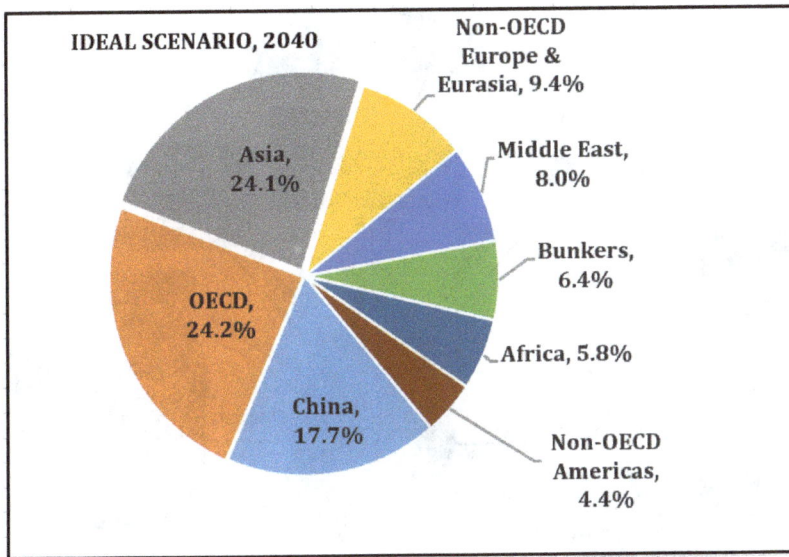

FIGURE 13.21 Projection of CO_2 emissions by region to 2040, Ideal Policies Scenario (IPS). *(eia.org, 2016).* (IPS, also known as 450S: based on policies needed to limit global temperature increase to 2°C).

13.8.2 Capture, use and sequestration of carbon dioxide

Carbon capture can reduce CO_2 emissions by 85-90% from large point emission sources, such as power generating plants, cement and steel plants and several technologies are already in commercial use. The most promising and feasible method is to remove carbon dioxide from large point sources such as coal and gas power plants. One major issue is the cost of capture and storage, estimated to use up 10-40% of the energy generated by the power plant, thus reducing operating efficiency significantly. Furthermore, configuring power plants with carbon capture and storage technology could increase capital and operating costs by 30-60%. The potential effects on the ecosystem of transporting and storing large quantities of carbon dioxide underground and in oceans is largely unknown. For example, massive leakage in a populated area could suffocate thousands of people and the effect of ocean storage on aquatic life could be devastating. Another problem is what to do with the captured gas. Currently, use in various industrial production processes such as carbonated drinks, refrigeration, oil and gas production, fire extinguishers, metal annealing, is very low. Much of the captured gas is stripped of sulphur and nitrogen oxides and particulate matter, and released into the atmosphere. Current technologies for use and sequestration are at early stages of development and the level of commercialization and proliferation by 2040 will largely determine the reductions in anthropogenic emissions that can be achieved over the next 2-3 decades. Perhaps the best option is to focus on technologies which can convert carbon dioxide to useful products and research in this area is intensive. Some of the potential uses which are currently under development are summarized in Figure 13.22.

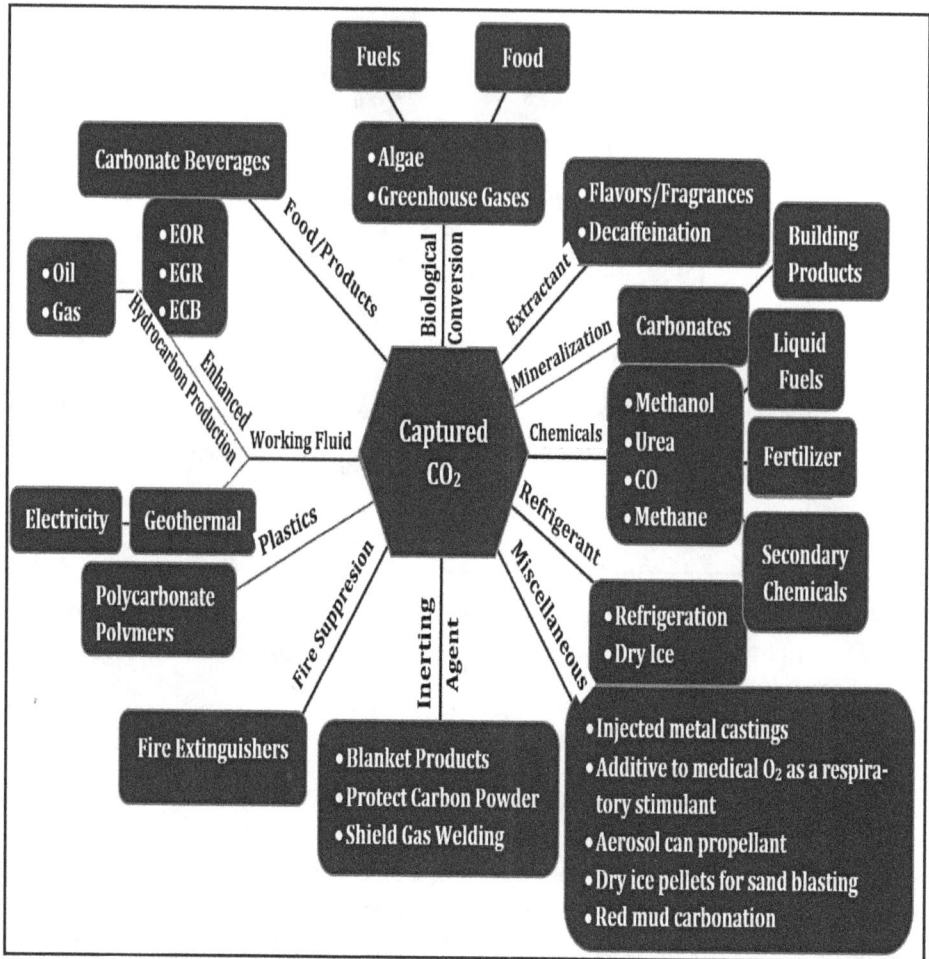

FIGURE 13.22 Current an potential uses of carbon dioxide. *(Damiani et al., 2011).*

REFERENCES

ABB (2014). "Energy Efficiency in the Power Grid." www.abb.com

Arya, A. And N. Hooda (2014). "Enhancing the Characteristics of Fly Ash Based Composite Material." International Journal for Research in Applied Science and Engineering Technology (IJRASET), Vol. 2 Issue VII, July 2014.

Blissett, R. S. And N. A. Rowson (2012). "A review of the multi-component utilisation of coal fly ash." *Fuel*, Volume 97, pp 1-23.

EES (2011). *Third survey of residential standby power consumption of Australian homes-2010.* Energy Efficient Strategies, www.energyrating.gov.au./wpcontent/uploads/Energy_Rating_Documents/Library/StandbyPower/

EIA (2016). *International Energy Outlook, 2016.* Energy Information Administration. www.eia.gov.

Hittinger, E. (2011). "Power consumption of video game consoles under realistic usage patterns." Carnegie Mellon Electricity Industry Center (CEIC) Working Paper. http://wpweb2.tepper.cmu.edu/ceic/pdfs/CEIC_11_01.pdf

IEA (2005). *Premium carbon products and organic chemicals from coal.* International Energy Agency, IEA Report PF 05-05, www.Iea.org.

IEA (2008). *Energy Efficiency Indicators for Public Electricity Production from Fossil Fuels.* International Energy Agency. www.iea.org.

IEA (2011). *25 Energy Efficiency Policy Recommendations.* International Energy Agency. www.iea.org.

IEA (2012). *World Energy Outlook.* International Energy Agency.www.iea.org.

IEA (2013). *Transition to Sustainable Buildings Strategies and Opportunities to 2050).* International Energy Agency, iea.org.

IEA (2014). *Energy Efficiency Indicators.* International Energy Agency. www.iea.org.

IEA (2016). *International Energy Outlook 2016.* International Energy Agency.www.iea.org.

IPCC (2005). *IPCC Special Report on Emissions Scenarios.* Intergovernmental Panel of Climate Change.

IPCC (2005). *IPCC Special Report on Emissions Scenarios.* Intergovernmental Panel of Climate Change.

ISES (2003). *Solar Energy for a Sustainable Future.* Proc. International Solar Energy Society ISES 2003), Goteborg, Sweden.

Joao Fernando Pereira Gomes (2013). *Carbon Dioxide Capture and Sequestration: An Integrated overview of Available Technologies.* Nova Science Publishers.

Jones, M. R., McCarthy, M. J. and A. McCarthy (2003). "Moving fly ash utilization in concrete forward: a UK perspective." Proc. of the 2003 International Ash Utilization Symposium, Center for Applied Energy Research, University of Kentucky, Paper # 113.

Leung, Y. C., Giorgio Caramanna, M. Maroto-Valer (2014). " An Overview of current status of carbon capture and storage technologies." *Renewable and Sustainable Energy Reviews,* Vol. 39, pp. 426-443.

McCarthy, M. J. and R. K. Dihr (2005). "Development of high volume fly ash cements for use in concrete construction." *Fuel,* Vol. 84, No. 11, 2005, pp. 1423-1432.

McKinsey, (2010). Energy efficiency, a compelling global resource. www. McKinsey.com.

NETL (2010a). *Overview of DOE's Gasification Program.* U. S. Department of Energy, National Energy Technology Laboratory, Pittsburgh. www.netl.doe.gov/technologies/coalpower/gasification/pubs.

NETL (2010c). *CCPI/Clean Coal Demonstrations Tidd PFBC Demonstration Project, Project Fact Sheet.* U. S. Department of Energy, National Energy Technology Laboratory, Pittsburgh, PA. www.netl.doe.gov/technologies/coalposer/gasification/pubs.

NRDC (2011). "Better viewing. Lower Energy Bills, and Less Pollution: Improving the Efficiency of Television Set-Top Boxes, NRDC." www.nrdc.org/energy/settopboxes.asp

OECD (2012) "OECD Internet Economy Outlook, 2012." http://www.oecd-library.org/science-and-technology/oecd-internet-economy-outlook-2012/oecd-ict-spending-2012_9789264086463-graph31-en.

Quaschning, Volker (2003). Solar thermal power plants. " Renewable Energy World 06/2003 pp. 109-113. www.DLR; volker-quaschning.de.

Reddy, P. J. (2014). *Clean Coal Technologies for Power Generation.* CRC Press.

Santos, Stanley (2012). "Direct Reduced Iron (DRI) an CO2 Capture." IEA Greenhouse Gas R&D Programme. *Meeting with European Commission,* July 13, 2012. www.ieaghg.org.

Sidique, R. And M. I. Khan (2011). *Supplementary Cementing Materials: Fly Ash* Volume 37 of the series Engineering Materials, pp. 1-66. Springer.

Wachsman, E. D. (2014). "Redox Power System's Revolutionary SOFC Technology: 25 Years of Persistence." https://www.hydrogen.energy.gov/pdfs/htac_apr14_14_wachsman.pdf.

WEC (2013). *World Energy Perspective. Energy Efficiency Policies-what works and what does not.* World Energy Council. www.wec.org.

WEF (2012) "The Global Information Technology Report 2012. World Economic Forum. www.reports.weforum.org/global-information-technology-2012/.

Wesche, K. (Ed.) (2005). *Fly Ash in Concrete: Properties and Performance.* Report of Technical Committee 67-FAB Use of Fly Ash in Building. RILEM ((The International Union of Testing and Research Laboratories for Materials and Structures). E and F. N. Spoon. Chapman & Hall (1991). Taylor & Francis e-Library, (2005).

COAL CONVERSION FACTORS

UNIT ABBREVIATIONS			
GCV	Gross calorific value	TWh	terawatt hour
NCV	Net calorific value	Mtoe	million tonnes of oil equivalent
kcal	kilocalorie	Mtce	million tonnes of coal equivalent
Btu	British thermal units	tCO$_2$	Carbon dioxide equivalent, tonnes
MBtu	million British thermal units	lt	long ton (imperial)
GW	gigawatt	st	short ton (US)
GWh	gigawatt hour	t	Tonne (metric)
kJ	kilojoule	psi	pounds per square inch
MJ	megajoule	kilo	watt x 10^3
TJ	terajoule	Megawatt	watt x 10^6
KJ	kilojoule	Gigawatt	watt x 10^9
kW	kilowatt	Terawatt	watt x 10^{12}
kWh	Kilowatt hour	Petawatt	watt x 10^{15}
MWh	megawatt hour		
GWh	gigagawatt hour		

CONVERSION FACTORS FOR MASS					
To:	kg	t	lt	st	lb
From:	Multiplication factor				
kilogramme(kg)	1	1.000x10^3	9.842x10^{-4}	1.102 x 10^{-3}	2.205
tonne (t)	1.000x10^3	1	9.842x10^{-1}	1.102	2.205x10^3
long ton (lt)	1.016x10^3	1.016	1	1.120	2.240x10^3
short ton (st)	9.072x10^2	9.072x10^{-1}	8.929x10^{-1}	1	2.000x10^3
pound (lb)	4.536x10^{-1}	4.536x10^{-4}	4.464x10^{-4}	5.000x10^{-4}	1

CONVERSION FACTORS FOR ENERGY					
To:	TJ	Gcal	Mtoe	Mbtu	GWh
From:	Multiplication factor				
TJ	1	2.388x10^2	2.388x10^{-5}	9.478x10^2	2.778x10^{-1}
Gcal	4.187x10^{-3}	1	1.000x10^{-7}	3.968	1.163x10^{-3}
Mtoe	4.187x10^4	1.000x10^7	1	3.968x10^7	1.163x10^4
Mbtu	1.055x10^{-3}	2.520x10^{-1}	2.520x10^{-8}	1	2.931x10^{-4}
GWh	3.600	8.598x10^2	8.598x10^{-5}	3.412x10^3	1

OTHER USEFUL CONVERSION FACTORS
1 atmosphere = 760 mmHg = 14.696 psi
1 calorie = 4.184 joules
1 joule = 0.239006 calorie
1 Btu = 1054.35 joules
1 Btu/lb = 1.8 x kcals/kg; 1 kcal/kg = 0.5556 x Btu/lb
1 Btu/lb = 0.429923J/g; 1 J/g = 2.326Btu/lb
1 kilogram =2.20462 pounds
1 pound = 0.45359 kilogram
1 tonne (metric) = 1000 kilograms = 2204.6 pounds = 0.984 long ton = 1.10 short ton
1 ton (long, imperial) = 2240 pounds = 1.016 tonne
From MJ/kg to kcal/kg multiply by 238.846
From MJ/kg to Btu/lb multiply by 429.923
From Btu/lb to MJ/kg, multiply by 0.002326
From kcal/kg to MJ/kg multiply by 0.004187
From Btu/lb to kcal/kg multiply by 0.5556
From kcal/kg to Btu/lb multiply by 1.8000
1 ton (short, US) = 2000 pounds = 907.185 kilograms = 0.907 tonne.

COAL CONVERSION FACTORS			
To:	**Air Dry**	**Dry Basis**	**As Received**
From:	**Multiplication factor**		
AR by:	$\dfrac{(100 - \text{Im}\%)}{(100 - \text{Tm}\%)}$	$\dfrac{100}{(100 - \text{Tm}\%)}$	-
AD by:	-	$\dfrac{100}{(100 - \text{Im}\%)}$	$\dfrac{(100 - \text{Tm}\%)}{(100 - \text{Im}\%)}$
DB by:	$\dfrac{(100 - \text{Im}\%)}{100}$	-	$\dfrac{(100 - \text{Tm}\%)}{100}$

DAF (dry, ash-free) = DB x $\dfrac{100}{(100-\text{A}\%)}$

AR = as received (includes total moisture, Tm)
AD = air-dried (includes inherent moisture Im only)
DD = dry basis (excludes all moisture)
DAF = dry, ash-free (excludes all moisture and ash)
A = ash content

DMMF = dry, mineral matter-free basis. Mineral matter and ash are distinctly different entities. Mineral matter consists of the various minerals contained in the coal. Ash is the organic solids remaining after the coal is completely burnt. Mineral matter content is usually higher than ash content because of the weight changes that take place during coal combustion, for example, loss of gaseous carbon dioxide from mineral carbonates, loss of water from silica minerals and loss of sulphur as gaseous sulphur dioxide from pyrites

DMMF = DB x $\dfrac{(100-\text{MMd}\%)}{100}$ where MMd is the mineral matter content on dry basis.

CALORIFIC VALUE OF COAL

Gross calorific value (GCV) of coal is the heating value under laboratory conditions. Net calorific value (NCV) or Net Effective Calories (NEC) is the available, useful heat value in combustion. The difference is essentially the latent heat of the water vapour produced.

EMPIRICAL FORMULA FOR ESTIMATION OF GCV

DULONG (1820)	GCV = (80.8 x C) + (344.6 x H) – (43.1 x O) + (25 x S)
SEYLER (1938)	GCV = (123.9 x C) + (388.1 x H) + (25 x O_2) - 4269
NEAVEL (1986)	GCV = (81.05 x C) + (316.4 x H) – (29.9 x O) + (23.9 x S) – (3.5 x Ash)
GIVEN (1986)	GCV = (78.3 x C) + (339.1 x H) – (33.0 x O) + (22.1 x S) + 152 C = carbon, H = hydrogen, O = oxygen, S = sulphur, all on % weight, dry basis

FUEL RATIO OF COAL

Fuel ratio = Fixed Carbon/Volatile Matter

Index

www.ingramcontent.com/pod-product-compliance
Lightning Source LLC
Chambersburg PA
CBHW080651220326
41598CB00033B/5172